Acoustic Waves Generated by Parametric Array Loudspeakers

Parametric array loudspeakers (PALs) are capable of generating highly directional audio beams from nonlinear interactions of intense airborne ultrasound waves. This unique capability holds great potential in audio engineering. This book introduces the physical principles of acoustics waves generated by PALs systematically, along with the commonly used and state-of-the-art numerical models, such as the Westervelt's model, the convolution directivity model, the Gaussian beam expansion method, and the spherical wave expansion method.

The book also delves into the properties of sound fields generated by PALs examining phenomena such as the reflection of acoustic waves generated by PALs from a surface, transmission through a thin partition, scattering by a rigid sphere, and propagation in rooms. It further introduces the steering and focusing of acoustic waves generated by PALs, as well as potential applications of PALs in active sound control. Issues related to hardware implementation, signal processing techniques, measurement, and safety issues are discussed.

The book is tailored to meet the needs of researchers in this field, as well as audio practitioners and acoustics engineers.

Jiaxin Zhong is Postdoctoral researcher in the Graduate Program in Acoustics at Pennsylvania State University, USA.

Xiaojun Qiu is Chief Scientist at Huawei. He is a Fellow of the Audio Engineering Society and the International Institute of Acoustics and Vibration. He is the author of *An Introduction to Virtual Sound Barriers* and co-author of *Active Control of Noise and Vibration*, 2nd edition, also published by CRC Press.

Acoustic Waves Generated by Parametric Array Loudspeakers

Jiaxin Zhong and Xiaojun Qiu

CRC Press is an imprint of the
Taylor & Francis Group, an **informa** business

Cover image: Jiaxin Zhong and Xiaojun Qiu

First edition published 2025
by CRC Press
2385 NW Executive Center Drive, Suite 320, Boca Raton FL 33431

and by CRC Press
4 Park Square, Milton Park, Abingdon, Oxon, OX14 4RN

CRC Press is an imprint of Taylor & Francis Group, LLC

Library of Congress Cataloging-in-Publication Data

Names: Zhong, Jiaxin (Acoustical engineer), author. | Qiu, Xiaojun, author.
Title: Acoustic waves generated by parametric array loudspeakers / Jiaxin Zhong and Xiaojun Qiu.
Description: First edition. | Boca Raton, FL : CRC Press, 2024. | Includes bibliographical references and index.
Identifiers: LCCN 2024006190 | ISBN 9781032408521 (hbk) | ISBN 9781032408538 (pbk) | ISBN 9781003354994 (ebk)
Subjects: LCSH: Loudspeakers. | Sound-waves--Mathematical models. | Acoustical engineering.
Classification: LCC TK5983 .Z47 2024 | DDC 620.2--dc23/eng/20240404
LC record available at https://lccn.loc.gov/2024006190

ISBN: 978-1-032-40852-1 (hbk)
ISBN: 978-1-032-40853-8 (pbk)
ISBN: 978-1-003-35499-4 (ebk)

DOI: 10.1201/9781003354994

Typeset in Nimbus Roman
by KnowledgeWorks Global Ltd.

Dedication

This book is dedicated to

Liu
Donna
Harry

Contents

Preface

Parametric array loudspeakers (PALs) are known for their capability of generating highly directional audio beams due to the nonlinear interactions of intense airborne ultrasound waves. Since the first successful prototype fabricated in 1983 by Yoneyama, they have demonstrated potentials of being used in many applications such as active noise control, sound reproduction, personal communication, museum exhibitions and service and multimedia booths, and measurement of the acoustic parameters of materials. This book is intended to introduce the physical principles of the acoustic waves generated by PALs systematically.

This book contains eight chapters. Chapter 1 introduces the physical basis for PALs, including the history, physical models and equations, and the theoretical framework of the quasilinear solution. Chapter 2 discusses the various numerical models for calculating acoustic waves generated by PALs. Not only the classical methods, such as the Westervelt's model and Gaussian beam expansion method, are discussed but also the recent developed methods such as the convolution directivity model and the spherical wave expansion method. Chapter 3 deals with the sound fields generated by a PAL in free space. Unlike the conventional loudspeakers, the front side sound fields of PALs are divided into three regions: the near field, the Westervelt far field, and the inverse-law far field, and the properties in each region are discussed. The back side of a non-baffled PAL is also modeled and analyzed.

Chapter 4 examines three typical physical phenomena for the acoustic waves generated by a PAL, which are the reflection from a surface, transmission through a thin partition, and the scattering by a rigid sphere. Chapter 5 deals with the wave propagations in a rectangular room with lightly damped walls. The physical modeling is based on the normal mode theory.

Chapter 6 discusses the steerable and focusing PALs, which enable the flexible manipulations of the acoustic waves. The physical limitations of steering the audio beam at a specific direction and improving the low frequency response by the beam focusing are analyzed. Chapter 7 introduces a typical application of PALs in active noise control systems. The feasibility and the limitation of using one or multiple PALs in generating a quiet zone are discussed. Chapter 8 introduces the implementation of PALs, where the hardware, signal processing techniques, measurement, and safety issues are discussed.

The website http://www.parametricarray.com provides supplementary materials for this book.

November 2023
Jiaxin Zhong and Xiaojun Qiu

Acronyms

2D	Two-Dimensional
3D	Three-Dimensional
AM	Amplitude Modulation
ANC	Active Noise Control
CDM	Convolution Directivity Method
CPU	Central Processing Unit
CWE	Cylindrical Wave Expansion
DFT	Discrete Fourier Transform
DFW	Difference-Frequency Wave
DIM	Direct Integration Method
DSB	Double Side Band
DSP	Digital Signal Processor
FEM	Finite Element Method
FDTD	Finite-Difference Time-Domain
FFT	Fast Fourier Transform
FPGA	Field Programmable Gate Array
FxLMS	Filtered-x Least Mean Square
GBE	Gaussian Beam Expansion
HATS	Head And Torso Simulator
HPBW	Half Power Beam Width
IL	Insertion Loss
ISM	Image Source Method
KZK	Khokhlov-Zabolotskaya-Kuznetsov
LDV	Laser Doppler Vibrometer
MEM	Modal Expansion Method
MOSFET	Metal-Oxide Semiconductor Field-Effect Transistor
PAA	Parametric Acoustic Array
PAL	Parametric Array Loudspeaker
PSL	Peak Sidelobe Level
RAM	Random Access Memory
RH	Relative Humidity
SA	Simulated Annealing
SPL	Sound Pressure Level
SRT	Square RooT
SSB	Single Side Band
SSS	Scattering of Sound by Sound
SWE	Spherical Wave Expansion

Symbols

ALPHABET LETTERS

a	aperture size of the source
a_x, a_y	aperture sizes of the source in x- and y-directions
$\mathcal{A}(\cdot)$	aperture area
$\mathcal{A}_0$	total area of the radiation surface
B/A	nonlinear parameter
c_0	linear speed of sound in air ($c_0 = 343\,\mathrm{m/s}$ is assumed)
$\mathscr{C}$	Bessel ($\mathscr{C} = J$) or Hankel ($\mathscr{C} = H$) functions
$\mathcal{D}(\cdot)$	directivity
$\mathscr{D}_{\mathrm{ab}}$	absorption distance, defined by Eq. (1.114)
$\mathscr{D}_{\mathrm{c}}$	critical distance, defined by Eq. (1.112)
$\mathscr{D}_{\mathrm{sf}}$	shock formation distance, defined by Eq. (1.110)
$\mathscr{D}_{\mathrm{R}}$	Rayleigh distance, defined by Eq. (1.111)
$E(t)$	envelope signal
f	frequency
f_1, f_2	ultrasound frequencies, $f_2 > f_1$
f_{a}	audio sound frequency, $f_{\mathrm{a}} = f_2 - f_1$
f_{u}	average ultrasound frequency $f_{\mathrm{u}} = (f_1 + f_2)/2$; carrier ultrasound frequency
$g_{\mathrm{2D}}, g_{\mathrm{3D}}$	2D and 3D Green's functions in free space
G	general Green's function
$\mathcal{G}$	auxiliary Gaussian type integral, defined by Eq. (2.104)
$\mathscr{G}$	Gaunt coefficients, defined by Eq. (B.15)
$\mathrm{h}_\ell(\cdot)$	spherical Hankel function of the first kind of ℓ-th order
h_{rel}	relative humidity; if not specified otherwise, the relative humidity is set as 50% in calculating the absorption coefficient as discussed in Appendix A
$\mathrm{H}(\cdot)$	Heaviside function, defined by Eq. (2.31)
$H_m(\cdot)$	Hankel function of the first kind of m-th order
i	imaginary unit, $\mathrm{i}^2 = -1$
$\mathrm{j}_\ell(\cdot)$	spherical Bessel function of ℓ-th order
$\mathcal{J}(\cdot)$	objective function
$J_m(\cdot)$	Bessel function of m-th order
$\mathrm{jinc}(\cdot)$	jinc function, defined by Eq. (2.84)
k	wavenumber
$\mathcal{L}$	Lagrangian density
$\mathcal{M}$	modulation index used in Sec. 8.3
$\mathcal{N}_{\mathrm{sf}}^{\mathrm{ab}}$	absorption-saturation number, defined by Eq. (1.115)

$\mathcal{N}_{\text{sf}}^{\text{R}}$	diffraction-saturation number, defined by Eq. (1.113)
$\mathbb{N}$	set of natural numbers
$\mathcal{O}(\cdot)$	big O notation
$p(\cdot)$	sound pressure
p_0	on-surface sound pressure amplitude
P	pressure
$P_\ell(\cdot)$	Legendre polynomial
$P_\ell^m(\cdot)$	associated Legendre polynomial
Pr	Prandtl number, defined by Eq. (1.47)
$q(\cdot)$	source density
$\mathbf{r} = (x, y, z)$	position vector in Cartesian coordinates
$\Re\{\cdot\}$	real part
$\hat{\mathbf{s}}$	unit vector indicating the beam direction of a steerable PAL
$\text{sinc}(\cdot)$	sinc function, defined by Eq. (2.82)
t	time
T	temperature; if not specified otherwise, the ambient temperature is set as $T_{\text{ref}} = 293.15\,\text{K}\,(20\,^\circ\text{C})$ in calculating the absorption coefficient as discussed in Appendix A
$u(t)$	the modulated signal fed into the ultrasonic emitter
$\mathbf{v}(\cdot)$	particle velocity
v_0	on-surface particle velocity amplitude
W_1, W_2, W_a	sound power at frequency f_i, $i = 1, 2, \text{a}$
$Y_\ell^m(\cdot,\cdot)$	spherical harmonic of ℓ-th degree and m-th order
$\mathbb{Z}$	set of integers

GREEK LETTERS

α	absorption coefficient
β	nonlinear coefficient, defined by Eq. (1.68)
δ	sound diffusivity, defined by Eq. (1.70)
$\delta(\cdot)$	Dirac delta function
$\delta_{nn'}$	Kronecker delta function
ε	Mach number
$\tilde{\varepsilon}$	a generic small parameter characterizing the smallness of both ε and η
η	viscosity parameter, defined by Eq. (1.45); power conversion efficiency (Chap. 3)
(η, ξ, φ)	oblate spheroidal coordinates used in Sec. 3.3
θ	zenithal angle in the spherical coordinate system
κ	thermal conductivity
λ	wavelength
μ_{b}	bulk viscosity coefficient
μ_{s}	shear viscosity coefficient
$\Pi(\cdot)$	rectangle function, defined by Eq. (2.11)
ρ	radial coordinate in polar or cylindrical coordinate systems
$\boldsymbol{\rho} = (x, y)$	position vector in 2D Cartesian (rectangular) coordinates
ρ_0	air density
σ	area mass density, used in Sec. 4.3
τ	retarded time
φ	azimuthal angle in polar, cylindrical, or spherical coordinate systems
Φ	velocity potential
ω	angular frequency

OTHER SYMBOLS

∇	gradient operator
∇^2	Laplacian operator
$*$	linear convolution
$\otimes$	tensor product
$\circledast$	spherical convolution

SUPERSCRIPTS AND SUBSCRIPTS

$\hat{(\cdot)}$	unit vector
$(\cdot)^*$	complex conjugate
$(\cdot)^{\mathrm{T}}$	matrix transpose
$(\cdot)_{\mathrm{I}}$	first-order component
$(\cdot)_{\mathrm{II}}$	second-order component
$(\cdot)_{\mathrm{a}}$	index for audio
$(\cdot)_{\mathrm{d}}$	beam steering
$(\cdot)_{\mathrm{s}}$	source
$(\cdot)_{\mathrm{u}}$	index for ultrasound
$(\cdot)_{\mathrm{v}}$	virtual source
$(\cdot)_x, (\cdot)_y, (\cdot)_z$	components in x-, y-, z-directions

1 Physical Basis for Parametric Array Loudspeakers (PALs)

1.1 INTRODUCTION

Acoustics deals with the propagation of mechanical waves in fluids or solids. Linear acoustics, which assumes small amplitudes of acoustic waves, has been extensively studied and applied successfully in various fields. This assumption allows linear superposition of acoustic waves, and simplifies the modeling and understanding of wave propagation. However, when the amplitude of acoustic waves becomes sufficiently large, the linear superposition is no longer valid, leading to *finite-amplitude acoustic waves*. Modeling acoustic waves with large amplitudes requires full systems of fundamental equations of fluid dynamics for acoustic waves in liquids and gases and elasticity for those in solids. These governing equations are nonlinear in nature, but they can be simplified to linear equations when the amplitude of acoustic waves is small. The traditional linearization is no longer applicable when acoustic waves have large amplitudes. The study of the finite amplitude acoustic waves results in a new branch of the acoustics—*nonlinear acoustics*.

Nonlinear acoustics encompasses a wide range of phenomena that exhibit unique behaviors not observed in linear counterparts. The origins of nonlinear acoustics can be traced back to the work of Euler in 1755, who derived the formulations for the conservation of mass and momentum. A detailed history of nonlinear acoustics prior to the 1930s can be found in [Enflo and Hedberg, 2002, Sec. 1.2; Hamilton and Blackstock, 2008, Chap. 1]. Nonlinear acoustics includes various fascinating topics and phenomena. For example, the occurrence of sonic booms, which are generated when an object moves faster than the speed of sound and produces a shock wave; and acoustic levitation, involves the use of intense sound waves to suspend and manipulate objects in the air. Medical ultrasound waves, widely used in diagnostic imaging and therapeutic applications, also fall under the purview of nonlinear acoustics.

This book specifically delves into the nonlinear phenomenon of *parametric acoustic arrays* (PAAs) and their applications in the air—*parametric array loudspeakers* (PALs). The subsequent sections, specifically Secs. 1.1.1 and 1.1.3, offer a thorough introduction of PAAs and PALs, respectively. Sections 1.1.2 and 1.1.4 delve into the common applications of PAAs and PALs, respectively. In Sec. 1.1.5, some commercial product examples of PALs that are currently available in the market are presented. Moving forward, Sec. 1.2 introduces the fundamental equations governing the wave propagation for PALs. These equations include the continuity equation (Sec. 1.2.1), Navier-Stokes equation (Sec. 1.2.2), Kirchhoff-Fourier

DOI: 10.1201/9781003354994-1

equation (Sec. 1.2.3), and the equation of state (Sec. 1.2.4), derived from the conservation of mass, momentum, energy, and the thermodynamic laws, respectively. Due to their complexity and nonlinearity, solving these equations is challenging. Hence, in Sec. 1.2.5, a second-order approximation approach is introduced. Based on these fundamental equations, Sec. 1.3 presents the commonly used nonlinear wave equations in this field, which include the general second-order nonlinear wave equation (Sec. 1.3.1), the Kuznetsov equation (Sec. 1.3.2), the Westervelt equation (Sec. 1.3.3), and the Khokhlov-Zabolotskaya-Kuznetsov (KZK) equation (Sec. 1.3.4). Section 1.4 introduces an essential technique in the realm of nonlinear acoustics and PALs, which is the perturbation analysis leading to the derivation of quasilinear solutions of the nonlinear wave equations. This technique is explored in both the time domain (Sec. 1.4.1) and the frequency domain (Sec. 1.4.2), offering valuable insights and analysis tools.

1.1.1 PARAMETRIC ACOUSTIC ARRAYS

One typical phenomenon of a finite-amplitude acoustic wave is that its waveform distorts as it propagates. For a harmonic acoustic wave with a frequency of f_1, an additional component with a frequency of $2f_1$ can be generated due to the distortion of the second-order nonlinearity. For two harmonic *primary waves* with frequencies f_1 and f_2 and $f_2 > f_1$, the *secondary waves* are generated at frequencies of $2f_1, 2f_2, f_1+f_2$, and f_2-f_1. Among them, the components of frequencies f_2-f_1 and f_1+f_2 are referred to as the *difference-frequency wave* (DFW) and the sum-frequency wave, respectively.

The generation of the additional components can be explained by simple mathematics. For a linear system, as illustrated in Fig. 1.1(a), the system response can be described by a linear function $H(x) = C_1 x + C_0$, where x is the system input, and C_1 and C_0 are constants. For an input with two harmonic waves, the input can be written as $x(t) = \cos(\omega_1 t) + \cos(\omega_2 t)$, where t is the time and the angular frequency $\omega_i = 2\pi f_i$, $i = 1, 2$. The output of the system $H(x) = C_1\cos(\omega_1 t) + C_1\cos(\omega_2 t) + C_0$ contains only the first-order components of frequencies f_1 and f_2. But for a second-order nonlinear system with a system response function of $H(x) = C_2 x^2 + C_1 x + C_0$, as shown in Fig. 1.1(b), the output can be written as

$$\begin{aligned} H(x) = \frac{C_2}{2}\cos(2\omega_1 t) + \frac{C_2}{2}\cos(2\omega_2 t) + C_1\cos(\omega_1 t) + C_1\cos(\omega_2 t) + C_0 + C_2 \\ + C_2\cos\left[(\omega_1+\omega_2)t\right] + C_2\cos\left[(\omega_2-\omega_1)t\right]. \end{aligned} \tag{1.1}$$

Equation (1.1) reveals that, in addition to the first-order components f_1 and f_2, the second-order components $2f_1, 2f_2, f_1+f_2, f_2-f_1$ are generated. In more intricate nonlinear systems, the output may encompass higher-order components, such as $3f_1, 3f_2, 2f_2-f_1, 4f_1, 4f_2$, and so forth. While PAAs and PALs typically leverage the second-order nonlinearity, various studies and applications explored the utilization of higher-order nonlinear components [Qian, 1995; Ma et al., 2006; Liu et al., 2007; Mitri, 2010; Garner, 2011; Garner and Steer, 2012a; Johnson and Steer, 2014; Garner and Steer, 2012b; Johnson, 2016].

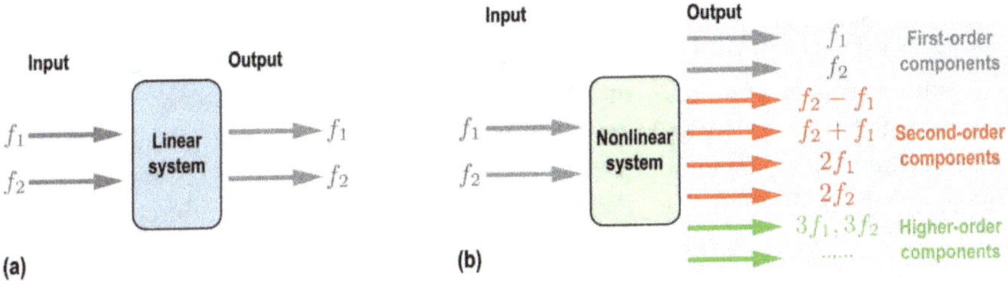

Figure 1.1 Outputs of (a) linear and (b) nonlinear systems with an input of two harmonic waves at frequencies f_1 and f_2 with $f_2 > f_1$.

Generation of additional components other than the first-order ones f_1 and f_2 usually refers to as the phenomenon of *scattering of sound by sound (SSS)* [Darvennes and Hamilton, 1990; Dimant et al., 2021]. This concept was firstly proposed by Ingard and Pridmore-Brown in 1956 [Ingard and Pridmore-Brown, 1956]. They calculated the scattered field based on the far-field solution of the Lighthill's equation and predicted that there exists a scattered field outside the overlap region of the two orthogonal primary beams. This prediction was verified by their experiments. However, Westervelt argued that there should not exist the scattered field outside the overlap region [Westervelt, 1957a; Westervelt, 1957b] and underwater experiments were conducted to verify this argument [Bellin and Beyer, 1960]. This contradictory conclusion was solved by Tjøtta's group with a new theory [Tjøtta and Tjøtta, 1987a; Tjøtta and Tjøtta, 1987b; Tjøtta and Tjøtta, 1988], where they pointed out that contradictory results were observed due to the inappropriate assumptions made in their theories. The scattered field beyond the overlap region does indeed exist; however, its magnitude is too small to measure due to the region being too small to induce a substantial generation of the DFW.

Enlightened by these findings, Westervelt introduced the concept of PAA in 1963 [Westervelt, 1963]. When two primary sound beams are collinear or nearly collinear, a stronger scattered secondary field is produced, attributed to the enlarged overlap region of the primary waves. Furthermore, it is observed that the DFW in the second-order components exhibits significant directionality, even when the difference frequency is very low. From this aspect, PAAs can be regarded as a specialized case of the SSS phenomenon. The origins of PAAs trace back to around 1951 in London when Westervelt was tasked with delivering a steak-and-kidney pie to the esteemed Captain H. J. Round. In Round's laboratory, as the transducer emitted sound waves, Westervelt noticed an exceptionally powerful, low-frequency sound with directional characteristics. This piqued his curiosity, leading him to question whether the sound originated from the transducer, his ears, his brain, or somewhere in between [Foote, 2021].

Underwater experiments were conducted to validate the concept of PAA in the 1960s [Bellin et al., 1960; Bellin and Beyer, 1962; Muir and Blue, 1969]. The word *parametric* originates from the parameter of nonlinearity in the acoustic fundamental equations, which is similar to the coefficient C_2 in Eq. (1.1) [Beyer, 1960; Foote,

2021]. Consequently, generating the DFW is also called the *parametric conversion* [Foote, 2021]. For a PAA, the primary wave beam acts like a distribution of virtual sources for the generation of DFW [Westervelt, 1960; Shi and Gan, 2010]. The source density of these virtual sources is proportional to the amplitude of the primary waves, which is exponentially attenuated along the propagation axis approximately due to the absorption in fluids [Berktay and Shooter, 1973]. These sources form a so-called end-fire array with a hemi-infinitely long length [Shi and Gan, 2010; Gan et al., 2012], which can produce low-frequency directional sound beams in the end-fire direction without side lobes [Bellin and Beyer, 1962; Mellen, 1976; Mellen, 1977; Tu et al., 2016]. This physical mechanism of generating the DFW results in the word *array* in the terminology of PAAs.

The directivity of the DFW is of the most interest, and its closed-form formula was initially provided by Westervelt in 1963, assuming collimated beams [Westervelt, 1963]. This formula, now known as the *Westervelt's directivity*, is applicable to the pure-tone DFW resulting from the excitation of two harmonic primary waves. In 1965, Berktay derived a closed-form solution in the time domain for the DFW along the propagation axis, termed the *Berktay's far-field model* [Berktay, 1965a]. This model is widely utilized for analyzing the performance of wideband DFWs. Experimental validation of this self-modulation process for transient signals was conducted several years later [Moffett et al., 1970; Moffett et al., 1971]. However, the theoretical capability for modeling PAAs is limited to the propagation axis and the far field. The development and applications of PAAs have been greatly facilitated by the well-known KZK equation in nonlinear acoustics [Novikov et al., 1987], derived via the Burgers' equation, encompassing diffraction, absorption, and nonlinearity. The KZK equation enables the modeling of wave propagation and analysis of performance in the near field. This valuable tool paved the way for exploring the feasibility of using PAAs in underwater applications from the 1960s [Zhou et al., 2020].

1.1.2 APPLICATIONS OF PARAMETRIC ACOUSTIC ARRAYS

Parametric sonar stands as one of the earliest and notable applications of PAAs [Berktay, 1965b; Browning et al., 2009; Ostrovsky, 2009; Foote, 2021]. It serves as a device capable of transmitting and receiving underwater DFWs. Parametric sonar offers three distinct advantages over conventional techniques: wide bandwidth, narrow beamwidth, and the absence of sidelobes. Its wide-ranging use includes applications such as measuring sea depth through echos [Nichols Jr., 1971], quantifying seabed properties in shelf waters [Hines, 1999], detecting buried objects [Trucco and Pescetto, 2000], sub-bottom profiling [Humphrey et al., 2008; Marchal and Cervenka, 2012; Qu et al., 2018], high-resolution monitoring of contaminated sediments [Guigné et al., 1991], ensonification for fish swimbladder-resonance spectroscopy, and measurements in confined volumes when low frequencies are required [Foote, 2021]. Furthermore, a parametric Doppler sonar, integrating the Doppler effect, was proposed and employed to measure ground speed and distance traveled at extreme ocean depths in the 1970s [Kritz, 1977].

Acoustic communication plays a critical role in underwater applications due to the rapid attenuation of optical waves in water, which poses significant challenges for long-range underwater optical communication. The PAA offers a distinct advantage in underwater communication due to its ability to generate low-frequency directional beams [Coates et al., 1996; Loubet et al., 1996; Kopp et al., 2000; Wiedmann et al., 2012; Campo-Valera et al., 2023; Yu et al., 2023]. An example of this advantage is demonstrated in the transmission of video signals, where the signal is processed and encoded before being transmitted by a parametric source with a relatively small aperture diameter of ten inches. The resulting DFW is received at the bottom of a lake, approximately 4000 m below the surface, and can be decoded and displayed on a screen [Browning et al., 2009]. For a comprehensive exploration of underwater applications utilizing PAAs, readers can refer to Zhou et al., 2020, which provides a comprehensive review of various uses and advancements in this field.

PAAs have found applications in the field of medical imaging, particularly in improving diagnostic image resolution through two distinct approaches. The first approach involves the utilization of two collinear primary waves transmitted through the tissue. The resulting DFW carries information about the tissue's nonlinearity. By receiving the DFW, it becomes possible to reconstruct a tomogram of the scanned tissue [Nakagawa et al., 1986; Cai et al., 1988; Zhang et al., 2001; Zhang et al., 2002; Nomura et al., 2018]. The second approach employs two non-collinear primary beams that intersect in an orthogonal or nearly orthogonal manner. In a homogeneous medium, the generated DFW does not propagate beyond the overlap area, resulting in no detectable DFW signal. However, when the tissue is inhomogeneous, such as encountering a skull, the sources within the overlap area effectively propagate to the far field and can be received [Li et al., 2020; Kim et al., 2022b]. To enhance the contrast between the target and surrounding tissue, microbubbles are introduced as contrast agents. These microbubbles are released into the bloodstream and can be excited by an external fundamental frequency wave. The energy of the second harmonic wave generated by the microbubbles can be detected to form an image. The narrow-beam characteristics of the parametric array make it a potential ultrasound source for exciting microbubbles [Vos et al., 2011].

PAAs have been applied in various other fields as well. In the realm of high-energy neutrino interaction, PAAs are employed to generate neutrino-like signals in the calibrator system [Ardid et al., 2009]. Additionally, PAAs find utility in non-destructive/noncontact testing by exciting different materials and detecting their responses. The responses are detected and the attenuation in the materials are extracted [Guigné et al., 1989]. This enables the extraction of attenuation information, which in turn allows for material classification [Kaduchak et al., 2000]. PAAs offer a non-contact method to measure the surface normal vibration of objects, eliminating the need for physical contact with the object itself [Mater et al., 1998]. PAAs have also demonstrated their potentials in improving spatial resolution for ultrasound ranging in both air and underwater settings [Lee et al., 2006; Lee et al., 2007; Lee et al., 2009]. By leveraging the unique characteristics of PAAs, such as narrow beam-forming, enhanced spatial resolution can be achieved, advancing ultrasound imaging capabilities. Overall, the versatility and diverse applications of PAAs highlight their significance in many fields, showcasing their potentials for innovation and advancing various scientific and technological domains.

1.1.3 PARAMETRIC ARRAY LOUDSPEAKERS

Besides the underwater and medical applications, PAAs are used in the air for generating highly directional audio beams, known as the *parametric array loudspeakers (PALs)*. The primary wave for a PAL is the airborne ultrasound while the DFW corresponds to the audio sound. The air acts like a low-pass filter to wideband sound signal, so the sum-frequency wave at $f_1 + f_2$ attenuates quickly due to the atmospheric absorption [ISO 9613-1, 1993]. Bennett and Blackstock first experimentally observed the parametric effect in the air in 1975 [Bennett and Blackstock, 1975]. In their experiment, a circular piston transducer transmitted two primary waves at 18.6 kHz and 23.6 kHz, and the anticipated detection of the DFW at 5 kHz was observed.

Subsequent to this study, Yoneyama et al. designed and fabricated the first PAL prototype, drawing inspiration from the principles of PAA [Yoneyama et al., 1983]. As shown in Fig. 1.2, they employed 547 PZT bimorph ultrasonic transducers with a center frequency of 40 kHz to produce audio sound spanning from 200 Hz to 20 kHz. The measurements revealed distinct directivity patterns for audio waves, aligning with the anticipated outcomes. Despite being referred to as the *audio spotlight* in their study, the more prevalent term in the literature is the PAL. Additionally, their findings illustrated that the demodulated audio sound exhibited a rate of decrease of 12 dB/octave as the frequency was halved.

Figure 1.2 Front view of the first PAL prototype, which consists of 547 PZT bimorph ultrasonic transducers with a center frequency of 40 kHz. Extracted from [Yoneyama et al., 1983, Fig. 2].

The illustration of the directional audio beam generated by a PAL is presented in Fig. 1.3. To enhance sound quality, the audio signal undergoes preprocessing and is subsequently amplitude-modulated to a pure-tone ultrasound signal at frequency f_{u}. The modulated signal receives amplification from the amplifier before being emitted. Following this procedure, the PAL emits only ultrasound by its ultrasonic emitters at an ultrasound frequency typically ranging from 40 kHz to 100 kHz, beyond the audible range for humans. The audio signal is automatically *demodulated* in the air due to the nonlinear effect. The emitted ultrasonic beam establishes a virtual end-fire array along the propagation axis, with the effective virtual array length typically determined by the *absorption distance*. This distance is defined as [Sparrow and Kim, 2002, Eq. (15)]

$$\mathscr{D}_{\mathrm{ab}} \equiv \frac{1}{\alpha_{\mathrm{u}}}, \tag{1.2}$$

where α_{u} is the absorption coefficient (in Np/m) of the ultrasound beam at frequency f_{u}. Beyond this absorption distance, the ultrasound beam experiences significant attenuation. The readers may refer to Appendix A for more details on absorption coefficients and distances. When the difference frequency falls within the human hearing range (20 Hz–20 kHz), the audible DFW, also known as the audio beam, is generated through the nonlinear interaction of ultrasound waves in the air.

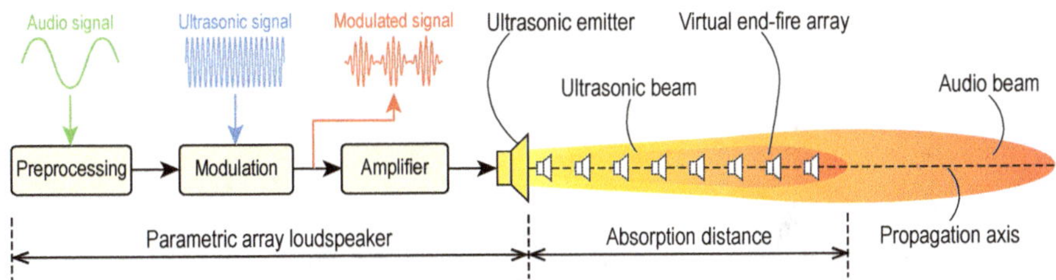

Figure 1.3 Demonstration of the generation of a directional audio beam by a PAL.

Figure 1.4 provides a comparison between the audio sound fields generated by a circular PAL and a circular conventional loudspeaker, both possessing a same aperture size of 100 mm at 1 kHz. The audio beam emitted by the PAL shows significantly enhanced concentration along the propagation axis. Figure 1.5 provides a comparison between far-field directivities generated by a PAL and a conventional loudspeaker, both achieving the same −6 dB angle. Figures 1.5(a), (b), and (c) demonstrate that in order to attain an equivalent level of directivity within the main lobe, the PAL dimensions can be, respectively, 7.25, 5.15, and 35 times smaller at these three cases. Therefore, the PAL outperforms in generating directional audio beams, particularly at lower frequencies and/or smaller aperture sizes.

While PALs possess the ability to generate narrow beams at low frequencies, they do exhibit a slight decrease in beam amplitude with distance, which can be a drawback in certain applications. For instance, the audio beam at 1 kHz radiated by a PAL with the dimensions of 0.6 m × 0.6 m attenuates only around 6 dB at the distance of 4 m along the propagation axis [Holosonics, 2019]. This implies that undesirable

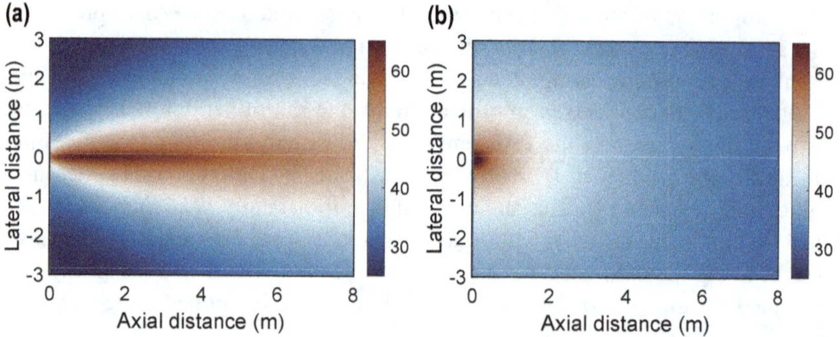

Figure 1.4 Sound pressure level (SPL) distribution at 1 kHz generated by a circular baffled (a) PAL and (b) conventional loudspeaker with the same radius of 100 mm (0.3 wavelengths). The average ultrasound frequency for the PAL is 40 kHz. The on-surface pressure amplitude is (a) $p_0 = 50$ Pa (125 dB) and (b) $p_0 = 0.1$ Pa (71 dB).

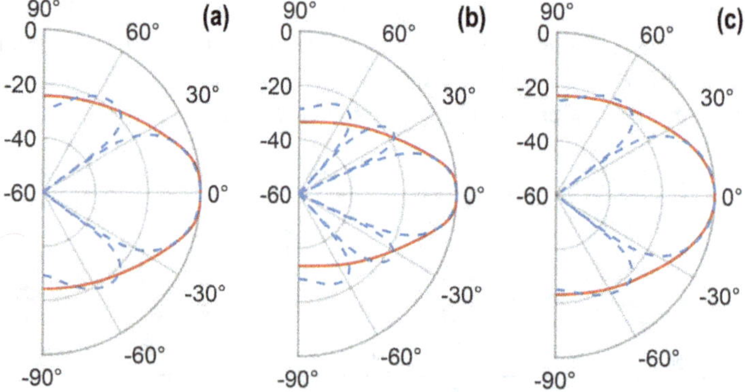

Figure 1.5 Far-field directivities (dB) generated by baffled circular PALs and conventional loudspeakers with the same −6 dB angle. (a) Audio frequency is 500 Hz; PAL radius is 100 mm; conventional loudspeaker radius is 725 mm. (b) Audio frequency is 1 kHz; PAL radius is 100 mm; conventional loudspeaker radius is 515 mm. (c) Audio frequency is 1 kHz; PAL radius is 10 mm; conventional loudspeaker radius is 350 mm. ——, PALs; - - - , conventional loudspeakers.

multiple reflections from walls and the floor can occur when PALs are used in enclosed spaces. To address this issue, a concept known as the *length-limited PAL* has been proposed, enabling the reproduction of audible sound only within a limited range directly in front of the source [Hedberg et al., 2010; Nomura et al., 2012; Lam et al., 2014; Skinner et al., 2019; Liang et al., 2020].

The fundamental concept is canceling the sound produced by one PAL in the far field through the utilization of another PAL operating at a distinct ultrasound frequency. This process generates audible sound with equivalent amplitude but in an opposing phase. While the demodulated audio sound in the near field is complicated and not entirely eliminated, it leads to the creation of an audible zone confined to the near field. Numerical modeling using the finite-difference time-domain method has been employed to simulate the sound propagation generated by an axisymmetric circular length-limited PAL [Nomura et al., 2012]. Figure 1.6 illustrates simulation results demonstrating the feasibility of implementing such a length-limited PAL, with audible sound rapidly decreasing beyond an axial distance of 2 m.

In a study conducted in 2019, pairs of commercially available PALs were used, and a series of measurements were performed to assess the practicality of a length-limited PAL [Skinner et al., 2019]. Overall, the concept of a length-limited PAL addresses the issue of slow beam attenuation over distance, providing a potential solution for mitigating undesirable reflections and optimizing sound propagation in specific applications [Shi and Gan, 2013; Lam et al., 2014; Wang et al., 2019]. However, it is important to note that the trade-off for achieving length limitation is a lower power conversion compared to conventional PALs. As a result, the length-limited PAL may not be able to generate high-volume audio sound.

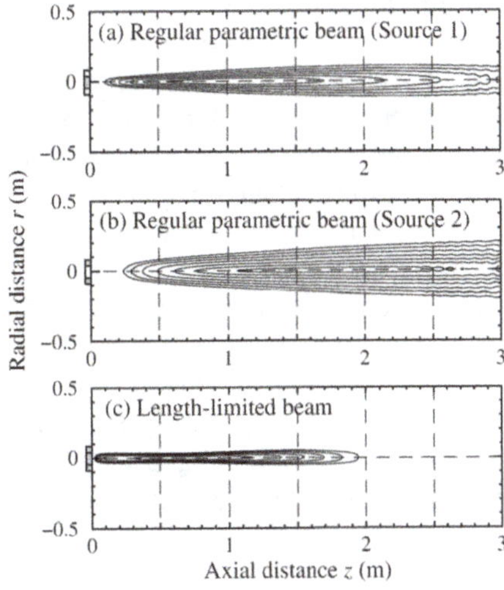

Figure 1.6 Contour lines indicate the SPL normalized by each maximum level from −6 to 0 dB in steps of 1 dB. The carrier frequencies for sources 1 and 2 are 77 kHz and 52 kHz, respectively. Extracted from [Nomura et al., 2012, Fig. 10].

Apart from the limitation of audio sound in the near field of the PAL, there have been studies on remote creation of an audio spot [Ye et al., 2008; Ji et al., 2009; Matsui et al., 2013; Iijima et al., 2019]. The fundamental concept involves the intersection of two high-intensity ultrasound beams in an orthogonal or nearly orthogonal manner, resulting in the localized generation of DFW within the overlapped area. These ultrasound beams are inaudible along their propagation path as they are ultrasonic waves. However, due to the nonlinear interactions between the beams, the sound becomes audible within the overlap region, creating a small locally audible zone. Unlike the conventional PALs, this locally generated audio sound is confined to the overlap region and does not propagate. This characteristic makes the PAL less susceptible to reflections caused by surrounding obstacles or walls. However, the locally generated DFW has a low amplitude and the waveform is more challenging to control. As a result, it is still hard to generate high-volume and high-quality audio sound remotely now with this technique.

1.1.4 APPLICATIONS OF PARAMETRIC ARRAY LOUDSPEAKERS

PALs offer unique capability to deliver highly isolated audio contents to a listener, creating a distinct listening zone while maintaining quiet in other regions. This feature allows PALs to function as virtual headphones, providing an immersive audio experience without the need to wear physical headphones. This characteristic sets PALs apart from conventional omnidirectional loudspeakers and opens up new possibilities in audio applications. This section briefly summarizes some applications for PALs, while more reviews can be found in [Gan et al., 2011; Gan et al., 2012; Reis, 2016; Mei et al., 2023].

Advertising and exhibition. PALs offer significant advantages in projecting product information to customers without disturbing other shoppers. They can be employed in various settings such as retail stores, where a PAL can generate a highly focused sound beam to deliver targeted audio content about a particular product. This allows interested customers to receive detailed information while maintaining privacy and minimizing disruptions to others. In scenarios like ATM kiosks in banks, conventional omnidirectional loudspeakers are typically used for voice playback. PALs can be utilized to create a tightly focused sound beam, enabling private conversations and enhancing the confidentiality of customer interactions. Furthermore, PALs find applications in museums and art galleries. The targeted audio delivery allows visitors in that area to fully immerse themselves in the exhibit experience, gaining detailed information. At the same time, by focusing the audio beam, PALs ensure that other visitors standing nearby are not disturbed or distracted by the audio content. This creates a more enjoyable and respectful environment for all museum and gallery visitors, allowing them to explore and appreciate the exhibits at their own pace [Ortega et al., 2021].

Entertainment. PALs can be integrated with a directional video display system to offer both directional audio and visual content [Pompei, 2002]. This system has the potential to be further developed, enabling multiple family members to share a single television set while each person can individually see and hear their preferred show. In the context of automotive entertainment, PALs can be installed in the cabin to enhance the audio experience. Research has shown that low-frequency audio content can be effectively heard by the intended listener while remaining unnoticed by other passengers, allowing for the creation of a PAL system that can reproduce localized audio fields in different seat regions, catering to individual preferences without disturbing others [Pompei, 2002]. In the gaming industry, PALs can be employed to deliver more immersive and realistic spatial audio, enhancing the overall gaming experience. Furthermore, PALs can be incorporated into mobile devices, such as smartphones and laptops, to provide a highly directional audio beam, thus improving the audio quality of mobile entertainment [Nakashima et al., 2005; Nakashima et al., 2007; Ahn et al., 2019; Nakagawa et al., 2019; Okano and Kajikawa, 2022].

Safety. PALs have proven to be effective in pedestrian warning systems designed to reduce bus-pedestrian accidents in dense urban areas. These systems utilize visual detection to identify pedestrians, particularly when buses are making turns at busy intersections. By employing PALs, a focused beam of sound can be directed specifically toward the detected target pedestrian, minimizing disruption and disturbance to surrounding areas [Burka et al., 2014]. In addition, PALs can be employed to assist visually impaired individuals in navigating streets and avoiding obstacles or hazardous areas [Gan et al., 2011]. By providing localized audio cues, PALs help guiding visually impaired individuals safely across streets and enhance their mobility. Moreover, PALs have showcased their effectiveness in mitigating behavioral and psychological symptoms associated with dementia, consequently alleviating the caregiving burden and offering tailored rehabilitation for an enhanced quality of life among residents [Nishiura et al., 2018]. It is worth noting, however, that there have been concerns regarding the potential misuse of PALs. Some research has demonstrated that PALs can be used to perform secret malicious voice commands on voice assistance systems such as Siri, Google Assistant, and Amazon Alexa [Iijima et al., 2019]. This highlights the importance of implementing appropriate security measures to prevent unauthorized access and misuse of PALs.

Multilingual teleconferencing, meeting, movie viewing, and broadcasting. PALs can enhance multilingual teleconferencing and meetings in large meeting rooms. By utilizing PALs, translations in different languages can be directed to specific PALs and projected to participants sitting in their preferred language zones. This eliminates the need for participants to wear headphones and creates a more seamless communication environment. Similarly, this approach can be applied to watching movies in the same room, where different languages can be simultaneously projected into defined tune-in zones, providing audio content in multiple languages [Gan et al., 2011]. This capability extends to public areas as well, such as broadcasting, where PALs can be utilized to deliver audio content in multiple languages. This ensures that individuals from diverse language backgrounds can access and enjoy the content without the need for separate audio devices or language-specific broadcasts.

Wave field manipulation. Active noise control (ANC) is a technique used to reduce noise in specific target regions by introducing additional loudspeakers, known as secondary sources, to create a quiet zone. When conventional omnidirectional loudspeakers are used as secondary sources, there can be an undesirable amplification of noise in other areas outside the intended quiet zone, known as the spillover effect [Guo et al., 1997; Kidner et al., 2006; Tanaka and Tanaka, 2010; Zhong et al., 2022e]. PALs offer a promising solution to effectively eliminate the spillover effect in ANC systems [Tanaka and Tanaka, 2010; Zhong et al., 2022e]. Additionally, the use of an array of PALs enables the manipulation of the wave field, creating a virtual sound source that provides a highly immersive spatial audio experience for the listener [Ochiai et al., 2017; Ogami et al., 2019]. PALs have been showcased as an alternative to realize sound field reproduction, particularly for high-frequency components (> 2 kHz), which are susceptible to spatial aliasing issues when conventional loudspeakers are employed [Donley et al., 2016; Donley, 2018].

Measurement of material parameters. PALs find applications in the measurement of material parameters, particularly in the accurate assessment of sound absorption, reflection, and transmission coefficients [Humphrey, 1985; Castagnède et al., 2006; Castagnède et al., 2008; Humphrey et al., 2008; Sugahara et al., 2019; Romanova et al., 2019]. Traditionally, precise measurements of these acoustic parameters require an ideal acoustic environment such as a full anechoic chamber or an impedance tube. However, in some cases, it is necessary to conduct measurements in situ, where materials cannot be cut or removed. In situ measurements using conventional omnidirectional loudspeakers can be challenging due to the reflections of acoustic waves from walls or obstacles, which can distort the measurement results. PALs offer a solution to improve the accuracy of in situ measurements. The DFW generated by PALs is quasi-planar along the propagation axis, aligning with the requirements for accurate measurements [ISO 10534-2, 2001]. Furthermore, the highly directional nature of PAL-generated waves eliminates the effects of reflections, ensuring more precise and reliable measurement results. By utilizing PALs in material parameter measurements, researchers and engineers can overcome the limitations of traditional methods and obtain more accurate data in complex and challenging environments.

Other applications. PALs have found applications in various fields beyond those mentioned above. One notable application is in the field of animal vocal communication studies [Johnson-Ulrich et al., 2023]. PALs have proven to be an effective tool in conducting experiments and studying animal behavior and communication patterns. Compared to conventional loudspeakers, PALs offer advantages in terms of their directional control, allowing researchers to precisely direct and manipulate sound stimuli for animal subjects. This enables more accurate and controlled experiments in studying animal vocalizations and their responses to specific auditory stimuli. Furthermore, the development of omnidirectional PALs has expanded their application in acoustic measurements. By installing an array of PALs on a spherical surface, an omnidirectional audio beam can be achieved, particularly at high frequencies [Sayin et al., 2013]. This offers advantages over traditional dodecahedron loudspeakers, making PALs more suitable for certain acoustic measurement tasks [Arnela et al., 2018; Arnela et al., 2020; Arnela et al., 2021; Arnela et al., 2022]. Finally, a study showed that the PAL improves speech understanding compared to a conventional loudspeaker in participants with mild to severe hearing loss [Mehta et al., 2015a; Mehta et al., 2015b]. One reason that participants experienced greater speech intelligibility with PALs may be due to the precise targeting of sound within a narrow beam.

1.1.5 COMMERCIAL PRODUCT EXAMPLES OF PARAMETRIC ARRAY LOUDSPEAKERS

Since the initial PAL prototype fabricated by Yoneyama et al. in 1983 [Yoneyama et al., 1983], there have been significant efforts in the industry to develop commercial products that combine low cost with high performance. It took approximately 16 years for PAL technology to be successfully commercialized. Dr. Joseph Pompei played a crucial role in this process [Pompei, 1999]. Dr. Pompei conducted fundamental research on PALs at the MIT Media Lab and went on to establish Holosonics, a company dedicated to commercializing this innovative technology. Currently, there are numerous companies worldwide that offer PALs, reflecting the growing interest and demand for this technology. Table 1.1 provides information regarding some of these products.

Table 1.1

Curated selection of commercial PAL products, accompanied by details on the corresponding companies and their countries of origin.

Model	Company	Country
Acouspade Speaker	Ultrasonic Audio Technologies	Slovenia
Audio Spotlight	Holosonics	USA
Focusonics	Neurotechnology	Lithuania
Focusound	Audfly Technology (Suzhou)	China
Hypersound	Turtle Beach Corporation	France
MSP-50E	Mistubishi Electric Engineering Company	Japan
Soundlazer	Richard Haberkern	USA

PALs have several disadvantages that limit their applications. Due to the second-order nonlinear nature of the physical mechanism of the PAL, the generated audio sound is generally weak. To produce audible sound at a level of 70 dB, PALs often need to emit ultrasound waves with amplitudes as high as 130 dB (re 20 μPa) because of the low power conversion efficiency from ultrasound to audio sound waves [Fenlon, 1978; Li et al., 2023]. This high power requirement can result in significant power consumption, reaching up to 50 Watts [Holosonics, 2019]. Although micromachined techniques can help reduce power consumption to several Watts, the associated cost remains prohibitive for widespread commercialization [Je et al., 2013; Ahn et al., 2019]. Moreover, prolonged exposure to such high-intensity ultrasound may result in health issues.

Another drawback is the inherent nonlinearity in the demodulation process of audio sound, leading to distortion and degradation of sound quality. The demodulated audio sound amplitude is approximately proportional to the square of its frequency, resulting in a poor low-frequency response for PALs, typically with a cutoff frequency around 1 kHz. This limitation makes it challenging for PALs to compete with conventional loudspeakers in reproducing low-frequency sounds. Efforts have been made to address these issues, such as using virtual bass enhancement techniques [Karnapi et al., 2002; Tao et al., 2021], and focusing the audio beam to enhance low-frequency performance [Ochiai et al., 2017; Ochiai and Hoshi, 2018; Zhong et al., 2022d]. Hybrid loudspeaker systems have also been proposed, incorporating conventional loudspeakers or subwoofers to handle low-frequency content alongside PALs to leverage their directional audio capabilities [Sugibayashi et al., 2012; Donley et al., 2016; Donley, 2018].

1.2 FUNDAMENTAL EQUATIONS FOR ACOUSTIC WAVES IN FLUIDS

Acoustic waves are mechanical waves that propagate in a medium. This section begins to derive the wave equations for PALs by obtaining the fundamental equations for acoustic waves in fluids. These fundamental equations encompass the continuity equation (Sec. 1.2.1), Navier-Stokes equation (Sec. 1.2.2), Kirchhoff-Fourier equation (Sec. 1.2.3), and the equation of state (Sec. 1.2.4). These equations are derived from the principles of conservation of mass, momentum, energy, and the laws of thermodynamics, respectively. Solving these equations directly can be challenging due to their complexity. To overcome this challenge, in Sec. 1.2.5, a second-order approximation approach is introduced. This approximation simplifies the equations, making them more manageable to solve and providing valuable insights into the behaviors of acoustic waves generated by PALs.

In order to model the acoustic waves generated by PALs, it is essential to consider a diffusive medium that incorporates the effects of viscosity and heat conduction. To describe the state of the fluid, an arbitrary fixed volume V with the boundary denoted as $\partial V \equiv \mathbf{S} \equiv S\mathbf{n}$ is considered here, where S represents the surface area and $\mathbf{n}$ denotes the unit vector pointing outward from the surface. At any given spatial point (described by Eulerian coordinates) $\mathbf{r} = (x, y, z)$ and time t, the state of the fluid is determined by six variables: three state variables including pressure, $P(\mathbf{r}, t)$, density, $\rho(\mathbf{r}, t)$, temperature, $T(\mathbf{r}, t)$; and three components of particle velocity, $\mathbf{v}(\mathbf{r}, t) = (v_x(\mathbf{r}, t), v_y(\mathbf{r}, t), v_z(\mathbf{r}, t))$. Hence, a system of six equations is necessary to solve for these functions and determine the state of the medium at any point $\mathbf{r}$ and time t.

Of these equations, five are derived from the principles of conservation, which include the conservation of mass, the conservation of momentum in the three Cartesian components, and the conservation of energy. The sixth equation is a state equation that relates the properties of the fluid. By considering these six equations, one can gain a comprehensive understanding of the behavior of the medium and effectively model the acoustic waves generated by PALs.

1.2.1 CONSERVATION OF MASS – CONTINUITY EQUATION

The conservation of the mass for a fluid implies that the increase rate of the mass within a volume V is equal to the net flow of mass through the boundary $\mathbf{S}$ caused by the fluid motion. This equality can be expressed in integral form as [Enflo and Hedberg, 2002, Eq. (2.1)]

$$\frac{\partial}{\partial t} \iiint_V \rho \mathrm{d}V = -\oiint_S \mathbf{j}_{\mathrm{m}} \cdot \mathrm{d}\mathbf{S}, \tag{1.3}$$

where the minus sign on the right-hand side denotes that the mass flows out of the surface, indicating a loss of the mass, and the mass flux is defined as

$$\mathbf{j}_{\mathrm{m}} = \rho \mathbf{v}. \tag{1.4}$$

Transformation of the surface integral in Eq. (1.3) gives

$$\iiint_V \left(\frac{\partial \rho}{\partial t} + \nabla \cdot \mathbf{j}_{\mathrm{m}} \right) \mathrm{d}V = 0. \tag{1.5}$$

The *divergence theorem* (also known as the *Gauss's theorem*) is used in the derivation, which states that the surface integral of a vector field **f** over a closed surface **S** (the flux through the surface) is equal to the volume integral of the divergence over the region inside the surface. The mathematical form is [Kreyszig et al., 2011, p. 453]

$$\iiint_V \nabla \cdot \mathbf{f} \mathrm{d}V = \oiint_S \mathbf{f} \cdot \mathrm{d}\mathbf{S}. \tag{1.6}$$

Since the volume V is arbitrary, the integrand in Eq. (1.5) must vanish to give

$$\frac{\partial \rho}{\partial t} + \nabla \cdot \mathbf{j}_{\mathrm{m}} = 0. \tag{1.7}$$

Using the *substantial time derivative* (also known as *material derivative*) [Enflo and Hedberg, 2002, Eq. (2.5)]

$$\frac{\mathrm{d}}{\mathrm{d}t} \equiv \frac{\partial}{\partial t} + \mathbf{v} \cdot \nabla, \tag{1.8}$$

Eq. (1.7) becomes [Enflo and Hedberg, 2002, Eq. (2.6)]

$$\frac{\mathrm{d}\rho}{\mathrm{d}t} + \rho \nabla \cdot \mathbf{v} = 0. \tag{1.9}$$

This is the *continuity equation* resulted from the conservation of the mass.

The derivation of the substantial time derivative operator, as presented in Eq. (1.8), follows the following steps. Consider an arbitrary fluid particle described in Eulerian coordinates, and let its initial position at time t be denoted by $\mathbf{r}$. After a small time interval of Δt, this particle undergoes displacement to a new position $\mathbf{r} + \Delta \mathbf{r}$. For any state variable ϕ associated with this particle, denoted as $\phi(\mathbf{r}, t)$ at location $\mathbf{r}$ and time t, the state variable undergoes a transformation to $\phi(\mathbf{r} + \Delta \mathbf{r}, t + \Delta t)$ after the time interval Δt. Consequently, the substantial time derivative of any state variable for this particle can be determined as follows

$$\begin{aligned}
\frac{\mathrm{d}\phi(\mathbf{r},t)}{\mathrm{d}t} &= \lim_{\Delta t \to 0} \frac{\phi(\mathbf{r}+\Delta\mathbf{r}, t+\Delta t) - \phi(\mathbf{r},t)}{\Delta t} \\
&= \lim_{\Delta t \to 0} \left[\frac{\phi(\mathbf{r}+\Delta\mathbf{r}, t+\Delta t) - \phi(\mathbf{r}+\Delta\mathbf{r},t)}{\Delta t} - \frac{\phi(\mathbf{r}+\Delta\mathbf{r},t) - \phi(\mathbf{r},t)}{\Delta t} \right] \\
&= \frac{\partial \phi}{\partial t} + \left(\lim_{\Delta t \to 0} \frac{\Delta \mathbf{r}}{\Delta t} \cdot \nabla \right) \phi \\
&= \left(\frac{\partial}{\partial t} + \mathbf{v} \cdot \nabla \right) \phi.
\end{aligned} \tag{1.10}$$

1.2.2 CONSERVATION OF MOMENTUM – NAVIER-STOKES EQUATIONS

The principle of conservation of momentum states that the rate of change of momentum within a volume V is equivalent to the net flow of momentum through the boundary $\mathbf{S}$ caused by the fluid motion, as well as the surface force exerted on the fluid. This equality can be expressed in an integral form as [Enflo and Hedberg, 2002, Eq. (2.7)]

$$\frac{\partial}{\partial t}\iiint_V \rho \mathbf{v} \mathrm{d}V = -\oiint_S \mathbf{J}\cdot \mathrm{d}\mathbf{S} + \oiint_S \sigma \cdot \mathrm{d}\mathbf{S}. \tag{1.11}$$

Here, the *momentum tensor* is defined as

$$\mathbf{J} = \rho \mathbf{v} \otimes \mathbf{v}, \tag{1.12}$$

where $\otimes$ denotes the tensor product, and $\rho\mathbf{v}$ is referred to as the *Cartesian momentum density component*. The *Cauchy stress tensor* σ can be expressed as the summation of a viscosity term (the *viscous stress*), τ, and a pressure term (*volumetric stress*), $-P\mathbf{I}$, as

$$\sigma = \tau - P\mathbf{I}, \tag{1.13}$$

where $\mathbf{I}$ is a unit tensor.

Writing the surface integrals in Eq. (1.11) as volume integrals by using the divergence theorem [Eq. (1.6)] and the arbitrariness of the volume V, it is obtained that

$$\frac{\partial(\rho \mathbf{v})}{\partial t} = \nabla \cdot (-\mathbf{J} + \sigma). \tag{1.14}$$

Using the continuity equation (1.7) and the identity,

$$\nabla \cdot (\mathbf{a} \otimes \mathbf{b}) = \mathbf{a}(\nabla \cdot \mathbf{b}) + \mathbf{a} \cdot \nabla \mathbf{b}, \tag{1.15}$$

with arbitrary vectors $\mathbf{a}$ and $\mathbf{b}$, Eq. (1.14) is transformed to the *Newton's second law of motion* as

$$\rho \frac{\mathrm{d}\mathbf{v}}{\mathrm{d}t} = \nabla \cdot \sigma, \tag{1.16}$$

where $\mathrm{d}\mathbf{v}/\mathrm{d}t$ corresponds to the acceleration of a mass element. Substituting Eq. (1.13) into Eq. (1.16) yields the *Cauchy momentum equation*

$$\rho \frac{\mathrm{d}\mathbf{v}}{\mathrm{d}t} = \nabla \cdot \tau - \nabla P. \tag{1.17}$$

Thermoviscous Fluids

In thermoviscous fluids, the viscous stress tensor is linearly related to the gradient of the velocity, $\nabla\mathbf{v}$, by [Cheng, 2019a, Eq. (6.1.5a)]

$$\tau_{ij} = c_{ijkl}\frac{\partial v_k}{\partial x_l}, \tag{1.18}$$

where c_{ijkl} has $3^4 = 81$ coefficients in general. In Eq. (1.18), the *Einstein's summation convention* is used for simplicity [Hassani, 2013, Sec. 13.2]. Specifically, when an index variable appears twice in a single term, it implies a summation of that term over all the values of the index. For three-dimensional (3D) physical problems, the summation is over the set $\{1,2,3\}$ for Cartesian coordinates (x_1, x_2, x_3). For instance,

$$c_i x_i \equiv c_1 x_1 + c_2 x_2 + c_3 x_3. \tag{1.19}$$

In the case of a homogeneous fluid, it can be demonstrated that only two coefficients, denoted as μ_1 and μ_2, are independent. As a result, the viscous stress tensor can be simplified to

$$\tau_{ij} = \mu_1 \left(\frac{\partial v_i}{\partial x_j} + \frac{\partial v_j}{\partial x_i} \right) + \mu_2 \frac{\partial v_k}{\partial x_k} \delta_{ij}, \tag{1.20}$$

where δ_{ij} is the *Kronecker delta function*. By defining

$$\begin{cases} \mu_1 = \mu_s, \\ \mu_2 = \mu_b - \frac{2}{3}\mu_s, \end{cases} \qquad \begin{cases} \mu_s = \mu_1, \\ \mu_b = \frac{2}{3}\mu_1 + \mu_2, \end{cases} \tag{1.21}$$

Eq. (1.20) is rewritten as

$$\tau_{ij} = \mu_s \left(\frac{\partial v_i}{\partial x_j} + \frac{\partial v_j}{\partial x_i} - \frac{2}{3} \frac{\partial v_k}{\partial x_k} \delta_{ij} \right) + \mu_b \frac{\partial v_k}{\partial x_k} \delta_{ij}, \quad i, j = 1, 2, 3. \tag{1.22}$$

Here, μ_s is denoted as the *shear viscosity coefficient*, also known as *dynamic viscosity*. In the case of air, at 25°C and 1 bar, $\mu_s \approx 1.85 \times 10^{-5}$ Pa · s [Shang et al., 2019]. On the other hand, μ_b is referred to as the *bulk viscosity coefficient*, which is also known as *volume viscosity* or *second viscosity*. Bulk viscosity accounts for losses attributed to the expansion and compression of fluids and quantifies the difference between mechanical and thermodynamic pressures. As a rough approximation, the bulk viscosity can be estimated as $\mu_b \approx 0.6\mu_s$ [Bruno et al., 2023].

It is convenient to introduce the *strain rate tensor* (also known as the *deformation velocity tensor*)

$$\mathbf{D} \equiv \frac{1}{2}\left[\nabla \mathbf{v} + (\nabla \mathbf{v})^{\mathrm{T}}\right] \quad \text{or} \quad D_{ij} \equiv \frac{1}{2}\left(\frac{\partial v_i}{\partial x_j} + \frac{\partial v_j}{\partial x_i} \right), \tag{1.23}$$

which has the symmetric property $D_{ij} = D_{ji}$. Equations (1.20) and (1.22) can then be written as

$$\boldsymbol{\tau} = 2\mu_1 \mathbf{D} + \mu_2 (\nabla \cdot \mathbf{v}) \mathbf{I} = 2\mu_s \mathbf{D} + \left(\mu_b - \frac{2}{3}\mu_s \right) (\nabla \cdot \mathbf{v}) \mathbf{I}. \tag{1.24}$$

Substituting Eq. (1.24) into the Cauchy momentum equation (1.17) yields [Enflo and Hedberg, 2002, Eq. (2.13)]

$$\rho \frac{\mathrm{d}\mathbf{v}}{\mathrm{d}t} = -\nabla P + \mu_s \nabla^2 \mathbf{v} + \left(\mu_b + \frac{\mu_s}{3} \right) \nabla (\nabla \cdot \mathbf{v}). \tag{1.25}$$

Equation (1.25) is known as the *Navier-Stokes equation* in the case of irrotational flow. In the special case where $\mu_b = \mu_s = 0$, Eq. (1.27), along with the continuity equation (1.9), constitutes *Euler's fundamental hydrodynamic equations.*

In the case of acoustic wave propagation, the fluid is assumed to be irrotational. Utilizing the vector Laplacian [Cheng, 2019a, Eq. (9.2.2)]

$$\nabla^2 \mathbf{a} = \nabla(\nabla \cdot \mathbf{a}) - \nabla \times (\nabla \times \mathbf{a}), \tag{1.26}$$

and since $\nabla \times \mathbf{v} = 0$ for irrotational fluids, Eq. (1.25) can be simplified to be the following form [Aanonsen et al., 1984, Eq. (2)]

$$\rho \frac{d\mathbf{v}}{dt} = -\nabla P + \left(\mu_b + \frac{4}{3}\mu_s\right)\nabla^2 \mathbf{v}. \tag{1.27}$$

1.2.3 CONSERVATION OF ENERGY—KIRCHHOFF-FOURIER EQUATION

The principle of conservation of energy states that the rate of reduction of energy within a fixed volume V, with boundary $\mathbf{S}$, contains the net internal energy flow from V through $\mathbf{S}$ due to the fluid motion, the net thermal flow out from V through $\mathbf{S}$, and the work per second done by the fluid within the volume on its surroundings.

The *specific energy* (the total energy per mass) is defined as

$$e = u + \frac{1}{2}v^2, \tag{1.28}$$

where u is the *specific internal energy*, and $v^2/2$ is the *specific kinetic energy*. The integral form of the conservation of energy can then be expressed as [Enflo and Hedberg, 2002, Eq. (2.14)]

$$\frac{\partial}{\partial t}\iiint_V \rho e dV = -\oiint_S \mathbf{j}_e \cdot d\mathbf{S} - \oiint_S \mathbf{v} \cdot (\boldsymbol{\sigma} \cdot d\mathbf{S}) + \oiint_S \mathbf{q} \cdot d\mathbf{S}, \tag{1.29}$$

where the *energy flux density*

$$\mathbf{j}_e \equiv \rho e \mathbf{v}, \tag{1.30}$$

the *thermal flux density* is

$$\mathbf{q} = -\kappa \nabla T, \tag{1.31}$$

and κ is the *thermal conductivity* (also known as the *heat conduction number*). In the air, $\kappa \approx 0.026\,\mathrm{W \cdot m^{-1} \cdot K^{-1}}$ at 25°C and 1 bar [Shang et al., 2019].

Similarly, by applying the divergence theorem (1.6) to transform the surface integrals in Eq. (1.29) into volume integrals, and considering the arbitrariness of the volume V, the conservation of energy equation can be expressed in its differential form as

$$\frac{\partial(\rho e)}{\partial t} = \nabla \cdot (-\mathbf{j}_e + \boldsymbol{\sigma} \cdot \mathbf{v} - \mathbf{q}). \tag{1.32}$$

By using the identity of the divergence of the product of a scalar field ψ and a vector field $\mathbf{a}$

$$\nabla \cdot (\psi \mathbf{a}) = \psi \nabla \cdot \mathbf{a} + \mathbf{a} \cdot \nabla \psi, \tag{1.33}$$

as well as the continuity equation (1.9), Eq. (1.32) can be simplified to

$$\rho \frac{\mathrm{d}e}{\mathrm{d}t} = \boldsymbol{\nabla} \cdot (\boldsymbol{\sigma} \cdot \mathbf{v} - \mathbf{q}). \tag{1.34}$$

From the first and second laws of thermodynamics, one has the relation

$$\mathrm{d}u = T\mathrm{d}s - P\mathrm{d}(\rho^{-1}) = T\mathrm{d}s + \frac{P}{\rho^2}\mathrm{d}\rho, \tag{1.35}$$

where s is the *specific entropy*. Therefore, it is obtained that

$$\frac{\mathrm{d}u}{\mathrm{d}t} = T\frac{\mathrm{d}s}{\mathrm{d}t} + \frac{P}{\rho^2}\frac{\mathrm{d}\rho}{\mathrm{d}t}. \tag{1.36}$$

Substituting Eqs. (1.36), (1.28), and (1.13) into Eq. (1.34) yields

$$\rho T\frac{\mathrm{d}s}{\mathrm{d}t} + \frac{P}{\rho}\left(\frac{\mathrm{d}\rho}{\mathrm{d}t} + \rho\boldsymbol{\nabla}\cdot\mathbf{v}\right) + \mathbf{v}\cdot\left(\rho\frac{\mathrm{d}\mathbf{v}}{\mathrm{d}t} + \boldsymbol{\nabla}P\right) = \boldsymbol{\nabla}\cdot(\boldsymbol{\tau}\cdot\mathbf{v} - \mathbf{q}), \tag{1.37}$$

where the identity [Cheng, 2019a, Eq. (9.2.2)]

$$(\mathbf{v}\cdot\boldsymbol{\nabla})\mathbf{v} = \boldsymbol{\nabla}v^2/2 - \mathbf{v}\times(\boldsymbol{\nabla}\times\mathbf{v}), \tag{1.38}$$

is used. By using the continuity equation (1.9) and the Cauchy momentum equation (1.17), Eq. (1.37) reduces to

$$\rho T\frac{\mathrm{d}s}{\mathrm{d}t} = \boldsymbol{\nabla}\cdot(\boldsymbol{\tau}\cdot\mathbf{v} - \mathbf{q}) - \mathbf{v}\cdot(\boldsymbol{\nabla}\cdot\boldsymbol{\tau}). \tag{1.39}$$

Substituting the viscous stress tensor Eq. (1.24) into Eq. (1.37), the entropy equation is obtained as [Enflo and Hedberg, 2002, Eq. (2.30); Hamilton and Blackstock, 2008, Eq. (3-3)]

$$\rho T\frac{\mathrm{d}s}{\mathrm{d}t} = \kappa\nabla^2 T + \frac{1}{2}\mu_{\mathrm{s}}\boldsymbol{\phi}:\boldsymbol{\phi} + \mu_{\mathrm{b}}(\boldsymbol{\nabla}\cdot\mathbf{v})^2, \tag{1.40}$$

where the colon symbol ":" represents the inner product of tensors, and a symmetric traceless tensor is introduced as

$$\boldsymbol{\phi} = 2\mathbf{D} - \frac{2}{3}(\boldsymbol{\nabla}\cdot\mathbf{v})\mathbf{I}, \tag{1.41}$$

which satisfies that $\phi_{ij} = \phi_{ji}$ and $\mathrm{Tr}(\boldsymbol{\phi}) = 0$. Equation (1.40) is known as the *Kirchhoff-Fourier equation*.

1.2.4 STATE EQUATION OF THE FLUID

The pressure, P, can be expressed as a function of the density, ρ, and the entropy per mass, s, as

$$P = P(\rho, s). \tag{1.42}$$

Suppose p, ρ', s' are small disturbance of P, ρ, s, respectively. The *state equation* of $P_0 + p = P(\rho_0 + \rho', s_0 + s')$ can be expanded as [Cheng, 2019a, Eq. (9.2.3d)]

$$p = \left(\frac{\partial P}{\partial \rho}\right)_{s,0} \rho' + \left(\frac{\partial P}{\partial s}\right)_{\rho,0} s' + \frac{1}{2}\left(\frac{\partial^2 P}{\partial \rho^2}\right)_{s,0} \rho'^2 + \frac{1}{2}\left(\frac{\partial^2 P}{\partial s^2}\right)_{\rho,0} s'^2 + \cdots, \tag{1.43}$$

where P_0, ρ_0, s_0 are corresponding equilibrium values.

1.2.5 SECOND-ORDER APPROXIMATION

It is challenging to directly solve the full system of fundamental equations (1.9), (1.27), (1.40), and (1.42). In the field of nonlinear acoustics, a common approach is to employ perturbation approximation, which aims to simplify the equations to second order in terms of the acoustic *Mach number*. The acoustic Mach number,

$$\varepsilon \equiv \frac{v_0}{c_0}, \tag{1.44}$$

defined as the ratio of the peak particle velocity v_0 to the small-signal speed of sound, serves as a small parameter. It is required that $\varepsilon \ll 1$ for this approximation to be valid. For instance, a typical value of $\varepsilon = 0.01$ corresponds to $v_0 = 3.43\,\mathrm{m/s}$, resulting in a pressure amplitude of $p_0 = \rho_0 c_0 v_0 = 1.42\,\mathrm{kPa}$ $(154\,\mathrm{dB}$ re $20\,\mu\mathrm{Pa})$.

In addition, a second small parameter is assumed as [Hamilton and Blackstock, 2008, p. 49]

$$\eta = \frac{\mu_s \omega}{\rho_0 c_0^2}, \tag{1.45}$$

where ω is a characteristic angular frequency. Physically, η measures the importance of viscosity in a plane progressive sound wave, relative to the fluctuating pressure. For a typical carrier ultrasound frequency for the PAL at 50 kHz, the value of η is approximately 10^{-5}. As a result, η can be considered a small parameter, and it is advisable to retain the first-order term of η, denoted as $\mathcal{O}(\eta)$, in the equations.

The corresponding small parameter associated with heat conduction is [Hamilton and Blackstock, 2008, p. 49]

$$\frac{\kappa \omega}{\rho_0 c_0^2 c_P} = \frac{\eta}{\mathrm{Pr}}. \tag{1.46}$$

Here, the *Prandtl number* is defined as [Hamilton and Blackstock, 2008, p. 49]

$$\mathrm{Pr} = \frac{\mu c_P}{\kappa}, \tag{1.47}$$

where c_P is the specific heat at constant pressure. The Prandtl number measures the ratio of momentum diffusivity to thermal diffusivity. It is $\mathcal{O}(1)$ in terms of ε and η.

Following [Hamilton and Blackstock, 2008, Chap. 3], this book considers both ε and η to be of comparable smallness. To simplify matters, a generic small parameter, denoted as $\tilde{\varepsilon}$, is introduced that characterizes the smallness of both ε and η. Throughout the subsequent sections, equations are derived valid up to order $\tilde{\varepsilon}^2$. Specifically, the terms of order $\varepsilon, \eta, \eta\varepsilon$, and ε^2 are retained in these derivations.

In the following analysis, it is assumed that

$$P = P_0 + p, \quad \rho = \rho_0 + \rho', \quad T = T_0 + T', \quad s = s_0 + s', \quad \mathbf{v} = \mathbf{v}_0 + \mathbf{v}', \tag{1.48}$$

where p, ρ', T', s' represent small disturbances relative to the equilibrium values P_0, ρ_0, T_0, s_0, respectively. It is noted that in the equilibrium state, the fluid is typically assumed to be static, leading to $\mathbf{v}_0 = \mathbf{0}$. Consequently, the disturbance of the velocity, denoted as $\mathbf{v}'$, is equivalent to the actual velocity, $\mathbf{v}$. In summary, the variables $p, \rho', T', s', \mathbf{v}$ are first-order terms $\mathcal{O}(\tilde{\varepsilon})$.

Continuity Equation

By substituting $\rho = \rho_0 + \rho'$ into the continuity equation (1.9) and assuming that the density is independent of spatial coordinates $\nabla \rho_0$, the equation up to the second order $\mathcal{O}(\varepsilon^2)$ is obtained as [Hamilton and Blackstock, 2008, Eq. (3-30)]

$$\frac{\partial \rho'}{\partial t} + \rho_0 \nabla \cdot \mathbf{v} = -\rho' \nabla \cdot \mathbf{v} - \mathbf{v} \cdot \nabla \rho'. \tag{1.49}$$

The first-order version of Eq. (1.49) is commonly used in modeling linear acoustics, and can be expressed as

$$\frac{\partial \rho'}{\partial t} + \rho_0 \nabla \cdot \mathbf{v} = 0. \tag{1.50}$$

Navier-Stokes Equation

By utilizing Eqs. (1.26) and (1.38), and taking into account that $\nabla \times \mathbf{v} = 0$ for irrotational flow, the Navier-Stokes equation (1.27) can be simplified by discarding the components smaller than second-order as [Hamilton and Blackstock, 2008, Eq. (3-32); Cheng, 2019a, Eq. (9.2.3b)]

$$\rho_0 \frac{\partial \mathbf{v}}{\partial t} + \nabla p = \left(\mu_\mathrm{b} + \frac{4}{3} \mu_\mathrm{s} \right) \nabla (\nabla \cdot \mathbf{v}) - \rho' \frac{\partial \mathbf{v}}{\partial t} - \frac{\rho_0}{2} \nabla v^2. \tag{1.51}$$

The first-order version of Eq. (1.51) is expressed as

$$\rho_0 \frac{\partial \mathbf{v}}{\partial t} = -\nabla p. \tag{1.52}$$

Kirchhoff-Fourier Equation

Similarly, under the second-order approximation, the Kirchhoff-Fourier equation (1.40) can be simplified to

$$\rho_0 T_0 \frac{\partial s'}{\partial t} = \kappa \nabla^2 T. \tag{1.53}$$

State Equation

Under the second-order approximation, the equation of state, as given by Eq. (1.42), is $P_0 + p = P(\rho_0 + \rho', s_0 + s')$, which can be simplified to

$$p \approx \rho' c_0^2 + \frac{1}{2} \frac{B}{A} \frac{c_0^2}{\rho_0} \rho'^2 + \left(\frac{\partial P}{\partial s} \right)_{\rho,0} s', \tag{1.54}$$

where the *nonlinear parameter* B/A is defined as

$$\frac{B}{A} \equiv \rho_0 \left(\frac{\partial^2 P}{\partial \rho^2}\right)_{s,0} \Big/ \left(\frac{\partial P}{\partial \rho}\right)_{s,0} = \frac{\rho_0}{c_0^2}\left(\frac{\partial^2 P}{\partial \rho^2}\right)_{s,0}. \tag{1.55}$$

Combining Eqs. (1.53) and (1.54) and utilizing the thermodynamic relations, the entropy s' can be removed to give [Cheng, 2019a, Eq. (9.2.4c)]

$$p = \rho' c_0^2 + \frac{B}{2A}\frac{c_0^2}{\rho_0}\rho'^2 - \kappa\left(\frac{1}{c_V} - \frac{1}{c_P}\right)\nabla \cdot \mathbf{v}, \tag{1.56}$$

where c_P and c_V are the specific heats at constant pressure and volume. It can be observed from Eq. (1.56) that the nonlinear parameter B/A is the coefficient of the second-order quantity ρ'^2. As a result, the magnitude of wave distortion, influenced by second-order nonlinearity, is determined by this nonlinear parameter.

The first-order version of Eq. (1.56) is expressed as

$$p = \rho' c_0^2. \tag{1.57}$$

Equations (1.49), (1.51), and (1.56) are main results obtained in this section. These equations are approximated to the second-order of small quantities and serve as the foundation for developing the single nonlinear wave equation in Sec. 1.3.

1.3 NONLINEAR EQUATIONS FOR PALS

In Sec. 1.2, the fundamental equations governing the propagation of acoustic waves in fluids, along with their second-order approximations, are derived. The objective of this section is to utilize these equations to derive a single second-order nonlinear wave equation that can be employed to solve for the sound field generated by PALs. This section presents commonly used nonlinear equations for modeling PALs, including the general second-order nonlinear wave equation (Sec. 1.3.1), the Kuznetsov equation (Sec. 1.3.2), the Westervelt equation (Sec. 1.3.3), and the KZK equation (Sec. 1.3.4). These equations provide valuable tools for studying and analyzing the behavior of PALs and their acoustic wave generation.

1.3.1 GENERAL SECOND-ORDER NONLINEAR WAVE EQUATION

The derivation of the general second-order nonlinear wave equation requires the following corollary: $\mathcal{O}(\varepsilon)$ acoustic relations can be substituted into any $\mathcal{O}(\varepsilon^2)$ terms, since the resulting errors are $\mathcal{O}(\varepsilon^3)$. As a result, the first term on the right-hand side of Eq. (1.49) can be rewritten as [Hamilton and Blackstock, 2008, Eq. (3-35)]

$$-\rho'\nabla \cdot \mathbf{v} = -\left(\frac{p}{c_0^2}\right)\left(-\frac{1}{\rho_0}\frac{\partial \rho'}{\partial t}\right) = \frac{p}{\rho_0 c_0^4}\frac{\partial p}{\partial t} = \frac{1}{2\rho_0 c_0^4}\frac{\partial p^2}{\partial t}. \tag{1.58}$$

Here, ρ' and $\nabla \cdot \mathbf{v}$ are substituted based on Eqs. (1.57) and (1.50), respectively. The second term on the right-hand side of Eq. (1.49) can be rewritten as

$$-\mathbf{v} \cdot \nabla \rho' = -\frac{1}{c_0^2}\mathbf{v} \cdot \nabla p = \frac{\rho_0}{c_0^2}\mathbf{v} \cdot \frac{\partial \mathbf{v}}{\partial t} = \frac{\rho_0}{2c_0^2}\frac{\partial v^2}{\partial t}. \tag{1.59}$$

Here, the first and second equal signs are derived using Eqs. (1.57) and (1.52), respectively.

Substituting Eqs. (1.58) and (1.59) into the second-order continuity equation (1.49), it is obtained that [Hamilton and Blackstock, 2008, Eq. (3-36)]

$$\frac{\partial \rho'}{\partial t} + \rho_0 \nabla \cdot \mathbf{v} = \frac{1}{\rho_0 c_0^4}\frac{\partial p^2}{\partial t} + \frac{1}{c_0^2}\frac{\partial \mathcal{L}}{\partial t}, \tag{1.60}$$

where the second-order *Lagrangian density* is defined as

$$\mathcal{L} = \frac{\rho_0 v^2}{2} - \frac{p^2}{2\rho_0 c_0^2}, \tag{1.61}$$

representing the subtraction of the potential energy from the kinetic energy. In fact, the Lagrangian density characterizes the local (non-cumulative) nonlinearity [Červenka and Bednařík, 2019; Zhong et al., 2021b].

The first term on the right-hand side of the second-order Navier-Stokes equation (1.51) can be rewritten as

$$\begin{aligned}\left(\mu_b + \frac{4}{3}\mu_s\right)\nabla(\nabla \cdot \mathbf{v}) &= \left(\mu_b + \frac{4}{3}\mu_s\right)\nabla\left(-\frac{1}{\rho_0}\frac{\partial \rho'}{\partial t}\right)\\ &= -\frac{1}{\rho_0 c_0^2}\left(\mu_b + \frac{4}{3}\mu_s\right)\nabla\frac{\partial p}{\partial t}.\end{aligned} \tag{1.62}$$

Here, the first and second equal signs are derived using Eqs. (1.50) and (1.57), respectively. The second term on the right-hand side of Eq. (1.51) can be rewritten as

$$\rho'\frac{\partial \mathbf{v}}{\partial t} = \left(\frac{p}{c_0^2}\right)\left(-\frac{\nabla p}{\rho_0}\right) = -\frac{\nabla p^2}{2\rho_0 c_0^2}, \tag{1.63}$$

where Eqs. (1.57) and (1.52) are used.

As a result, substituting Eqs. (1.62) and (1.63) into Eq. (1.51) results in [Hamilton and Blackstock, 2008, Eq. (3-37)]

$$\rho_0\frac{\partial \mathbf{v}}{\partial t} + \nabla p = -\frac{1}{\rho_0 c_0^2}\left(\mu_b + \frac{4}{3}\mu_s\right)\nabla\frac{\partial p}{\partial t} - \nabla \mathcal{L}. \tag{1.64}$$

It is interesting to note that $\mathcal{L} = 0$ for progressive plane waves due to the plane wave relation $p = \rho_0 c_0 v$ at first order. In this particular scenario, the momentum equation (1.64) is linear, thus making no contribution to the nonlinearity at second order [Hamilton and Blackstock, 1990].

By using Eq. (1.57), Eq. (1.56) can be rewritten as

$$\rho' = \frac{p}{c_0^2} - \frac{B}{2A}\frac{p^2}{\rho_0 c_0^4} + \frac{\kappa}{c_0^2}\left(\frac{1}{c_V} - \frac{1}{c_P}\right)\nabla\cdot\mathbf{v}. \tag{1.65}$$

By using Eqs. (1.50) and (1.57), the third term on the right-hand side of Eq. (1.65) can be expressed with respect to p, and Eq. (1.65) is rewritten as [Hamilton and Blackstock, 2008, Eq. (3-40)]

$$\rho' = \frac{p}{c_0^2} - \frac{B}{2A}\frac{p^2}{\rho_0 c_0^4} - \frac{\kappa}{\rho_0 c_0^4}\left(\frac{1}{c_V} - \frac{1}{c_P}\right)\frac{\partial p}{\partial t}. \tag{1.66}$$

Using Eq. (1.66) to eliminate ρ' in Eq. (1.60), it is obtained that [Aanonsen et al., 1984, Eq. (6); Tjøtta and Tjøtta, 1987b, Eq. (5)]

$$\frac{\partial p}{\partial t} + \rho_0 c_0^2 \nabla\cdot\mathbf{v} = \frac{\beta}{\rho_0 c_0^2}\frac{\partial p^2}{\partial t} + \frac{\kappa}{\rho_0 c_0^2}\left(\frac{1}{c_V} - \frac{1}{c_P}\right)\frac{\partial^2 p}{\partial t^2} + \frac{\partial \mathcal{L}}{\partial t}, \tag{1.67}$$

where the *nonlinear coefficient* is defined as [Lighthill, 1978, Sec. I.13]

$$\beta \equiv 1 + \frac{B}{2A}, \tag{1.68}$$

which has a typical value of 1.2 in the air.

Subtracting the time derivative of Eq. (1.67) from the divergence of Eq. (1.64) to eliminate $\mathbf{v}$, and substituting the first-order wave equation $\left(\nabla^2 - c_0^{-2}\,\partial^2/\partial t^2\right)p = \mathcal{O}(\tilde{\varepsilon}^2)$, into the viscosity term lead to [Aanonsen et al., 1984, Eq. (9); Hamilton and Blackstock, 2008, Eq. (3-41)]

$$\left(\nabla^2 - \frac{1}{c_0^2}\frac{\partial^2}{\partial t^2}\right)p + \frac{\delta}{c_0^4}\frac{\partial^3 p}{\partial t^3} = -\frac{\beta}{\rho_0 c_0^4}\frac{\partial^2 p^2}{\partial t^2} - \left(\nabla^2 + \frac{1}{c_0^2}\frac{\partial^2}{\partial t^2}\right)\mathcal{L}, \tag{1.69}$$

where the *sound diffusivity* is defined as [Hamilton and Blackstock, 2008, Eq. (3-42)]

$$\delta \equiv \frac{1}{\rho_0}\left[\kappa\left(\frac{1}{c_V} - \frac{1}{c_P}\right) + \left(\mu_{\mathrm{b}} + \frac{4}{3}\mu_{\mathrm{s}}\right)\right]. \tag{1.70}$$

Equation (1.69) is known as the *general second-order nonlinear wave equation*. It is the full wave equation subject to the second-order $\mathcal{O}(\tilde{\varepsilon}^2)$.

Equation (1.69) is the most common form of the second-order nonlinear wave equation, but it does not account for the relaxation effect of fluids. The inclusion of relaxation effects in the wave equation can be derived similarly [Hamilton and Bilbao, 2021], but for simplicity, the derivation is not presented here. To incorporate the relaxation effects, Eq. (1.69) is modified as

$$\begin{aligned}\left(\nabla^2 - \frac{1}{c_0^2}\frac{\partial^2}{\partial t^2}\right)p + \frac{\delta}{c_0^4}\frac{\partial^3 p}{\partial t^3} - \frac{2}{\pi c_0^2}\sum_{n=1}^{\mathcal{N}}(\alpha_n\lambda)_{\mathrm{m}}\frac{\partial^2 p_n}{\partial t^2}\\ = -\frac{\beta}{\rho_0 c_0^4}\frac{\partial^2 p^2}{\partial t^2} - \left(\nabla^2 + \frac{1}{c_0^2}\frac{\partial^2}{\partial t^2}\right)\mathcal{L}.\end{aligned} \tag{1.71}$$

The third term on the left-hand side accounts for the relaxation effects of $\mathcal{N}$ molecules in the air, such as nitrogen and oxygen. In this term, $(\alpha_n\lambda)_\mathrm{m}$ is the maximum absorption per wavelength due to relaxation effects from the n-th molecule. The variable $p_n = p_n(\mathbf{r},t)$ is the solution of the equation $\partial p_n/\partial t + (p_n - p)/\tau_n = 0$ associated with the n-th molecule,and τ_n is the relaxation time [Pierce, 2019; Hamilton and Bilbao, 2021].

1.3.2 KUZNETSOV EQUATION

It is challenging to obtain numerical results of the general second-order nonlinear wave equation (1.69) due to the complexity of evaluating the spatial second derivatives of the Lagrangian density. To address this issue, an alternative approach involves introducing the velocity potential Φ, which satisfies the relation

$$\mathbf{v} = \boldsymbol{\nabla}\Phi. \tag{1.72}$$

By substituting Eqs. (1.72) and (1.62) into Eq. (1.64), and eliminating the gradient operator for all terms, it is obtained that

$$p = -\rho_0 \frac{\partial \Phi}{\partial t} + \left(\mu_\mathrm{b} + \frac{4}{3}\mu_\mathrm{s}\right)\boldsymbol{\nabla}^2\Phi - \mathcal{L} + \mathcal{O}(\tilde{\varepsilon}^3). \tag{1.73}$$

The above equation is valid up to the second-order $\mathcal{O}(\tilde{\varepsilon}^2)$. It is clear that the first-order approximation is simplified to

$$p = -\rho_0 \frac{\partial \Phi}{\partial t} + \mathcal{O}(\tilde{\varepsilon}^2). \tag{1.74}$$

Substituting Eqs. (1.72) and (1.74) into the definition of the Lagrangian density Eq. (1.61), it can be expressed in terms of the velocity potential as [Cheng, 2019a, Eq. (9.2.44a)]

$$\mathcal{L} = \frac{\rho_0}{2}(\boldsymbol{\nabla}\Phi)^2 - \frac{\rho_0}{2c_0^2}\left(\frac{\partial \Phi}{\partial t}\right)^2 + \mathcal{O}(\tilde{\varepsilon}^3). \tag{1.75}$$

Substituting Eqs. (1.72) and (1.74) into Eq. (1.67), it is expressed in terms of the velocity potential and up to the second-order as

$$\frac{\partial p}{\partial t} + \rho_0 c_0^2 \boldsymbol{\nabla}^2\Phi = \frac{\rho_0\beta}{c_0^2}\frac{\partial}{\partial t}\left(\frac{\partial \Phi}{\partial t}\right)^2 - \frac{\kappa}{c_0^2}\left(\frac{1}{c_V} - \frac{1}{c_P}\right)\frac{\partial^3\Phi}{\partial t^3} + \frac{\partial \mathcal{L}}{\partial t} + \mathcal{O}(\tilde{\varepsilon}^3). \tag{1.76}$$

Subtracting the time derivative of Eq. (1.73) from Eq. (1.76), and substituting the first-order wave equation $\left(\boldsymbol{\nabla}^2 - c_0^{-2}\,\partial^2/\partial t^2\right)\Phi = \mathcal{O}(\tilde{\varepsilon}^2)$ into the viscosity term lead to

$$\left(\nabla^2 - \frac{1}{c_0^2}\frac{\partial^2}{\partial t^2}\right)\Phi + \frac{\delta}{c_0^4}\frac{\partial \Phi}{\partial t} = \frac{\beta}{c_0^4}\frac{\partial}{\partial t}\left(\frac{\partial \Phi}{\partial t}\right)^2 + \frac{2}{\rho_0 c_0^2}\frac{\partial \mathcal{L}}{\partial t}. \tag{1.77}$$

Substituting the Lagrangian density given by Eq. (1.75) into Eq. (1.77) yields [Enflo and Hedberg, 2002, Eq. (2.60); Červenka and Bednařík, 2019, Eq. (5)]

$$\left(\nabla^2 - \frac{1}{c_0^2}\frac{\partial^2}{\partial t^2}\right)\Phi + \frac{\delta}{c_0^4}\frac{\partial \Phi}{\partial t} = \frac{1}{c_0^2}\frac{\partial}{\partial t}\left[(\nabla\Phi)^2 + \frac{\beta - 1}{c_0^2}\left(\frac{\partial \Phi}{\partial t}\right)^2\right], \tag{1.78}$$

which is known as the *Kuznetsov equation.* Another derivation of Eq. (1.78), based on the fundamental equations, is presented in Enflo and Hedberg, 2002, Sec. 2.2. It is important to note that the Kuznetsov equation (1.78) is equivalent to the general second-order nonlinear wave equation (1.69) under the second-order approximation. However, the Kuznetsov equation is often preferred in numerical computations due to its avoidance of second-order spatial derivatives, which can be computationally expensive and susceptible to numerical discretization errors [Kagawa et al., 1992; Červenka and Bednařík, 2019].

1.3.3 WESTERVELT EQUATION

The Westervelt equation can be obtained by neglecting the Lagrangian density in the general second-order nonlinear wave equation (1.69). In the case of progressive plane waves, the relationship between sound pressure and particle velocity is expressed as $p = \rho_0 c_0 v$. Upon substituting this relation into Eq. (1.61), it becomes evident that the Lagrangian density is zero, $\mathcal{L} = 0$. In practical applications of PALs, the ultrasound beams are highly collimated, behaving akin to quasi-planar waves. Consequently, the Lagrangian density can typically be ignored at $\mathcal{O}(\tilde{\varepsilon}^2)$, introducing only negligible errors, except in proximity to the source locations [Červenka and Bednařík, 2019; Zhong et al., 2021b]. As a result, the *Westervelt equation* is formulated as [Hamilton and Blackstock, 2008, Eq. (3-46)]

$$\left(\nabla^2 - \frac{1}{c_0^2}\frac{\partial^2}{\partial t^2}\right)p + \frac{\delta}{c_0^4}\frac{\partial^3 p}{\partial t^3} = -\frac{\beta}{\rho_0 c_0^4}\frac{\partial^2 p^2}{\partial t^2}. \tag{1.79}$$

It can been alternatively derived from the Lighthill's equations, as demonstrated in Westervelt's seminal work [Westervelt, 1963].

Consider the equality [Cheng, 2019a, Eq. (9.2.44b)]

$$\left(\nabla^2 - \frac{1}{c_0^2}\frac{\partial^2}{\partial t^2}\right)\Phi^2 = 2\left[(\nabla\Phi)^2 - \frac{1}{c_0^2}\left(\frac{\partial \Phi}{\partial t}\right)^2\right] + 2\Phi\left(\nabla^2 - \frac{1}{c_0^2}\frac{\partial^2}{\partial t^2}\right)\Phi. \tag{1.80}$$

It is observed from Eq. (1.80) that the first term on the right-hand side possesses the form of Lagrangian density presented in Eq. (1.75). Meanwhile, the second term contains the first-order wave equation, making it of third-order $\mathcal{O}(\tilde{\varepsilon}^3)$. Hence, the Lagrangian density can also be expressed as [Aanonsen et al., 1984, Eq. (10)]

$$\mathcal{L} = \frac{\rho_0}{4}\left(\nabla^2 - \frac{1}{c_0^2}\frac{\partial^2}{\partial t^2}\right)\Phi^2 + \mathcal{O}(\tilde{\varepsilon}^3). \tag{1.81}$$

Inserting Eq. (1.81) into the general second-order nonlinear wave equation (1.69) results in [Hamilton and Blackstock, 2008, Eq. (3-45)]

$$\left(\nabla^2 - \frac{1}{c_0^2}\frac{\partial^2}{\partial t^2}\right)\tilde{p} + \frac{\delta}{c_0^4}\frac{\partial^3 \tilde{p}}{\partial t^3} = -\frac{\beta}{\rho_0 c_0^4}\frac{\partial^2 \tilde{p}^2}{\partial t^2}. \tag{1.82}$$

Here, the auxiliary pressure variable is introduced as [Hamilton and Blackstock, 2008, Eq. (3-44)]

$$\tilde{p} \equiv p + \frac{\rho_0}{4}\left(\boldsymbol{\nabla}^2 + \frac{1}{c_0^2}\frac{\partial^2}{\partial t^2}\right)\Phi^2. \tag{1.83}$$

It is clear that $\tilde{p} = p + \mathcal{O}(\tilde{\varepsilon}^2)$. This first-order relation is used in the derivation of Eq. (1.82). Interestingly, Eq. (1.82) has the same form of the Westervelt equation (1.79), hence being termed the *generalized Westervelt equation* [Tjøtta and Tjøtta, 1987b]. When $\tilde{p}$ is substituted with p, Eq. (1.82) reduces to Eq. (1.79).

It is noted that, by utilizing Eqs. (1.81) and (1.75), the auxiliary pressure in Eq. (1.83) can be rewritten as [Červenka and Bednařík, 2022, Eq. (10)]

$$\begin{aligned}\tilde{p} &= p + \mathcal{L} + \frac{\rho_0}{2c_0^2}\frac{\partial^2 \Phi^2}{\partial t^2} + \mathcal{O}(\tilde{\varepsilon}^3), \\ &= p + \frac{\rho_0}{2}(\boldsymbol{\nabla}\Phi)^2 - \frac{\rho_0}{2c_0^2}\left(\frac{\partial \Phi}{\partial t}\right)^2 + \mathcal{O}(\tilde{\varepsilon}^3).\end{aligned} \tag{1.84}$$

Similarly, inserting Eq. (1.81) into the Kuznetsov equation (1.77) results in

$$\left(\boldsymbol{\nabla}^2 - \frac{\partial^2}{\partial t^2}\right)\tilde{\Phi} + \frac{\delta}{c_0^4}\frac{\partial^3 \tilde{\Phi}}{\partial t^3} = -\frac{\beta}{c_0^4}\frac{\partial}{\partial t}\left(\frac{\partial \tilde{\Phi}}{\partial t}\right)^2. \tag{1.85}$$

Here, the auxiliary potential variable is introduced as [Tjøtta and Tjøtta, 1987b, Eq. (11); Červenka and Bednařík, 2022, Eq. (6)]

$$\tilde{\Phi} \equiv \Phi - \frac{1}{2c_0^2}\frac{\partial \Phi^2}{\partial t}. \tag{1.86}$$

It is clear that $\tilde{\Phi} = \Phi + \mathcal{O}(\tilde{\varepsilon}^2)$. The first-order relation is used in the derivation of Eq. (1.85). Equation (1.85) has the same form of Eq. (1.82), and is also referred to as the generalized Westervelt equation [Tjøtta and Tjøtta, 1987b]. Upon substituting Eq. (1.73) into Eq. (1.84) and comparing the result against the time derivative of Eq. (1.86), it is observed that

$$\tilde{p} = -\rho_0\frac{\partial \tilde{\Phi}}{\partial t} + \left(\mu_\mathrm{b} + \frac{4}{3}\mu_\mathrm{s}\right)\boldsymbol{\nabla}^2\Phi. \tag{1.87}$$

In scenarios where the viscosity is significantly smaller, denoted as $\eta \ll \varepsilon$, the second term on the right-hand side of Eq. (1.87) becomes negligible. Consequently, it simplifies to Červenka and Bednařík, 2022, Eq. (8).

1.3.4 KHOKHLOV-ZABOLOTSKAYA-KUZNETSOV EQUATION

To derive the KZK equation, several assumptions are made. First, it is assumed that a baffled source is positioned at the plane $z = 0$ in 3D space, with the positive z-axis indicating the direction of beam radiation. Second, an important assumption is that $ka \gg 1$, where k is the wavenumber and a represents the aperture size of the source. This assumption ensures that the beam is highly directional and focused along its radiation axis, and that the wavefronts are approximately planar.

The slow scale in the direction of propagation is chosen as (z_1, τ), where $z_1 = \varepsilon z$ and the *retarded time* $\tau = t - z/c_0$. The scale for the transverse coordinates is chosen as $x_1 = \varepsilon^\nu x$ and $y_1 = \varepsilon^\nu y$, where the exponent ν will be determined by the fact that the effects of absorption, nonlinearity, and the diffraction are at the same order. Consequently, the sound pressure can be expressed as

$$p = p(x_1, y_1, z_1, \tau); \; (x_1, y_1, z_1) = (\varepsilon^\nu x, \varepsilon^\nu y, \varepsilon z); \; \tau = t - \frac{z}{c_0}. \tag{1.88}$$

The Laplacian operator can be written as

$$\nabla^2 = \varepsilon^{2\nu}\left(\frac{\partial^2}{\partial x_1^2} + \frac{\partial^2}{\partial y_1^2}\right) + \varepsilon^2 \frac{\partial^2}{\partial z_1^2} - \varepsilon \frac{2}{c_0}\frac{\partial^2}{\partial z_1 \partial \tau} + \frac{1}{c_0^2}\frac{\partial^2}{\partial \tau^2}. \tag{1.89}$$

Substituting Eq. (1.89) into the Westervelt equation (1.79) yields

$$\varepsilon^{2\nu}\left(\frac{\partial^2}{\partial x_1^2} + \frac{\partial^2}{\partial y_1^2}\right)p + \varepsilon^2 \frac{\partial^2 p}{\partial z_1^2} - \varepsilon \frac{2}{c_0}\frac{\partial^2 p}{\partial z_1 \partial \tau} = -\frac{\delta}{c_0^4}\frac{\partial^3 p}{\partial \tau^3} - \frac{\beta}{\rho_0 c_0^4}\frac{\partial^2 p^2}{\partial \tau^2}. \tag{1.90}$$

The first term on the left-hand side in Eq. (1.90) accounts for the diffraction. It is observed that when $\nu = 1/2$, the diffraction is of the same order as the nonlinearity and absorption. The second term on the left-hand side in Eq. (1.90) can be discarded as it is a third-order $\mathcal{O}(\varepsilon^3)$ term. Transformation of Eq. (1.90) from the slow scale (x_1, y_1, z_1) back to (x, y, z) yields [Hamilton and Blackstock, 2008, Eq. (3-65)]

$$\frac{\partial^2 p}{\partial z \partial \tau} - \frac{c_0}{2}\nabla_\perp^2 p - \frac{\delta}{2c_0^3}\frac{\partial^3 p}{\partial \tau^3} = \frac{\beta}{2\rho_0 c_0^3}\frac{\partial^2 p^2}{\partial \tau^2}. \tag{1.91}$$

where $\nabla_\perp^2 \equiv \partial^2/\partial x^2 + \partial^2/\partial y^2$. Equation (1.91) is known as the *KZK equation.*

The acronym 'KZK' comes from the surnames of three physicists—Khokhlov, Zabolotskaya, and Kuznetsov—whose collaborative efforts led to the development of the equation between 1969 and 1971. The KZK equation can be viewed as the *parabolic approximation* of the Westervelt equation (1.79) because the elliptic (Westervelt) equation is simplified to the parabolic (KZK) equation [Filippi et al., 1999, Sec. 5.6]. It is also known as the *paraxial approximation* of the Westervelt equation because it is valid only in the paraxial region, which is roughly determined by $\rho \ll z$ [Gu and Jing, 2015]. In the absence of the diffraction effect ($\nabla_\perp^2 p = 0$), the KZK equation reduces to the Burgers equation[Hamilton and Blackstock, 2008, Sec. 3-7]. Note that the Burgers equation does not capture the diffraction behavior necessary for accurately modeling PALs. Therefore, it is not discussed in the context of PAL modeling in this book. The relations among different governing equations are presented in Fig. 1.7.

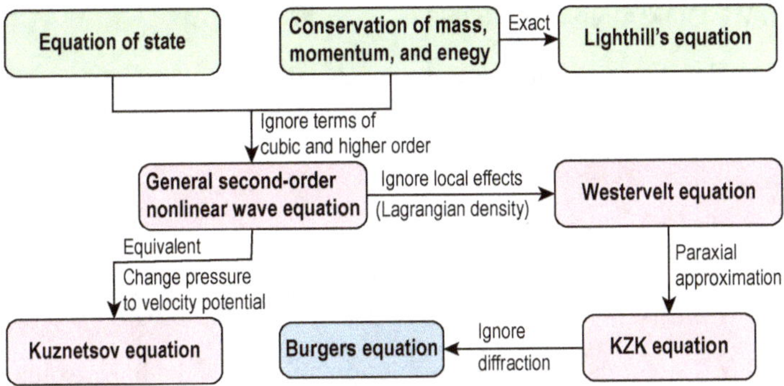

Figure 1.7 Relations among various nonlinear equations governing the acoustic waves generated by PALs.

1.4 QUASILINEAR SOLUTION

Perturbation techniques have been proven effective for solving the nonlinear wave equations such as Eqs. (1.69), (1.78), (1.79), and (1.91). In practical applications involving PALs, it is common to control ultrasound levels for safety reasons, resulting in *weak nonlinearity*. This characteristic of weak nonlinearity is crucial for applications of perturbation analysis. In this book, the focus is on second-order nonlinearity by neglecting higher-order nonlinear terms. Under this assumption, the acoustic state variables such as pressure p, velocity potential Φ, and particle velocity components $\mathbf{v}$ are all of the order $\mathcal{O}(\tilde{\varepsilon})$ or smaller. Therefore, it is possible to expand any state variable ϕ in a power series relative to ε as

$$\phi(\mathbf{r},t,\tilde{\varepsilon}) = \tilde{\varepsilon}\phi_{\mathrm{I}}(\mathbf{r},t) + \tilde{\varepsilon}^2\phi_{\mathrm{II}}(\mathbf{r},t) + \mathcal{O}(\tilde{\varepsilon}^3) \tag{1.92}$$

where ϕ_{I} is the linear solution (first-order approximation) of the wave equation. The subsequent term ϕ_{II} represents a small correction to the linear solution ϕ_{I} and serves as the second-order approximation. By definition, the linear solution corresponds to the limit as $\tilde{\varepsilon} \to 0$. Consequently, the first-order term on the right-hand side of the expansion Eq. (1.92) is the linear approximation, and Eq. (1.92) is referred to as the *quasilinear solution*.

The basic idea of a perturbation analysis is that the expansion should be valid for any value of $\tilde{\varepsilon}$ that is sufficiently small. Therefore, when the expansion Eq. (1.92) is substituted into the wave equations, the coefficients associated with each order of magnitude should match with each other. In the following texts, the quasilinear solutions in the time-domain and the frequency-domain of different governing wave equations are presented in Secs. 1.4.1 and 1.4.2, respectively. The quasilinear solutions offer an approximation to the true solutions of the wave equations by taking into account the weak nonlinearity and allowing for analysis and characterization of sound fields generated by PALs. Finally, the validity of the quasilinear approximation is discussed in Sec. 1.4.3.

1.4.1 TIME-DOMAIN SOLUTIONS

Substituting Eq. (1.92) into the general second-order nonlinear wave equation (1.69) and after discarding the high order terms, two coupled equations are obtained as

$$\begin{cases} \left(\nabla^2 - \frac{1}{c_0^2}\frac{\partial^2}{\partial t^2}\right)p_{\mathrm{I}} + \frac{\delta}{c_0^4}\frac{\partial^3 p_{\mathrm{I}}}{\partial t^3} = 0, \\ \left(\nabla^2 - \frac{1}{c_0^2}\frac{\partial^2}{\partial t^2}\right)p_{\mathrm{II}} + \frac{\delta}{c_0^4}\frac{\partial^3 p_{\mathrm{II}}}{\partial t^3} = -\frac{\beta}{\rho_0 c_0^4}\frac{\partial^2 p_{\mathrm{I}}^2}{\partial t^2} - \left(\nabla^2 + \frac{1}{c_0^2}\frac{\partial^2}{\partial t^2}\right)\mathcal{L}_{\mathrm{I}}. \end{cases} \tag{1.93}$$

Equation (1.93) suggests that the first-order pressure p_{I} is the linear solution, while this solution as well as the Lagrangian density $\mathcal{L}_{\mathrm{I}}$ then become a source term driving the second-order equation for solving the second-order pressure p_{II}.

Substituting Eq. (1.92) into the Kuznetsov equation (1.78) and after discarding the higher-order terms, two coupled equations are obtained as [Hamilton and Blackstock, 2008, Eqs. (10-6) and (10-7)]

$$\begin{cases} \left(\nabla^2 - \frac{1}{c_0^2}\frac{\partial^2}{\partial t^2}\right)\Phi_{\mathrm{I}} + \frac{\delta}{c_0^4}\frac{\partial \Phi_{\mathrm{I}}}{\partial t} = 0 \\ \left(\nabla^2 - \frac{1}{c_0^2}\frac{\partial^2}{\partial t^2}\right)\Phi_{\mathrm{II}} + \frac{\delta}{c_0^4}\frac{\partial \Phi_{\mathrm{II}}}{\partial t} = \frac{1}{c_0^2}\frac{\partial}{\partial t}\left[(\boldsymbol{\nabla}\Phi_{\mathrm{I}})^2 + \frac{\beta-1}{c_0^2}\left(\frac{\partial \Phi_{\mathrm{I}}}{\partial t}\right)^2\right]. \end{cases} \tag{1.94}$$

It shows that the first-order potential Φ_{I} is the linear solution, while this solution as well as its gradient then become a source term driving the second-order equation for solving the second-order potential Φ_{II}.

Substituting Eq. (1.92) into the Westervelt equation (1.79) and after discarding the higher-order terms, two coupled equations are obtained as [Westervelt, 1963, Eq. (7)]

$$\begin{cases} \left(\nabla^2 - \frac{1}{c_0^2}\frac{\partial^2}{\partial t^2}\right)p_{\mathrm{I}} + \frac{\delta}{c_0^4}\frac{\partial^3 p_{\mathrm{I}}}{\partial t^3} = 0, \\ \left(\nabla^2 - \frac{1}{c_0^2}\frac{\partial^2}{\partial t^2}\right)p_{\mathrm{II}} + \frac{\delta}{c_0^4}\frac{\partial^3 p_{\mathrm{II}}}{\partial t^3} = -\rho_0\frac{\partial q_{\mathrm{II}}}{\partial t}, \end{cases} \tag{1.95}$$

where the *source density (function)* for the virtual audio source is [Westervelt, 1963, Eq. (8)]

$$q_{\mathrm{II}} \equiv \frac{\beta}{\rho_0^2 c_0^4}\frac{\partial p_{\mathrm{I}}^2}{\partial t}. \tag{1.96}$$

The difference between Eqs. (1.95) and (1.93) is that the Lagrangian density is removed on the right-hand side of the second-order wave equation in (1.95).

Substituting Eq. (1.92) into the KZK equation (1.91) and after discarding the high-order terms, two coupled equations are obtained as [Averkiou et al., 1993,

Eqs. (11–12); Cheng, 2019a, Eq. (9.4.6b)]

$$\begin{cases} \dfrac{\partial^2 p_{\mathrm{I}}}{\partial z \partial \tau} - \dfrac{c_0}{2}\nabla_{\perp}^2 p_{\mathrm{I}} - \dfrac{\delta}{2c_0^3}\dfrac{\partial^3 p_{\mathrm{I}}}{\partial \tau^3} = 0, \\ \dfrac{\partial^2 p_{\mathrm{II}}}{\partial z \partial \tau} - \dfrac{c_0}{2}\nabla_{\perp}^2 p_{\mathrm{II}} - \dfrac{\delta}{2c_0^3}\dfrac{\partial^3 p_{\mathrm{II}}}{\partial \tau^3} = \dfrac{\beta}{2\rho_0 c_0^3}\dfrac{\partial^2 p_{\mathrm{I}}^2}{\partial \tau^2}. \end{cases} \tag{1.97}$$

1.4.2 FREQUENCY-DOMAIN SOLUTIONS

The time-domain solutions given by Eqs (1.93), (1.94), (1.95), and (1.97) are computationally expensive to solve. In certain scenarios where only the pure-tone audio sound field is of interest, a simplified approach can be used. For example, the audio sound field is generated by two ultrasonic waves at frequencies f_1 and f_2 with $f_2 > f_1$. By considering the first-order approximation, the sound pressure can be written as a sum of terms representing the individual ultrasonic waves

$$p_{\mathrm{I}}(\mathbf{r},t) = p_1(\mathbf{r})\mathrm{e}^{-\mathrm{i}\omega_1 t} + p_2(\mathbf{r})\mathrm{e}^{-\mathrm{i}\omega_2 t}, \tag{1.98}$$

where i is the imaginary unit, and $p_i(\mathbf{r})$ is the spatial pressure field at the frequency f_i, $i = 1,2$. The second-order sound field contains four components as demonstrated by Eq. (1.1)

$$p_{\mathrm{II}}(\mathbf{r},t) = p_{11}(\mathbf{r})\mathrm{e}^{-2\mathrm{i}\omega_1 t} + p_{22}(\mathbf{r})\mathrm{e}^{-2\mathrm{i}\omega_2 t} + p_{+}(\mathbf{r})\mathrm{e}^{-\mathrm{i}(\omega_1+\omega_2)t} + p_{\mathrm{a}}(\mathbf{r})\mathrm{e}^{-\mathrm{i}\omega_{\mathrm{a}} t}, \tag{1.99}$$

corresponding to the waves at frequencies of $2f_1, 2f_2, f_1 + f_2$, and $f_2 - f_1$. The DFW component is the audio sound with the frequency of $f_{\mathrm{a}} = f_2 - f_1$.

For the first-order solution in the general second-order nonlinear wave equation (1.93), when considering only the harmonic sound field, it reduces to two homogeneous Helmholtz equations, while the second-order solution in it reduces to an inhomogeneous Helmholtz equation as

$$\begin{cases} (\nabla^2 + k_i^2)\, p_i = 0, \\ (\nabla^2 + k_{\mathrm{a}}^2)\, p_{\mathrm{a}} = \mathrm{i}\rho_0 \omega_{\mathrm{a}} q_{\mathrm{a}}, \end{cases} \tag{1.100}$$

where $i = 1,2$, and the source density for the virtual audio source is [Červenka and Bednařík, 2019, Eq. (10)]

$$q_{\mathrm{a}}(\mathbf{r}) = \frac{\beta\omega_{\mathrm{a}}}{\mathrm{i}\rho_0^2 c_0^4} p_1^*(\mathbf{r}) p_2(\mathbf{r}) - (\boldsymbol{\nabla} - k_{\mathrm{a}}^2)\left[\frac{\rho_0}{2}\mathbf{v}_1^*(\mathbf{r})\cdot\mathbf{v}_2(\mathbf{r}) - \frac{p_1^*(\mathbf{r})p_2(\mathbf{r})}{2\rho_0 c_0^2}\right], \tag{1.101}$$

where the superscript '*' is the complex conjugate, $\mathbf{v}_i$ is the particle velocity at ultrasound frequency f_i, $i = 1,2$. When considering harmonic sound fields, the dissipative process including the classical thermoviscosity and relaxation is simplified by introducing a complex wavenumber $k_i = \omega_i / c_0 + \mathrm{i}\alpha(\omega_i)$ as shown in Eq. (1.100), where

$i = 1, 2$ and $i = \mathrm{a}$ represent the indices for both ultrasound and audio sound waves, respectively. The sound absorption coefficient takes the general form of [Pierce, 2019, Eq. (10.8.11)]

$$\alpha(\omega) = \frac{\delta\omega^2}{2c_0^3} + \sum_{n=1}^{\mathcal{N}} \frac{(\alpha_n\lambda)_{\mathrm{m}}}{\pi c_0} \frac{\omega^2\tau_n}{1 + (\omega\tau_n)^2}. \tag{1.102}$$

The empirical formula of Eq. (1.102) obtained by fitting the experimental data can be found in Appendix A.

The particle velocity of ultrasound in Eq. (1.101) can be derived by applying the first-order approximation of the Navier-Stokes equation (1.52) in the frequency domain, expressed as

$$\mathbf{v}(\mathbf{r}) = \frac{\nabla p(\mathbf{r})}{\mathrm{i}\rho_0\omega}. \tag{1.103}$$

Similarly, for the velocity potential involved in the Kuznetsov equation, Eq. (1.94) reduces to [Červenka and Bednařík, 2019, Eq. (12)]

$$\begin{cases} (\nabla^2 + k_i^2)\Phi_i = 0 \\ (\nabla^2 + k_{\mathrm{a}}^2)\Phi_{\mathrm{a}} = \dfrac{\omega_{\mathrm{a}}}{\mathrm{i}c_0^2}[\mathbf{v}_1^* \cdot \mathbf{v}_2 + (\beta - 1)k_1^* k_2 \Phi_1^* \Phi_2]. \end{cases} \tag{1.104}$$

Once the two coupled perturbation equations are linearly solved, the pressure field of the audio sound is obtained by [Červenka and Bednařík, 2019, Eq. (14)]

$$p_{\mathrm{a}}(\mathbf{r}) = \mathrm{i}\rho_0\omega_{\mathrm{a}}\Phi_{\mathrm{a}}(\mathbf{r}) - \frac{\rho_0}{2}\mathbf{v}_1^*(\mathbf{r}) \cdot \mathbf{v}_2(\mathbf{r}) + \frac{\rho_0}{2c_0^2}\Phi_1^*(\mathbf{r})\Phi_2(\mathbf{r}). \tag{1.105}$$

For the quasilinear solution of the Westervelt equation in the time domain, Eq. (1.95) reduces to the same form as Eq. (1.100) in the frequency domain, but the source density Eq. (1.101) is modified as

$$q_{\mathrm{a}}(\mathbf{r}) = \frac{\beta\omega_{\mathrm{a}}}{\mathrm{i}\rho_0^2 c_0^4} p_1^*(\mathbf{r}) p_2(\mathbf{r}). \tag{1.106}$$

The frequency-domain solution of the Westervelt equation is easier to solve compared to the general second-order nonlinear wave equation because the evaluation of the velocity field is not required in the source density function. However, the Westervelt equation is less accurate due to neglecting the local effects characterized by the Lagrangian density. To account for the local effects and to simplify the calculation, an algebraic correction to the audio sound pressure obtained using the Westervelt equation was proposed as [Červenka and Bednařík, 2022, Eq. (22)]

$$\tilde{p}_{\mathrm{a}}(\mathbf{r}) = p_{\mathrm{a}}(\mathbf{r}) - \left[\frac{\rho_0}{2}\mathbf{v}_1^*(\mathbf{r}) \cdot \mathbf{v}_2(\mathbf{r}) - \left(\frac{\omega_1}{\omega_2} + \frac{\omega_2}{\omega_1} - 1\right)\frac{p_1^*(\mathbf{r})p_2(\mathbf{r})}{2\rho_0 c_0^2}\right]. \tag{1.107}$$

For the KZK equation, by utilizing the assumption that $k_i a \gg 1$, where a is the aperture size, the harmonic solution can be written as

$$p_i(\mathbf{r}, \tau) = p_i(\mathbf{r})\mathrm{e}^{-\mathrm{i}\omega_i\tau}, \quad i = 1, 2, \mathrm{a}. \tag{1.108}$$

Then the quasilinear solution of the KZK equation Eq. (1.97) reduces to [Hamilton and Blackstock, 2008, Eq. (8-4)]

$$\begin{cases} \dfrac{\partial p_i}{\partial z} + \dfrac{1}{2\mathrm{i}k_i}\nabla_\perp^2 p_i + \dfrac{\delta\omega_i^2}{2c_0^3} p_i = 0 \\ \dfrac{\partial p_\mathrm{a}}{\partial z} + \dfrac{1}{2\mathrm{i}k_\mathrm{a}}\nabla_\perp^2 p_\mathrm{a} + \dfrac{\delta\omega_\mathrm{a}^2}{2c_0^3} p_\mathrm{a} = \dfrac{\beta\omega_\mathrm{a}}{2\rho_0 c_0^3} p_1^* p_2. \end{cases} \tag{1.109}$$

Equations (1.100), (1.104), and (1.109) are the main results of this section. These equations can be solved by utilizing the Green's function. In Chap. 2, detailed numerical models are introduced, providing practical implementation and solutions of these equations.

1.4.3 VALIDITY OF THE QUASILINEAR APPROXIMATION

As presented in this section, the quasilinear approximation serves to simplify the nonlinear problem into two linear problems, leading to a substantial simplification of both analysis and computation. However, it is crucial to assess the validity of the quasilinear approximation.

In the realm of linear acoustics, the sound pressure amplitude at a field location increases proportionally with the increase of the on-surface pressure amplitude, p_0, or the acoustic Mach number, ε, as defined by Eq. (1.44). Nevertheless, in the realm of nonlinear acoustics at large p_0 values, the nonlinearity becomes strong. Consequently, the waveform undergoes significant distortion, and *shock waves* form during the wave propagation [Hamilton and Blackstock, 2008, Sec. 8.4]. The wave energy is lost at the shock fronts, thereby imposing an upper bound on how much sound can be transmitted to a given distance. In such scenarios, one has to directly solve the nonlinear wave equations presented in Sec. 1.3 to obtain numerical solutions.

In a 1D lossless medium, finite-amplitude waves can be modeled by planar waves, and shock waves are formed at the distance of [Marchal and Cervenka, 2004, Eq. (10); Hamilton and Blackstock, 2008, Eq. (4.22)]

$$\mathscr{D}_\mathrm{sf}(\omega_\mathrm{u}) \equiv \frac{\rho_0 c_0^3}{\beta\omega_\mathrm{u} p_0} = \frac{1}{\beta\varepsilon k_\mathrm{u}}, \tag{1.110}$$

which is referred to as the *shock formation distance*. Here, ω_u is the angular frequency of the carrier ultrasound, $k_\mathrm{u} = \omega_\mathrm{u}/c_0$ is the real wavenumber of the ultrasound, ε is the acoustic Mach number as defined by Eq. (1.44), and $p_0 = \rho_0 c_0 v_0$ and v_0 are the on-surface pressure and particle velocity amplitudes, respectively.

In a 3D lossless medium, the wave attenuates due to the diffraction, while the diffraction features of the wave is characterized by the source aperture size and operating frequency. Specifically, in the case of a circular baffled source, the amplitude of sound pressure decreases inversely with the propagation distance in the region beyond the *Rayleigh distance* [Foote, 2014]

$$\mathscr{D}_\mathrm{R}(\omega_\mathrm{u}) \equiv \frac{\mathcal{A}_0}{\lambda_\mathrm{u}} = \frac{\pi a^2}{\lambda_\mathrm{u}} = \frac{1}{2}k_\mathrm{u}a^2, \tag{1.111}$$

where $\mathcal{A}_0$ is the radiation area and λ_u is the wavelength of ultrasound. For a circular source with a radius of a, $\mathcal{A}_0 = \pi a^2$, and the Rayleigh distance has the form of $\mathscr{D}_R(\omega_u) = \pi a^2/\lambda_u = \pi\mathscr{D}_c(\omega_u)$. Here, the *critical distance* is defined as

$$\mathscr{D}_c(\omega_u) = a^2/\lambda_u, \tag{1.112}$$

where the on-axis sound pressure amplitude reaches its final maximum.

For convenience, a dimensionless parameter can be defined as [Averkiou et al., 1993]

$$\mathcal{N}_{sf}^{R}(\omega) \equiv \frac{\mathscr{D}_R(\omega)}{\mathscr{D}_{sf}(\omega)}, \tag{1.113}$$

which is referred to as the *diffraction-saturation number* in this book because it signifies the ratio of diffractive effects to saturation effects. Within the Rayleigh distance, the wave experiences minor decay. If the Rayleigh distance $\mathscr{D}_R$ exceeds the shock formation distance $\mathscr{D}_{sf}$, implying that $\mathcal{N}_{sf}^{R} > 1$, shock wave can be formed within the Rayleigh distance. However, if the Rayleigh distance $\mathscr{D}_R$ is shorter than the nonlinear shock formation distance $\mathscr{D}_{sf}$, implying that $\mathcal{N}_{sf}^{R} < 1$, the sound pressure amplitude falls below p_0 before shock waves can form, making shock wave formation impossible [Hamilton et al., 1997].

Figure 1.8 illustrates the diffraction-saturation number at various on-surface pressure amplitudes, source radii, and ultrasound frequencies. It is clear that the diffraction-saturation number increases with the increase of these parameters. This trend suggests that the quasilinear approximation becomes invalid at higher values of these parameters. For example, as depicted in Fig. 1.8(a), in the case of a circular

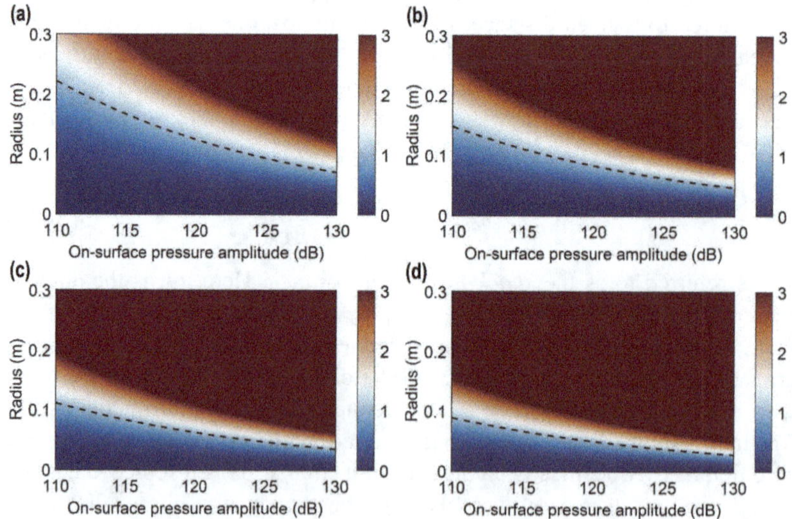

Figure 1.8 The diffraction-saturation number, defined by Eq. (1.113), at various source radii and on-surface pressure amplitudes. The ultrasound frequency is (a) 40 kHz, (b) 60 kHz, (c) 80 kHz, and (d) 100 kHz. - - - , the ratio is equal to 1.

PAL with a source radius of 0.1 m and a carrier ultrasound frequency of 40 kHz, the maximum on-surface pressure amplitude is $p_0 = 43.3\,\text{Pa}$ $(123.7\,\text{dB})$ for a diffraction-saturation number less than 1 ($\mathcal{N}_{\text{sf}}^{\text{R}} < 1$).

It is worth noting that the above analysis pertains to lossless media. In the context of PALs, the ultrasound attenuation due to the absorption by air cannot be neglected. The sound pressure decays exponentially with distance, expressed as $\exp[-\alpha(\omega_{\text{u}})r]$, where $\alpha(\omega_{\text{u}})$ is the absorption coefficient at angular frequency ω_{u} in air (see Appendix A for details), and r is the propagation distance. Beyond the *absorption distance*, defined as

$$\mathscr{D}_{\text{ab}}(\omega_{\text{u}}) = \frac{1}{\alpha(\omega_{\text{u}})}, \tag{1.114}$$

the sound pressure is greatly attenuated. Similarly, it is convenient to define a dimensionless number as [Averkiou et al., 1993; Hamilton, 2016, Eq. (4)]

$$\mathcal{N}_{\text{sf}}^{\text{ab}}(\omega_{\text{u}}) \equiv \frac{\mathscr{D}_{\text{ab}}(\omega_{\text{u}})}{\mathscr{D}_{\text{sf}}(\omega_{\text{u}})} = \frac{\beta\omega_{\text{u}}p_0}{\alpha(\omega_{\text{u}})\rho_0 c_0^3}, \tag{1.115}$$

which is known as the *Gol'dberg number*. In this book, it is referred to as the *absorption-saturation number* because it signifies the absorptive effects to saturation effects. It is noted that Eq. (1.115) is independent of the source aperture size.

If $\mathcal{N}_{\text{sf}}^{\text{ab}} < 1$, shock wave will not form either, thus the quasilinear approximation could also be valid. Therefore, in the lossy medium, either the Rayleigh distance or absorption distance is less than the shock formation distance ($\mathcal{N}_{\text{sf}}^{\text{R}} < 1$ or $\mathcal{N}_{\text{sf}}^{\text{ab}} < 1$), the shock wave will not be formed so the quasilinear approximation holds. In other words, the quasilinear approximation is valid as long as the shock formation distance is larger than the absorption distance ($\mathscr{D}_{\text{sf}} > \mathscr{D}_{\text{ab}}$ or $\mathcal{N}_{\text{sf}}^{\text{ab}} < 1$), even when the shock formation distance is less than the Rayleigh distance ($\mathscr{D}_{\text{sf}} < \mathscr{D}_{\text{R}}$ or $\mathcal{N}_{\text{sf}}^{\text{R}} > 1$).

Figure 1.9 illustrates the absorption-saturation number at various ultrasound frequencies and on-surface pressure amplitudes. It is evident the absorption-saturation number increases with the increase of the on-surface pressure amplitude, while the effects of the ultrasound frequency on this number are more complicated. For example, in the case of a PAL with a carrier ultrasound frequency of 40 kHz, the maximum on-surface pressure amplitude is $p_0 = 24.6\,\text{Pa}$ $(118.8\,\text{dB})$ for a absorption-saturation number less than 1 ($\mathcal{N}_{\text{sf}}^{\text{ab}} < 1$).

1.5 SUMMARY

This chapter provides a comprehensive overview of the physical basis for PALs. It begins in Sec. 1.1 by introducing the concepts of PAAs (Sec. 1.1.1) and PALs (Sec. 1.1.3), along with their applications and commercial examples in Secs. 1.1.2, 1.1.4, and 1.1.5. Moving forward, Sec. 1.2 delves into the fundamental equations governing acoustic waves in fluids. This includes the conservation of mass (continuity equation, Sec. 1.2.1), conservation of momentum (Navier-Stokes equations, Sec. 1.2.2), conservation of energy (Kirchhoff-Fourier equation, Sec. 1.2.3), and the

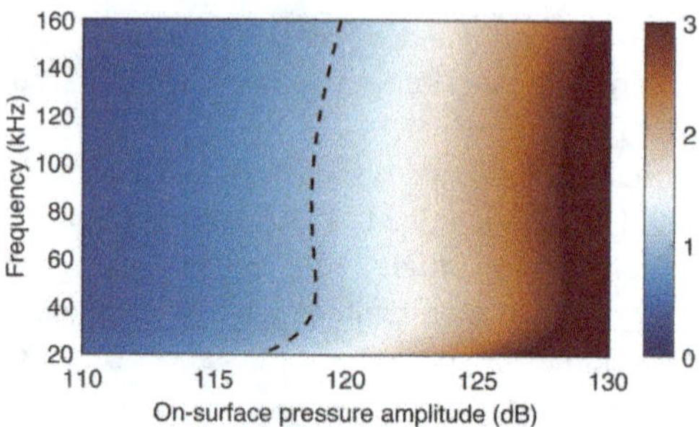

Figure 1.9 The absorption-saturation number, defined by Eq. (1.115), at various ultrasound frequencies and on-surface pressure amplitudes. The temperature is 293.15 K (20°C), and the relative humidity is 50%. - - - , the absorption-saturation number is equal to 1.

state equation (Sec. 1.2.4). The concept of second-order approximation is introduced in Sec. 1.2.5 to approximate these equations.

This chapter then explores nonlinear equations specifically applicable to PALs in Sec. 1.3, where the general second-order nonlinear wave equation (Sec. 1.3.1), the Kuznetsov equation (Sec. 1.3.2), the Westervelt equation (Sec. 1.3.3), and the KZK equation (Sec. 1.3.4) are presented. In order to simplify the analysis, the concept of quasilinear solutions is introduced in Sec. 1.4, and both time-domain (Sec. 1.4.1) and frequency-domain (Sec. 1.4.2) solutions under the quasilinear approximation are provided. The validity of the quasilinear approximation is discussed in Sec. 1.4.3. This chapter serves as a foundation for understanding the physical principles and mathematical models underlying PALs, paving the way for more detailed numerical models and analysis to be discussed in subsequent chapters.

2 Numerical Models for Parametric Array Loudspeakers

2.1 INTRODUCTION

The generation of audio sound by a PAL entails a nonlinear process, adding complexity to its numerical modeling compared to conventional loudspeakers, which exclusively involve linear processes. The theoretical analysis of nonlinear interactions, as presented in the previous chapter, requires intricate mathematical treatments that lack simple analytical solutions. Throughout this book, two fundamental assumptions, commonly adopted in the literature, are utilized to simplify these calculations. Firstly, only second-order nonlinearity is considered, with higher-order nonlinearities being neglected. Secondly, the quasilinear approximation is applied. These two assumptions enable the decomposition of the nonlinear governing equations into two linear equations, effectively converting nonlinear models into linear ones. Within this framework, the ultrasound field is modeled as linear radiation, while the audio sound field is governed by an inhomogeneous linear wave equation, with a source density determined by the ultrasound field. This chapter summarizes various numerical models developed under this framework, providing insights into their utility and applicability.

The general second-order nonlinear wave equation, as represented by Eq. (1.71), is the most accurate equation when considering second-order nonlinearity. However, even with the quasilinear approximation, this equation lacks an analytical solution. It can be solved using the finite element method (FEM) [Kagawa et al., 1992]. In this method, the spatial domain is discretized into six-node triangular elements, and the Galerkin procedure is applied to obtain numerical solutions for sound fields. It is worth noting that due to the equation complexity and the substantial computational demands, the study conducted in Kagawa et al., 1992 focused solely on a two-dimensional (2D) model. The challenges encountered when employing FEM arise from the necessity to numerically evaluate both spatial and temporal derivatives of the Lagrangian density, as well as dealing with the fact that the region of interest greatly exceeds the ultrasound wavelength.

The Kuznetsov equation, as expressed by Eq. (1.78), is fundamentally equivalent to the second-order nonlinear wave equation. The primary distinction lies in the utilization of the velocity potential instead of sound pressure. However, the Kuznetsov equation is computationally more straightforward to address due to its avoidance of the need to evaluate spatial and temporal derivatives of the Lagrangian density. In a study conducted by [Červenka and Bednařík, 2019], the FEM was successfully

DOI: 10.1201/9781003354994-2

applied to solve the Kuznetsov equation, with the obtained numerical results demonstrating equivalence to those derived from the second-order nonlinear wave equation. Another approach for calculating the audio sound field based on the Kuznetsov equation is through an integration method, as also detailed in Červenka and Bednařík, 2019.

When the Lagrangian density, which characterizes nonlinear local effects, is disregarded, the second-order nonlinear wave equation simplifies to the Westervelt equation, as represented by Eq. (1.79). Typically, the quasilinear solution of the Westervelt equation is achieved using an integration method, which shares similarities with the approach employed for the Kuznetsov equation [Červenka and Bednařík, 2019; Zhong et al., 2020c]. In this book, such integration methods are referred to as the *direct integration method* (DIM), as elaborated in Sec. 2.2. Compared to the FEM, the DIM holds advantages because it is relatively easier to implement using numerical integration techniques and demands less memory. In the DIM, the ultrasound field is initially determined via the well-known Rayleigh integral. Subsequently, the source density of the virtual audio source is computed based on the calculated ultrasound pressure and particle velocities. Finally, the audio sound field is obtained through a three-fold integral across the entire spatial domain. It is important to note that the Green's function in the integrand of these calculations exhibits high oscillation, which can lead to slow convergence in numerical integrations. Therefore, additional techniques or simplifications are often required to alleviate the computational burden associated with the DIM.

In certain applications, particularly those where only far-field directivity is of concern, the directivity serves as a crucial indicator for describing the radiation pattern of audio sound produced by a PAL. The concept of directivity is a straightforward means to convey this radiation pattern. The Westervelt's seminal work introduced a closed-form expression for audio beam directivity, which is now known as the *Westervelt's directivity* [Westervelt, 1963]. This expression illustrates that audio sound directivity exhibits no sidelobes, being solely influenced by the audio frequency and the ultrasound attenuation coefficient associated with atmospheric absorption. However, discrepancies between predictions based on the Westervelt's directivity and experimental measurements have been observed, which are attributed to the assumption that ultrasound beams are perfectly collimated and that all nonlinear interactions occur over a limited distance [Shi and Kajikawa, 2015a].

Berktay and Leahy made advancements in improving the accuracy of the Westervelt's directivity by considering the aperture factor and wave shape [Berktay, 1965a; Berktay and Leahy, 1974]. However, their models are primarily suitable for piston-type sources with uniform velocity profiles on their radiation surfaces. The product directivity model, which approximates audio sound directivity as the product of the directivities of two ultrasonic waves, offers a more convenient approach for calculating the directivity of steerable PALs [Gan et al., 2006]. While this method predicts the main lobe location accurately, it tends to underestimate the sidelobe levels due to the omission of audio sound wave diffraction effects. Some enhancement was achieved by employing an equivalent Gaussian source array [Shi and Gan, 2012].

Presently, the most precise and computationally efficient numerical model for calculating far-field directivity is the *convolution directivity method* (CDM) [Shi and Kajikawa, 2015a; Guasch and Sánchez-Martín, 2018; Shi et al., 2022; Zhong et al., 2023d]. The core concept behind the CDM is that a PAL emits an infinite number of ultrasonic beams into the air, with audio sound generated in each direction possessing a beam pattern corresponding to the Westervelt's directivity, weighted by the amplitude of ultrasonic waves in that direction. Consequently, the total audio directivity emerges as a sum of all contributing audio beam patterns, resulting in a convolution of the Westervelt's directivity and ultrasonic directivities. The CDM has demonstrated improved agreement with measurements and stands as a potent tool for directivity control in steerable PALs [Shi et al., 2015; Zhong et al., 2023b]. Additionally, it finds utility in predicting harmonic distortion arising from modulation algorithms [Shi et al., 2022]. Further details on the CDM are presented in Sec. 2.3.

Predicting audio sound behavior in the near field presents greater challenges than assessing far-field directivity. Early investigations focused on solving the Khokhlov-Zabolotskaya-Kuznetsov (KZK) equation, which can be considered as a paraxial approximation of the Westervelt equation, achieved by approximating the second-order derivative of sound pressure with respect to the propagation direction using a first-order derivative. One approach to obtaining the quasilinear solution of the KZK equation involves utilizing the Hankel transformation [Hamilton and Blackstock, 2008, Sec. 8.2]. Another method is based on finite-difference time domain (FDTD) methods, which offer the advantage of calculating transient sound fields without the requirement of the quasilinear approximation. However, FDTD methods are susceptible to issues like numerical reflections and dispersions and entail significant computational demands, particularly for locations distant from the PAL [Nomura et al., 2012].

The *Gaussian beam expansion* (GBE) method has gained prominence due to its superior attributes. The GBE approximates the velocity profile on the radiation source using a series of Gaussian functions and is introduced in Sec. 2.4. This method models ultrasound as emanating from multiple Gaussian sources [Wen and Breazeale, 1988]. Leveraging the paraxial approximation employed in the KZK equation, it derives closed-form expressions for the radiation from each Gaussian source, thereby substantially enhancing computational efficiency. While the KZK equation accounts for diffraction, absorption, and nonlinearity, its accuracy is confined primarily to the paraxial region, typically within 20 degrees from the transducer axis. Studies have demonstrated that predictions based on the KZK equation tend to be inaccurate near the PAL, with diminishing accuracy observed for small aperture sizes and lower audio and ultrasound frequencies [Červenka and Bednařík, 2013].

Partial wave expansion methods were introduced to reduce computational costs without necessitating additional approximations like the paraxial approximation in the GBE methods [Zhong et al., 2020c]. In the context of a 2D radiation problem, where one dimension of the PAL vastly exceeds the wavelength, the *cylindrical wave expansion* (CWE) method becomes applicable and is elucidated in Sec. 2.5 [Zhong et al., 2021a]. On the other hand, for a 3D radiation problem in scenarios where audio sound is generated by a circular PAL, the *spherical wave expansion* (SWE) method method comes into play, as detailed in Sec. 2.6 [Zhong et al., 2020c; Zhong et al., 2021b; Zhong et al., 2022b]. The core principle behind these partial wave expansion methods is the representation of the Green's function in terms of partial waves under cylindrical (polar) and spherical coordinate systems. By utilizing the symmetrical characteristics, numerical computations using the partial wave expansion methods are notably more computationally efficient than those relying on the DIM.

In addition to the methods elucidated in Secs. 2.2 to 2.6, Sec. 2.7 provides a concise overview of other modeling approaches. These encompass the FEM [Kagawa et al., 1992; Dirkse, 2014; Červenka and Bednařík, 2019; Červenka and Bednařík, 2021], the operator splitting method [Yuldashev and Khokhlova, 2011], and the FDTD method [Nomura et al., 2012; Jimenez et al., 2021]. Furthermore, this section introduces the utilization of the k-Wave software for modeling PALs [Treeby and Cox, 2010]. Widely recognized for its proficiency in simulating medical acoustic waves, this software is extensively employed in the field.

2.2 DIRECT INTEGRATION METHOD (DIM)

2.2.1 KIRCHHOFF-HELMHOLTZ INTEGRAL EQUATION

In a given domain, represented as V and bounded by $S \equiv \partial V$, the linear sound pressure in the frequency domain follows the Helmholtz equation

$$\left(\nabla^2 + k^2\right) p(\mathbf{r}) = \mathrm{i}\rho_0 \omega q(\mathbf{r}), \quad \mathbf{r} \in V. \tag{2.1}$$

Here, i is the imaginary unit, ρ_0 is the medium density, $\omega = 2\pi f$ is the angular frequency, f is the frequency, k is the wavenumber, $\mathbf{r}$ is the field location, and $q(\mathbf{r})$ denotes the source density. Within this domain, the Green's function is governed by

$$\left(\nabla^2 + k^2\right) G(\mathbf{r};\mathbf{r}') = -\delta(\mathbf{r}-\mathbf{r}'), \quad \mathbf{r},\mathbf{r}' \in V \tag{2.2}$$

where $\delta(\cdot)$ is the Dirac delta function and $\mathbf{r}'$ is the source location.

An identity can be established within the volume V as

$$\begin{aligned} G(\mathbf{r};\mathbf{r}')(\nabla^2 + k^2)p(\mathbf{r}) - p(\mathbf{r})(\nabla^2 + k^2)G(\mathbf{r};\mathbf{r}') \\ = \nabla \cdot \left[G(\mathbf{r};\mathbf{r}')\nabla p(\mathbf{r}) - p(\mathbf{r})\nabla G(\mathbf{r};\mathbf{r}')\right]. \end{aligned} \tag{2.3}$$

Substituting Eqs. (2.1) and (2.2) into Eq. (2.3), conducting a volume integral over the domain V, and applying the divergence theorem on the right-hand side, result in

$$\begin{aligned} \mathrm{i}\rho_0\omega \iiint_V G(\mathbf{r};\mathbf{r}')q(\mathbf{r})\mathrm{d}^3\mathbf{r} + \iiint_V p(\mathbf{r})\delta(\mathbf{r}-\mathbf{r}')\mathrm{d}^3\mathbf{r} \\ = \oiint_S \left[G(\mathbf{r};\mathbf{r}')\nabla p(\mathbf{r}) - p(\mathbf{r})\nabla G(\mathbf{r};\mathbf{r}')\right] \cdot \mathrm{d}\mathbf{S}. \end{aligned} \tag{2.4}$$

Here, it is important to note that ∇ is applied with respect to $\mathbf{r}$ and the vector $\mathrm{d}\mathbf{S}$ points outward the boundary. By exchanging the symbols $\mathbf{r}$ and $\mathbf{r}'$, it is obtained that [Cheng, 2019b, Eq. (2.5.6a)]

$$\begin{aligned} p(\mathbf{r}) = -\mathrm{i}\rho_0\omega \iiint_V G(\mathbf{r};\mathbf{r}')q(\mathbf{r}')\mathrm{d}^3\mathbf{r}' \\ + \oiint_S \left[G(\mathbf{r};\mathbf{r}')\nabla p(\mathbf{r}') - p(\mathbf{r}')\nabla G(\mathbf{r};\mathbf{r}')\right] \cdot \mathrm{d}\mathbf{S}', \quad \mathbf{r} \in V. \end{aligned} \tag{2.5}$$

In Eq. (2.5), it is worth highlighting that ∇ is applied with respect to $\mathbf{r}'$ as a result of symbol exchange. Additionally, the reciprocal property for Green's function, $G(\mathbf{r};\mathbf{r}') = G(\mathbf{r}';\mathbf{r})$, is utilized. Equation (2.5) is commonly referred to as the *Kirchhoff-Helmholtz integral equation (KHIE)*, serving as the cornerstone for addressing linear radiation problems.

In practical applications of the KHIE, it is common to select the Green's function in a manner that satisfies the boundary condition $\nabla G(\mathbf{r};\mathbf{r}') = 0$ on the boundary S, where $\mathbf{r}' \in S$. Additionally, the gradient of sound pressure at the boundary can be expressed as $\nabla p(\mathbf{r}') = \mathrm{i}\rho_0\omega\mathbf{v}(\mathbf{r}')$, utilizing Eq. (1.103). Consequently, the KHIE given by Eq. (2.5) can be reformulated as

$$p(\mathbf{r}) = -\mathrm{i}\rho_0\omega \iiint_V G(\mathbf{r};\mathbf{r}')q(\mathbf{r}')\mathrm{d}^3\mathbf{r}' - \mathrm{i}\rho_0\omega \oiint_S G(\mathbf{r};\mathbf{r}')\mathrm{d}Q_S, \quad \mathbf{r} \in V. \tag{2.6}$$

In Eq. (2.6), it is clear that the effect of vibrations on the boundary can be effectively represented as a source with a volume velocity distribution of $\mathrm{d}Q_S \equiv -\mathbf{v}(\mathbf{r}') \cdot \mathrm{d}\mathbf{S}'$. When there are no vibrations present on the boundary, Eq. (2.6) is solely contributed by the source density, and it can be expressed as follows

$$p(\mathbf{r}) = -\mathrm{i}\rho_0\omega \iiint_V G(\mathbf{r};\mathbf{r}')q(\mathbf{r}')\mathrm{d}^3\mathbf{r}', \quad \mathbf{r} \in V. \tag{2.7}$$

2.2.2 TWO-DIMENSIONAL MODEL

2.2.2.1 Problem Description

PALs can take on various shapes such as rectangular configurations. When one side of a rectangular PAL significantly exceeds the wavelength, while the other side does not, it becomes possible to approximate the radiated sound field as a 2D radiation emitting from a finite-sized line source that is baffled [Mellow and Karkkainen, 2011; Poletti, 2019]. This approximation has found application in modeling the radiation from linear phased array PALs [Shi and Kajikawa, 2015a; Zhong et al., 2021a; Zhong et al., 2023d].

Figure 2.1 depicts a schematic representation of the 2D physical model for PAL radiation. Rectangular (x, y) and polar (ρ, φ) coordinate systems are established with their origin, O, at the centroid of the PAL, and ρ and φ representing the radial and polar angle coordinates, respectively. The positive x-axis corresponds to the radiation direction ($\varphi = 0$). The length of the PAL along the y-axis is $2a$, while the baffled region extends across $|y| > a$. In the baffled region, the particle velocity of acoustic waves is zero.

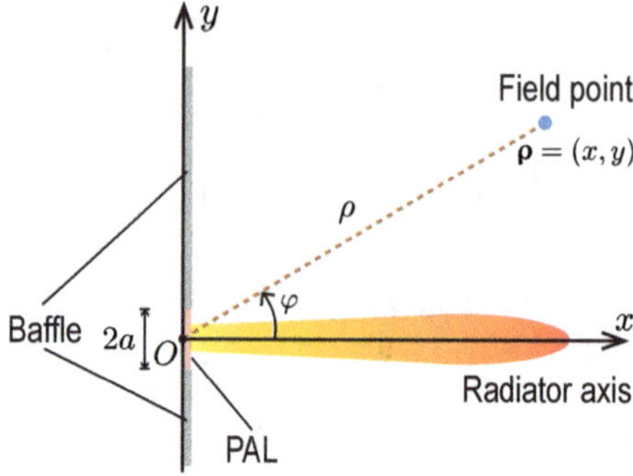

Figure 2.1 Sketch of the 2D physical model for a baffled PAL. Both the baffle and the PAL are located at $x = 0$.

In this model, the boundary condition for ultrasound on the radiation surface $x = x_s$ is given as

$$v_{\mathrm{I},x}(\boldsymbol{\rho}_s, t) = v_{1,x}(\boldsymbol{\rho}_s)\mathrm{e}^{-\mathrm{i}\omega_1 t} + v_{2,x}(\boldsymbol{\rho}_s)\mathrm{e}^{-\mathrm{i}\omega_2 t}, \tag{2.8}$$

where the source point $\boldsymbol{\rho}_s = (x_s, y_s)$, t is the time, $\omega_i = 2\pi f_i$ is the angular frequency at frequency f_i with $i = 1, 2$, and $v_{i,x}$ is the particle velocity in x-direction at frequency f_i. Without loss of generality, it is assumed that $x_s = 0$.

2.2.2.2 Velocity Profile

The velocity profile of ultrasound as shown in Eq. (2.8) determines the radiation of a PAL. The piston line source is a frequently employed model due to its simplicity. In this model, it is assumed that the vibration of each point on the radiation surface is in-phase with the same amplitude. For a piston line source within a 2D radiation model, the velocity profile can be expressed as

$$v_{i,x}(\boldsymbol{\rho}_s) = v_0 \mathcal{A}(\boldsymbol{\rho}_s), \tag{2.9}$$

which is uniform across the entire radiation surface. Here, v_0 is a constant with units of m/s, the aperture area

$$\mathcal{A}(\boldsymbol{\rho}_s) = \Pi\left(\frac{y_s}{2a}\right), \tag{2.10}$$

and the *rectangle function* $\Pi(\xi)$ is defined as

$$\Pi(\xi) = \begin{cases} 1, & |\xi| < \dfrac{1}{2}, \\ 0, & \text{otherwise}. \end{cases} \tag{2.11}$$

Apodized sources are more realistic in applications [Assaad and Rouvaen, 1998; Schmerr Jr, 2014; Zhong et al., 2023d]. For instance, a line source apodized by a cosine window represents a vibration mode of a membrane with fixed edges. The velocity profile for a line source apodized by a cosine window in a 2D radiation model is expressed as

$$v_{i,x}(\boldsymbol{\rho}_\mathrm{s}) = v_0 \cos\left(\frac{\pi y_\mathrm{s}}{2a}\right) \mathcal{A}(\boldsymbol{\rho}_\mathrm{s}). \tag{2.12}$$

The concept of a 2D steerable PAL has garnered significant attention in the realm of directivity control (see Chap. 6 for more details), which aims to steer the highly directional audio beam into a desired direction without necessitating mechanical source rotation [Pompei, 2002; Tan et al., 2004; Gan et al., 2006; Shi, 2013; Akahori et al., 2014; Kashiwase and Kondo, 2014; Shi and Kajikawa, 2015a; Nakagawa et al., 2019; Zhong et al., 2021a; Okano and Kajikawa, 2022]. In the context of a 2D radiation model, the velocity profile for a steerable line source can be expressed as

$$v_{i,x}(\boldsymbol{\rho}_\mathrm{s}) = v_0 \mathrm{e}^{\mathrm{i}\Re(k_i)\boldsymbol{\rho}_\mathrm{s}\cdot\hat{\mathbf{s}}} \mathcal{A}(\boldsymbol{\rho}_\mathrm{s}) = v_0 \mathrm{e}^{\mathrm{i}\Re(k_i) y_\mathrm{s} \sin\varphi_\mathrm{d}} \mathcal{A}(\boldsymbol{\rho}_\mathrm{s}), \tag{2.13}$$

where $\Re(\cdot)$ takes the real part of the argument, k_i is the complex wavenumber at frequency f_i,

$$\hat{\mathbf{s}} = (\cos\varphi_\mathrm{d}, \sin\varphi_\mathrm{d}), \tag{2.14}$$

is the unit vector indicating the beam direction, and φ_d is the steering angle.

2.2.2.3 Ultrasound Field

In the framework of the quasilinear solution, as illustrated in Sec. 1.4.2, the ultrasound field is determined by the boundary condition given by Eq. (2.8). Consider the volume $V = \{(x,y)\,|\,x \geq x_\mathrm{s}\}$, and select the Green's function with a rigid baffle as $G(\boldsymbol{\rho};\boldsymbol{\rho}') = 2g_{2\mathrm{D}}(\boldsymbol{\rho},\boldsymbol{\rho}',\omega)$, where the 2D Green's function in free space is expressed as [Poletti, 2019, Eq. (6)]

$$g_{2\mathrm{D}}(\boldsymbol{\rho},\boldsymbol{\rho}',\omega) \equiv \frac{\mathrm{i}}{4} H_0(k|\boldsymbol{\rho}-\boldsymbol{\rho}'|), \tag{2.15}$$

where $H_0(\cdot)$ is the Hankel function of the first kind of order 0, and the Euclidean distance between points $\boldsymbol{\rho}$ and $\boldsymbol{\rho}'$ is

$$|\boldsymbol{\rho}-\boldsymbol{\rho}'| = \sqrt{(x-x')^2+(y-y')^2}. \tag{2.16}$$

On the boundary $x = x_\mathrm{s}$, it is clear that $-\mathbf{v}_i(\boldsymbol{\rho}_\mathrm{s})\cdot\mathrm{d}\mathbf{S} = v_{i,x}(\boldsymbol{\rho}_\mathrm{s})\mathrm{d}y_\mathrm{s}$. Consequently, by using Eq. (2.6), the ultrasound pressure at a field location $\boldsymbol{\rho} = (x,y)$ can be obtained by a Rayleigh-like integral

$$p_i(\boldsymbol{\rho}) = -2\mathrm{i}\rho_0\omega_i \int_{-\infty}^{\infty} v_{i,x}(\boldsymbol{\rho}_\mathrm{s}) g_{2\mathrm{D}}(\boldsymbol{\rho},\boldsymbol{\rho}_\mathrm{s},\omega_i)\mathrm{d}y_\mathrm{s}, \tag{2.17}$$

By substituting Eq. (2.15) into Eq. (2.17), the ultrasound pressure is obtained as [Poletti, 2019, Eq. (37); Schmerr Jr, 2014, Eq. (3.96)]

$$p_i(\boldsymbol{\rho}) = \frac{\rho_0\omega_i}{2} \int_{-\infty}^{\infty} v_{i,x}(\boldsymbol{\rho}_\mathrm{s}) H_0(k_i|\boldsymbol{\rho}-\boldsymbol{\rho}_\mathrm{s}|)\mathrm{d}y_\mathrm{s}. \tag{2.18}$$

It is worth noting that through the utilization of the image source method (ISM), the radiation from a baffled line source can also be interpreted as the radiation stemming from a volume source, characterized by a source density given by

$$q_i(\boldsymbol{\rho}_s) = 2v_{i,x}(\boldsymbol{\rho}_s)\delta(x - x_s). \tag{2.19}$$

In this scenario, the region of interest encompasses the entire space, defined as $V = \{(x,y) \mid -\infty < x, y < \infty\}$. The total surface velocity is then

$$Q_i \equiv \iint_V q_i(\boldsymbol{\rho}_s)\mathrm{d}^2\boldsymbol{\rho}_s = 2\int_{-\infty}^{\infty} v_{i,x}(\boldsymbol{\rho}_s)\mathrm{d}y_s. \tag{2.20}$$

By substituting Eq. (2.19) into the KHIE given by Eq. (2.6), while selecting the free space Green's function $G(\boldsymbol{\rho};\boldsymbol{\rho}_s) = g_{2\mathrm{D}}(\boldsymbol{\rho},\boldsymbol{\rho}_s,\omega_i)$, only the first term on the right-hand side of Eq. (2.6) is retained. Consequently, the resulting expression for the ultrasound pressure is identical to Eq. (2.18).

The components of the velocity field of ultrasound can be determined by substituting Eq. (2.18) into Eq. (1.103) as

$$\mathbf{v}_i(\boldsymbol{\rho}) = \frac{\mathrm{i}}{2}\int_{-\infty}^{\infty} v_{i,x}(\boldsymbol{\rho}_s)\frac{\boldsymbol{\rho}-\boldsymbol{\rho}_s}{|\boldsymbol{\rho}-\boldsymbol{\rho}_s|}H_1(k_i|\boldsymbol{\rho}-\boldsymbol{\rho}_s|)k_i\mathrm{d}y_s, \tag{2.21}$$

where $H_1(\cdot)$ is the Hankel function of the first kind of order 1, and the derivative relation given by Eq. (B.22) is used.

In the far field when $\rho \to \infty$, Eq. (2.18) can be simplified by employing the limiting form of Hankel function given by Eq. (B.20), resulting in

$$p_i(\boldsymbol{\rho}) = \frac{\rho_0\omega_i}{\sqrt{2\mathrm{i}\pi k_i}}\int_{-\infty}^{\infty}\frac{v_{i,x}(\boldsymbol{\rho}_s)}{\sqrt{|\boldsymbol{\rho}-\boldsymbol{\rho}_s|}}\mathrm{e}^{\mathrm{i}k_i|\boldsymbol{\rho}-\boldsymbol{\rho}_s|}\mathrm{d}y_s. \tag{2.22}$$

The 2D direct integration method (DIM) evaluates the integral expressions given by Eqs. (2.18), (2.21), and (2.22) directly with numerical integration techniques to obtain the ultrasound field with a specified velocity profile $v_{i,x}(\boldsymbol{\rho}_s)$. Commonly employed numerical integration techniques include Newton-Cotes methods, Romberg integration, adaptive quadrature, and Gaussian quadrature [Sauer, 2012, Chap. 5; Chapra and Canale, 2015, Chap. 21; Zhong et al., 2020c].

2.2.2.4 Audio Sound Field

As discussed in Sec. 1.4, in the framework of the quasilinear solution, the audio sound can be considered as the radiation from a virtual source with a source density of Eq. (1.106). In the framework of the 2D radiation model, the source density at virtual source location $\boldsymbol{\rho}_v = (x_v, y_v)$ is expressed as

$$q_a(\boldsymbol{\rho}_v) = \frac{\beta\omega_a}{\mathrm{i}\rho_0^2c_0^4}p_1^*(\boldsymbol{\rho}_v)p_2(\boldsymbol{\rho}_v), \tag{2.23}$$

where $\omega_a = 2\pi f_a$ is the audio angular frequency at frequency f_a.

In the context of a baffled line source model with the baffle located at $x = x_s$, each virtual source at $\boldsymbol{\rho}_v$ has a corresponding image source at $(2x_s - x_v, y_v)$ with an identical source density of $q_a(\boldsymbol{\rho}_v)$. By setting $q_a(2x_s - x_v, y_v) = q_a(\boldsymbol{\rho}_v)$, the region of interest is selected as the entire space, defined as $V = \{(x_v, y_v) \mid -\infty < x_v, y_v < \infty\}$. The audio sound pressure can then be determined based on the KHIE in Eq. (2.7) as

$$p_a(\boldsymbol{\rho}) = -i\rho_0\omega_a \iint_{-\infty}^{\infty} q_a(\boldsymbol{\rho}_v) g_{2D}(\boldsymbol{\rho}, \boldsymbol{\rho}_v, \omega_a) d^2\boldsymbol{\rho}_v. \tag{2.24}$$

By substituting Eqs. (2.15) and (2.23) into Eq. (2.24), the audio sound in free space is rewritten as

$$p_a(\boldsymbol{\rho}) = \frac{\beta\omega_a^2}{4i\rho_0 c_0^4} \iint_{-\infty}^{\infty} p_1^*(\boldsymbol{\rho}_v) p_2(\boldsymbol{\rho}_v) H_0(k_a|\boldsymbol{\rho} - \boldsymbol{\rho}_v|) d^2\boldsymbol{\rho}_v. \tag{2.25}$$

In some scenarios, it is more convenient to express Eq. (2.25) in polar coordinates as

$$p_a(\boldsymbol{\rho}) = \frac{\beta\omega_a^2}{4i\rho_0 c_0^4} \int_0^{2\pi} \int_0^{\infty} p_1^*(\boldsymbol{\rho}_v) p_2(\boldsymbol{\rho}_v) H_0(k_a|\boldsymbol{\rho} - \boldsymbol{\rho}_v|) \rho_v d\rho_v d\varphi_v. \tag{2.26}$$

In the far field when $\rho \to \infty$, Eq. (2.25) can be simplified by employing the limiting form of Hankel function given by Eq. (B.20), resulting in

$$p_a(\boldsymbol{\rho}) = \frac{\beta\omega_a^2}{2i\rho_0 c_0^4 \sqrt{2i\pi k_a}} \iint_{-\infty}^{\infty} \frac{p_1^*(\boldsymbol{\rho}_v) p_2(\boldsymbol{\rho}_v)}{|\boldsymbol{\rho} - \boldsymbol{\rho}_v|} e^{ik_a|\boldsymbol{\rho} - \boldsymbol{\rho}_v|} d^2\boldsymbol{\rho}_v. \tag{2.27}$$

The 2D DIM evaluates the integral expression given by Eqs. (2.25) and (2.27) directly with numerical integration techniques to obtain the audio sound field, where the ultrasound pressure $p_i(\boldsymbol{\rho}_v)$ in the integrand is obtained with Eq. (2.18). To include the local effects shown by Eq. (1.107), the ultrasonic particle velocity is required which can be obtained by evaluating Eq. (2.21) directly. The exponential term within the integrals in Eqs. (2.25) and (2.27) is highly oscillatory throughout the integration domain, leading to slow convergence during numerical integration computations.

2.2.3 THREE-DIMENSIONAL MODEL

2.2.3.1 Problem Description

In comparison to the 2D model, the 3D model offers greater accuracy when it comes to simulating a PAL, albeit at the expense of increased computational resources. As illustrated in Fig. 2.2, this 3D physical model depicts a baffled circular PAL with a radius of a. It is worth noting that while a circular PAL is illustrated here, the model developed in this section is versatile and can be applied to PALs of other shapes, including rectangular configurations. For convenient references, Cartesian (x, y, z), cylindrical (ρ, φ, z), and spherical (r, θ, φ) coordinate systems are established with their origin, O, at the centroid of the PAL and the positive z-axis pointing to the radiation direction, where r, θ, and φ are the radial, zenithal, and azimuthal coordinates, respectively.

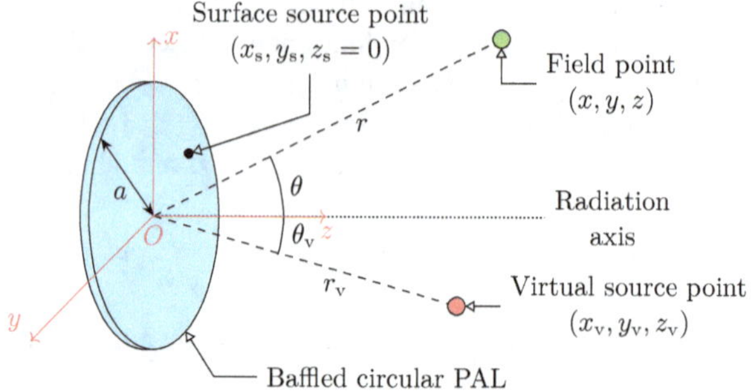

Figure 2.2 Sketch of the 3D physical model for a baffled circular PAL. Both the baffle and the PAL are located at $z = 0$. Extracted from [Zhong et al., 2020c, Fig. 1].

In the 3D model, the boundary condition on the radiation surface $z = z_s$ is given as

$$v_{I,z}(\mathbf{r}_s, t) = v_{1,z}(\mathbf{r}_s)e^{-i\omega_1 t} + v_{2,z}(\mathbf{r}_s)e^{-i\omega_2 t}, \tag{2.28}$$

where $v_{i,z}(\mathbf{r}_s)$ is the particle velocity in the z-direction at the source point $\mathbf{r}_s = (x_s, y_s, z_s)$ and frequency of f_i, $i = 1, 2$. Without loss of generality, it is assumed that the radiation surface is located on the plane $z = 0$, indicating that $z_s = 0$.

2.2.3.2 Velocity Profile

The velocity profile shown in Eq. (2.28) determines the radiation of a PAL. The velocity profile for a circular piston source with a radius of a is

$$v_{i,z}(\mathbf{r}_s) = v_0 \mathcal{A}(\mathbf{r}_s), \tag{2.29}$$

where the aperture area is

$$\mathcal{A}(\mathbf{r}_s) = \mathrm{H}(a - \rho_s), \tag{2.30}$$

and $\mathrm{H}(\xi)$ is the Heaviside function which is defined as

$$\mathrm{H}(\xi) = \begin{cases} 1, & \xi > 0, \\ 0, & \text{otherwise.} \end{cases} \tag{2.31}$$

The circular source apodized by a quadratic profile has the form of

$$v_{i,z}(\mathbf{r}_s) = v_0 \left[1 - \left(\frac{\rho_s}{a}\right)^2\right]^m \mathcal{A}(\mathbf{r}_s), \tag{2.32}$$

where $m = 0, 1, 2, \ldots$ is the order number. The degenerated quadratic profile with an order of $m = 0$ represents a piston source given by Eq. (2.32). Higher order quadratic profiles are more realistic in applications [Greenspan, 1979; Aarts and Janssen, 2009; Červenka and Bednařík, 2015; Červenka and Bednařík, 2019]. For example, the profiles of order $m = 1$ and 2 correspond to a simply supported source and a clamped source, respectively.

A 3D steerable PAL means that the directional audio beam can be steered in both azimuthal φ and zenithal θ directions (see Chap. 6 for more details), which is more flexible than a 2D steerable PAL as shown by Eq. (2.13). For a steerable source in a 3D model, the steering direction is assumed to be determined by a unit vector of

$$\hat{\mathbf{s}} = (\cos\varphi_\mathrm{d}\sin\theta_\mathrm{d}, \sin\varphi_\mathrm{d}\sin\theta_\mathrm{d}, \cos\theta_\mathrm{d}), \tag{2.33}$$

where θ_d and φ_d are the steering zenithal and azimuthal angles, respectively. The velocity profile for a circular steerable PAL is then

$$\begin{aligned} v_{i,z}(\mathbf{r}_\mathrm{s}) &= v_0 \exp(\mathrm{i}\Re k_i \mathbf{r}_\mathrm{s}\cdot\hat{\mathbf{s}})\mathcal{A}(\mathbf{r}_\mathrm{s}) \\ &= v_0 \exp[\mathrm{i}\Re k_i(x_\mathrm{s}\cos\varphi_\mathrm{d}\sin\theta_\mathrm{d} + y_\mathrm{s}\sin\varphi_\mathrm{d}\sin\theta_\mathrm{d})]\mathcal{A}(\mathbf{r}_\mathrm{s}). \end{aligned} \tag{2.34}$$

Here, the aperture area for a circular PAL is given by Eq. (2.30).

Rectangular PALs are widely used in applications. The velocity profile for a rectangular piston PAL with side lengths of $2a_x$ and $2a_y$ in x- and y-directions, respectively, has the same form of Eq. (2.29) while the aperture area is modified as

$$\mathcal{A}(\mathbf{r}_\mathrm{s}) = \Pi\left(\frac{x_\mathrm{s}}{2a_x}\right)\Pi\left(\frac{y_\mathrm{s}}{2a_y}\right). \tag{2.35}$$

The velocity profile for a rectangular and steerable PAL is same as Eq. (2.34) while the aperture area is given by Eq. (2.35).

2.2.3.3 Ultrasound Field

Consider the volume $V = \{(x,y,z)\,|\,-\infty < x,y < \infty, z_\mathrm{s} \le z < \infty\}$, and select the Green's function with a rigid baffle as $G(\mathbf{r};\mathbf{r}') = 2g_\mathrm{3D}(\mathbf{r},\mathbf{r}',\omega)$, where the 3D Green's function in free space is expressed as

$$g_\mathrm{3D}(\mathbf{r},\mathbf{r}',\omega) \equiv \frac{\mathrm{e}^{\mathrm{i}k|\mathbf{r}-\mathbf{r}'|}}{4\pi|\mathbf{r}-\mathbf{r}'|}. \tag{2.36}$$

The Euclidean distance between two points located at $\mathbf{r}$ and $\mathbf{r}'$ is

$$|\mathbf{r}-\mathbf{r}'| = \sqrt{(x-x')^2+(y-y')^2+(z-z')^2}. \tag{2.37}$$

On the boundary $z = z_\mathrm{s}$, it is clear that $-\mathbf{v}_i(\mathbf{r}_\mathrm{s})\cdot\mathrm{d}\mathbf{S} = v_{i,z}(\mathbf{r}_\mathrm{s})\mathrm{d}x_\mathrm{s}\mathrm{d}y_\mathrm{s}$. Consequently, by using Eq. (2.6), the ultrasound pressure at a field location $\mathbf{r} = (x,y,z)$ can be obtained by the *Rayleigh integral* (also known as the *Rayleigh-Sommerfeld integral*) as

$$p_i(\mathbf{r}) = -2\mathrm{i}\rho_0\omega_i \iint_{-\infty}^{\infty} v_{i,z}(\mathbf{r}_\mathrm{s})g_\mathrm{3D}(\mathbf{r},\mathbf{r}_\mathrm{s},\omega_i)\mathrm{d}^2\mathbf{r}_\mathrm{s}, \tag{2.38}$$

By substituting Eq. (2.36) into Eq. (2.38), the ultrasound pressure is rewritten as

$$p_i(\mathbf{r}) = \frac{\rho_0\omega_i}{2\pi\mathrm{i}} \iint_{-\infty}^{\infty} v_{i,z}(\mathbf{r}_\mathrm{s})\frac{\mathrm{e}^{\mathrm{i}k_i|\mathbf{r}-\mathbf{r}_\mathrm{s}|}}{|\mathbf{r}-\mathbf{r}_\mathrm{s}|}\mathrm{d}x_\mathrm{s}\,\mathrm{d}y_\mathrm{s}. \tag{2.39}$$

It is worth noting that through the utilization of the image source method, the radiation from a baffled planar source can also be interpreted as the radiation stemming from a volume source, characterized by a source density given by

$$q_i(\mathbf{r}_s) = 2v_{i,z}(\mathbf{r}_s)\delta(z - z_s). \tag{2.40}$$

In this scenario, the region of interest encompasses the entire space, defined as $V = \{(x,y,z) \mid -\infty < x,y,z < \infty\}$. The total volume velocity is then

$$Q_i \equiv \iiint_V q_i(\mathbf{r}_s)\mathrm{d}^3\mathbf{r}_s = 2\iint_{-\infty}^{\infty} v_{i,z}(\mathbf{r}_s)\mathrm{d}x_s\mathrm{d}y_s. \tag{2.41}$$

It is note worthing that when the profile is uniform, as given by Eq. (2.29), the total volume velocity is simplified to

$$Q_i = 2v_0\mathcal{A}_0, \tag{2.42}$$

where $\mathcal{A}_0$ is the total area of the radiation surface.

By substituting Eq. (2.40) into the KHIE given by Eq. (2.6), while selecting the free space Green's function $G(\mathbf{r};\mathbf{r}_s) = g_{3\mathrm{D}}(\mathbf{r},\mathbf{r}_s,\omega_i)$, only the first term on the right-hand side of Eq. (2.6) is retained. Consequently, the resulting expression for the ultrasound pressure is identical to Eq. (2.39).

The components of the velocity field of ultrasound can be determined by substituting Eq. (2.39) into Eq. (1.103) as

$$\mathbf{v}_i(\mathbf{r}) = \frac{1}{2\pi}\iint_{-\infty}^{\infty} v_{i,z}(\mathbf{r}_s)\frac{(\mathbf{r}-\mathbf{r}_s)(1-\mathrm{i}k_i|\mathbf{r}-\mathbf{r}_s|)}{|\mathbf{r}-\mathbf{r}_s|^3}\mathrm{e}^{\mathrm{i}k_i|\mathbf{r}-\mathbf{r}_s|}\,\mathrm{d}x_s\,\mathrm{d}y_s\,. \tag{2.43}$$

The 3D DIM evaluates the integral expressions given by Eqs. (2.39) and (2.43) directly with numerical integration techniques to obtain the ultrasound field.

2.2.3.4 Audio Sound Field

Under the framework of the quasilinear solution, the audio sound can be considered as the radiation from a virtual source. Similar to Eq. (2.23), the source density at virtual source location $\mathbf{r}_v = (x_v, y_v, z_v)$ is

$$q_a(\mathbf{r}_v) = \frac{\beta\omega_a}{\mathrm{i}\rho_0^2 c_0^4}p_1^*(\mathbf{r}_v)p_2(\mathbf{r}_v). \tag{2.44}$$

In the context of a baffled planar source model with the baffle located at $z = z_s$, each virtual source at $\mathbf{r}_v$ has a corresponding image source at $(x_v, y_v, 2z_s - z_v)$ with an identical source density of $q_a(\mathbf{r}_v)$. By setting $q_a(x_v, y_v, 2z_s - z_v) = q_a(\mathbf{r}_v)$, the region of interest is selected as the entire space, defined as $V = \{(x_v,y_v,z_v) \mid -\infty < x_v,y_v,z_v < \infty\}$. The audio sound pressure is then obtained as

$$p_a(\mathbf{r}) = -\mathrm{i}\rho_0\omega_a\iiint_{-\infty}^{\infty} q_a(\mathbf{r}_v)g_{3\mathrm{D}}(\mathbf{r},\mathbf{r}_v,\omega_a)\mathrm{d}^3\mathbf{r}_v. \tag{2.45}$$

By substituting Eqs. (2.36) and (2.44) into Eq. (2.45), the audio sound pressure in free space is obtained as

$$p_{\mathrm{a}}(\mathbf{r}) = -\frac{\beta\omega_{\mathrm{a}}^2}{4\pi\rho_0 c_0^4}\iiint_{-\infty}^{\infty}\frac{p_1^*(\mathbf{r}_{\mathrm{v}})p_2(\mathbf{r}_{\mathrm{v}})}{|\mathbf{r}-\mathbf{r}_{\mathrm{v}}|}\mathrm{e}^{\mathrm{i}k_{\mathrm{a}}|\mathbf{r}-\mathbf{r}_{\mathrm{v}}|}\mathrm{d}^3\mathbf{r}_{\mathrm{v}}. \tag{2.46}$$

The 3D DIM evaluates the integral expression given by Eq. (2.46) directly with numerical integration techniques to obtain the audio sound field, where the ultrasound pressure $p_i(\mathbf{r}_{\mathrm{v}})$ in the integrand is obtained according to Eq. (2.39). To account for the local effects demonstrated in Eq. (1.107), it becomes necessary to calculate the ultrasonic particle velocity, which can be obtained by evaluating Eq. (2.43) directly. Both 2D and 3D DIMs provide accurate results within the quasilinear approximation. However, the integrands in both cases involve highly oscillatory Green's functions, which result in slow convergence during numerical integration.

2.2.4 WESTERVELT'S SOLUTIONS FOR COLLIMATED BEAMS

This section underscores the increased complexity in modeling PALs compared to linear acoustic sources, even when employing the quasilinear approximation. In earlier studies, an assumption was made of a collimated beam, signifying that ultrasound waves are highly directional and confined within the aperture region. The solutions developed in the collimated beam assumption are frequently referred to as the *Westervelt's solution* in the literature [Westervelt, 1963; Berktay, 1965a]. Subsequent Secs. 2.2.4.1 and 2.2.4.2 present 2D and 3D variants of the Westervelt's solutions, respectively.

2.2.4.1 Two-Dimensional Model

The schematic representation of the 2D collimated beam model is depicted in Fig. 2.3. For a line source with a half-width of a, the aperture area is determined by Eq. (2.10). In the collimated beam assumption, ultrasound waves are confined solely

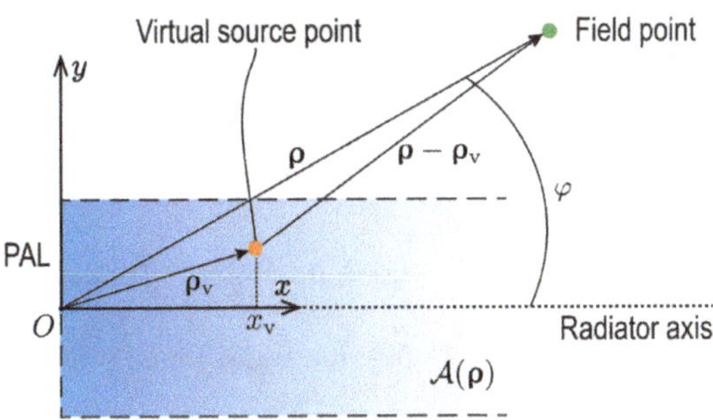

Figure 2.3 Sketch of a PAL radiating collimated beams within the aperture area $\mathcal{A}(\rho)$ in a 2D physical model. The PAL is located at $x = 0$.

to the aperture region $\mathcal{A}(\boldsymbol{\rho})$ and behave like plane waves. As a result, the ultrasound pressure is approximated by

$$p_{\mathrm{I}}(\mathbf{r},t) = \mathcal{A}(\boldsymbol{\rho}) p_0 \left[\mathrm{e}^{\mathrm{i}(k_1 x - \omega_1 t)} + \mathrm{e}^{\mathrm{i}(k_2 x - \omega_2 t)} \right], \tag{2.47}$$

where $p_0 = \rho_0 c_0 v_0$ is the on-surface sound pressure amplitude. Equation (2.47) reveals that the ultrasound pressure undergoes exponential attenuation along the x-axis, characterized by the exponent $\exp(-\alpha_i x)$, which results from substituting $k_i = \omega_i / c_0 + \mathrm{i}\alpha_i$ into the equation.

The audio source density is obtained by substituting Eq. (2.47) into Eq. (2.23) as

$$q_{\mathrm{a}}(\boldsymbol{\rho}_{\mathrm{v}}) = \mathcal{A}(\boldsymbol{\rho}_{\mathrm{v}}) \frac{\beta p_0^2 \omega_{\mathrm{a}}}{\mathrm{i} \rho_0^2 c_0^4} \mathrm{e}^{\mathrm{i}(k_2 - k_1^*) x_{\mathrm{v}}}. \tag{2.48}$$

Substituting Eq. (2.48) into Eq. (2.24) and performing the integration in the y-direction, the audio sound pressure is obtained as

$$p_{\mathrm{a}}(\boldsymbol{\rho}) = \frac{\beta a p_0^2 \omega_{\mathrm{a}}^2}{2\mathrm{i} \rho_0 c_0^4} \int_0^\infty H_0(k_{\mathrm{a}} |\boldsymbol{\rho} - \boldsymbol{\rho}_{\mathrm{v}}|) \mathrm{e}^{\mathrm{i}(k_2 - k_1^*) x_{\mathrm{v}}} \mathrm{d}x_{\mathrm{v}}. \tag{2.49}$$

It is worth noting that in the collimated beam model, the presence of the image of the audio virtual source resulting from the rigid baffle is often neglected, leading to the omission of the integration term from $-\infty$ to 0 in Eq. (2.49).

In the far field $\rho \to \infty$, the limiting form of Hankel function given by Eq. (B.20) can be used to further simplify Eq. (2.49) as

$$p_{\mathrm{a}}(\boldsymbol{\rho}) = \frac{\beta a p_0^2 \omega_{\mathrm{a}}^2}{\mathrm{i} \rho_0 c_0^4} \frac{1}{\sqrt{2\mathrm{i}\pi k_{\mathrm{a}}}} \int_0^\infty \frac{\exp\left[\mathrm{i}(k_2 - k_1^*) x_{\mathrm{v}} + \mathrm{i} k_{\mathrm{a}} |\boldsymbol{\rho} - \boldsymbol{\rho}_{\mathrm{v}}|\right]}{\sqrt{|\boldsymbol{\rho} - \boldsymbol{\rho}_{\mathrm{v}}|}} \mathrm{d}x_{\mathrm{v}}. \tag{2.50}$$

The distance between field location located at $\boldsymbol{\rho}$ and virtual source location $\boldsymbol{\rho}_{\mathrm{v}}$ can be approximated in the far field as

$$\begin{aligned} |\boldsymbol{\rho} - \boldsymbol{\rho}_{\mathrm{v}}| &= \rho \sqrt{1 - 2\frac{\boldsymbol{\rho} \cdot \boldsymbol{\rho}_{\mathrm{v}}}{\rho^2} + \frac{\rho_{\mathrm{v}}^2}{\rho^2}} = \rho - \hat{\boldsymbol{\rho}} \cdot \boldsymbol{\rho}_{\mathrm{v}} + \mathcal{O}\left(\frac{\rho_{\mathrm{v}}^2}{\rho}\right) \\ &\approx \rho - \hat{\boldsymbol{\rho}} \cdot \boldsymbol{\rho}_{\mathrm{v}} = \rho - \rho_{\mathrm{v}} \cos(\varphi - \varphi_{\mathrm{v}}) \\ &\approx \rho - x_{\mathrm{v}} \cos\varphi, \end{aligned} \tag{2.51}$$

where the term $\rho_{\mathrm{v}} \sin\varphi \sin\varphi_{\mathrm{v}}$ is dropped in the last step because $|\sin\varphi| \ll |\cos\varphi|$ at small angles.

Substituting Eq. (2.51) into Eq. (2.50), the audio sound pressure in the far field is expressed as

$$p_{\mathrm{a}}(\boldsymbol{\rho}) = \frac{-\beta a p_0^2 \omega_{\mathrm{a}}^2}{\alpha_{\mathrm{t}} \rho_0 c_0^4} \frac{\exp(\mathrm{i} k_{\mathrm{a}} \rho)}{\sqrt{2\mathrm{i}\pi k_{\mathrm{a}} \rho}} \mathcal{D}_{\mathrm{W}}(\varphi). \tag{2.52}$$

Here, $\mathcal{D}_\mathrm{W}(\varphi)$ is known as the *Westervelt's directivity* [Westervelt, 1963], which is expressed as [Zhong et al., 2023d, Eq. (23)]

$$\begin{aligned}\mathcal{D}_\mathrm{W}(\varphi) &\equiv \int_0^\infty \exp\left[\mathrm{i}(k_2 - k_1^* - k_\mathrm{a}\cos\varphi)x_\mathrm{v}\right]\mathrm{d}x_\mathrm{v} \\ &= \frac{\mathrm{i}\alpha_\mathrm{t}}{k_2 - k_1^* - k_\mathrm{a}\cos\varphi} = \frac{1}{1 - 2\mathrm{i}k_\mathrm{a}\alpha_\mathrm{t}^{-1}\sin^2(\varphi/2)},\end{aligned} \tag{2.53}$$

where the total absorption coefficient is defined as

$$\alpha_\mathrm{t} \equiv \alpha_1 + \alpha_2 - \alpha_\mathrm{a}. \tag{2.54}$$

The Westervelt's directivity shown in Eq. (2.53) is normalized at $\varphi = 0$ such that $\mathcal{D}_\mathrm{W}(0) = 1$.

An important observation is that the Westervelt's directivity is solely influenced by the audio frequency and the absorption distance of ultrasound, exhibiting independence from the aperture area. Furthermore, it is evident that the magnitude of the Westervelt's directivity Eq. (2.53), denoted as $|\mathcal{D}_\mathrm{W}(\varphi)|$, is a monotonic function within the interval $\varphi \in [0, \pi/2]$ or $\varphi \in [-\pi/2, 0]$. This observation implies that the Westervelt's directivity magnitude has no local maxima except at the main lobe $\varphi = 0$. As a result, the demodulated audio beam displays no sidelobes. It is important to note, however, that the assumption of a collimated beam is made in the derivation of Westervelt's directivity. In practical PALs, this assumption is sometimes invalid and sidelobes can be generated in the audio sound directivity [Zhong et al., 2023d].

By setting $\left|\mathcal{D}_\mathrm{W}(\varphi_\mathrm{hp})\right|^2 = 1/2$, the half power beam width (HPBW) can be determined as

$$\varphi_\mathrm{hp} = 2\arcsin\frac{1}{2\sqrt{k_\mathrm{a}\alpha_\mathrm{t}^{-1}}} \approx \frac{1}{\sqrt{k_\mathrm{a}\alpha_\mathrm{t}^{-1}}}. \tag{2.55}$$

Here, the approximation $\arcsin\xi \approx \xi$ is employed, reflecting the fact that the HPBW is generally exceptionally small, owing due to the sharp directivity of the generated audio beam. In certain literature, the absorption coefficient at the carrier ultrasound frequency f_u is commonly denoted as α_u. Additionally, the absorption distance is related by $\mathscr{D}_\mathrm{ab} = 1/\alpha_\mathrm{u}$ [refer to Eq. (A.7)]. In such scenarios, $\alpha_\mathrm{t} \approx 2\alpha_\mathrm{u} = 2\mathscr{D}_\mathrm{ab}$. Consequently, Eq. (2.55) can also be expressed as [Westervelt, 1963, Eq. (15)]

$$\varphi_\mathrm{hp} \approx \frac{2}{\sqrt{k_\mathrm{a}\alpha_\mathrm{u}^{-1}}} = \frac{2}{\sqrt{k_\mathrm{a}\mathscr{D}_\mathrm{ab}}}. \tag{2.56}$$

2.2.4.2 Three-Dimensional Model

The illustration of the 3D collimated wave model is presented in Fig. 2.4. It is important to emphasize that while a circular PAL is depicted in Fig. 2.4, the model can be applied to PALs of arbitrary shapes, including rectangles. The aperture area is determined by Eqs. (2.30) and (2.35) for circular and rectangular PALs, respectively.

Similar to Eq. (2.47), the ultrasound pressure field is approximated by

$$p_{\mathrm{I}}(\mathbf{r},t) = \mathcal{A}(\mathbf{r})p_0\left[\mathrm{e}^{\mathrm{i}(k_1z-\omega_1t)} + \mathrm{e}^{\mathrm{i}(k_2z-\omega_2t)}\right]. \tag{2.57}$$

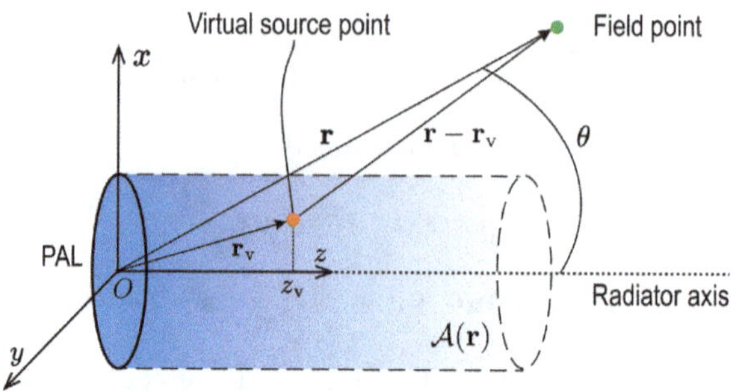

Figure 2.4 Sketch of a circular PAL radiating collimated beams within the aperture area $\mathcal{A}(\mathbf{r})$ in a 3D physical model. The PAL is located at $z = 0$.

The audio source density is obtained by substituting Eq. (2.57) into Eq. (1.106) as

$$q_{\mathrm{a}}(\mathbf{r}_\mathrm{v}) = \mathcal{A}(\mathbf{r}_\mathrm{v})\frac{\beta p_0^2\omega_\mathrm{a}}{\mathrm{i}\rho_0^2c_0^4}\mathrm{e}^{\mathrm{i}(k_2-k_1^*)z_\mathrm{v}}. \tag{2.58}$$

Substituting Eq. (2.58) into Eq. (2.24) and performing the integration on x- and y-directions, the ultrasound pressure is obtained as

$$p_{\mathrm{a}}(\mathbf{r}) = \frac{\beta p_0^2\mathcal{A}_0\omega_\mathrm{a}^2}{4\mathrm{i}\pi\rho_0c_0^4}\int_0^\infty \frac{\exp\left[\mathrm{i}(k_2-k_1^*)z_\mathrm{v}+\mathrm{i}k_\mathrm{a}|\mathbf{r}-\mathbf{r}_\mathrm{v}|\right]}{|\mathbf{r}-\mathbf{r}_\mathrm{v}|}\mathrm{d}^3\mathbf{z}_\mathrm{v}, \tag{2.59}$$

where $\mathcal{A}_0$ is the total area of the radiation surface. Similar to Eq. (2.49), in the collimated beam model, the integration domain ranging from $-\infty$ to 0 is neglected in Eq. (2.59).

The distance between field location $\mathbf{r}$ and virtual source location $\mathbf{r}_\mathrm{v}$ can be approximated in the far field ($r_\mathrm{v} \ll r$) as

$$\begin{aligned}|\mathbf{r}-\mathbf{r}_\mathrm{v}| &= r\sqrt{1-2\frac{\mathbf{r}\cdot\mathbf{r}_\mathrm{v}}{r^2}+\frac{r_\mathrm{v}^2}{r^2}} = r-\hat{\mathbf{r}}\cdot\mathbf{r}_\mathrm{v}+\mathcal{O}\left(\frac{r_\mathrm{v}^2}{r}\right)\\ &\approx r-\hat{\mathbf{r}}\cdot\mathbf{r}_\mathrm{v} = r-r_\mathrm{v}\cos\gamma,\end{aligned} \tag{2.60}$$

where cosine of the angle between $\mathbf{r}$ and $\mathbf{r}_\mathrm{v}$ is

$$\cos\gamma \equiv \cos\langle\hat{\mathbf{r}},\hat{\mathbf{r}}_\mathrm{v}\rangle = r_\mathrm{v}\cos\theta_\mathrm{v}\cos\theta + r_\mathrm{v}\sin\theta_\mathrm{v}\sin\theta\cos(\varphi-\varphi_\mathrm{v}). \tag{2.61}$$

The 3D Green's function in Eq. (2.59) can then be approximated by

$$\frac{\mathrm{e}^{\mathrm{i}k_\mathrm{a}|\mathbf{r}-\mathbf{r}_\mathrm{v}|}}{|\mathbf{r}-\mathbf{r}_\mathrm{v}|} \approx \frac{1}{r}\mathrm{e}^{\mathrm{i}k_\mathrm{a}(r-r_\mathrm{v}\cos\gamma)}. \tag{2.62}$$

In the Westervelt's models, when the term $\rho_v \sin\theta \cos(\varphi - \varphi_v)$ is omitted due to the condition $|\sin\theta| \ll |\cos\theta|$ for small angles, Eq. (2.61) is further simplified to

$$r_v \cos\gamma = z_v \cos\theta + \rho_v \sin\theta \cos(\varphi - \varphi_v) \approx z_v \cos\theta. \tag{2.63}$$

Substituting Eqs. (2.62) and (2.63) into Eq. (2.59), a closed-form solution of the audio sound pressure in the far field is obtained as

$$p_a(\mathbf{r}) = \frac{i\beta p_0^2 \mathcal{A}_0 \omega_a^2}{4\pi\alpha_t \rho_0 c_0^4 r} e^{ik_a r} \mathcal{D}_W(\theta), \tag{2.64}$$

where the same Westervelt's directivity is obtained as that in Eq. (2.52).

2.2.5 REMARKS

The reader is introduced to both 2D and 3D DIMs in this section. In the 2D DIM, the ultrasound pressure and particle velocity are computed using a one-fold Rayleigh-like integral as illustrated in Eqs. (2.18) and (2.21). Subsequently, the audio sound pressure is derived through a three-fold integral after substituting the ultrasound pressure into Eq. (2.25). In contrast, the 3D DIM involves a two-fold Rayleigh integral to compute the ultrasound pressure and particle velocity, as shown by Eqs. (2.39) and (2.43). Subsequently, the audio sound pressure is determined through a more complicated five-fold integral after substituting the ultrasound pressure into Eq. (2.46).

Both the 2D and 3D DIMs utilize numerical integration techniques to calculate the audio sound field. However, the integrands in both cases involve highly oscillatory Green's functions, resulting in slow convergence during numerical integration. Consequently, the DIM is infrequently employed in the literature due to its substantial computational demands. Nevertheless, it lays the foundation for other simplified methods introduced in subsequent sections. Building upon the DIMs, the Westervelt's solutions in the collimated beam assumption are presented in Sec. 2.2.4. The Westervelt's solutions in the far field encompass the Westervelt's directivity, as described by Eq. (2.53), which represents a closed-form solution applicable to both 2D and 3D radiation models. An important observation is that the Westervelt's directivity is solely influenced by the audio frequency and the absorption distance of ultrasound, exhibiting independence from the aperture area. In addition, the Westervelt's directivity reveals that the demodulated audio sound exhibit no sidelobes.

2.3 CONVOLUTION DIRECTIVITY MODEL (CDM)

2.3.1 TWO-DIMENSIONAL MODEL

In the far field when $\rho \to \infty$, the ultrasound pressure given by Eq. (2.18) can be written as [Schmerr Jr, 2014, Eq. (2.54); Zhong et al., 2023d, Eq. (17)]

$$p_i(\rho) = \sqrt{\frac{2}{i\pi k_i \rho_v}} p_0 k_i a e^{ik_i\rho} \mathcal{D}_i(\varphi), \tag{2.65}$$

where the coefficient is chosen to ensure the ultrasound directivity $\mathcal{D}_i(\varphi) = 1$ at the main lobe. The essence of the CMDs is to approximate the exact ultrasound field Eq. (2.18) by the inward-extrapolated far-field pressure given by Eq. (2.65). The source density of audio sound is then obtained by substituting Eq. (2.65) into Eq. (2.23) yielding

$$q_{\mathrm{a}}(\boldsymbol{\rho}_{\mathrm{v}}) = \frac{2\beta p_0^2 k_{\mathrm{a}} \sqrt{k_1 k_2} a^2}{\mathrm{i}\pi \rho_0^2 c_0^3 \rho_{\mathrm{v}}} \mathrm{e}^{\mathrm{i}(k_2 - k_1^*)\rho_{\mathrm{v}}} \mathcal{D}_1^*(\varphi_{\mathrm{v}}) \mathcal{D}_2(\varphi_{\mathrm{v}}). \tag{2.66}$$

Substituting Eq. (2.66) into the audio sound pressure given by Eq. (2.26) yields

$$\begin{aligned} p_{\mathrm{a}}(\boldsymbol{\rho}) = \frac{\beta p_0^2 k_{\mathrm{a}}^2 \sqrt{k_1 k_2} a^2}{2\pi \mathrm{i} \rho_0 c_0^2} \int_0^{2\pi} \int_0^{\infty} \mathcal{D}_1^*(\varphi_{\mathrm{v}}) \mathcal{D}_2(\varphi_{\mathrm{v}}) \mathrm{e}^{\mathrm{i}(k_2 - k_1^*)\rho_{\mathrm{v}}} \\ \times H_0(k_{\mathrm{a}} |\boldsymbol{\rho} - \boldsymbol{\rho}_{\mathrm{v}}|) \rho_{\mathrm{v}} \mathrm{d}\rho_{\mathrm{v}} \mathrm{d}\varphi_{\mathrm{v}}. \end{aligned} \tag{2.67}$$

Similar to Eq. (2.52), Eq. (2.67) can be approximated in the far field by

$$\begin{aligned} p_{\mathrm{a}}(\boldsymbol{\rho}) = \frac{\beta p_0^2 k_{\mathrm{a}}^2 a^2}{\mathrm{i}\pi \rho_0 c_0^2} \sqrt{\frac{k_1 k_2}{2\pi \mathrm{i} k_{\mathrm{a}} \rho}} \mathrm{e}^{\mathrm{i}k_{\mathrm{a}}\rho} \int_0^{2\pi} \int_0^{\infty} \mathcal{D}_1^*(\varphi_{\mathrm{v}}) \mathcal{D}_2(\varphi_{\mathrm{v}}) \\ \times \mathrm{e}^{\mathrm{i}\left[k_2 - k_1^* - k_{\mathrm{a}} \cos(\varphi - \varphi_{\mathrm{v}})\right]\rho_{\mathrm{v}}} \mathrm{d}\rho_{\mathrm{v}} \mathrm{d}\varphi_{\mathrm{v}}. \end{aligned} \tag{2.68}$$

By using the definition of the Westervelt's directivity given by Eq. (2.53), Eq. (2.68) is simplified to

$$p_{\mathrm{a}}(\boldsymbol{\rho}) = \frac{\beta p_0^2 k_{\mathrm{a}}^2 a^2}{\pi \mathrm{i} \alpha_t \rho_0 c_0^2} \sqrt{\frac{k_1 k_2}{2\pi \mathrm{i} k_{\mathrm{a}} \rho}} \mathrm{e}^{\mathrm{i}k_{\mathrm{a}}\rho} \mathcal{D}_{\mathrm{a}}(\varphi), \tag{2.69}$$

where the audio sound directivity is written as [Zhong et al., 2023d, Eq. (25)]

$$\mathcal{D}_{\mathrm{a}}(\varphi) = (\mathcal{D}_1^* \mathcal{D}_2 * \mathcal{D}_{\mathrm{W}})(\varphi) = \int_0^{2\pi} \mathcal{D}_1^*(\varphi_{\mathrm{v}}) \mathcal{D}_2(\varphi_{\mathrm{v}}) \mathcal{D}_{\mathrm{W}}(\varphi - \varphi_{\mathrm{v}}) \mathrm{d}\varphi_{\mathrm{v}}. \tag{2.70}$$

Here, "$*$" denotes the convolution operator while the superscript "$*$" denotes the complex conjugate.

Equation (2.70) defines the *direct convolution directivity model* (direct CDM), illustrating that the directivity of audio sound is a convolution of the Westervelt's directivity and the product directivity of ultrasound. A geometrically similar form was proposed in Shi and Kajikawa, 2015a, Eq. (21), albeit with the omission of the phase of the Westervelt's directivity, a simplification that can lead to inaccurate predictions, as discussed in Guasch and Sánchez-Martín, 2018. In contrast to Shi and Kajikawa, 2015a, Eq. (21), the CDM presented in Eq. (2.70) within this section is rigorously derived based on the DIMs introduced in Sec. 2.2. A critical step in the direct CDM involves approximating the ultrasound field by its inward-extrapolated far-field pressure, as depicted in Eq. (2.65). It is important to note that this approximation holds true only in the far field, beyond the Rayleigh distance [Foote, 2014]. Consequently, the nonlinear interactions of ultrasonic waves in the near field cannot be accurately captured, resulting in an imprecise prediction of the far-field directivity of audio sound.

The accuracy of the audio sound directivity obtained using the direct CDM given by Eq. (2.70) can be improved by taking into account the aperture factor of audio sound, and the directivity is modified as [Zhong et al., 2023d, Eq. (26)]

$$\tilde{\mathcal{D}}_{\mathrm{a}}(\varphi) = \mathcal{D}_{\mathrm{A}}(\varphi)\mathcal{D}_{\mathrm{a}}(\varphi) = \mathcal{D}_{\mathrm{A}}(\varphi)(\mathcal{D}_1^*\mathcal{D}_2 * \mathcal{D}_{\mathrm{W}})(\varphi), \tag{2.71}$$

where the effective directivity $\mathcal{D}_{\mathrm{A}}(\varphi)$ is the far-field directivity of an audio source with a same aperture size and a velocity profile of $v_{\mathrm{a},x}(\boldsymbol{\rho}_{\mathrm{s}}) = v_{1,x}^*(\boldsymbol{\rho}_{\mathrm{s}})v_{2,x}(\boldsymbol{\rho}_{\mathrm{s}})/v_0$. The audio sound directivity obtained using Eq. (2.71) is termed the *modified DCM* in this book.

2.3.2 THREE-DIMENSIONAL MODEL

In the 3D model shown in Fig. 2.2, the ultrasound pressure given by Eq. (2.38) in the far field, $r \to \infty$, is written as [Zhong et al., 2023d, Eq. (27)]

$$p_i(\mathbf{r}) = \frac{\rho_0 \omega_i \mathcal{A}_0}{2\pi \mathrm{i} r} \mathrm{e}^{\mathrm{i}k_i r}\mathcal{D}_i(\theta,\varphi), \tag{2.72}$$

where the coefficient is chosen to ensure the ultrasound directivity $\mathcal{D}_i(\theta,\varphi) = 1$ at the main lobe. The source density of the audio sound is then obtained by substituting Eq. (2.72) into Eq. (2.44) as

$$q_{\mathrm{a}}(\mathbf{r}_{\mathrm{v}}) = \frac{\beta \mathcal{A}_0^2 \omega_1 \omega_2 \omega_i}{4\mathrm{i}\pi^2 c_0^4 r_{\mathrm{v}}^2}\mathcal{D}_1^*(\theta_{\mathrm{v}},\varphi_{\mathrm{v}})\mathcal{D}_2(\theta_{\mathrm{v}},\varphi_{\mathrm{v}})\mathrm{e}^{\mathrm{i}(k_2-k_1^*)r_{\mathrm{v}}}. \tag{2.73}$$

Substituting Eq. (2.73) into Eq. (2.46) yields

$$\begin{aligned} p_{\mathrm{a}}(\mathbf{r}) = -\frac{\beta \mathcal{A}_0^2 \omega_1 \omega_2 \omega_i}{16\mathrm{i}\pi c_0^4} \int_0^{2\pi}\int_0^{\pi}\int_0^{\infty} \mathcal{D}_1^*(\theta_{\mathrm{v}},\varphi_{\mathrm{v}})\mathcal{D}_2(\theta_{\mathrm{v}},\varphi_{\mathrm{v}}) \\ \times \mathrm{e}^{\mathrm{i}(k_2-k_1^*)r_{\mathrm{v}}} \frac{\mathrm{e}^{\mathrm{i}k_{\mathrm{a}}|\mathbf{r}-\mathbf{r}_{\mathrm{v}}|}}{|\mathbf{r}-\mathbf{r}_{\mathrm{v}}|} \sin\theta_{\mathrm{v}} \mathrm{d}r_{\mathrm{v}}\mathrm{d}\theta_{\mathrm{v}}\mathrm{d}\varphi_{\mathrm{v}} \end{aligned}. \tag{2.74}$$

In the far field when $r \to \infty$, the 3D Green's function in Eq. (2.74) can be approximated by its limiting form Eq. (2.62). Substituting Eq. (2.62) into Eq. (2.74) with the definition of the Westervelt's directivity given by Eq. (2.53) yields

$$p_{\mathrm{a}}(\mathbf{r}) = -\frac{\beta p_0^2 \omega_1 \omega_2 \omega_{\mathrm{a}}^2 a^4}{16\pi \alpha_{\mathrm{t}} c_0^4 r} \mathrm{e}^{\mathrm{i}k_{\mathrm{a}} r}\mathcal{D}_{\mathrm{a}}(\theta,\varphi), \tag{2.75}$$

where the audio sound directivity is written as [Zhong et al., 2023d, Eq. (34)]

$$\begin{aligned} \mathcal{D}_{\mathrm{a}}(\theta,\varphi) &= (\mathcal{D}_1^*\mathcal{D}_2 \circledast \mathcal{D}_{\mathrm{W}})(\theta,\varphi) \\ &= \int_0^{2\pi}\int_0^{\pi} \mathcal{D}_1^*(\theta_{\mathrm{v}},\varphi_{\mathrm{v}})\mathcal{D}_2(\theta_{\mathrm{v}},\varphi_{\mathrm{v}})\mathcal{D}_{\mathrm{W}}(\gamma)\sin\theta_{\mathrm{v}}\mathrm{d}\theta_{\mathrm{v}}\mathrm{d}\varphi_{\mathrm{v}}. \end{aligned} \tag{2.76}$$

Here, the functions $\mathcal{D}_1^*(\theta,\varphi)\mathcal{D}_2(\theta,\varphi)$ and $\mathcal{D}_\mathrm{W}(\gamma)$ are defined on a unit sphere with the zenithal and azimuthal angles of θ and φ, respectively. Consequently, the directivity of audio sound shown by Eq. (2.76) can be conceptualized as a spherical convolution [Roddy and McEwen, 2021] between these functions, where "$\circledast$" denotes the spherical convolution operator. Similar to Eq. (2.71), the 3D modified CDM is derived by considering the aperture factor of audio sound as

$$\tilde{\mathcal{D}}_\mathrm{a}(\theta,\varphi) = \mathcal{D}_\mathrm{A}(\theta,\varphi)(\mathcal{D}_1^*\mathcal{D}_2 \circledast \mathcal{D}_\mathrm{W})(\theta,\varphi), \tag{2.77}$$

where the effective directivity $\mathcal{D}_\mathrm{A}(\theta,\varphi)$ is the far-field directivity of an audio source with the same aperture size and a velocity profile of $v_{\mathrm{a},z}(\mathbf{r}_\mathrm{s}) = v_{1,z}^*(\mathbf{r}_\mathrm{s})v_{2,z}(\mathbf{r}_\mathrm{s})/v_0$.

The geometric interpretation of the direct CDM given by Eq. (2.76) is that the PAL emits infinitely many ultrasonic waves in all directions and the audio sound directivity in each direction $(\theta_\mathrm{v},\varphi_\mathrm{v})$ has a form of the Westervelt's directivity weighted by ultrasonic directivity $\mathcal{D}_1^*(\theta_\mathrm{v},\varphi_\mathrm{v})\mathcal{D}_2(\theta_\mathrm{v},\varphi_\mathrm{v})$. The argument of the Westervelt's directivity γ is the included angle between the direction of the field and virtual source locations, i.e., (θ,φ) and $(\theta_\mathrm{v},\varphi_\mathrm{v})$. A similar form of Eq. (2.76) has been derived in Guasch and Sánchez-Martín, 2018, Eq. (20) in a geometric way, where γ is chosen as the included angle between the zenithal angles θ and θ_v. This leads to inaccurate predictions because the convolution in the azimuthal direction is not included.

2.3.3 THREE-DIMENSIONAL AXISYMMETRIC MODEL

In the 3D axisymmetric model, the velocity profile of ultrasound $v_{i,z}(\mathbf{r}_\mathrm{s})$ is independent of the azimuthal angle φ_s. Consequently, the audio sound directivity is axisymmetric and is independent of φ. By setting $\varphi = 0$ in Eq. (2.61) and using the Westervelt's directivity given by Eq. (2.53), the integration with respect to the azimuthal angle φ_v can be simplified to (see [Zhong et al., 2023d, Appendix] for detailed derivations)

$$\begin{aligned}\frac{1}{2\pi}\int_0^{2\pi} &\frac{\mathrm{d}\varphi_\mathrm{v}}{1 - \mathrm{i}k_\mathrm{a}\alpha_\mathrm{t}^{-1}(1-\cos\theta\cos\theta_\mathrm{v} - \sin\theta\sin\theta_\mathrm{v}\cos\varphi_\mathrm{v})} \\ &= \sqrt{\mathcal{D}_\mathrm{W}(\theta-\theta_\mathrm{v})\mathcal{D}_\mathrm{W}(\theta+\theta_\mathrm{v})}.\end{aligned} \tag{2.78}$$

The 3D axisymmetric direct CDM is then obtained by substituting Eq. (2.78) into Eq. (2.76) as [Zhong et al., 2023d, Eq. (37)]

$$\begin{aligned}\mathcal{D}_\mathrm{a}(\theta) &= \int_0^{\pi} \mathcal{D}_1^*(\theta_\mathrm{v})\mathcal{D}_2(\theta_\mathrm{v})\sqrt{\mathcal{D}_\mathrm{W}(\theta-\theta_\mathrm{v})\mathcal{D}_\mathrm{W}(\theta+\theta_\mathrm{v})}\sin\theta_\mathrm{v}\mathrm{d}\theta_\mathrm{v} \\ &= [\mathcal{D}_1^*(\theta)\mathcal{D}_2(\theta)] * \sqrt{\mathcal{D}_\mathrm{W}(\theta)\mathcal{D}_\mathrm{W}(-\theta)}\end{aligned}. \tag{2.79}$$

Here, the property $\mathcal{D}_\mathrm{W}(-\theta-\theta_\mathrm{v}) = \mathcal{D}_\mathrm{W}(\theta+\theta_\mathrm{v})$ is used because the Westervelt's directivity Eq. (2.53) is an even function. It is worth noting that the integration domain is from 0 to π in Eq. (2.79), while it is from 0 to 2π in the 2D CDM given by Eq. (2.70).

The 3D axisymmetric improved CDM is then [Zhong et al., 2023d, Eq. (38)]

$$\tilde{\mathcal{D}}_{\mathrm{a}}(\theta) = \mathcal{D}_{\mathrm{A}}(\theta)\left\{[\mathcal{D}_1^*(\theta)\mathcal{D}_2(\theta)] * \sqrt{\mathcal{D}_{\mathrm{W}}(\theta)\mathcal{D}_{\mathrm{W}}(-\theta)}\right\}, \tag{2.80}$$

where the effective directivity $\mathcal{D}_{\mathrm{A}}(\theta)$ is the far-field directivity of an axisymmetric audio source with the same aperture size and an axisymmetric velocity profile.

The geometric interpretation of Eq. (2.79) is that the PAL emits infinitely many ultrasonic waves in all directions and the audio sound directivity in each zenithal direction θ_{v} has a form of $\sqrt{\mathcal{D}_{\mathrm{W}}(\theta-\theta_{\mathrm{v}})\mathcal{D}_{\mathrm{W}}(\theta+\theta_{\mathrm{v}})}$ weighted by the ultrasonic directivity $\mathcal{D}_1^*(\theta_{\mathrm{v}})\mathcal{D}_2(\theta_{\mathrm{v}})$. According to Eq. (2.78), it is clear that $\sqrt{\mathcal{D}_{\mathrm{W}}(\theta-\theta_{\mathrm{v}})\mathcal{D}_{\mathrm{W}}(\theta+\theta_{\mathrm{v}})}$ represents an effective Westervelt's directivity generated by a highly collimated spherical cone source on the surface $\theta = \theta_{\mathrm{v}}$. A 3D axisymmetric direct CDM was given in Guasch and Sánchez-Martín, 2018, Eq. (22) similar to Eq. (2.79) in this book, but the Westervelt's directivity given by Eq. (2.53) is used, which is less accurate than $\sqrt{\mathcal{D}_{\mathrm{W}}(\theta-\theta_{\mathrm{v}})\mathcal{D}_{\mathrm{W}}(\theta+\theta_{\mathrm{v}})}$ used in Eq. (2.79).

2.3.4 DIRECTIVITIES OF TYPICAL SOURCES

To derive the far-field directivity of the audio sound produced by a PAL using the introduced CDM, it is evident that both Westervelt's directivity and the directivity of ultrasound are required for the calculations. The Westervelt's directivity is obtained in Sec. 2.2.4 and given as Eq. (2.53). The ultrasound source is conventionally modeled as a baffled source in linear acoustics. The directivities of typical sources are summarized here for reference.

For a line piston source with a length of $2a$ and a velocity profile described by Eq. (2.9), the directivity is expressed as

$$\mathcal{D}(\varphi) = \mathrm{sinc}(\Re ka \sin\varphi). \tag{2.81}$$

Here, the *sinc function* is defined as

$$\mathrm{sinc}\,\xi \equiv \frac{\sin\xi}{\xi}. \tag{2.82}$$

For a circular piston source with a radius of a and a velocity profile described by Eq. (2.29), the directivity is expressed as

$$\mathcal{D}(\theta,\varphi) = \mathrm{jinc}(\Re ka \sin\theta). \tag{2.83}$$

Here, the *jinc function* is defined as

$$\mathrm{jinc}\xi \equiv \begin{cases} \dfrac{2J_1(\xi)}{\xi}, & \xi \neq 0, \\ 1, & \xi = 0, \end{cases} \tag{2.84}$$

where $J_1(\cdot)$ is the Bessel function of order 1.

For a rectangular piston source with side lengths of $2a_x$ and $2a_y$ in x and y directions, respectively, the velocity profile is described by Eq. (2.35). Consequently, the directivity is expressed as

$$\mathcal{D}(\theta,\varphi) = \operatorname{sinc}(\Re k a_x \sin\theta\cos\varphi)\operatorname{sinc}(\Re k a_y \sin\theta\sin\varphi). \tag{2.85}$$

2.3.5 REMARKS

In this section, the reader is introduced to both the 2D and 3D CDMs. The derived formulas are expressed as linear and spherical convolutions of ultrasound directivity and the Westervelt's directivity for 2D and 3D models, as illustrated in Eqs. (2.70) and (2.76), respectively. In cases where the ultrasound velocity profile exhibits axisymmetry in a 3D model, the expression can be further simplified, as demonstrated in Eq. (2.79). To enhance prediction accuracy, the derived formulas are adjusted by multiplying them with an effective directivity factor resulting from the aperture characteristics of the audio sound, as demonstrated in Eqs. (2.71), (2.77), and (2.80).

The essence of the CDMs is to approximate the ultrasound pressure in the near field using its inward-extrapolated far-field pressure, making them unsuitable for predicting audio sound in the near field. The strength of CDMs lies in their capability to predict the far-field directivity of audio sound generated by a PAL with an acceptable accuracy, while maintaining a relatively low computational load [Zhong et al., 2023d]. However, it is important to acknowledge that CDMs lack the ability to capture intricate nonlinear interactions in the near field. This limitation adversely affects the prediction accuracy of CDMs, even for the calculation of far-field directivities. A comprehensive analysis of the accuracy of CDMs can be found in Zhong et al., 2023d.

2.4 GAUSSIAN BEAM EXPANSION (GBE) METHOD

2.4.1 GBE COEFFICIENTS AND PARAXIAL APPROXIMATIONS

The idea of the *Gaussian beam expansion* (GBE) method is to approximate the velocity profile of ultrasound by a sum of a set of Gaussian functions. For an arbitrary profile function, it is approximated by N Gaussian functions as [Wen and Breazeale, 1988, Eq. (18)]

$$u(\xi) \approx \sum_{n=1}^{N} A_n \exp\left(-B_n \xi^2\right), \tag{2.86}$$

where A_n and B_n are called the *GBE coefficients*. Because the Gaussian functions are not a complete orthogonal set, the approximation equal sign is used in Eq. (2.86).

The GBE coefficients are commonly determined using various optimization methods. For example, the GBE coefficients can be obtained by minimizing the cost function [Wen and Breazeale, 1988, Eq. (20)]

$$\mathcal{J}(A_1,B_1,\ldots,A_N,B_N)=\int_{-\infty}^{\infty}\left|u(\xi)-\sum_{n=1}^{N}A_n\exp(-B_n\xi^2)\right|^2\mathrm{d}\xi. \tag{2.87}$$

The cost function can also be chosen as [Červenka and Bednařík, 2015, Eq. (7)]

$$\mathcal{J}(A_1,B_1,\ldots,A_N,B_N)=\int_{-\infty}^{\infty}\left|u(\xi)-\sum_{n=1}^{N}A_n\exp(-B_n\xi^2)\right|^2\xi\mathrm{d}\xi, \tag{2.88}$$

where a weighting factor ξ is included to improve the convergence.

Equations (2.87) and (2.88) represent the cost functions with respect to the profile in the spatial domain. The GBE coefficients can also be obtained in the k-space (spatial frequency domain) [Sha et al., 2003; Kim et al., 2006; Ding and Xu, 2006]. By introducing the Fourier transform pair,

$$\begin{cases} u(k_\xi)=\int_{-\infty}^{\infty}u(\xi)\mathrm{e}^{-\mathrm{i}k_\xi\xi}\mathrm{d}\xi, \\ u(\xi)=\dfrac{1}{2\pi}\int_{-\infty}^{\infty}u(k_\xi)\mathrm{e}^{\mathrm{i}k_\xi\xi}\mathrm{d}k_\xi, \end{cases} \tag{2.89}$$

the profile in the k-space is approximated by N Gaussian functions as

$$u(k_\xi)\approx\sum_{n=1}^{N}A_n^k\exp\left(-B_n^k k_\xi^2\right), \tag{2.90}$$

where A_n^k and B_n^k are GBE coefficients in the k-space. They are obtained by minimizing the cost function [Sha et al., 2003, Eq. (13)]

$$\mathcal{J}(A_1^k,B_1^k,\ldots,A_N^k,B_N^k)=\int_{-\infty}^{\infty}\left|u(k_\xi)-\sum_{n=1}^{N}A_n^k\exp\left(-B_n^k k_\xi^2\right)\right|^2\mathrm{d}k_\xi. \tag{2.91}$$

By substituting Eqs. (2.86) and (2.90) into Eq. (2.89) and using the integral identity [Ding et al., 2003, Eq. (5); Gradshteyn and Ryzhik, 2015, Eq. (3.323-2)]

$$\int_{-\infty}^{\infty}\exp\left(-\alpha^2x^2\pm\beta x\right)\mathrm{d}x=\frac{\sqrt{\pi}}{\alpha}\exp\left(\frac{\beta^2}{4\alpha^2}\right),\quad\Re(\alpha^2)>0, \tag{2.92}$$

it is found that the GBE coefficients in the spatial domain are related to those in the k-space by

$$\begin{cases} A_n^k=A_n\sqrt{\dfrac{\pi}{B_n}}, \\ B_n^k=\dfrac{1}{4B_n}. \end{cases} \tag{2.93}$$

Table 2.1

GBE coefficients with $N = 10$ Gaussian functions for a piston source. Copied from [Wen and Breazeale, 1988, Table I].

n	A_n	B_n
1	11.428 + 0.95175i	4.0697 + 0.22726i
2	0.06002 − 0.08013i	1.1531 − 20.933i
3	−4.2743 − 8.5562i	4.4608 + 5.1268i
4	1.6576 + 2.7015i	4.3521 + 14.997i
5	−5.0418 + 3.2488i	4.5443 + 10.003i
6	1.1227 − 0.68854i	3.8478 + 20.078i
7	−1.0106 − 0.26955i	2.5280 − 10.310i
8	−2.5974 + 3.2202i	3.3197 − 4.8008i
9	−0.14840 − 0.31193i	1.9002 − 15.820i
10	−0.20850 − 0.23851i	2.6340 + 25.009i

It is clear that the values of GBE coefficients depend on the specific profile. For the most commonly used piston source, the profile has the form of

$$u(\xi) = \Pi(\xi/2), \tag{2.94}$$

where the rectangle function is defined by Eq. (2.11). A set of GBE coefficients with $N = 10$ Gaussian functions for this profile is given in Table 2.1. More sets can be found in [Huang and Breazeale, 1999; Sha et al., 2003; Ding and Zhang, 2004; Kim et al., 2006; Liu et al., 2008; Červenka et al., 2009; Liu and Yang, 2010; Červenka and Bednařík, 2013; Červenka and Bednařík, 2015; Johnson, 2016].

Using Gaussian functions to represent velocity profiles offers the advantage of simplifying the calculation of sound pressure, allowing for a closed-form solution when using the paraxial approximation. In the 2D radiation model depicted in Fig. 2.1, for field locations near the radiation axis, where the condition $|y-y'| \ll |x-x'|$ holds, the distance between the field and source locations has the expansion of

$$\begin{aligned} |\rho-\rho'| &= \sqrt{(x-x')^2+(y-y')^2} = |x-x'|\sqrt{1+\frac{(y-y')^2}{(x-x')^2}} \\ &= |x-x'| + \frac{(y-y')^2}{2|x-x'|} + \mathcal{O}\left[\frac{(y-y')^4}{|x-x'|^3}\right], \quad \text{as } \frac{|y-y'|}{|x-x'|} \to 0. \end{aligned} \tag{2.95}$$

Truncating Eq. (2.95) to the second-order term, the 2D Green's function given by Eq. (2.15) can be approximated by

$$\begin{aligned} g_{2D}(\boldsymbol{\rho}, \boldsymbol{\rho}', \omega) &\approx \sqrt{\frac{\mathrm{i}}{8\pi k|\boldsymbol{\rho}-\boldsymbol{\rho}'|}} \mathrm{e}^{\mathrm{i}k|\boldsymbol{\rho}-\boldsymbol{\rho}'|} \\ &\approx \sqrt{\frac{\mathrm{i}}{8\pi k|x-x'|}} \exp\left(\mathrm{i}k\left[|x-x'| + \frac{(y-y')^2}{2|x-x'|}\right]\right), \end{aligned} \tag{2.96}$$

where the limiting form of Hankel function given by Eq. (B.20) is used.

For the 3D radiation model as illustrated in Fig. 2.2, similar expansion can be obtained in the condition $[(x-x')^2 + (y-y')^2] \ll (z-z')^2$ as

$$\begin{aligned} |\mathbf{r}-\mathbf{r}'| &= |z-z'|\sqrt{1 + \frac{(x-x')^2 + (y-y')^2}{(z-z')^2}} \\ &= |z-z'| + \frac{(x-x')^2 + (y-y')^2}{2|z-z'|} + \mathcal{O}\left(\frac{[(x-x')^2 + (y-y')^2]^2}{2|z-z'|^3}\right). \end{aligned} \tag{2.97}$$

Truncating Eq. (2.97) to the second term, the 3D Green's function given by Eq. (2.36) can be approximated by [Červenka and Bednařík, 2013]

$$g_{3D}(\mathbf{r}, \mathbf{r}', \omega) \approx \frac{1}{4\pi|z-z'|} \exp\left(\mathrm{i}k\left[|z-z'| + \frac{(x-x')^2 + (y-y')^2}{2|z-z'|}\right]\right). \tag{2.98}$$

2.4.2 TWO-DIMENSIONAL MODEL

2.4.2.1 Uniform Profile

2.4.2.1.1 Ultrasound Field

For a piston line source with a length of $2a$ in 2D space, a uniform profile is assumed as given by Eq. (2.10). Substituting Eq. (B.20), the limiting form of the Hankel function at large arguments $|k_i\rho| \to \infty$, into Eq. (2.18), the ultrasound pressure field is obtained as

$$p_i(\boldsymbol{\rho}) \approx \frac{\rho_0 \omega_i}{\sqrt{2\pi \mathrm{i} k_i}} \int_{-\infty}^{\infty} \frac{v_{i,y}(\boldsymbol{\rho}_\mathrm{s})}{\sqrt{|\boldsymbol{\rho}-\boldsymbol{\rho}_\mathrm{s}|}} \mathrm{e}^{\mathrm{i}k_i|\boldsymbol{\rho}-\boldsymbol{\rho}_\mathrm{s}|} \mathrm{d}y_\mathrm{s}. \tag{2.99}$$

By approximating the velocity profile as shown by Eqs. (2.9–2.10) using the Gaussian functions in Eq. (2.86), it is obtained that

$$v_{i,x}(\boldsymbol{\rho}_\mathrm{s}) = v_0 \Pi\left(\frac{y_\mathrm{s}}{2a}\right) \approx v_0 \sum_{n=1}^{N} A_n \exp\left(-B_n \frac{y_\mathrm{s}^2}{a^2}\right). \tag{2.100}$$

Substituting Eq. (2.100) into Eq. (2.99) yields

$$p_i(\boldsymbol{\rho}) \approx \sum_{n=1}^{N} \frac{\rho_0 \omega_i v_0 A_n}{\sqrt{2\pi \mathrm{i} k_i}} \int_{-\infty}^{\infty} \frac{1}{\sqrt{|\boldsymbol{\rho}-\boldsymbol{\rho}_\mathrm{s}|}} \exp\left(-B_n \frac{y_\mathrm{s}^2}{a^2} + \mathrm{i}k_i|\boldsymbol{\rho}-\boldsymbol{\rho}_\mathrm{s}|\right) \mathrm{d}y_\mathrm{s}. \tag{2.101}$$

Substituting the paraxial approximation Eq. (2.96) into Eq. (2.101) and introducing the following variables

$$\tilde{y} \equiv \frac{y}{a}, \quad \tilde{x}_i \equiv \frac{x}{\mathscr{D}_{\mathrm{R},i}}, \quad \mathscr{D}_{\mathrm{R},i} \equiv \frac{1}{2}k_i a^2, \tag{2.102}$$

the ultrasound pressure field is obtained as

$$\begin{aligned} p_i(\rho) &= \frac{p_0}{\sqrt{\mathrm{i}\pi|\tilde{x}_i|}} \mathrm{e}^{\mathrm{i}k_i|x|} \sum_{n=1}^{N} A_n \int_{-\infty}^{\infty} \exp\left[-B_n\tilde{y}_{\mathrm{s}}^2 + \mathrm{i}\frac{(\tilde{y}-\tilde{y}_{\mathrm{s}})^2}{|\tilde{x}_i|}\right] \mathrm{d}\tilde{y}_{\mathrm{s}} \\ &= p_0 \mathrm{e}^{\mathrm{i}k_i|x|} \sum_{n=1}^{N} \mathcal{G}(A_n, B_n, \tilde{y}, |\tilde{x}_i|). \end{aligned} \tag{2.103}$$

For convenience, the domain of the sound field $x \geq 0$ is extended to the entire space $-\infty < x < \infty$. Because the source is baffled, the sound field is symmetric about $x = 0$, indicating that $p_i(x,y) = p_i(-x,y)$. This results in the expression of $|x|$ in Eq. (2.103) instead of x.

The auxiliary Gaussian type integral in Eq. (2.103) is defined as

$$\mathcal{G}(A_n, B_n, \tilde{\xi}, \tilde{\zeta}_i) \equiv \frac{A_n}{\sqrt{\mathrm{i}\pi\tilde{\zeta}_i}} \int_{-\infty}^{\infty} \exp\left[\mathrm{i}\frac{(\tilde{\xi}-\tilde{\xi}_{\mathrm{s}})^2}{\tilde{\zeta}_i} - B_n\tilde{\xi}_{\mathrm{s}}^2\right] \mathrm{d}\tilde{\xi}_{\mathrm{s}}. \tag{2.104}$$

By using the integral Eq. (2.92), Eq. (2.104) can be simplified to a closed-form solution of

$$\mathcal{G}(A_n, B_n, \tilde{\xi}, \tilde{\zeta}_i) = \frac{A_n}{\sqrt{1+\mathrm{i}B_n\tilde{\zeta}_i}} \exp\left(-\frac{B_n\tilde{\xi}^2}{1+\mathrm{i}B_n\tilde{\zeta}_i}\right). \tag{2.105}$$

By substituting Eq. (2.105) into Eq. (2.103), the ultrasound pressure reduces to a closed-form of [Schmerr Jr, 2014, Eq. (3.102)]

$$p_i(\rho) = p_0 \mathrm{e}^{\mathrm{i}k_i|x|} \sum_{n=1}^{N} \frac{A_n}{\sqrt{1+\mathrm{i}B_n|\tilde{x}_i|}} \exp\left(-\frac{B_n\tilde{y}^2}{1+\mathrm{i}B_n|\tilde{x}_i|}\right). \tag{2.106}$$

Equation (2.106) shows that if a Gaussian velocity profile is generated on $x = 0$ and in the paraxial approximation, its wavefront remains a Gaussian profile traveling in the medium.

2.4.2.1.2 Audio Sound Field

Substituting Eq. (2.106) into Eq. (2.23), the GBE of the source density for audio sound is expressed as

$$q_{\mathrm{a}}(\rho_{\mathrm{v}}) = \frac{\beta\omega_{\mathrm{a}}}{\mathrm{i}c_0^2} \mathrm{e}^{\mathrm{i}(k_2-k_1^*)|x_{\mathrm{v}}|} \sum_{n_1,n_2}^{N} A_{n_1,n_2} \exp\left(-B_{n_1,n_2}\tilde{y}_{\mathrm{v}}^2\right), \tag{2.107}$$

where the coefficients are defined as

$$\begin{cases} A_{n_1,n_2} \equiv \dfrac{A_{n_1}^* A_{n_2}}{\sqrt{(1+\mathrm{i}B_{n_1}|\tilde{x}_{\mathrm{v},1}|)^*(1+B_{n_2}|\tilde{x}_{\mathrm{v},2}|)}}, \\ B_{n_1,n_2} \equiv \dfrac{B_{n_1}^*}{(1+\mathrm{i}B_{n_1}|\tilde{x}_{\mathrm{v},1}|)^*} + \dfrac{B_{n_2}}{1+\mathrm{i}B_{n_2}|\tilde{x}_{\mathrm{v},2}|}. \end{cases} \tag{2.108}$$

Equation (2.107) shows that the audio source density has the form of Gaussian functions. Therefore, it is intuitively to simplify the audio sound pressure using the paraxial approximation. Specifically, substituting Eq. (2.107) into Eq. (2.24), the audio sound pressure is

$$\begin{aligned} p_{\mathrm{a}}(\boldsymbol{\rho}) = \frac{p_0^2 \beta \omega_{\mathrm{a}}^2}{4\mathrm{i}\rho_0 c_0^4} \iint_{-\infty}^{\infty} & \mathrm{e}^{\mathrm{i}(k_2-k_1^*)|x_{\mathrm{v}}|} \\ & \times \sum_{n_1,n_2=1}^{N} A_{n_1,n_2} \mathrm{e}^{-B_{n_1,n_2}\tilde{y}_{\mathrm{v}}^2} H_0(k_{\mathrm{a}}|\boldsymbol{\rho}-\boldsymbol{\rho}_{\mathrm{v}}|)\mathrm{d}x_{\mathrm{v}}\mathrm{d}y_{\mathrm{v}}. \end{aligned} \tag{2.109}$$

Similar to Eq. (2.95), for the field locations close to the radiation axis satisfying $|y-y_{\mathrm{v}}| \ll |x-x_{\mathrm{v}}|$, the audio sound pressure is simplified to

$$\begin{aligned} p_{\mathrm{a}}(\boldsymbol{\rho}) = \frac{p_0^2 \beta \omega_{\mathrm{a}}}{2\mathrm{i}\rho_0 c_0^3} \sum_{n_1,n_2=1}^{N} A_{n_1,n_2} \int_{-\infty}^{\infty} & \mathcal{G}(A_{n_1,n_2}, B_{n_1,n_2}, \tilde{y}, |\tilde{x}_{\mathrm{a}}|) \\ & \times \mathrm{e}^{\mathrm{i}\left[(k_2-k_1^*)|x_{\mathrm{v}}|+k_{\mathrm{a}}|x-x_{\mathrm{v}}|\right]}\mathrm{d}x_{\mathrm{v}}, \end{aligned} \tag{2.110}$$

where the closed-form of the Gaussian type integral $\mathcal{G}$ is given by Eq. (2.105), and the variables are introduced as

$$\tilde{x}_{\mathrm{a}} \equiv \frac{x-x_{\mathrm{v}}}{\mathscr{D}_{\mathrm{R,a}}}, \quad \mathscr{D}_{\mathrm{R,a}} \equiv \frac{1}{2}k_{\mathrm{a}}a^2. \tag{2.111}$$

It is noted that in some literature [Červenka and Bednařík, 2013, Eq. (11)], the integration domain in Eq. (2.110) is constrained to $(0,x)$. The GBE expression Eq. (2.110) contains only one-fold integral which can be computed using various numerical integration techniques. It is clear that the GBE is far more computationally efficient compared to the 2D DIM [Eq. (2.27)] which entails a three-fold integral after substituting the ultrasound pressure expression. However, it is noted that the paraxial approximation is assumed in the GBE making it less accurate.

2.4.2.2 Steerable Profile

2.4.2.2.1 Ultrasound Field

When the PAL has a steerable profile given by Eq. (2.13) with a steering angle of φ_{d}, a rotated coordinate system $Ox'y'$ is constructed as shown in Fig. 2.5, which is obtained by rotating the original coordinate system Oxy anti-clockwise by an angle of $\varphi_d \in (-\pi/2, \pi/2)$. The coordinate of a field location $\boldsymbol{\rho}$ in the unprimed coordinate system can be transformed to $\boldsymbol{\rho}'$ in the primed one by using a rotation matrix as

$$\boldsymbol{\rho}' = \mathbf{R}\boldsymbol{\rho}, \quad \mathbf{R} \equiv \begin{bmatrix} \cos\varphi_d & \sin\varphi_d \\ -\sin\varphi_d & \cos\varphi_d \end{bmatrix}. \tag{2.112}$$

Because $\mathbf{R}$ is orthonormal, $\mathbf{R}^{-1} = \mathbf{R}^{\mathrm{T}}$. The inverse transform from $\boldsymbol{\rho}'$ to $\boldsymbol{\rho}$ is obtained as $\boldsymbol{\rho} = \mathbf{R}^{\mathrm{T}}\boldsymbol{\rho}'$. In the primed coordinates, the beam direction is denoted by the unit vector $\hat{\mathbf{s}}' = (1, 0)$. By using the inverse transformation, the beam direction in the unprimed coordinates can be obtained as $\hat{\mathbf{s}} = (\cos\varphi_d, \sin\varphi_d)$.

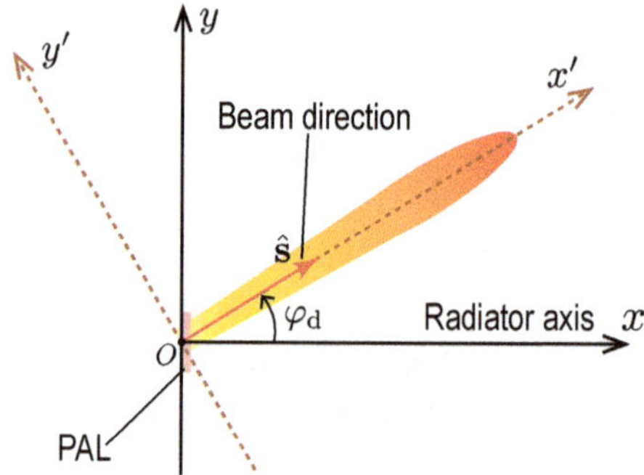

Figure 2.5 Sketch of a steerable PAL in 2D space, where φ_d is the steering angle. The PAL is located at $x = 0$.

Similar to Eq. (2.100), the steerable profile given by Eq. (2.13) can be approximated by multiple Gaussian functions as

$$v_{i,x}(\boldsymbol{\rho}_s) \approx v_0 e^{ik_i y_s \sin\varphi_d} \sum_{n=1}^{N} A_n \exp\left(-B_n \frac{y_s^2}{a^2}\right). \tag{2.113}$$

The distance between a field location $\boldsymbol{\rho}'$ and a source point $\boldsymbol{\rho}_s'$ in the primed coordinate system is expressed as

$$|\boldsymbol{\rho}' - \boldsymbol{\rho}_s'| = \sqrt{(x' - y_s\sin\varphi_d)^2 + (y' - y_s\cos\varphi_d)^2}. \tag{2.114}$$

When the steering angle is small, the term $y_s^2 \sin^2\varphi_d$ inside the square root in Eq. (2.114) can be neglected. Additionally, assuming the paraxial approximation that $|x'| \gg |y'|, |y_s|$, an approximation similar to Eq. (2.96) is obtained as

$$\frac{e^{ik_i|\boldsymbol{\rho}' - \boldsymbol{\rho}_s'|}}{\sqrt{ik_i|\boldsymbol{\rho}' - \boldsymbol{\rho}_s'|}} \approx \frac{1}{\sqrt{ik_i|x'|}} \exp\left(ik_i\left[|x'| + \frac{(y' - y_s\cos\varphi_d)^2}{2|x'|} - y_s\sin\varphi_d\right]\right). \tag{2.115}$$

Substituting Eqs. (2.113) and (2.115) into Eq. (2.99), the ultrasound pressure field is obtained as

$$p_i(\boldsymbol{\rho}) = p_0 e^{ik_i|x'|} \sum_{n=1}^{N} \mathcal{G}(A_n/\cos\varphi_d, B_n, \tilde{y}', |\tilde{x}_i'|), \tag{2.116}$$

which is similar to Eq. (2.103), and the following variable definitions are introduced

$$\tilde{y}' \equiv \frac{y'}{a}, \quad \tilde{x}_i' \equiv \frac{x'}{\mathscr{D}_{\mathrm{R},i}}, \quad \mathscr{D}_{\mathrm{R},i} \equiv \frac{1}{2} k_i a^2 \cos^2 \varphi_{\mathrm{d}}. \tag{2.117}$$

Equation (2.116) is the GBE of the ultrasound pressure generated by a baffled line source 2D with a steerable profile. This expression is similar to Eq. (2.103) that is for a source with a uniform profile.

2.4.2.2.2 *Audio Sound Field*

The audio sound pressure in the primed coordinate system can be obtained similar to Eq. (2.110) as

$$\begin{aligned} p_{\mathrm{a}}(\rho') = \frac{p_0^2 \beta \omega_{\mathrm{a}}}{2\mathrm{i}\rho_0 c_0^3} \sum_{n_1,n_2=1}^{N} \int_{-\infty}^{\infty} \mathcal{G}(A_{n_1,n_2}, B_{n_1,n_2}, \tilde{y}', |\tilde{x}_{\mathrm{a}}'|) \\ \times \mathrm{e}^{\mathrm{i}\left[(k_2 - k_1^*)|x_{\mathrm{v}}'| + k_{\mathrm{a}}|x - x_{\mathrm{v}}'|\right]} \mathrm{d}x_{\mathrm{v}}', \end{aligned} \tag{2.118}$$

where the coefficients are defined in Eq. (2.108) while A_{n_i} is replaced by $A_{n_i}/\cos\varphi_{\mathrm{d}}$. In addition, the following dimensionless variables are introduced as

$$\tilde{x}_{\mathrm{a}}' \equiv \frac{x' - x_{\mathrm{v}}'}{\mathscr{D}_{\mathrm{R,a}}}, \quad \mathscr{D}_{\mathrm{R,a}} \equiv \frac{1}{2} k_{\mathrm{a}} a^2 \cos^2 \varphi_{\mathrm{d}}. \tag{2.119}$$

It is worth noting that, the integration domain in Eq. (2.118) is commonly constrained to $(0, x')$, assuming only the forward propagation of the radiation from the virtual source.

2.4.3 CIRCULAR PAL WITH A UNIFORM PROFILE

2.4.3.1 Ultrasound field

For a circular PAL with a uniform profile given by Eq. (2.29), the velocity profile is approximated by Gaussian functions as

$$v_{i,z}(\mathbf{r}_{\mathrm{s}}) \approx v_0 \sum_{n=1}^{N} A_n \exp\left(-B_n \frac{\rho_{\mathrm{s}}^2}{a^2}\right). \tag{2.120}$$

By using the paraxial approximation Eq. (2.98) and Eq. (2.120), the ultrasound pressure can be obtained as

$$p_i(\mathbf{r}) = \frac{p_0 k_i}{2\mathrm{i}\pi|z|} \mathrm{e}^{\mathrm{i}k_i|z|} \sum_{n=1}^{N} A_n \iint_{-\infty}^{\infty} \exp\left[\mathrm{i}k_i \frac{(x - x_{\mathrm{s}})^2 + (y - y_{\mathrm{s}})^2}{2|z|} - B_n \frac{x_{\mathrm{s}}^2 + y_{\mathrm{s}}^2}{a^2}\right] \mathrm{d}x_{\mathrm{s}} \mathrm{d}y_{\mathrm{s}}. \tag{2.121}$$

The following dimensionless variables are defined to simplify the calculation

$$\tilde{x} \equiv \frac{x}{a}, \quad \tilde{y} \equiv \frac{y}{a}, \quad \tilde{\rho} \equiv \frac{\rho}{a}, \quad \tilde{z}_i \equiv \frac{z}{\mathscr{D}_{\mathrm{R},i}}, \tag{2.122}$$

where the Rayleigh distance for the ultrasound is defined as

$$\mathscr{D}_{\mathrm{R},i} = \frac{1}{2}k_i a^2. \tag{2.123}$$

The ultrasound pressure given by Eq. (2.121) is then simplified to

$$\begin{aligned} p_i(\mathbf{r}) &= \frac{p_0}{\mathrm{i}\pi|\tilde{z}_i|}\mathrm{e}^{\mathrm{i}k_i|z|}\sum_{n=1}^{N} A_n \int_{-\infty}^{\infty}\int_{-\infty}^{\infty} \exp\left[\mathrm{i}\frac{(\tilde{x}-\tilde{x}_\mathrm{s})^2+(\tilde{y}-\tilde{y}_\mathrm{s})^2}{|\tilde{z}_i|} - B_n(\tilde{x}_\mathrm{s}^2+\tilde{y}_\mathrm{s}^2)\right]\mathrm{d}\tilde{x}_\mathrm{s}\mathrm{d}\tilde{y}_\mathrm{s} \\ &= p_0\mathrm{e}^{\mathrm{i}k_i|z|}\sum_{n=1}^{N}\mathcal{G}\left(\sqrt{A_n}, B_n, \tilde{x}, |\tilde{z}_i|\right)\mathcal{G}\left(\sqrt{A_n}, B_n, \tilde{y}, |\tilde{z}_i|\right). \end{aligned} \tag{2.124}$$

Substituting Eq. (2.105) into Eq. (2.124), the ultrasound pressure reduces to a closed-form of

$$p_i(\mathbf{r}) = p_0\mathrm{e}^{\mathrm{i}k_i|z|}\sum_{n=1}^{N}\frac{A_n}{1+\mathrm{i}B_n|\tilde{z}_i|}\exp\left(-\frac{B_n\tilde{\rho}^2}{1+\mathrm{i}B_n|\tilde{z}_i|}\right). \tag{2.125}$$

In comparison to the Rayleigh integral Eq. (2.39), it is evident that the GBE expression Eq. (2.125) is more computationally efficient since it does not necessitate numerical integrations.

2.4.3.2 Audio Sound Field

Substituting the ultrasound pressure Eq. (2.125) into Eq. (2.44), the GBE of the source density of the audio sound is

$$q_\mathrm{a}(\mathbf{r}_\mathrm{v}) = \frac{\beta\omega_\mathrm{a}}{\mathrm{i}c_0^2}\mathrm{e}^{\mathrm{i}(k_2-k_1^*)|z_\mathrm{v}|}\sum_{n_1,n_2=1}^{N} A_{n_1,n_2}\mathrm{e}^{-B_{n_1,n_2}\tilde{\rho}_\mathrm{v}^2}, \tag{2.126}$$

where the coefficients are defined as

$$A_{n_1,n_2} \equiv \frac{A_{n_1}^* A_{n_2}}{(1+\mathrm{i}B_{n_1}|\tilde{z}_{\mathrm{v},1}|)^*(1+\mathrm{i}B_{n_2}|\tilde{z}_{\mathrm{v},2}|)}, \tag{2.127}$$

$$B_{n_1,n_2} \equiv \frac{B_{n_1}^*}{(1+\mathrm{i}B_{n_1}|\tilde{z}_{\mathrm{v},1}|)^*} + \frac{B_{n_2}}{1+\mathrm{i}B_{n_2}|\tilde{z}_{\mathrm{v},2}|}. \tag{2.128}$$

The audio sound pressure is obtained by substituting Eq. (2.126) into Eq. (2.45) as

$$p_\mathrm{a}(\mathbf{r}) = -\frac{\beta p_0^2\omega_\mathrm{a}^2}{4\pi\rho_0 c_0^4}\sum_{n_1,n_2=1}^{N}\iiint_{-\infty}^{\infty}\frac{A_{n_1,n_2}}{|\mathbf{r}-\mathbf{r}_\mathrm{v}|}\mathrm{e}^{\mathrm{i}k_\mathrm{a}|\mathbf{r}-\mathbf{r}_\mathrm{v}|-B_{n_1,n_2}\tilde{\rho}_\mathrm{v}^2+\mathrm{i}(k_2-k_1^*)|z_\mathrm{v}|}\mathrm{d}^3\mathbf{r}_\mathrm{v}. \tag{2.129}$$

By using the paraxial approximation Eq. (2.98), the audio sound pressure given by Eq. (2.129) reduced to

$$\begin{aligned} p_\mathrm{a}(\mathbf{r}) = \frac{\beta\rho_0\omega_\mathrm{a}}{2\mathrm{i}c_0}\sum_{n_1,n_2=1}^{N}\int_{-\infty}^{\infty}&\mathrm{e}^{\mathrm{i}(k_2-k_1^*)|z_\mathrm{v}|+\mathrm{i}k_\mathrm{a}|z-z_\mathrm{v}|} \\ &\mathcal{G}(A_{n_1,n_2}, B_{n_1,n_2}, \tilde{x}, |\tilde{z}_\mathrm{a}|)\mathcal{G}(A_{n_1,n_2},, B_{n_1,n_2}\tilde{y}, |\tilde{z}_\mathrm{a}|)\mathrm{d}z_\mathrm{v}, \end{aligned} \tag{2.130}$$

where the dimensionless variable is introduced as

$$\tilde{z}_{\mathrm{a}} = \frac{z - z_{\mathrm{v}}}{\mathscr{D}_{\mathrm{R,a}}}, \tag{2.131}$$

with the Rayleigh distance for audio sound $\mathscr{D}_{\mathrm{R,a}} = k_{\mathrm{a}} a^2/2$.

Using the closed-form of the function $\mathcal{G}$ given by Eq. (2.105), Eq. (2.130) reduces to

$$\begin{aligned} p_{\mathrm{a}}(\mathbf{r}) = {} & \frac{\beta p_0^2 \omega_{\mathrm{a}}}{2\mathrm{i}\rho_0 c_0^3} \int_{-\infty}^{\infty} \mathrm{e}^{\mathrm{i}(k_2 - k_1^*)|z_{\mathrm{v}}| + \mathrm{i}k_{\mathrm{a}}|z - z_{\mathrm{v}}|} \\ & \times \sum_{n_1, n_2 = 1}^{N} \frac{A_{n_1,n_2}}{1 + \mathrm{i}B_{n_1,n_2}|\tilde{z}_{\mathrm{a}}|} \exp\left(-\frac{B_{n_1,n_2}}{1 + \mathrm{i}B_{n_1,n_2}|\tilde{z}_{\mathrm{a}}|} \tilde{\rho} \right) \mathrm{d}z_{\mathrm{v}}. \end{aligned} \tag{2.132}$$

Equation (2.132) is the GBE of the audio sound pressure generated by a circular PAL with a uniform velocity profile. It contains a one-fold integral and a two-fold summation which is more computationally efficient than the five-fold integral in the DIM shown in Eq. (2.46). It is worth noting that the integration domain Eq. (2.132) is commonly constrained to $(0, z)$, assuming only the forward propagation of the radiation from the virtual source.

2.4.4 RECTANGULAR PAL

2.4.4.1 Uniform Profile

2.4.4.1.1 Ultrasound Field

For a rectangular PAL with side lengths of $2a_x$ and $2a_y$ in x- and y-directions, the velocity profile on the radiation surface can be expanded as the superposition of N^2 Gaussian functions as [Yang et al., 2005, Eq. (8); Ji and Yang, 2019, Eq. (5)]

$$v_{i,z}(\mathbf{r}_{\mathrm{s}}) = v_0 \sum_{m_i=1}^{N} \sum_{n_i=1}^{N} A_{m_i} A_{n_i} \exp\left(-B_{m_i} \frac{x_{\mathrm{s}}^2}{a_x^2} - B_{n_i} \frac{y_{\mathrm{s}}^2}{a_y^2} \right). \tag{2.133}$$

By using the paraxial approximation given by Eq. (2.98), the ultrasound pressure can be obtained by the sum of the pressure generated by these Gaussian functions,

$$\begin{aligned} p_i(\mathbf{r}) = {} & \frac{\rho_0 \omega_i}{2\mathrm{i}\pi|z|} \mathrm{e}^{\mathrm{i}k_i|z|} \sum_{m_i, n_i = 1}^{N} A_{m_i} A_{n_i} \\ & \times \int_{-\infty}^{\infty} \int_{-\infty}^{\infty} \exp\left[\mathrm{i}k_i \frac{(x - x_{\mathrm{s}})^2 + (y - y_{\mathrm{s}})^2}{2|z|} - B_{m_i} \frac{x_{\mathrm{s}}^2}{a_x^2} - B_{n_i} \frac{y_{\mathrm{s}}^2}{a_y^2} \right] \mathrm{d}x_{\mathrm{s}} \mathrm{d}y_{\mathrm{s}}. \end{aligned} \tag{2.134}$$

To simplify the derivation, the dimensionless variables are introduced as

$$\tilde{x} = \frac{x}{a_x}, \quad \tilde{y} = \frac{y}{a_y}, \quad \tilde{z}_{i,x} = \frac{z}{\mathscr{D}_{\mathrm{R},i,x}}, \quad \tilde{z}_{i,y} = \frac{z}{\mathscr{D}_{\mathrm{R},i,y}}, \tag{2.135}$$

where the Rayleigh distances in x and y directions are defined as $\mathscr{D}_{\mathrm{R},i,x} = k_i a_x^2/2$ and $\mathscr{D}_{\mathrm{R},i,y} = k_i a_y^2/2$, respectively. Equation (2.134) is then rewritten as

$$\begin{aligned} p_i(\mathbf{r}) = p_0 \mathrm{e}^{\mathrm{i}k_i|z|} \sum_{m_i,n_i=1}^{N} \frac{A_{m_i}}{\sqrt{\mathrm{i}\pi|\tilde{z}_{i,x}|}} \int_{-\infty}^{\infty} \exp\left[\mathrm{i}\frac{(\tilde{x}-\tilde{x}_\mathrm{s})^2}{|\tilde{z}_{i,x}|} - B_{m_i}\tilde{x}_\mathrm{s}^2\right] \mathrm{d}\tilde{x}_\mathrm{s} \\ \times \frac{A_{n_i}}{\sqrt{\mathrm{i}\pi|\tilde{z}_{i,y}|}} \int_{-\infty}^{\infty} \exp\left[\mathrm{i}\frac{(\tilde{y}-\tilde{y}_\mathrm{s})^2}{|\tilde{z}_{i,y}|} - B_{n_i}\tilde{y}_\mathrm{s}^2\right] \mathrm{d}\tilde{y}_\mathrm{s}. \end{aligned} \tag{2.136}$$

By using the integral given by Eq. (2.104), the ultrasound pressure can be written in a compact form of [Zhuang et al., 2023, Eq. (11)]

$$p_i(\mathbf{r}) = p_0 \mathrm{e}^{\mathrm{i}k_i|z|} \sum_{m_i,n_i=1}^{N} \mathcal{G}(A_{m_i}, B_{m_i}, \tilde{x}, |\tilde{z}_{i,x}|)\mathcal{G}(A_{n_i}, B_{n_i}\tilde{y}, |\tilde{z}_{i,y}|). \tag{2.137}$$

By using the closed-form of $\mathcal{G}$ given by Eq. (2.105), the ultrasound pressure is obtained as

$$\begin{aligned} p_i(\mathbf{r}) = p_0 \mathrm{e}^{\mathrm{i}k_i|z|} \sum_{m_i,n_i=1}^{N} \frac{A_{m_i}A_{n_i}}{\sqrt{1+\mathrm{i}B_{m_i}|\tilde{z}_{i,x}|}\sqrt{1+\mathrm{i}B_{n_i}|\tilde{z}_{i,y}|}} \\ \times \exp\left(-\frac{B_{m_i}\tilde{x}^2}{1+\mathrm{i}B_{m_i}|\tilde{z}_{i,x}|} - \frac{B_{n_i}\tilde{y}^2}{1+\mathrm{i}B_{n_i}|\tilde{z}_{i,y}|}\right). \end{aligned} \tag{2.138}$$

Equation (2.138) is the GBE of the ultrasound generated by a baffled rectangular piston source. It is clear that Eq. (2.138) is expressed in a closed-form which is more computationally efficient when compared to the two-fold Rayleigh integral given by Eq. (2.38).

2.4.4.1.2 *Audio Sound Field*

Substituting Eq. (2.138) into Eq. (2.44), the GBE of the source density of the audio sound is

$$q_\mathrm{a}(\mathbf{r}_\mathrm{v}) = \frac{\beta p_0^2 \omega_\mathrm{a}}{\mathrm{i}\rho_0^2 c_0^4} \mathrm{e}^{\mathrm{i}k_\mathrm{a}|z_\mathrm{v}|} \sum_{m_1,m_2,n_1,n_2=1}^{N} A_{x,m}A_{y,n} \exp\left(-B_{x,m}\tilde{x}_\mathrm{v}^2 - B_{y,n}\tilde{y}_\mathrm{v}^2\right), \tag{2.139}$$

where the coefficients are defined as

$$A_{\xi,m} = \frac{A_{m_1}^* A_{m_2}}{\sqrt{1+\mathrm{i}B_{m_1}|\tilde{z}_{\xi,1}|}^* \sqrt{1+\mathrm{i}B_{m_2}|\tilde{z}_{\xi,2}|}}, \tag{2.140}$$

$$B_{\xi,m} = \left(\frac{B_{m_1}}{1+\mathrm{i}B_{m_1}|\tilde{z}_{\xi,1}|}\right)^* + \frac{B_{m_2}}{1+\mathrm{i}B_{m_2}|\tilde{z}_{\xi,2}|}, \quad \xi = x, y. \tag{2.141}$$

The dimensionless variables are introduced as

$$\tilde{z}_{x,\mathrm{a}} = \frac{z - z_\mathrm{v}}{\mathscr{D}_{\mathrm{R,a},x}}, \quad \tilde{z}_{y,\mathrm{a}} = \frac{z - z_\mathrm{v}}{\mathscr{D}_{\mathrm{R,a},y}}, \quad \tilde{x}_\mathrm{s} = \frac{x_\mathrm{s}}{a_x}, \quad \tilde{y}_\mathrm{s} = \frac{y_\mathrm{s}}{a_y}, \tag{2.142}$$

where the Rayleigh distance for audio sound in x- and y- directions are $\mathscr{D}_{\mathrm{R,a},x} = k_\mathrm{a}a_x^2/2$ and $\mathscr{D}_{\mathrm{R,a},y} = k_\mathrm{a}a_y^2/2$, respectively.

Equation (2.139) shows that the source density has the same form as the velocity profile given by Eq. (2.133). Therefore, the volume integral for the virtual audio source can be simplified in a form like that for the Rayleigh integral. After substituting Eqs. (2.139) and (2.133) into Eq. (2.45), the audio sound pressure is written as

$$\begin{aligned} p_\mathrm{a}(\mathbf{r}) = {} & \frac{\beta \omega_\mathrm{a} p_0^2}{2\mathrm{i}\rho_0 c_0^3} \int_{-\infty}^{\infty} \mathrm{e}^{\mathrm{i}\left[(k_2 - k_1^*)|z_\mathrm{v}| + k_\mathrm{a}|z - z_\mathrm{v}|\right]} \\ & \times \sum_{m_1, m_2 = 1}^{N} \int_{-\infty}^{\infty} \frac{A_{x,m}}{\sqrt{\mathrm{i}\pi |\tilde{z}_{x,\mathrm{a}}|}} \exp\left[\mathrm{i}\frac{(\tilde{x} - \tilde{x}_\mathrm{v})^2}{|\tilde{z}_{x,\mathrm{a}}|} - B_{x,m}\tilde{x}_\mathrm{v}^2\right] \mathrm{d}\tilde{x}_\mathrm{v} \\ & \times \sum_{n_1, n_2 = 1}^{N} \int_{-\infty}^{\infty} \frac{A_{y,n}}{\sqrt{\mathrm{i}\pi |\tilde{z}_{y,\mathrm{a}}|}} \exp\left[\mathrm{i}\frac{(\tilde{y} - \tilde{y}_\mathrm{v})^2}{|\tilde{z}_{y,\mathrm{a}}|} - B_{y,m}\tilde{y}_\mathrm{v}^2\right] \mathrm{d}\tilde{y}_\mathrm{v} \mathrm{d}z_\mathrm{v}. \end{aligned} \tag{2.143}$$

By using the integral given by Eq. (2.104), the audio sound pressure is simplified to [Zhuang et al., 2023, Eq. (18)]

$$\begin{aligned} p_\mathrm{a}(\mathbf{r}) = {} & \frac{\beta p_0^2 \omega_\mathrm{a}}{2\mathrm{i}\rho_0 c_0^3} \int_{-\infty}^{\infty} \mathrm{e}^{\mathrm{i}(k_2 - k_1^*)|z_\mathrm{v}| + \mathrm{i}k_\mathrm{a}|z - z_\mathrm{v}|} \\ & \times \left[\sum_{m_1, m_2 = 1}^{N} \mathcal{G}(A_{x,m}, B_{x,m}, \tilde{x}, |\tilde{z}_{x,\mathrm{a}}|)\right] \left[\sum_{n_1, n_2 = 1}^{N} \mathcal{G}(A_{y,n}, B_{y,n}, \tilde{y}, |\tilde{z}_{y,\mathrm{a}}|)\right] \mathrm{d}z_\mathrm{v}, \end{aligned} \tag{2.144}$$

where the closed-form solution of the Gaussian type integral $\mathcal{G}$ is given by Eq. (2.105). It is worth noting that the integration domain Eq. (2.144) is commonly constrained to $(0, z)$, assuming only the forward propagation of the radiation from the virtual source.

2.4.4.2 Steerable Profile

When the rectangular PAL has a steerable profile given by Eq. (2.34), a Cartesian coordinate system $Oxyz$ is established with the origin O at the center of the PAL as shown in Fig. 2.6. A rotated coordinate system $O'x'y'z'$ is established with the origin O' at the center of the PAL. The positive z'-axis coincides with the beam direction $\hat{\mathbf{s}}$. The rotated coordinates can be obtained by rotating the original coordinates twice. The first step is to rotate the original coordinate system around the x-axis clockwise through an angle of $\theta_y \in (-\pi/2, \pi/2)$, and the y'-axis is then determined. The second step is to rotate the coordinate system around the y'-axis counterclockwise through an angle of $\theta_x \in (-\pi/2, \pi/2)$. The positive value of θ_x (or θ_y) represents a rotation

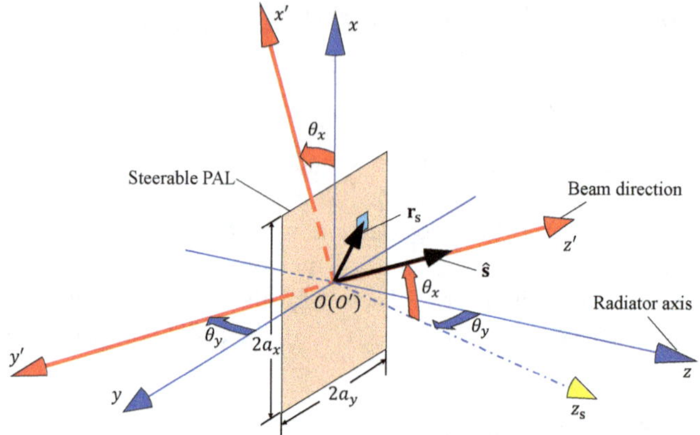

Figure 2.6 The sketch of a rectangular steerable PAL. The PAL is located at $z = 0$. Extracted from [Zhuang et al., 2023, Fig. 1].

from the positive z_s (or z) axis to the positive x (or y) axis, whereas the negative value corresponds to a rotation toward the minus x (or y) axis. The coordinate of a field location $\mathbf{r}$ can be transformed to $\mathbf{r}'$ by using a rotation matrix $\mathbf{R}$ as

$$\mathbf{r}' = \mathbf{R}\mathbf{r}. \tag{2.145}$$

The rotation matrix $\mathbf{R}$ can be obtained by using two elemental rotation matrices,

$$\mathbf{R} = \mathbf{R}_y(\theta_x)\mathbf{R}_x(-\theta_y) = \begin{bmatrix} \cos\theta_x & -\sin\theta_x \sin\theta_y & -\sin\theta_x \cos\theta_y \\ 0 & \cos\theta_y & -\sin\theta_y \\ \sin\theta_x & \cos\theta_x \sin\theta_y & \cos\theta_x \cos\theta_y \end{bmatrix}, \tag{2.146}$$

where

$$\mathbf{R}_x(-\theta_y) = \begin{bmatrix} 1 & 0 & 0 \\ 0 & \cos\theta_y & -\sin\theta_y \\ 0 & \sin\theta_y & \cos\theta_y \end{bmatrix}, \quad \mathbf{R}_y(\theta_x) = \begin{bmatrix} \cos\theta_x & 0 & -\sin\theta_x \\ 0 & 1 & 0 \\ \sin\theta_x & 0 & \cos\theta_x \end{bmatrix}. \tag{2.147}$$

By using Eq. (2.145), the source point in the primed coordinate system is

$$\mathbf{r}'_s \equiv (x'_s, y'_s, z'_s) = (x_s \cos\theta_x - y_s \sin\theta_x \sin\theta_y, y_s \cos\theta_y, x_s \sin\theta_x + y_s \cos\theta_x \sin\theta_y), \tag{2.148}$$

where $z_s = 0$ is used.

Because $\mathbf{R}$ is an orthogonal matrix, $\mathbf{R}^{-1} = \mathbf{R}^{\mathrm{T}}$, where $\mathbf{R}^{-1}$ is the inverse matrix of $\mathbf{R}$, and the superscript "T" denotes the transpose. The inverse transformation from $\mathbf{r}'$ to $\mathbf{r}$ can be expressed as

$$\mathbf{r} = \mathbf{R}^{-1}\mathbf{r}' = \mathbf{R}^{\mathrm{T}}\mathbf{r}'. \tag{2.149}$$

In the primed coordinates, the beam direction is denoted by the vector $\hat{\mathbf{s}}' = (0, 0, 1)$. By using the inverse transformation in Eq. (2.149), the beam direction in the original coordinates can be obtained as

$$\hat{\mathbf{s}} = (\sin\theta_x, \cos\theta_x \sin\theta_y, \cos\theta_x \cos\theta_y). \tag{2.150}$$

2.4.4.2.1 Ultrasound Field

The velocity profile given by Eq. (2.34) can be expanded using a summation of N^2 Gaussian functions

$$v_{i,z}(\mathbf{r}_s) \approx v_0 \exp(\mathrm{i}k_i \mathbf{r}_s \cdot \hat{\mathbf{s}}) \sum_{m_i,n_i=1}^{N} A_{m_i} A_{n_i} \exp\left(-B_{m_i}\frac{x_s^2}{a_x^2} - B_{n_i}\frac{y_s^2}{a_y^2}\right) \tag{2.151}$$

The distance between $\mathbf{r}_v$ and $\mathbf{r}_s$ can be transformed to that expressed in the rotated coordinates using Eq. (2.148) as

$$|\mathbf{r}'_v - \mathbf{r}'_s| = \sqrt{(x'_v - x_s\cos\theta_x + y_s\sin\theta_x\sin\theta_y)^2 + (y'_v - y_s\cos\theta_y)^2 + (z'_v - z'_s)^2}. \tag{2.152}$$

The ultrasound beam can be treated as a steered collimated beam, so $|\mathbf{r}'_v - \mathbf{r}'_s|$ can be approximated by using the paraxial approximation Eq. (2.97). After neglecting $y_s \sin\theta_x \sin\theta_y$, it is obtained that the approximation as

$$|\mathbf{r}'_v - \mathbf{r}'_s| \approx |z'_v - z'_s| + \frac{(x'_v - x_s\cos\theta_x)^2 + (y'_v - y_s\cos\theta_y)^2}{2|z'_v - z'_s|}. \tag{2.153}$$

Similar to Eqs. (2.135) and (2.142), the following dimensionless variables are introduced in the rotated coordinates

$$\begin{aligned} &\tilde{x}'_v = \frac{x'_v}{a_x\cos\theta_x}, \quad \tilde{y}'_v = \frac{y'_v}{a_y\cos\theta_y}, \quad \tilde{z}'_{v,i,x} = \frac{z'_v}{\mathscr{D}'_{R,i,x}}, \quad \tilde{z}'_{v,i,y} = \frac{z'_v}{\mathscr{D}'_{R,i,y}}, \\ &\tilde{x}_s = \frac{x_s}{a_x}, \quad \tilde{y}_s = \frac{y_s}{a_y}, \end{aligned} \tag{2.154}$$

where the Rayleigh distance $\mathscr{D}'_{R,i,x} = k_i a_x^2 \cos^2\theta_x/2$, $\mathscr{D}'_{R,i,y} = k_i a_y^2 \cos^2\theta_y/2$. Here the key concept is to project the sides a_x and a_y onto the primed x and y axes, respectively. When the rotation angles $\theta_x = \theta_y = 0$, these substitutions reduce to those given by Eqs. (2.135) and (2.142).

Because of the sharp directivity of the ultrasound, the ultrasound in the region $z'_v < z'_s$ can be ignored, $|z'_v - z'_s|$ can then be approximated by $z'_v - z'_s$ and can be approximated by z'_v when it is the denominator in Eq. (2.153). By substituting Eqs. (2.151) and (2.153) into Eq. (2.39), the Rayleigh integral for calculating the ultrasound can be written as

$$\begin{aligned} p_i(\mathbf{r}'_v) = p_0 e^{\mathrm{i}k_i|z'_v|} &\sum_{m_i=1}^{N} \frac{A_{m_i}/\cos\theta_x}{\sqrt{\mathrm{i}\pi|\tilde{z}'_{v,i,x}|}} \int_{-\infty}^{\infty} \exp\left[\mathrm{i}\frac{(\tilde{x}'_v - \tilde{x}_s)^2}{|\tilde{z}'_{v,i,x}|} - B_{m_i}\tilde{x}_s^2\right] \mathrm{d}\tilde{x}_s \\ &\times \sum_{n_i=1}^{N} \frac{A_{n_i}/\cos\theta_y}{\sqrt{\mathrm{i}\pi|\tilde{z}'_{v,i,y}|}} \int_{-\infty}^{\infty} \exp\left[\mathrm{i}\frac{(\tilde{y}'_v - \tilde{y}_s)^2}{|\tilde{z}'_{v,i,y}|} - B_{n_i}\tilde{y}_s^2\right] \mathrm{d}\tilde{y}_s. \end{aligned} \tag{2.155}$$

By using the integral formula Eq. (2.104), Eq. (2.155) can be simplified to [Zhuang et al., 2023, Eq. (28)]

$$p_i(\mathbf{r}'_{\mathrm{v}}) = p_0 \mathrm{e}^{\mathrm{i}k_i z'_{\mathrm{v}}} \sum_{m_i=1}^{N} \mathcal{G}\left(\frac{A_{m_i}}{\cos\theta_x}, B_{m_i}, \tilde{x}'_{\mathrm{v}}, \left|\tilde{z}'_{\mathrm{v},i,x}\right|\right) \sum_{n_i=1}^{N} \mathcal{G}\left(\frac{A_{n_i}}{\cos\theta_y}, B_{n_i}, \tilde{y}'_{\mathrm{v}}, \left|\tilde{z}'_{\mathrm{v},i,y}\right|\right), \tag{2.156}$$

which is the ultrasound pressure in rotated coordinates. The closed-form solution of $\mathcal{G}$ is given by Eq. (2.105). This expression reduces to Eq. (2.137) when the steering angles $\theta_x = \theta_y = 0$.

2.4.4.2.2 Audio Sound Field

By substituting the ultrasound pressure from Eq. (2.156) into Eq. (2.44), the virtual audio source density in the rotated coordinates can be expressed as

$$q_{\mathrm{a}}(\mathbf{r}'_{\mathrm{v}}) = \frac{\beta\omega_{\mathrm{a}} p_0^2}{\mathrm{i}\rho_0^2 c_0^4} \mathrm{e}^{\mathrm{i}(k_2-k_1^*)|z'_{\mathrm{v}}|} \sum_{m_1,m_2,n_1,n_2=1}^{N} A'_{x,m} A'_{y,n} \exp\left(-B'_{x,m}\tilde{x}'^2_{\mathrm{v}} - B'_{y,n}\tilde{y}'^2_{\mathrm{v}}\right), \tag{2.157}$$

where the primed coefficients are defined as

$$\begin{cases} A'_{\xi,m} & \equiv \left(\dfrac{A_{m_1}/\cos\theta_x}{\sqrt{1+\mathrm{i}B_{m_1}\left|\tilde{z}'_{\mathrm{v},1,\xi}\right|}}\right)^* \dfrac{A_{m_2}/\cos\theta_y}{\sqrt{1+\mathrm{i}B_{m_2}\left|\tilde{z}'_{\mathrm{v},2,\xi}\right|}}, \\ B'_{\xi,m} & \equiv \left(\dfrac{B_{m_1}}{1+\mathrm{i}B_{m_1}\left|\tilde{z}'_{\mathrm{v},1,\xi}\right|}\right)^* + \dfrac{B_{m_2}}{1+\mathrm{i}B_{m_2}\left|\tilde{z}'_{\mathrm{v},2,\xi}\right|}, (\xi = x, y), \end{cases} \tag{2.158}$$

The dimensionless variables are introduced as

$$\tilde{z}'_{\mathrm{a},x} = \frac{z'-z'_{\mathrm{v}}}{\mathscr{D}'_{\mathrm{R,a},x}}, \quad \tilde{z}'_{\mathrm{a},y} = \frac{z'-z'_{\mathrm{v}}}{\mathscr{D}'_{\mathrm{R,a},y}}, \tag{2.159}$$

where the Rayleigh distances in x- and y- directions are $\mathscr{D}'_{\mathrm{R,a},x} = k_{\mathrm{a}} a_x^2 \cos^2\theta_x/2$ and $\mathscr{D}'_{\mathrm{R,a},y} = k_{\mathrm{a}} a_y^2 \cos^2\theta_y/2$, respectively. By substituting Eq. (2.157) into Eq. (2.46) and using the paraxial approximation Eq. (2.98), the audio sound pressure in the primed coordinates system can be obtained similar to Eq. (2.144) as

$$\begin{aligned} p_{\mathrm{a}}(\mathbf{r}') = & \frac{\beta\omega_{\mathrm{a}} p_0^2}{2\mathrm{i}\rho_0 c_0^3} \int_{-\infty}^{\infty} \mathrm{e}^{\mathrm{i}\left[k_{\mathrm{a}}|z'-z'_{\mathrm{v}}|+(k_2-k_1^*)|z'_{\mathrm{v}}|\right]} \\ & \times \left[\sum_{m_1,m_2=1}^{N} G\left(A'_{x,m}, B'_{x,m}, \tilde{x}', \tilde{z}_{x,\mathrm{a}}\right)\right]\left[\sum_{n_1,n_2=1}^{N} G\left(A'_{y,n}, B'_{y,n}, \tilde{y}', \tilde{z}_{y,\mathrm{a}}\right)\right] \mathrm{d}z'_{\mathrm{v}}, \end{aligned} \tag{2.160}$$

which is the main result for the steerable non-paraxial GBE. The audio sound in the original coordinate can be obtained by substituting inverse transformation [Eq. (2.149)] into Eq. (2.160).

2.4.5 REMARKS

In this section, the reader is introduced to the GBE method. The central idea behind the GBE method is approximating the source velocity profile with multiple Gaussian functions, as demonstrated in Eq. (2.86). By employing this technique, the radiation emitted by a source with a Gaussian profile becomes amenable to a closed-form solution in the paraxial approximation, as elucidated in Eqs. (2.96) and (2.98), thereby significantly simplifying sound field calculations.

Within the context of the 2D radiation model, the ultrasound pressure fields generated by PALs with uniform and steerable profiles are initially transformed into closed-form expressions, detailed in Eqs. (2.106) and (2.116), respectively. Subsequently, the GBEs for audio sound pressure are presented as Eqs. (2.110) and (2.118). Notably, the GBEs stand out in computational efficiency, requiring merely a one-fold integral as opposed to the three-fold integral necessitated by the DIMs. This efficiency becomes more pronounced in the 3D radiation model, where circular and rectangular PALs, both uniform and steerable, result in ultrasound pressure expressions represented by Eqs. (2.125), (2.138), and (2.156), respectively. Correspondingly, the GBEs for audio sound pressure are given as Eqs. (2.132), (2.144), and (2.160), respectively. The contrast with the DIMs is clear: while the latter entails a cumbersome five-fold integral, the GBE emerges as a significantly more computationally efficient alternative, particularly effective in the near field for audio sound prediction.

Nonetheless, the derivation of the GBEs hinges on paraxial approximations for both ultrasound and audio sound. In the 3D radiation model, the paraxial approximation applied to ultrasound simplifies the two-fold Rayleigh integral into a sum of finite Gaussian functions. This simplification comes at the cost of accuracy outside the paraxial region, leading to diminished prediction accuracy for source density at virtual points located beyond this region. Similarly, when the paraxial approximation is extended to audio sound, it streamlines the three-fold integral into a simpler one-fold integral and involves summations of finite Gaussian functions. This negatively affects the precision of audio sound pressure predictions as well. The situation becomes more pronounced in the near field, where the field location also serves as a virtual source location in numerical integration. Consequently, the shortcomings of the paraxial approximation are magnified, resulting in inaccurate predictions of audio sound pressure in proximity to the PAL, especially when dealing with low audio frequencies or small aperture sizes.

Recent advancements have led to the development of a non-paraxial GBE method which is introduced to enhance the accuracy of audio sound calculations [Zhuang et al., 2023]. This innovative approach eliminates the strict requirement for the field location to reside exclusively within the paraxial region, although the virtual source location must still adhere to paraxial constraints. Apart from the continuous profiles discussed in this chapter, the GBEs have been applied in scenarios involving phased array PALs characterized by discrete profiles based on the (conventional) paraxial [Ye et al., 2010] and non-paraxial [Zhu et al., 2023] GBE methods.

2.5 CYLINDRICAL WAVE EXPANSION (CWE) METHOD

The CWE method aims to simplify the calculation of the 2D DIM without resorting to additional approximations. It is initially proposed in Zhong et al., 2021a. The essence of the CWE method lies in representing the Green's function in the polar coordinate system. Section 2.5.1 provides an overview of the framework of the CWE method, while Secs. 2.5.2 and 2.5.3 derive the CWEs of the ultrasound and audio sound fields, respectively.

2.5.1 FRAMEWORK OF THE CWE METHOD

The 2D Green's function Eq. (2.15) in free space can be represented in polar coordinates as [Poletti, 2019, Eq. (6)]

$$g_{2\mathrm{D}}(\boldsymbol{\rho},\boldsymbol{\rho}',\omega) = \frac{\mathrm{i}}{4}H_0(k|\boldsymbol{\rho}-\boldsymbol{\rho}'|) = \frac{\mathrm{i}}{4}\sum_{m=-\infty}^{\infty} J_m(k\rho_<)H_m(k\rho_>)\mathrm{e}^{\mathrm{i}m(\varphi-\varphi')} \tag{2.161}$$

where $J_m(\cdot)$ is the Bessel function of order m, $H_m(\cdot)$ is the Hankel function of order m, $\rho_< = \min(\rho,\rho')$, and $\rho_> = \max(\rho,\rho')$.

For an arbitrary source density $q(\boldsymbol{\rho})$, substituting Eq. (2.161) into Eq. (2.24) yields

$$p(\boldsymbol{\rho}) = \frac{\rho_0\omega}{4}\sum_{m=-\infty}^{\infty} \mathcal{R}_m(\rho,k)\mathrm{e}^{\mathrm{i}m\varphi}. \tag{2.162}$$

It can be found in Eq. (2.162) that the radial (ρ) and angular (φ) coordinates of fields points are separated in two components. The angular component has the form of exponential functions, $\mathrm{e}^{\mathrm{i}m\varphi}$, and the radial component is

$$\mathcal{R}_m(\rho,k) \equiv \mathcal{Q}_m(0,\rho;J)H_m(k\rho) + \mathcal{Q}_m(\rho,\infty;H)J_m(k\rho), \tag{2.163}$$

where the multipole moment of the source density is defined as

$$\mathcal{Q}_m(\rho_1,\rho_2;\mathscr{C}) \equiv \iint_{\rho_1\le\rho_2} q_\mathrm{a}(\boldsymbol{\rho}')\mathscr{C}_m(k\rho')\mathrm{e}^{-\mathrm{i}m\varphi'}\mathrm{d}^2\boldsymbol{\rho}'. \tag{2.164}$$

Here, the symbol $\mathscr{C}$ can be Bessel ($\mathscr{C} = J$) or Hankel ($\mathscr{C} = H$) functions.

Equations (2.162) and (2.164) indicate that when the source density is separable in polar coordinates, the calculation of the sound pressure can be simplified as the radial and angular components can be calculated separately. Assume that the source density has the separable form of

$$q_\mathrm{a}(\boldsymbol{\rho}) = q_0 q_\rho(\rho) q_\varphi(\varphi), \tag{2.165}$$

where q_0 is a constant. The multipole moment of the source density Eq. (2.164) can then be simplified to

$$\mathcal{Q}_m(\rho_1,\rho_2;\mathscr{C}) = q_0\left[\int_{\rho_1}^{\rho_2} q_\rho(\rho')\mathscr{C}_m(k\rho')\rho'\mathrm{d}\rho'\right]\left[\int_0^{2\pi} q_\varphi(\varphi')\mathrm{e}^{-\mathrm{i}m\varphi'}\mathrm{d}\varphi'\right] \tag{2.166}$$

It is essential to emphasize that the derivation of the CWE for sound pressure, as expressed in Eq. (2.162), does not rely on any additional approximations or assumptions.

2.5.2 ULTRASOUND FIELD

Based on the derivation in Sec. 2.5.1, the CWE of ultrasound pressure can be expressed as the form of Eq. (2.162) as

$$p_i(\boldsymbol{\rho}) = \frac{\rho_0 \omega_i}{4} \sum_{m_i=-\infty}^{\infty} \mathcal{R}_{m_i}(\rho, k_i) \mathrm{e}^{\mathrm{i} m_i \varphi}. \tag{2.167}$$

Here, the radial component is obtained using Eq. (2.163).

In the polar coordinate system, the source density of ultrasound, described by Eq. (2.19), can be expressed as

$$q_i(\boldsymbol{\rho}_\mathrm{s}) = 2 v_{i,x}(\boldsymbol{\rho}_\mathrm{s}) \frac{\mathrm{H}(a-\rho_\mathrm{s})}{\rho_\mathrm{s}} \left[\delta\left(\varphi_\mathrm{s} - \frac{\pi}{2}\right) + \delta\left(\varphi_\mathrm{s} + \frac{\pi}{2}\right)\right]. \tag{2.168}$$

Here, ρ_s and φ_s represent radial and angular coordinates of the source point $\boldsymbol{\rho}_\mathrm{s}$. It is evident that Eq. (2.168) is separable in the polar coordinate system. Substituting Eq. (2.168) into Eq. (2.166) results in

$$\begin{aligned} Q_{m_i}(\rho_1, \rho_2; \mathscr{C}) = 2 \int_{\rho_1}^{\rho_2} & \left[\mathrm{i}^{-m_i} v_{i,x}(\rho_\mathrm{s}, \pi/2) + \mathrm{i}^{m_i} v_{i,x}(\rho_\mathrm{s}, -\pi/2)\right] \\ & \times \mathrm{H}(a-\rho_\mathrm{s}) \mathscr{C}_{m_i}(k_i \rho_\mathrm{s}) \mathrm{d}\rho_\mathrm{s}. \end{aligned} \tag{2.169}$$

Substituting Eq. (2.169) into the radial component Eq. (2.163), the expressions for ultrasound pressure take different forms in the interior ($\rho < a$) and exterior ($\rho > a$) regions. In the interior region ($\rho < a$), the ultrasound pressure is given by

$$p_i(\boldsymbol{\rho}) = \frac{\rho_0 \omega_i}{4} \sum_{m_i=-\infty}^{\infty} [Q_{m_i}(0, \rho; J) H_{m_i}(k\rho) + Q_{m_i}(\rho, a; H) J_{m_i}(k_i \rho)] \mathrm{e}^{\mathrm{i} m_i \varphi}, \tag{2.170}$$

where the multipole moments are

$$Q_{m_i}(0, \rho; J) = 2 \int_0^\rho \left[\mathrm{i}^{-m_i} v_{i,x}(\rho_\mathrm{s}, \pi/2) + \mathrm{i}^{m_i} v_{i,x}(\rho_\mathrm{s}, -\pi/2)\right] J_{m_i}(k_i \rho_\mathrm{s}) \mathrm{d}\rho_\mathrm{s}, \tag{2.171}$$

$$Q_{m_i}(\rho, a; H) = 2 \int_\rho^a \left[\mathrm{i}^{-m_i} v_{i,x}(\rho_\mathrm{s}, \pi/2) + \mathrm{i}^{m} v_{i,x}(\rho_\mathrm{s}, -\pi/2)\right] H_{m_i}(k_i \rho_\mathrm{s}) \mathrm{d}\rho_\mathrm{s}. \tag{2.172}$$

In the exterior region ($\rho > a$), the expression for ultrasound pressure is given as

$$p_i(\boldsymbol{\rho}) = \frac{\rho_0 \omega_i}{4} \sum_{m_i=-\infty}^{\infty} Q_{m_i}(0, a; J) H_{m_i}(k_i \rho) \mathrm{e}^{\mathrm{i} m_i \varphi}, \tag{2.173}$$

where the multipole moments are

$$Q_{m_i}(0, a; J) = 2 \int_0^a \left[\mathrm{i}^{-m_i} v_{i,x}(\rho_\mathrm{s}, \pi/2) + \mathrm{i}^{m_i} v_{i,x}(\rho_\mathrm{s}, -\pi/2)\right] J_{m_i}(k_i \rho_\mathrm{s}) \mathrm{d}\rho_\mathrm{s}. \tag{2.174}$$

In the far field ($\rho \to \infty$), the Hankel function in Eq. (2.173) can be approximated using its limiting form Eq. (B.20). Consequently, the ultrasound pressure is simplified to

$$p_i(\rho) = \frac{\rho_0 \omega_i}{2\sqrt{2\mathrm{i}\pi k_i \rho}} \mathrm{e}^{\mathrm{i}k_i\rho} \sum_{m_i=-\infty}^{\infty} \mathrm{i}^{-m_i} \mathcal{Q}_{m_i}(0,a;J) \mathrm{e}^{\mathrm{i}m_i\varphi}. \tag{2.175}$$

Equations (2.170) and (2.173) provide insights into the physical interpretation of the CWE of ultrasound. For an exterior field location ($\rho > a$), the ultrasound field consists of infinitely many outgoing cylindrical waves characterized by $H_{m_i}(k_i\rho)\mathrm{e}^{\mathrm{i}m_i\varphi}$, with the weight of each cylindrical wave determined by the aperture effects as $\mathcal{Q}_{m_i}(0,a;J)$. In contrast, for an interior location point ($\rho < a$), the ultrasound field is a superposition of both converging or outgoing cylindrical waves characterized by $J_{m_i}(k_i\rho)\mathrm{e}^{\mathrm{i}m_i\varphi}$ or $H_{m_i}(k_i\rho)\mathrm{e}^{\mathrm{i}m_i\varphi}$, respectively. The outgoing wave contributes from the source inside the circle with a radius equal to the radial coordinate of the field location ($0 \le \rho_\mathrm{s} < \rho$), while the weight of the converging wave contributes from the source outside this circle ($\rho < \rho_\mathrm{s} \le a$). Consequently, the weights of converging and outgoing waves are $\mathcal{Q}_{m_i}(0,\rho;J)$ and $\mathcal{Q}_{m_i}(\rho,a;H)$, respectively.

The components of the particle velocity field $\mathbf{v}_i(\boldsymbol{\rho})$ in the polar coordinate system can be obtained by substituting Eq. (2.167) into Eq. (1.103) as

$$\begin{cases} v_{i,\rho}(\boldsymbol{\rho}) = \dfrac{1}{\mathrm{i}\rho_0\omega_i}\dfrac{\partial p_i(\boldsymbol{\rho})}{\partial \rho} = \displaystyle\sum_{m_i=-\infty}^{\infty} \frac{\partial \mathcal{R}_{m_i}(\rho,k_i)}{4\mathrm{i}\partial\rho}\mathrm{e}^{\mathrm{i}m_i\varphi}, \\ v_{i,\varphi}(\boldsymbol{\rho}) = \dfrac{1}{\mathrm{i}\rho_0\omega_i\rho}\dfrac{\partial p_i(\boldsymbol{\rho})}{\partial \varphi} = \displaystyle\sum_{m_i=-\infty}^{\infty} \frac{\mathcal{R}_{m_i}(\rho,k_i)}{4\rho} m_i \mathrm{e}^{\mathrm{i}m_i\varphi}. \end{cases} \tag{2.176}$$

Equations (2.167) and (2.176) represent the CWE of ultrasound pressure and particle velocity, respectively. Notably, these expressions can be calculated separately in polar coordinates, saving computational resources for a large number of field locations. In contrast, the integrals presented in the DIM, as described by Eqs. (2.18), (2.21), and (2.22), require evaluation at each field location since they are not separable in the rectangular coordinate system. It is worth mentioning that while multipole moments ($\mathcal{Q}_{m_i}$) involving integrals are necessary for the CWE calculations, these moments only need to be computed once for all exterior points, as shown in Eq. (2.173). Although calculating multipole moments for interior field locations demands more computational resources, as shown in Eq. (2.170), the ultrasound pressure in exterior field locations dominates the generation of audio sound, and the calculation of the ultrasound field in the interior region results in negligible computational cost.

2.5.3 AUDIO SOUND FIELD

The source density for audio sound at virtual source location $\boldsymbol{\rho}_\mathrm{v}$ in the polar coordinate system is derived by substituting Eq. (2.167) into Eq. (2.23), resulting in

$$q_\mathrm{a}(\boldsymbol{\rho}_\mathrm{v}) = \frac{\beta\omega_\mathrm{a}\omega_1\omega_2}{16\mathrm{i}c_0^4} \sum_{m_1,m_2=-\infty}^{\infty} \mathcal{R}_{m_1}^*(\rho_\mathrm{v},k_1)\mathcal{R}_{m_2}(\rho_\mathrm{v},k_2)\mathrm{e}^{\mathrm{i}(m_2-m_1)\varphi_\mathrm{v}}. \tag{2.177}$$

It is evident that the source density described by Eq. (2.177) can be separated into radial and angular components in the polar coordinate system. Building on the derivation in Sec. 2.5.1, the audio sound pressure can be expressed using the CWE given by Eq. (2.162) as

$$p_{\mathrm{a}}(\boldsymbol{\rho}) = \frac{\rho_0 \omega_{\mathrm{a}}}{4} \sum_{m_{\mathrm{a}}=-\infty}^{\infty} \mathcal{R}_{m_{\mathrm{a}}}(\rho, k_{\mathrm{a}}) \mathrm{e}^{\mathrm{i} m_{\mathrm{a}} \varphi}. \tag{2.178}$$

By applying the radial component expression in Eq. (2.163), the audio sound pressure Eq. (2.178) can also be represented as

$$p_{\mathrm{a}}(\boldsymbol{\rho}) = \frac{\rho_0 \omega_{\mathrm{a}}}{4} \sum_{m_{\mathrm{a}}=-\infty}^{\infty} [\mathcal{Q}_{m_{\mathrm{a}}}(0, \rho; J) H_{m_{\mathrm{a}}}(k_{\mathrm{a}} \rho) + \mathcal{Q}_{m_{\mathrm{a}}}(\rho, \infty; H) J_{m_{\mathrm{a}}}(k_{\mathrm{a}} \rho)] \mathrm{e}^{\mathrm{i} m_{\mathrm{a}} \varphi}. \tag{2.179}$$

The multipole moments for audio sound can be obtained by substituting Eq. (2.177) into Eq. (2.166), resulting in

$$\mathcal{Q}_{m_{\mathrm{a}}}(0, \rho; J) = \frac{\pi \beta \omega_{\mathrm{a}} \omega_1 \omega_2}{8 \mathrm{i} c_0^4} \sum_{m_1=-\infty}^{\infty} \int_0^{\rho} \mathcal{R}_{m_1}^*(\rho_{\mathrm{v}}, k_1) \mathcal{R}_{m_2}(\rho_{\mathrm{v}}, k_2) J_{m_{\mathrm{a}}}(k_{\mathrm{a}} \rho_{\mathrm{v}}) \rho_{\mathrm{v}} \mathrm{d}\rho_{\mathrm{v}}, \tag{2.180}$$

$$\mathcal{Q}_{m_{\mathrm{a}}}(\rho, \infty; H) = \frac{\pi \beta \omega_{\mathrm{a}} \omega_1 \omega_2}{8 \mathrm{i} c_0^4} \sum_{m_1=-\infty}^{\infty} \int_{\rho}^{\infty} \mathcal{R}_{m_1}^*(\rho_{\mathrm{v}}, k_1) \mathcal{R}_{m_2}(\rho_{\mathrm{v}}, k_2) H_{m_{\mathrm{a}}}(k_{\mathrm{a}} \rho_{\mathrm{v}}) \rho_{\mathrm{v}} \mathrm{d}\rho_{\mathrm{v}}. \tag{2.181}$$

In these equations, the summation over m_2 is eliminated, retaining only terms where $m_2 = m_1 + m_{\mathrm{a}}$, due to the relationship

$$\int_0^{2\pi} \mathrm{e}^{\mathrm{i}(m_2 - m_1 - m_{\mathrm{a}})\varphi_{\mathrm{v}}} \mathrm{d}\varphi_{\mathrm{v}} = 2\pi \delta_{m_2, m_1 + m_{\mathrm{a}}}. \tag{2.182}$$

Equation (2.178) represents the CWE of audio sound pressure generated by a PAL in the 2D model. When compared to the DIM presented in Eq. (2.24), Eq. (2.178) for the audio sound pressure is noteworthy for its separability within the polar coordinate system. This inherent separability significantly reduces the computational cost when calculating audio sound pressure at numerous field locations. Besides, no additional approximations are applied in this derivation.

Equation (2.179) sheds light on the physical interpretation of the CWE for audio sound. In this context, audio sound is understood as a superposition of numerous converging and outgoing cylindrical waves characterized by $J_{m_{\mathrm{a}}}(k_{\mathrm{a}} \rho) \mathrm{e}^{\mathrm{i} m_{\mathrm{a}} \varphi}$ and $H_{m_{\mathrm{a}}}(k_{\mathrm{a}} \rho) \mathrm{e}^{\mathrm{i} m_{\mathrm{a}} \varphi}$, respectively. The magnitude of each cylindrical wave's contribution is determined by $\mathcal{Q}_{m_{\mathrm{a}}}(0, \rho; J)$ for converging waves and $\mathcal{Q}_{m_{\mathrm{a}}}(\rho, \infty; H)$ for outgoing waves. To clarify, the outgoing waves originate from the virtual audio source inside an infinitely long cylinder with a radius corresponding to the radial coordinate of the field location, while the converging waves originate from sources outside this cylinder.

Far-Field Approximation

In the far field when $\rho \to \infty$, only the outgoing wave components remain in the radial component. Consequently, by substituting the limiting form of the Hankel function from Eq. (B.20) into Eq. (2.179), the audio sound pressure can be simplified to

$$p_\mathrm{a}(\rho) = \frac{\rho_0 \omega_\mathrm{a}}{2\sqrt{2\mathrm{i}\pi k_\mathrm{a}\rho}} \mathrm{e}^{\mathrm{i}k_\mathrm{a}\rho} \sum_{m_\mathrm{a}=-\infty}^{\infty} \mathrm{i}^{-m_\mathrm{a}} Q_{m_\mathrm{a}}(0,\infty;J) \mathrm{e}^{\mathrm{i}m_\mathrm{a}\varphi}. \tag{2.183}$$

2.5.4 REMARKS

This section introduces the cylindrical wave expansion (CWE) method, with Eq. (2.178) describing the CWE of audio sound pressure. In contrast to the DIM discussed in Sec. 2.2, the CWE method represents sound fields as a summation of contributions from multiple cylindrical waves. This representation, being separable in the polar coordinate system, offers a computationally efficient alternative to DIM. Importantly, the CWE method simplifies the 2D radiation model without the need for additional approximations like the paraxial approximation. Consequently, it provides accurate results in both near and far fields. Moreover, the flexibility of the CWE method allows for arbitrary adjustments to the ultrasound velocity profile. This versatility makes it suitable for analyzing sound fields generated by linear phased array PALs (for specific examples, refer to Chap. 6).

2.6 SPHERICAL WAVE EXPANSION (SWE) METHOD

The *spherical wave expansion (SWE)* method provides a means for calculating sound fields generated by a circular PAL without relying on additional paraxial approximations. The fundamental idea behind the SWE method lies in expressing the Green's function in the spherical coordinate system. The framework of the SWE method is introduced in Sec. 2.6.1. Initially proposed for a circular PAL with an axisymmetric profile in [Zhong et al., 2020c; Zhong et al., 2021b], the method is detailed in Sec. 2.6.2. Subsequently, an extension was developed to accommodate a circular PAL with a non-axisymmetric profile, utilizing Zernike polynomials [Zhong et al., 2022b], and this extension is presented in Sec. 2.6.3.

2.6.1 FRAMEWORK OF THE SWE METHOD

The 3D Green's function in free space [Eq. (2.36)] can be expressed in the spherical coordinate system (r,θ,φ) as [Zhong et al., 2020c, Eq. (9)]

$$g_\mathrm{3D}(\mathbf{r},\mathbf{r}',\omega) = \mathrm{i}k \sum_{\ell=0}^{\infty} \mathrm{j}_\ell(kr_<)\mathrm{h}_\ell(kr_>) \sum_{m=-\ell}^{\ell} Y_\ell^m(\theta,\varphi) Y_\ell^{m,*}(\theta',\varphi'), \tag{2.184}$$

where $r_< = \min(r,r')$, $r_> = \max(r,r')$, $\mathrm{j}_\ell(\cdot)$ and $\mathrm{h}_\ell(\cdot)$ are spherical Bessel and Hankel functions, respectively, and $Y_\ell^m(\cdot)$ is the spherical harmonic function.

Consider an arbitrary source characterized by a source density of $q(\mathbf{r})$. The resultant sound field produced by this source is determined through the KHIE as presented in Eq. (2.7). Upon substituting Eq. (2.184) into Eq. (2.7), the expression becomes

$$p(\mathbf{r}) = \rho_0 \omega k \sum_{\ell=0}^{\infty} \sum_{m=-\ell}^{\ell} \mathcal{R}_\ell^m(r,k) Y_\ell^m(\theta,\varphi). \tag{2.185}$$

Equation (2.185) can be viewed as the 3D counterpart of Eq. (2.162). Notably, Eq. (2.185) reveals a separation of the radial (r) and angular (θ,φ) coordinates into two distinct components. The angular component has the form of spherical harmonics, $Y_\ell^m(\theta,\varphi)$, while the radial component is expressed as

$$\mathcal{R}_\ell^m(r,k) \equiv \mathcal{Q}_\ell^m(0,r;\mathrm{j})\mathrm{h}_\ell(kr) + \mathcal{Q}_\ell^m(r,\infty;\mathrm{h})\mathrm{j}_\ell(kr), \tag{2.186}$$

where the multipole moment of the source density is defined as [Bilbao and Ahrens, 2020, Eq. (19)]

$$\mathcal{Q}_\ell^m(r_1,r_2;\mathscr{S}) \equiv \iiint_{r_1 \le r' \le r_2} q(\mathbf{r}')\mathscr{S}_\ell(kr')Y_\ell^{m,*}(\theta',\varphi')\mathrm{d}^3\mathbf{r}'. \tag{2.187}$$

Here, the symbol $\mathscr{S}$ can represent either spherical Bessel ($\mathscr{S} = \mathrm{j}$) or spherical Hankel ($\mathscr{S} = \mathrm{h}$) functions. It is crucial to emphasize that the derivation of the SWE for sound pressure, as expressed in Eq. (2.185), is accomplished without reliance on additional approximations or assumptions.

Equations (2.185) and (2.187) indicate that when the source density is separable in spherical coordinates, the calculation of the sound pressure can be simplified as the radial and angular components can be calculated independently. Consider the source density in the separable form of

$$q_\mathrm{a}(\boldsymbol{\rho}) = q_0 q_r(r) q_\theta(\theta) q_\varphi(\varphi), \tag{2.188}$$

where q_0 is a constant. The multipole moment of the source density Eq. (2.187) can then be simplified to

$$\begin{aligned}\mathcal{Q}_\ell^m(r_1,r_2;\mathscr{S}) = q_0 &\left[\int_{r_1}^{r_2} q_r(r')\mathscr{S}_\ell(kr')r'^2\mathrm{d}r'\right] \\ &\times \left[\int_0^{2\pi}\int_0^{\pi} q_\theta(\theta')q_\varphi(\varphi')Y_\ell^{m,*}(\theta',\varphi')\sin\theta'\mathrm{d}\theta'\mathrm{d}\varphi'\right].\end{aligned} \tag{2.189}$$

Suppose the source density in the azimuthal direction is characterized by a complex exponential function

$$q_\varphi(\varphi) = \mathrm{e}^{\mathrm{i}m_0\varphi}, \quad m_0 \in \mathbb{Z}. \tag{2.190}$$

The multipole moment Eq. (2.189) can be further simplified to

$$\begin{aligned}\mathcal{Q}_\ell^m(r_1,r_2;\mathscr{S}) = 2\pi\delta_{mm_0} q_0 &\left[\int_{r_1}^{r_2} q_r(r')\mathscr{S}_\ell(kr')r'^2\mathrm{d}r'\right] \\ &\times \left[\int_0^{\pi} q_\theta(\theta')Y_\ell^{m_0}(\theta',0)\sin\theta'\mathrm{d}\theta'\right].\end{aligned} \tag{2.191}$$

Here, Eq. (B.8) is used, and δ_{mm_0} is the Kronecker delta function. Substituting Eq. (2.191) into the sound pressure Eq. (2.185) eliminates the summation over m, resulting in

$$p(\mathbf{r}) = \rho_0 \omega k \sum_{\ell=0}^{\infty} \mathcal{R}_{\ell+|m_0|}^{m_0}(r,k) Y_{\ell+|m_0|}^{m_0}(\theta,\varphi). \tag{2.192}$$

Here, the original index ℓ is replaced with $\ell+|m_0|$ to ensure that the resulting degree, $\ell+|m_0|$, consistently exceeds the absolute value of order, $|m_0|$.

Substituting Eq. (2.186) into Eq. (2.192), the sound pressure can be rewritten as

$$\begin{aligned} p(\mathbf{r}) = \rho_0 \omega k \sum_{\ell=0}^{\infty} \Big[& \mathcal{Q}_{\ell+|m_0|}^{m_0}(0,r;\mathrm{j}) \mathrm{h}_{\ell+|m_0|}(kr) \\ & + \mathcal{Q}_{\ell+|m_0|}^{m_0}(r,\infty;\mathrm{h}) \mathrm{j}_{\ell+|m_0|}(kr) \Big] Y_{\ell+|m_0|}^{m_0}(\theta,\varphi). \end{aligned} \tag{2.193}$$

Far-Field Approximation

In the far field when $r \to \infty$, it is observed from Eq. (2.186) that only the outgoing wave retains since $\lim_{r\to\infty} \mathcal{Q}_{\ell+|m_0|}^{m_0}(r,\infty;\mathrm{h}) \to 0$. By using the limiting form of the spherical Hankel function given by Eq. (B.26), the radial component Eq. (2.186) can be simplified to

$$\mathcal{R}_{\ell}^{m_0}(r,k) = \mathcal{Q}_{\ell}^{m_0}(0,\infty;\mathrm{j}) \mathrm{i}^{-\ell-|m_0|-1} \frac{\mathrm{e}^{\mathrm{i}kr}}{kr}. \tag{2.194}$$

By substituting Eq. (2.194) into Eq. (2.192), the sound pressure in the far field is simplified to

$$p(\mathbf{r}) = \frac{\rho_0 \omega}{r} \mathrm{e}^{\mathrm{i}kr} \sum_{\ell=0}^{\infty} \mathrm{i}^{-\ell-|m_0|-1} \mathcal{Q}_{\ell_i+|m_0|}^{m_0}(0,\infty;\mathrm{j}) Y_{\ell_i+|m_0|}^{m_0}(\theta,\varphi). \tag{2.195}$$

Particle Velocity Field

The ultrasonic particle velocity can be obtained by substituting Eq. (2.192) into the linear relation Eq. (1.103) as

$$\begin{cases} v_r(\mathbf{r}) = \dfrac{1}{\mathrm{i}\rho_0\omega} \dfrac{\partial p(\mathbf{r})}{\partial r} = -\mathrm{i}k \displaystyle\sum_{\ell=0}^{\infty} \dfrac{\partial \mathcal{R}_{\ell+|m_0|}^{m_0}(r,k)}{\partial r} Y_{\ell+|m_0|}^{m_0}(\theta,\varphi), \\ v_\theta(\mathbf{r}) = \dfrac{1}{\mathrm{i}\rho_0\omega r} \dfrac{\partial p(\mathbf{r})}{\partial \theta} = -\mathrm{i}k \displaystyle\sum_{\ell=0}^{\infty} \dfrac{\mathcal{R}_{\ell+|m_0|}^{m_0}(r,k)}{r} \dfrac{\partial Y_{\ell+|m_0|}^{m_0}(\theta,\varphi)}{\partial \theta}, \\ v_\varphi(\mathbf{r}) = \dfrac{1}{\mathrm{i}\rho_0\omega r \sin\theta} \dfrac{\partial p(\mathbf{r})}{\partial \varphi} = -\mathrm{i}k \dfrac{\mathcal{R}_{\ell+|m_0|}^{m_0}(r,k)}{r\sin\theta} \dfrac{\partial Y_{\ell+|m_0|}^{m_0}(\theta,\varphi)}{\partial \varphi}. \end{cases} \tag{2.196}$$

Here, v_r, v_θ, v_φ represent the radial, zenithal, and azimuthal components of the particle velocity field $\mathbf{v}$. In Eq. (2.196), the derivative of the spherical harmonics with respect to the zenithal (θ) and azimuthal (φ) angles can be obtained by Eqs. (B.6) and (B.7), respectively.

Equations (2.192) and (2.196) represent the SWEs for sound pressure and particle velocity, respectively, emanating from a circular source characterized by source density expressions Eqs. (2.188) and (2.190). For a special case when $m_0 = 0$, the resulting sound fields exhibit axisymmetry in the azimuthal direction. Equation (2.192) elucidates the physical interpretation of the SWE for sound fields. The sound pressure is the superposition of an infinite series of outgoing and converging spherical waves, defined by $\mathrm{h}_{\ell+|m_0|}(kr)Y_{\ell+|m_0|}^{m_0}(\theta,\varphi)$ and $\mathrm{j}_{\ell+|m_0|}(kr)Y_{\ell+|m_0|}^{m_0}(\theta,\varphi)$, respectively. The weight of each outgoing and converging spherical wave is characterized by $Q_{\ell+|m_0|}^{m_0}(0,r;\mathrm{j})$ and $Q_{\ell+|m_0|}^{m_0}(r,\infty;\mathrm{h})$, respectively. These outgoing and converging spherical waves correspond to the exterior [Williams, 1999, Sec. 6.7] and interior [Williams, 1999, Sec. 6.8] problems in the spherical coordinate system, respectively. In simpler terms, the outgoing wave originates from the source located inside a sphere with a radius equal to the radial coordinate of the field location, while the converging wave originates from the source outside this sphere.

2.6.2 AXISYMMETRIC PROFILE

This section derives the SWE of the sound field generated by a PAL featuring an axisymmetric velocity profile. The SWE for the quasilinear solution of the Westervelt equation was initially introduced in Zhong et al., 2020c and later extended to incorporate local effects in Zhong et al., 2021b. The ultrasound velocity profile is assumed to exhibit axisymmetry, implying that $v_{i,z}(\mathbf{r}_\mathrm{s})$ is independent of the azimuthal angle φ_s over the radiation surface. Here, the index $i = 1,2$ denotes the ultrasound at frequency f_i. For simplicity, the velocity profile is represented as

$$v_{i,z}(\mathbf{r}_\mathrm{s}) = v_0 u_i(r_\mathrm{s})\mathrm{H}(a - r_\mathrm{s}), \tag{2.197}$$

in the subsequent derivations, where r_s is the radial coordinate of the source point in the spherical coordinate system. Here, the source is assumed to be situated in the plane $z = 0$, setting the source coordinate in z-direction as $z_\mathrm{s} = 0$. Examples encompass circular sources with uniform and quadratic profiles, as depicted by Eqs. (2.29) and (2.32), respectively.

2.6.2.1 Ultrasound Field

In the spherical coordinate system, the equivalent source density for the ultrasound velocity profile, as defined in Eq. (2.197), can be expressed using Eq. (2.40) as

$$q_i(\mathbf{r}_\mathrm{s}) = \frac{2v_0}{r_\mathrm{s}} u_i(r_\mathrm{s})\mathrm{H}(a - r_\mathrm{s})\delta\left(\theta_\mathrm{s} - \frac{\pi}{2}\right). \tag{2.198}$$

Here, r_s and θ_s denote the radial and zenithal coordinates of the source point $\mathbf{r}_\mathrm{s}$. It is evident that Eq. (2.198) exhibits separability in the spherical coordinate system.

Building on the derivation in Sec. 2.6.1, the SWE for ultrasound pressure can be expressed in the form of Eq. (2.192) as

$$p_i(\mathbf{r}) = \rho_0 \omega_i k_i \sum_{\ell_i=0}^{\infty} \mathcal{R}_{\ell_i}^0(r, k_i) Y_{\ell_i}^0(\theta, \varphi), \tag{2.199}$$

where the condition $m_0 = 0$ is used for the axisymmetric profile.

The multipole moments are derived by substituting Eq. (2.198) into Eq. (2.191) as

$$\mathcal{Q}_{\ell_i}^0(r_1, r_2; \mathscr{S}) = 4\pi v_0 Y_{\ell_i}^0(\pi/2, 0) \int_{r_1}^{r_2} \mathrm{H}(a - r_s) u_i(r_s) \mathscr{S}_{\ell_i}(k_i r_s) r_s \mathrm{d}r_s. \tag{2.200}$$

According to Eq. (B.10), $Y_{\ell_i}^0(\pi/2, 0)$ is non-zero only when ℓ_i is even. By substituting the original ℓ_i with $2\ell_i$ to eliminate the odd terms, Eq. (2.199) can be reformulated as

$$p_i(\mathbf{r}) = \rho_0 \omega_i k_i \sum_{\ell_i=0}^{\infty} \mathcal{R}_{2\ell_i}^0(r, k_i) Y_{2\ell_i}^0(\theta, \varphi). \tag{2.201}$$

Upon substituting Eq. (2.200) into the radial component Eq. (2.186), the expressions for ultrasound pressure Eq. (2.201) take different forms in the interior ($r < a$) and exterior ($r > a$) regions. In the interior region ($r < a$), the ultrasound pressure is given by

$$p_i(\mathbf{r}) = \rho_0 \omega_i k_i \sum_{\ell_i=0}^{\infty} \left[\mathcal{Q}_{2\ell_i}^0(0, r; \mathrm{j}) \mathrm{h}_{2\ell_i}(k_i r) + \mathcal{Q}_{2\ell_i}^0(r, \infty; \mathrm{h}) \mathrm{j}_{2\ell_i}(k_i r)\right] Y_{2\ell_i}^0(\theta, \varphi), \tag{2.202}$$

where the multipole moments are obtained through Eq. (2.200) as

$$\mathcal{Q}_{2\ell_i}^0(0, r; \mathrm{j}) = 4\pi v_0 Y_{2\ell_i}^0(\pi/2, 0) \int_0^r u_i(r_s) \mathrm{j}_{2\ell_i}(k_i r_s) r_s \mathrm{d}r_s, \tag{2.203}$$

$$\mathcal{Q}_{2\ell_i}^0(r, a; \mathrm{h}) = 4\pi v_0 Y_{2\ell_i}^0(\pi/2, 0) \int_r^a u_i(r_s) \mathrm{h}_{2\ell_i}(k_i r_s) r_s \mathrm{d}r_s. \tag{2.204}$$

In the exterior region ($r > a$), the expression for ultrasound pressure is given by

$$p_i(\mathbf{r}) = \rho_0 \omega_i k_i \sum_{\ell_i=0}^{\infty} \mathcal{Q}_{2\ell_i}^0(0, a; \mathrm{j}) \mathrm{h}_{2\ell_i}(k_i r) Y_{2\ell_i}^0(\theta, \varphi), \tag{2.205}$$

where the multiple moments are

$$\mathcal{Q}_{2\ell_i}^0(0, a; \mathrm{j}) = 4\pi v_0 Y_{2\ell_i}^0(\pi/2, 0) \int_0^a u_i(r_s) \mathrm{j}_{2\ell_i}(k_i r_s) r_s \mathrm{d}r_s. \tag{2.206}$$

Equations (2.202) and (2.205) offer valuable insights into the physical interpretation of the SWE for ultrasound. For an exterior field location ($r > a$), the ultrasound field consists of an infinite series of outgoing spherical waves characterized by

$\mathrm{h}_{2\ell_i}(k_i r)Y_{2\ell_i}^0(\theta,\varphi)$, with the weight of each wave determined by the aperture effects, specifically $Q_{2\ell_i}^0(0,a;\mathrm{j})$. In contrast, for an interior location ($r < a$), the ultrasound field is a superposition of converging and outgoing spherical waves characterized by $\mathrm{j}_{2\ell_i}(k_i r)Y_{2\ell_i}^0(\theta,\varphi)$ and $\mathrm{h}_{2\ell_i}(k_i r)Y_{2\ell_i}^0(\theta,\varphi)$, respectively. The outgoing wave contributes from the source inside the sphere with a radius equal to the radial coordinate of the field location ($0 \le r_s < r$), while the weight of the converging wave contributes from the source outside this sphere ($r < r_s \le a$). Consequently, the weights of converging and outgoing waves are $Q_{2\ell_i}^0(0,r;\mathrm{j})$ and $Q_{2\ell_i}^0(\rho,a;\mathrm{h})$, respectively.

Far-Field Approximation

In the far field when $r \to \infty$, the spherical Hankel function in Eq. (2.205) can be approximated using the its limiting form Eq. (B.26). Consequently, the ultrasound pressure is simplified to

$$p_i(\mathbf{r}) = \frac{\rho_0 \omega_i}{r} \mathrm{e}^{\mathrm{i}k_i r} \sum_{\ell_i=0}^{\infty} (-1)^{\ell_i} \mathrm{i}^{-1} Q_{2\ell_i}^0(0,a;\mathrm{j}) Y_{2\ell_i}^0(\theta,\varphi). \tag{2.207}$$

Particle Velocity Field

Similar to Eq. (2.196), the components of the particle velocity field $\mathbf{v}_i(\boldsymbol{\rho})$ in the spherical coordinate system can be obtained by substituting Eq. (2.201) into Eq. (1.103) as

$$\begin{cases} v_{i,r}(\mathbf{r}) = -\mathrm{i}k_i \displaystyle\sum_{\ell_i=0}^{\infty} \frac{\partial \mathcal{R}_{2\ell_i}^0(r,k_i)}{\partial r} Y_{2\ell_i}^0(\theta,\varphi) \\ v_{i,\theta}(\mathbf{r}) = -\mathrm{i}k_i \displaystyle\sum_{\ell_i=0}^{\infty} \frac{\mathcal{R}_{2\ell_i}^0(r,k_i)}{r} \frac{\partial Y_{2\ell_i}^0(\theta,\varphi)}{\partial \theta}, \\ v_{i,\varphi}(\mathbf{r}) = 0. \end{cases} \tag{2.208}$$

Equations (2.201) and (2.208) represent the SWEs for ultrasound pressure and particle velocity, respectively. Notably, these expressions allow for separate computations in spherical coordinates, offering efficiency advantages when dealing with a substantial number of field locations. In contrast, the integrals presented in the DIM, described by Eqs. (2.39) and (2.43), necessitate evaluation at each field location as they are not separable in the Cartesian coordinate system.

It is worth mentioning that while multipole moments ($Q_{2\ell_i}^0$) involving integrals are required for SWE calculations, these moments only need to be computed once for all exterior points, as shown in Eq. (2.205). Despite the increased computational demand for calculating multipole moments in the interior field locations, as shown in Eq. (2.202), the contribution of ultrasound pressure in exterior field locations significantly outweighs that in the interior region. Consequently, the computation of the ultrasound field in the interior region incurs negligible computational cost in the context of generating audible sound.

2.6.2.2 Audio Sound Field

By substituting the SWE of ultrasound pressure given by Eq. (2.201) into the source density of the audio sound given by Eq. (2.44), the SWE of the source density at the virtual source location $\mathbf{r}_\mathrm{v}$ is obtained as

$$q_\mathrm{a}(\mathbf{r}_\mathrm{v}) = \frac{\beta\omega_\mathrm{a}\omega_1^2\omega_2^2}{\mathrm{i}c_0^6} \sum_{\ell_1,\ell_2=0}^{\infty} \mathcal{R}_{2\ell_1}^{0,*}(r_\mathrm{v},k_1)\mathcal{R}_{2\ell_2}^{0}(r_\mathrm{v},k_2)Y_{2\ell_1}^{0}(\theta_\mathrm{v},\varphi_\mathrm{v})Y_{2\ell_2}^{0}(\theta_\mathrm{v},\varphi_\mathrm{v}). \tag{2.209}$$

The source density described by Eq. (2.209) is separable into radial and angular components in the spherical coordinate system. Additionally, the source density is independent of the azimuthal angle, φ_v. Building on the derivation in Sec. 2.6.1, the audio sound pressure can be expressed using the SWE given by Eq. (2.201) as

$$p_\mathrm{a}(\mathbf{r}) = \rho_0\omega_\mathrm{a}k_\mathrm{a} \sum_{\ell_\mathrm{a}=0}^{\infty} \mathcal{R}_{\ell_\mathrm{a}}^{0}(r,k_\mathrm{a})Y_{\ell_\mathrm{a}}^{0}(\theta,\varphi). \tag{2.210}$$

The multiple moments in the radial component can be obtained by substituting Eq. (2.209) into Eq. (2.189) as

$$\begin{aligned} Q_{\ell_\mathrm{a}}^{0}(r_1,r_2;\mathscr{S}) = {}& \frac{\beta\omega_\mathrm{a}\omega_1^2\omega_2^2}{\mathrm{i}c_0^6} \sum_{\ell_1,\ell_2=0}^{\infty} \int_0^\infty \mathscr{S}_{\ell_\mathrm{a}}(k_\mathrm{a}r_\mathrm{a})\mathcal{R}_{2\ell_1}^{0,*}(r_\mathrm{v},k_1)\mathcal{R}_{2\ell_2}^{0}(r_\mathrm{v},k_2)r_\mathrm{v}^2\mathrm{d}r_\mathrm{v} \\ & \times \int_0^{2\pi}\int_0^{\pi} Y_{2\ell_1}^{0,*}(\theta_\mathrm{v},\varphi_\mathrm{v})Y_{2\ell_2}^{0}(\theta_\mathrm{v},\varphi_\mathrm{v})Y_{\ell_\mathrm{a}}^{0,*}(\theta_\mathrm{v},\varphi_\mathrm{v})\sin\theta_\mathrm{v}\mathrm{d}\theta_\mathrm{v}\mathrm{d}\varphi_\mathrm{v}. \end{aligned} \tag{2.211}$$

By using Eqs. (B.5) and (B.9), the triple integral with respect to the spherical harmonics in Eq. (2.211) can be simplified to

$$I(2\ell_1,2\ell_2,\ell_\mathrm{a};0,0,0) = \mathscr{G}(2\ell_1,0;2\ell_2,0;\ell_\mathrm{a}), \tag{2.212}$$

where I is defined by Eq. (B.9), and $\mathscr{G}$ is the Gaunt coefficient defined by Eq. (B.15). According to Eqs. (B.16) and (B.17), the Gaunt coefficient in Eq. (2.212) vanishes unless $2\ell_1+2\ell_2+\ell_\mathrm{a}$ is an even integer. Therefore, only the terms in Eqs. (2.210–2.211) for even ℓ_a are retained. By substituting the original ℓ_a with $2\ell_\mathrm{a}$ to eliminate the odd terms, the multipole moments Eq. (2.211) are simplified to

$$\begin{aligned} Q_{2\ell_\mathrm{a}}^{0}(r_1,r_2;\mathscr{S}) = {}& \frac{\beta\omega_\mathrm{a}\omega_1^2\omega_2^2}{\mathrm{i}c_0^6} \sum_{\ell_1,\ell_2=0}^{\infty} \mathscr{G}(2\ell_1,0;2\ell_2,0;2\ell_\mathrm{a}) \\ & \times \int_0^\infty \mathscr{S}_{2\ell_\mathrm{a}}(k_\mathrm{a}r_\mathrm{v})\mathcal{R}_{2\ell_1}^{0,*}(r_\mathrm{v},k_1)\mathcal{R}_{2\ell_2}^{0}(r_\mathrm{v},k_2)r_\mathrm{v}^2\mathrm{d}r_\mathrm{v}. \end{aligned} \tag{2.213}$$

The audio sound pressure Eq. (2.210) is rewritten as

$$p_\mathrm{a}(\mathbf{r}) = \rho_0\omega_\mathrm{a}k_\mathrm{a} \sum_{\ell_\mathrm{a}=0}^{\infty} \mathcal{R}_{2\ell_\mathrm{a}}^{0}(r,k_\mathrm{a})Y_{2\ell_\mathrm{a}}^{0}(\theta,\varphi). \tag{2.214}$$

By applying the radial component expression in Eq. (2.186), the audio sound pressure Eq. (2.214) can also be represented as

$$p_{\mathrm{a}}(\mathbf{r}) = \rho_0 \omega_{\mathrm{a}} k_{\mathrm{a}} \sum_{\ell_{\mathrm{a}}=0}^{\infty} \left[Q_{2\ell_{\mathrm{a}}}^0(0,r;\mathrm{j}) \mathrm{h}_{2\ell_{\mathrm{a}}}(k_{\mathrm{a}}r) + Q_{2\ell_{\mathrm{a}}}^0(r,\infty;\mathrm{h}) \mathrm{j}_{2\ell_{\mathrm{a}}}(k_{\mathrm{a}}r) \right] Y_{2\ell_{\mathrm{a}}}^0(\theta,\varphi). \tag{2.215}$$

Equation (2.214) represents the SWE of the audio sound pressure generated by a circular PAL with an axisymmetric profile. Equation (2.215) elucidates the physical meaning of the SWE for the audio sound field. The audio sound is the superposition of an infinite series of outgoing and converging spherical waves characterized by $\mathrm{h}_{2\ell_{\mathrm{a}}}(k_{\mathrm{a}}r) Y_{2\ell_{\mathrm{a}}}^0(\theta,\varphi)$ or $\mathrm{j}_{2\ell_{\mathrm{a}}}(k_{\mathrm{a}}r) Y_{2\ell_{\mathrm{a}}}^0(\theta,\varphi)$, respectively. The weight of of each converging and outgoing spherical wave is $Q_{2\ell_{\mathrm{a}}}^0(0,r;\mathrm{j})$ and $Q_{2\ell_{\mathrm{a}}}^0(r,\infty;\mathrm{h})$, respectively. In simpler terms, the outgoing wave emanates from the virtual audio source within the sphere, with a radius equal to the radial coordinate of the field location, while the converging wave originates from a source outside this sphere.

Far-Field Approximation

In the far field when $r \to \infty$, it is observed from Eq. (2.215) that only the outgoing wave retains since $\lim_{r\to\infty} Q_{2\ell_{\mathrm{a}}}^0(r,\infty;\mathrm{h}) \to 0$. By using the limiting form of the spherical Hankel function given by Eq. (B.26), the radial component for audio sound in Eq. (2.214) reduces to

$$\mathcal{R}_{2\ell_{\mathrm{a}}}^0(r,k_{\mathrm{a}}) = \frac{(-1)^{\ell_{\mathrm{a}}} Q_{2\ell_{\mathrm{a}}}^0(0,\infty;\mathrm{j})}{\mathrm{i}k_{\mathrm{a}}r} \mathrm{e}^{\mathrm{i}k_{\mathrm{a}}r}. \tag{2.216}$$

The audio sound is obtained by substituting Eq. (2.216) into Eq. (2.215) as

$$p_{\mathrm{a}}(\mathbf{r}) = \frac{\rho_0 \omega_{\mathrm{a}}}{\mathrm{i}r} \mathrm{e}^{\mathrm{i}k_{\mathrm{a}}r} \sum_{\ell_{\mathrm{a}}=0}^{\infty} (-1)^{\ell_{\mathrm{a}}} Q_{2\ell_{\mathrm{a}}}^0(0,\infty;\mathrm{j}) Y_{2\ell_{\mathrm{a}}}^0(\theta,\varphi). \tag{2.217}$$

2.6.3 NON-AXISYMMETRIC PROFILE

The SWE method introduced in Sec. 2.6.2 is specifically designed for the analysis of a circular PAL featuring an axisymmetric profile as defined by Eq. (2.197). The focus of this section is to broaden the applicability of the SWE approach, extending its capabilities to address the radiation emanating from a circular PAL with a non-axisymmetric profile. Examples of such non-axisymmetric profiles include steerable PALs, which electronically adjust highly directional audio beams toward a specified direction. This extension technique, initially presented in Zhong et al., 2022b, involves decomposing any arbitrary velocity profile into *Zernike polynomials*, which are orthogonal and form a complete set over a unit circle.

2.6.3.1 Ultrasound Field

An arbitrary velocity profile can be expressed as a summation of Zernike circular polynomials

$$v_{i,z}(\mathbf{r}_s) = v_0 \sum_{n_i=0}^{\infty} \sum_{m_i=-n_i}^{n_i}{}' M_{n_i}^{m_i} R_{n_i}^{m_i}(r_s/a) e^{\mathrm{i} m_i \varphi_s}, \tag{2.218}$$

where the prime over the summation sign indicates that the summation is performed for terms only with even $n_i - m_i$, and $R_{n_i}^{m_i}(r_s/a)$ is the *radial Zernike polynomial* of degree n_i and azimuthal order m_i defined by Eq. (B.30). Here, the argument of the radial Zernike polynomial is chosen as r_s/a to ensure it ranges from 0 to 1. In Eq. (2.218), $M_{n_i}^{m_i}$ are the *Zernike moments* defined by Eq. (B.34). They have the form of

$$M_{n_i}^{m_i} = \frac{n_i+1}{\pi} \int_0^{2\pi} \int_0^1 \frac{v_{i,z}(\mathbf{r}_s)}{v_0} R_{n_i}^{m_i}(\tilde{r}_s) e^{-\mathrm{i} m_i \varphi_s} \tilde{r}_s \mathrm{d}\tilde{r}_s \mathrm{d}\varphi_s, \tag{2.219}$$

where the normalized radial coordinate is defined as $\tilde{r}_s \equiv r_s/a$.

By using the Zernike expansion for the velocity profile given by Eq. (2.218), the ultrasound can be represented as the superposition of the radiation from each Zernike mode as

$$p_i(\mathbf{r}) = \sum_{n_i=0}^{\infty} \sum_{m_i=-n_i}^{n_i}{}' M_{n_i}^{m_i} p_{n_i}^{m_i}(\mathbf{r}), \tag{2.220}$$

where $p_{n_i}^{m_i}(\mathbf{r})$ is the ultrasound pressure radiated by the source with a Zernike profile

$$v_{n_i}^{m_i}(\mathbf{r}_s) = v_0 R_{n_i}^{m_i}(r_s/a) e^{\mathrm{i} m_i \varphi_s} \mathrm{H}(a - r_s). \tag{2.221}$$

The equivalent source density Eq. (2.198) is modified as

$$q_i(\mathbf{r}_s) = \frac{2v_0}{r_s} R_{n_i}^{m_i}(r_s/a) \mathrm{H}(a - r_s) \delta\left(\theta_s - \frac{\pi}{2}\right) e^{\mathrm{i} m_i \varphi_s}. \tag{2.222}$$

Based on the framework developed in Sec. 2.6.1, the ultrasound pressure $p_{n_i}^{m_i}(\mathbf{r})$ can be obtained by Eq. (2.192) as

$$p_{n_i}^{m_i}(\mathbf{r}) = \rho_0 \omega_i k_i \sum_{\ell_i=0}^{\infty} \mathcal{R}_{\ell_i+|m_i|}^{m_i}(r, k_i) Y_{\ell_i+|m_i|}^{m_i}(\theta, \varphi). \tag{2.223}$$

The multipole moments are obtained by substituting Eq. (2.222) into Eq. (2.191) as

$$\begin{aligned} Q_{\ell_i+|m_i|}^{m_i}(r_1, r_2; \mathscr{S}) &= 4\pi v_0 Y_{\ell_i+|m_i|}^{m_i}(\pi/2, 0) \\ &\quad \times \int_{r_1}^{r_2} \mathrm{H}(a - r_s) R_{n_i}^{m_i}(r_s/a) \mathscr{S}_{\ell_i+|m_i|}(k_i r_s) r_s \mathrm{d}r_s. \end{aligned} \tag{2.224}$$

According to Eq. (B.11), $Y_{\ell_i+|m_i|}^{m_i}(\pi/2, 0)$ is non-zero only when $\ell_i + |m_i| - m_i$ is even. By substituting the original $\ell_i + |m_i|$ with $2\ell_i + |m_i|$ to eliminate the odd terms, the ultrasound pressure Eq. (2.223) can be written as

$$p_{n_i}^{m_i}(\mathbf{r}) = \rho_0 \omega_i k_i \sum_{\ell_i=0}^{\infty} \mathcal{R}_{2\ell_i+|m_i|}^{m_i}(r, k_i) Y_{2\ell_i+|m_i|}^{m_i}(\theta, \varphi). \tag{2.225}$$

It is clear Eq. (2.201) is a special case of Eq. (2.225) by substituting $m_i = 0$ into it.

Similar to Eq. (2.202), in the interior region ($r < a$), the ultrasound pressure is given by

$$\begin{aligned} p_i(\mathbf{r}) = \rho_0 \omega_i k_i \sum_{\ell_i=0}^{\infty} \Big[& Q_{2\ell_i+|m_i|}^{m_i}(0, r; \mathrm{j}) \mathrm{h}_{2\ell_i+|m_i|}(k_i r) \\ & + Q_{2\ell_i+|m_i|}^{m_i}(r, \infty; \mathrm{h}) \mathrm{j}_{2\ell_i+|m_i|}(k_i r) \Big] Y_{2\ell_i+|m_i|}^{m_i}(\theta, \varphi), \end{aligned} \tag{2.226}$$

where the multipole moments are obtained through Eq. (2.200) as

$$Q_{2\ell_i+|m_i|}^{m_i}(0, r; \mathrm{j}) = 4\pi v_0 Y_{2\ell_i+|m_i|}^{m_i}(\pi/2, 0) \int_0^r R_{n_i}^{m_i}(r_s/a) \mathrm{j}_{2\ell_i+|m_i|}(k_i r_s) r_s \mathrm{d} r_s, \tag{2.227}$$

$$Q_{2\ell_i+|m_i|}^{m_i}(r, a; \mathrm{h}) = 4\pi v_0 Y_{2\ell_i+|m_i|}^{m_i}(\pi/2, 0) \int_r^a R_{n_i}^{m_i}(r_s/a) \mathrm{h}_{2\ell_i+|m_i|}(k_i r_s) r_s \mathrm{d} r_s. \tag{2.228}$$

In the exterior region ($r > a$), the expression for ultrasound pressure is given as

$$p_i(\mathbf{r}) = \rho_0 \omega_i k_i \sum_{\ell_i=0}^{\infty} Q_{2\ell_i+|m_i|}^{m_i}(0, a; \mathrm{j}) \mathrm{h}_{2\ell_i+|m_i|}(k_i r) Y_{2\ell_i+|m_i|}^{m_i}(\theta, \varphi), \tag{2.229}$$

where the multiple moments are

$$Q_{2\ell_i+|m_i|}^{m_i}(0, a; \mathrm{j}) = 4\pi v_0 Y_{2\ell_i+|m_i|}^{m_i}(\pi/2, 0) \int_0^a \mathrm{j}_{2\ell_i+|m_i|}(k_i r_s) r_s \mathrm{d} r_s. \tag{2.230}$$

Far-Field Approximation

In the far field when $r \to \infty$, the spherical Hankel function in Eq. (2.229) can be approximated using the its limiting form Eq. (B.26). Consequently, the ultrasound pressure is simplified to

$$p_{n_i}^{m_i}(\mathbf{r}) = \frac{\rho_0 \omega_i}{r} \mathrm{e}^{\mathrm{i} k_i r} \sum_{\ell_i=0}^{\infty} (-1)^{\ell_i} \mathrm{i}^{-|m_i|-1} Q_{2\ell_i+|m_i|}^{m_i}(0, a; \mathrm{j}) Y_{2\ell_i+|m_i|}^{m_i}(\theta, \varphi). \tag{2.231}$$

Particle Velocity Field

Similar to Eq. (2.220), the ultrasound particle velocity can be expressed as

$$\mathbf{v}_i(\mathbf{r}) = \sum_{n_i=0}^{\infty} \sum_{m_i=-n_i}^{n_i}{}' M_{n_i}^{m_i} \mathbf{v}_{n_i}^{m_i}(\mathbf{r}), \tag{2.232}$$

where $\mathbf{v}_{n_i}^{m_i}(\mathbf{r})$ is the ultrasound particle velocity resulted from the source with a Zernike profile given by Eq. (2.221).

The ultrasound particle velocity is then obtained by substituting Eq. (2.225) into the linear relation Eq. (1.103) as

$$\begin{cases} v_{n_i,r}^{m_i}(\mathbf{r}) = -\mathrm{i}k_i \sum_{\ell_i=0}^{\infty} \dfrac{\partial \mathcal{R}_{2\ell_i+|m_i|}^{m_i}(r,k_i)}{\partial r} Y_{2\ell_i+|m_i|}^{m_i}(\theta,\varphi), \\ v_{n_i,\theta}^{m_i}(\mathbf{r}) = -\mathrm{i}k_i \sum_{\ell_i=0}^{\infty} \dfrac{\mathcal{R}_{2\ell_i+|m_i|}^{m_i}(r,k_i)}{r} \dfrac{\partial Y_{2\ell_i+|m_i|}^{m_i}(\theta,\varphi)}{\partial \theta}, \\ v_{n_i,\varphi}^{m_i}(\mathbf{r}) = -\mathrm{i}k_i \dfrac{\mathcal{R}_{2\ell_i+|m_i|}^{m_i}(r,k_i)}{r\sin\theta} \dfrac{\partial Y_{2\ell_i+|m_i|}^{m_i}(\theta,\varphi)}{\partial \varphi}. \end{cases} \tag{2.233}$$

Equations (2.225) and (2.233) represent the SWEs of the ultrasound pressure and particle velocity, respectively, generated by a circular source with a non-axisymmetric profile. Specifically, this profile is separable as shown by Eq. (2.221). It is observed that the ultrasound fields given by Eqs. (2.201) and (2.208) are special cases of Eqs. (2.225) and (2.233), respectively, at the degenerated Zernike mode $m_i = n_i = 0$.

Equations (2.226) and (2.229) provide valuable insights into the physical interpretation of the SWE for ultrasound. For an exterior field location ($r > a$), the ultrasound field consists of an infinite series of outgoing spherical waves characterized by $\mathrm{h}_{2\ell_i+|m_i|}(k_i r) Y_{2\ell_i+|m_i|}^{m_i}(\theta,\varphi)$, with the weight of each spherical wave determined by the aperture effects as $Q_{2\ell_i+|m_i|}^{m_i}(0,a;\mathrm{j})$. In contrast, for an interior location point ($r < a$), the ultrasound field is a superposition of both converging and outgoing spherical waves characterized by $\mathrm{j}_{2\ell_i+|m_i|}(k_i r) Y_{2\ell_i+|m_i|}^{m_i}(\theta,\varphi)$ and $\mathrm{h}_{2\ell_i+|m_i|}(k_i r) Y_{2\ell_i+|m_i|}^{m_i}(\theta,\varphi)$, respectively. The outgoing wave contributes from the source inside the sphere with a radius equal to the radial coordinate of the field location ($0 \le r_\mathrm{s} < r$), while the weight of the converging wave contributes from the source outside this sphere ($r < r_\mathrm{s} \le a$). Consequently, the weights of converging and outgoing waves are $Q_{2\ell_i+|m_i|}^{m_i}(0,r;\mathrm{j})$ and $Q_{2\ell_i+|m_i|}^{m_i}(r,a;\mathrm{h})$, respectively.

2.6.3.2 Audio Sound Field

The source density contributed by the ultrasound with the Zernike profile of mode (n_i, m_i) at f_i is obtained by substituting Eq. (2.225) into Eq. (2.44), as

$$\begin{aligned} q_\mathrm{a}^{n,m}(\mathbf{r}_\mathrm{v}) = \frac{\beta\omega_\mathrm{a}\omega_1^2\omega_2^2}{\mathrm{i}c_0^6} \sum_{\ell_1,\ell_2=0}^{\infty} & \mathcal{R}_{2\ell_1+|m_1|}^{m_1,*}(r_\mathrm{v},k_1)\mathcal{R}_{2\ell_2+|m_2|}^{m_2}(r_\mathrm{v},k_2) \\ & \times Y_{2\ell_1+|m_1|}^{m_1,*}(\theta_\mathrm{v},\varphi_\mathrm{v}) Y_{2\ell_2+|m_2|}^{m_2}(\theta_\mathrm{v},\varphi_\mathrm{v}) \end{aligned} \tag{2.234}$$

where the index sets $n = (n_1, n_2)$ and $m = (m_1, m_2)$ are used for simplicity.

It is evident that the source density described by Eq. (2.234) can be separated into radial and angular components in the spherical coordinate system. By using Eq. (B.5) and the definition of the spherical harmoncis Eq. (B.3), it is clear that the azimuthal component in Eq. (2.234) is $e^{i(m_2-m_1)\varphi_v}$. Building on the derivation in Sec. 2.6.1, the audio sound pressure at mode (n,m) can then be expressed using the SWE given by Eq. (2.192) as

$$p_a^{n,m}(\mathbf{r}) = \rho_0\omega_a k_a \sum_{\ell_a=0}^{\infty} \mathcal{R}_{\ell_a+|m_a|}^{m_a}(r,k_a) Y_{\ell_a+|m_a|}^{m_a}(\theta,\varphi), \tag{2.235}$$

where the azimuthal order for audio sound is

$$m_a = m_2 - m_1. \tag{2.236}$$

The multiple moments can be obtained by substituting Eq. (2.234) into Eq. (2.191) as

$$\begin{aligned} Q_{\ell_a+|m_a|}^{m_a}(r_1,r_2;\mathscr{S}) = {} & \frac{\beta\omega_a\omega_1^2\omega_2^2}{ic_0^6} \\ & \times \sum_{\ell_1,\ell_2=0}^{\infty} I(2\ell_1+|m_1|, 2\ell_2+|m_2|, \ell_a+|m_a|; -m_1, m_2, -m_a) \\ & \times \int_0^{\infty} \mathscr{S}_{\ell_a+|m_a|}(k_a r_a) \mathcal{R}_{2\ell_1+|m_1|}^{m_1,*}(r_v,k_1) \mathcal{R}_{2\ell_2+|m_2|}^{m_2}(r_v,k_2) r_v^2 dr_v. \end{aligned} \tag{2.237}$$

Here, the integral $I(\cdot)$ involving triple spherical harmonics is defined by Eqs. (B.9) and (B.5) is used. By using Eq. (B.9), the triple integral in Eq. (2.237) can be simplified to

$$\begin{aligned} & I(2\ell_1+|m_1|, 2\ell_2+|m_2|, \ell_a+|m_a|; -m_1, m_2, -m_a) \\ & \quad = (-1)^{m_a} \mathscr{G}(2\ell_1+|m_2|, -m_1; 2\ell_2+|m_2|, m_2; \ell_a+|m_a|), \end{aligned} \tag{2.238}$$

where $\mathscr{G}$ is the Gaunt coefficient defined by Eq. (B.15). According to Eqs. (B.16) and (B.17), the Gaunt coefficient in Eq. (2.212) vanishes unless $2\ell_1+|m_1|+2\ell_2+|m_2|+\ell_a+|m_a|$ is an even integer. Therefore, only the terms in Eqs. (2.235) and (2.237) for even ℓ_a are retained. By substituting the original ℓ_a with $2\ell_a$ to eliminate the odd terms, the multipole moments Eq. (2.237) are simplified to

$$\begin{aligned} Q_{2\ell_a+|m_a|}^{m_a}(r_1,r_2;\mathscr{S}) = {} & \frac{\beta\omega_a\omega_1^2\omega_2^2}{ic_0^6}(-1)^{m_a} \\ & \times \sum_{\ell_1,\ell_2=0}^{\infty} \mathscr{G}(2\ell_1+|m_1|, m_1; 2\ell_2+|m_2|, m_2; 2\ell_a+|m_a|) \\ & \times \int_0^{\infty} \mathscr{S}_{2\ell_a+|m_a|}(k_a r_v) \mathcal{R}_{2\ell_1}^{0,*}(r_v,k_1) \mathcal{R}_{2\ell_2}^{0}(r_v,k_2) r_v^2 dr_v. \end{aligned} \tag{2.239}$$

The audio sound pressure Eq. (2.235) is rewritten as

$$p_{\mathrm{a}}^{n,m}(\mathbf{r}) = \rho_0 \omega_{\mathrm{a}} k_{\mathrm{a}} \sum_{\ell_{\mathrm{a}}=0}^{\infty} \mathcal{R}_{2\ell_{\mathrm{a}}+|m_{\mathrm{a}}|}^{m_{\mathrm{a}}}(r, k_{\mathrm{a}}) Y_{2\ell_{\mathrm{a}}+|m_{\mathrm{a}}|}^{m_{\mathrm{a}}}(\theta, \varphi). \tag{2.240}$$

By applying the radial component expression in Eq. (2.186), the audio sound pressure Eq. (2.240) can also be represented as

$$\begin{aligned} p_{\mathrm{a}}^{n,m}(\mathbf{r}) = \rho_0 \omega_{\mathrm{a}} k_{\mathrm{a}} \sum_{\ell_{\mathrm{a}}=0}^{\infty} \Big[& Q_{2\ell_{\mathrm{a}}+|m_{\mathrm{a}}|}^{m_{\mathrm{a}}}(0, r; \mathrm{j}) \mathrm{h}_{2\ell_{\mathrm{a}}+|m_{\mathrm{a}}|}(k_{\mathrm{a}} r) \\ & + Q_{2\ell_{\mathrm{a}}+|m_{\mathrm{a}}|}^{m_{\mathrm{a}}}(r, \infty; \mathrm{h}) \mathrm{j}_{2\ell_{\mathrm{a}}+|m_{\mathrm{a}}|}(k_{\mathrm{a}} r) \Big] Y_{2\ell_{\mathrm{a}}+|m_{\mathrm{a}}|}^{m_{\mathrm{a}}}(\theta, \varphi). \end{aligned} \tag{2.241}$$

Equation (2.241) is the SWE of the audio sound pressure generated by a circular PAL with a Zernike profile. The audio sound is the superposition of an infinite series of outgoing and converging spherical waves characterized by $\mathrm{h}_{2\ell_{\mathrm{a}}+|m_{\mathrm{a}}|}(k_{\mathrm{a}} r) Y_{2\ell_{\mathrm{a}}+|m_{\mathrm{a}}|}^{m_{\mathrm{a}}}(\theta, \varphi)$ or $\mathrm{j}_{2\ell_{\mathrm{a}}+|m_{\mathrm{a}}|}(k_{\mathrm{a}} r) Y_{2\ell_{\mathrm{a}}+|m_{\mathrm{a}}|}^{m_{\mathrm{a}}}(\theta, \varphi)$, respectively. The weight of each converging or outgoing spherical wave is $Q_{2\ell_{\mathrm{a}}+|m_{\mathrm{a}}|}^{m_{\mathrm{a}}}(0, r; \mathrm{j})$ and $Q_{2\ell_{\mathrm{a}}+|m_{\mathrm{a}}|}^{m_{\mathrm{a}}}(r, \infty; \mathrm{h})$, respectively. In other words, the outgoing wave contributes from the virtual audio source inside the sphere with a radius of the radial coordinate of the field location, while the converging wave contributes from the source outside this sphere.

After calculating the audio sound pressure at (n, m) mode given by Eq. (2.240) or Eq. (2.241), the total audio sound pressure can then be obtained as [Zhong et al., 2022b, Eq. (36)]

$$p_{\mathrm{a}}(\mathbf{r}) = \sum_{n=0}^{\infty} \sum_{m=-n}^{n}{}' M_{n_1}^{m_1,*} M_{n_2}^{m_2} p_{\mathrm{a}}^{n,m}(\mathbf{r}). \tag{2.242}$$

In numerical computations, $p_{\mathrm{a}}^{n,m}$ can be calculated first and saved as a table. Then the total audio sound pressure with a specific velocity profile can be obtained by using this table and Zernike moments $M_{n_i}^{m_i}$ obtained by Eq. (2.219).

Far-Field Approximation

In the far field when $r \to \infty$, it is observed from Eq. (2.241) that only the outgoing wave retains since $\lim_{r\to\infty} Q_{2\ell_{\mathrm{a}}+|m_{\mathrm{a}}|}^{m_{\mathrm{a}}}(r, \infty; \mathrm{h}) \to 0$. By using the limiting form of the spherical Hankel function given by Eq. (B.26), the radial component for audio sound in Eq. (2.240) reduces to

$$\mathcal{R}_{2\ell_{\mathrm{a}}+|m_{\mathrm{a}}|}^{m_{\mathrm{a}}}(r, k_{\mathrm{a}}) = \frac{(-1)^{\ell_{\mathrm{a}}} Q_{2\ell_{\mathrm{a}}+|m_{\mathrm{a}}|}^{m_{\mathrm{a}}}(0, \infty; \mathrm{j})}{\mathrm{i} k_{\mathrm{a}} r} \mathrm{e}^{\mathrm{i} k_{\mathrm{a}} r}. \tag{2.243}$$

The audio sound is obtained by substituting Eq. (2.243) into Eq. (2.240) to give

$$p_{\mathrm{a}}^{n,m}(\mathbf{r}) = \frac{\rho_0 \omega_{\mathrm{a}}}{\mathrm{i} r} \mathrm{e}^{\mathrm{i} k_{\mathrm{a}} r} \sum_{\ell_{\mathrm{a}}=0}^{\infty} (-1)^{\ell_{\mathrm{a}}} Q_{2\ell_{\mathrm{a}}+|m_{\mathrm{a}}|}^{m_{\mathrm{a}}}(0, \infty; \mathrm{j}) Y_{2\ell_{\mathrm{a}}+|m_{\mathrm{a}}|}^{m_{\mathrm{a}}}(\theta, \varphi). \tag{2.244}$$

2.6.4 REMARKS

The reader is introduced to the SWE method in this section. The SWE of the audio sound pressure generated by a circular PAL with an axisymmetric profile is given by Eq. (2.214). For a circular PAL with a non-axisymmetric profile, the profile is decomposed into a set of Zernike polynomials weighted by Zernike moments as shown in Eq. (2.218). The audio sound pressure generated by a circular PAL with a Zernike profile can be calculated first by the SWE given by Eq. (2.240) and saved as a table. The radiation from a PAL with an arbitrary profile can then be obtained by Eq. (2.242) after using this table and the corresponding Zernike moments.

The SWE developed in this chapter provides a framework for efficient and accurate calculation of the audio sound field generated by a PAL without any additional approximations. The radiation from a PAL which has other shapes instead of the circular shape can also be calculated using the SWE method. For example for a square PAL with a side length of $2a$, the square can be considered as a sphere with a radius of $\sqrt{2}a$ and setting the velocity profile outside the square to zero. However, the required number of Zernike modes increases rapidly because it is difficult to approximate the abrupt change on the edge of the square. It is noted that the fundamental mechanism of the SWE using Zernike modes is that any loudspeaker membrane can be considered being a vibrating object, which can be described in a modal manner, and the Zernike mode is chosen in this section. There are other options, such as the Bessel-trigonometric functions [Geddes and Lee, 2001; Geddes, 2010].

2.7 OTHER METHODS

In addition to the methods presented in this chapter, there are alternative approaches proposed in literature to calculate the audio sound field generated by a PAL. For example, the finite element method (FEM) has seen significant development in the realm of linear acoustics [Reddy, 2019]. It offers the advantage of handling fields with arbitrary boundary shapes and heterogeneous media and can be extended to nonlinear acoustics as well.

There are two primary ways to employ FEM for calculating the audio sound field produced by a PAL. The first approach involves directly solving the nonlinear wave equations using FEM, which demands substantial computational resources due to the inherent complexity of these equations. The second is to convert the nonlinear wave equation into two coupled linear wave equations using the quasilinear approximation, as illustrated in Sec. 1.4. Subsequently, the linear FEM is used to compute the first-order ultrasound field. This ultrasound field is then utilized as virtual sources to calculate the audio sound pressure. Given that both processes are linear, there is a significant reduction in computational cost.

As an example, in Kagawa et al., 1992, FEM was employed to solve both the Westervelt and Kuznetsov equations under conditions of weak nonlinearity. This approach was also implemented using the commercial software COMSOL Multiphysics [Červenka and Bednařík, 2019]. For instance, it has been effectively used to simulate the behavior of a PAL lensed by a gradient-index phononic crystal, which includes intricate scattering structures [Červenka and Bednařík, 2021]. However, the computational cost can be quite substantial, particularly because the ultrasound wavelength (e.g., 8.6 mm at 40 kHz) is often significantly smaller than the region of interest (e.g., several meters). Consequently, the literature has typically considered only 1D [Dirkse, 2014], 2D [Kagawa et al., 1992], and 3D axisymmetric [Červenka and Bednařík, 2019; Červenka and Bednařík, 2021] cases.

Operator splitting methods (OSMs) have been introduced as a means to solve the Westervelt equation [Yuldashev and Khokhlova, 2011]. The fundamental concept behind OSMs involves breaking down different operators governing various physical processes within the nonlinear wave equation. This division enables the application of distinct numerical methods to each operator, thereby enhancing the efficiency of the overall solving process. When tackling the Westervelt equation, the diffraction, nonlinear, and absorption operators are separated and computed using the angular spectrum method, Runge-Kutta method, and the exact solution for each harmonic in the frequency-domain, respectively [Jimenez et al., 2021]. Other numerical techniques, such as the finite-difference time-domain (FDTD) method, have also been employed [Nomura et al., 2012]. However, the proposed OSMs do not incorporate local effects, resulting in diminished prediction accuracy, especially in the near field.

The k-Wave software, originally designed for nonlinear acoustic simulations in the time domain within arbitrarily heterogeneous media [Treeby and Cox, 2010], primarily focuses on biomedical applications [Treeby et al., 2010; Robertson et al., 2017; Suomi et al., 2018]. Nevertheless, it is based on the Westervelt equation [Treeby et al., 2012], thus offering the potential to simulate audio sound pressure generated by a PAL. Recent applications have utilized k-Wave to calculate the audio sound field produced by a PAL [Shi et al., 2022; Okano and Kajikawa, 2022]. However, k-Wave does not account for local effects, leading to prediction inaccuracies in the near field. Additionally, the k-Wave relies on time-domain methods, which can be computationally expensive when dealing with large-scale problems.

2.8 SUMMARY

Calculating the audio sound field generated by a PAL is a challenging task, primarily due to the intricate nonlinear processes involved. This chapter summarized several commonly used numerical models in literatures. These models include the direct integration method (DIM) in Sec. 2.2, the convolution directivity model (CDM) in Sec. 2.3, the Gaussian beam expansion (GBE) method in Sec. 2.4, the cylindrical wave expansion (CWE) method in Sec. 2.5, the spherical wave expansion (SWE) method in Sec. 2.6, and various other methods in Sec. 2.7. To facilitate a comprehensive understanding, a comparative analysis of these methods is provided in Table 2.2.

In the DIMs, the ultrasound field is represented by a one-fold integral in a 2D physical model and a two-fold integral in a 3D physical model. On the other hand, the audio sound pressure involves a three-fold integral in a 2D physical model and a more complex five-fold integral in a 3D physical model. The mechanism behind the DIMs is to directly compute the ultrasound and audio sound field using numerical

Table 2.2
Comparison of computational methods for calculating the audio sound pressure generated by a PAL.

Model	Dimension	Profile	Accuracy	Computational cost
DIM	2D, 3D	Arbitrary	Full field	Very large
CDM	2D, 3D	Arbitrary	Inverse-law far field	Very small
GBE	2D, 3D	Arbitrary	Paraxial field, high frequencies, without local effects	Small
CWE	2D	Arbitrary	Full field	Moderate
Axisymmetric SWE	3D	Axisymmetric	Full field	Moderate
General SWE	3D	Arbitrary	Full field	Large
FEM	2D, 3D	Arbitrary	Full field	Very large
OSM	2D, 3D	Arbitrary	Without local effects	Large
k-Wave	2D, 3D	Arbitrary	Without local effects	Large

integration techniques, such as Gauss-Legendre quadrature. However, this method faces a challenge due to the highly oscillatory nature of the Green's function within the integrand. As a result, numerical evaluations within the DIMs tend to converge rather slowly. Consequently, the DIMs find limited practical applications due to its slow convergence rate. Nonetheless, it serves as a foundational framework upon which other numerical models are built.

The CDMs stand out as the most computationally efficient model, approximating the ultrasound field through the inward extrapolation of its far-field pressure. These derived expressions are subsequently formulated as linear and spherical convolutions of ultrasound directivity and the Westervelt's directivity for 2D and 3D models, respectively. While the CDMs are known for their computational efficiency, it is important to note that their applicability is limited to the calculation of the far-field directivity of the audio sound generated by a PAL.

The GBE methods simplify the source velocity profiles by employing multiple Gaussian functions, streamlining the integral calculations required for both ultrasound and audio sound computation, when assuming the paraxial approximation. It is able to calculate the audio sound pressure in the near field. Nonetheless, it is crucial to acknowledge that the derivation of the GBEs relies on the paraxial approximation, which can result in less accurate predictions near the PAL or beyond the paraxial region. This diminished accuracy becomes more pronounced when dealing with small aperture sizes and low audio and ultrasound frequencies.

The CWE method is equivalent to the 2D DIM without additional approximations but with a relatively low computational load. Its core technique involves converting the 2D Green's function into cylindrical waves. On the other hand, the SWE method is equivalent to a 3D DIM, also without additional approximations. It achieves this by rigorously transforming the 3D DIM into spherical waves, leveraging the spherical representations of 3D Green's functions. The SWE method is computationally efficient when dealing with a circular PAL featuring an axisymmetric profile. However, for PALs with non-axisymmetric profiles, these profiles are decomposed into multiple Zernike modes. As the number of required Zernike modes increases, the computational demands of the SWE method also rise accordingly.

In conclusion, there exist several methods for computing the sound fields produced by a PAL, each with its own set of advantages and drawbacks. It is essential to recognize that the approaches introduced in this chapter are applicable exclusively within the framework of the quasilinear approximation, assuming weak nonlinearity. In practical scenarios, substantial excitations, such as on-surface pressure amplitudes exceeding 130 dB, may be encountered to enhance the frequency response of the PAL [Ji et al., 2011; Zhong et al., 2022d]. In such situations, the nonlinearity becomes strong, challenging the validity of the quasilinear approximation, as discussed in Sec. 1.4.3. In these cases, a direct solution of the nonlinear wave equations presented in Sec. 1.3 is necessary to obtain numerical solutions.

3 Sound Fields Generated by PALs in Free Space

3.1 INTRODUCTION

A parametric array loudspeaker (PAL) generates audio sound through second-order nonlinear interactions among intense ultrasonic waves, distinguishing it from conventional loudspeakers where the linear acoustics is utilized. Consequently, PALs generate audio sound fields with distinct characteristics. This chapter delves into the analysis of audio sound fields created by PALs in free space.

Figure 3.1 illustrates the sound field regions for both a conventional loudspeaker and a PAL in free space. In the case of a conventional loudspeaker, the distinction between the near and far fields lies in whether the sound pressure amplitude follows an inverse relationship with the propagating distance [Foote, 2014; Pierce, 2019]. This characteristic leads to the term *inverse-law far field* for the far-field region. In this far-field region, the sound pressure produced by a conventional loudspeaker can be expressed as follows

$$p_{\mathrm{a}}(\mathbf{r}) = \frac{\rho_0 \omega_{\mathrm{a}} Q_0}{4\pi \mathrm{i} r} \mathrm{e}^{\mathrm{i}kr} \mathcal{D}_{\mathrm{a}}(\theta, \varphi). \tag{3.1}$$

Here, i is the imaginary unit, Q_0 represents an equivalent volume velocity, ω_{a} is the angular frequency, r is the radial propagating distance, and $\hat{\mathbf{r}} = (\theta, \varphi)$ denote the direction with zenithal (θ) and azimuthal (φ) angles, with $\mathcal{D}_{\mathrm{a}}(\hat{\mathbf{r}})$ representing the directivity.

Calculating the sound pressure in the near field is more complex and cannot be easily solved through a closed-form solution like Eq. (3.1). Instead, when the source is baffled, it necessitates a numerical evaluation of the Rayleigh integral, as defined by Eq. (2.38). The transition from the near field to the far field is denoted by R_0 in Fig. 3.1(a). This transition distance is typically determined by the Rayleigh distance, expressed as [Foote, 2014]

$$\mathscr{D}_{\mathrm{R}}(\omega) \equiv \frac{\mathcal{A}_0}{\lambda}, \tag{3.2}$$

where $\mathcal{A}_0$ is the radiation area, λ is the wavelength at frequency f, and $\omega = 2\pi f$ is the angular frequency. For a circular source with a radius of a, $\mathcal{A}_0 = \pi a^2$, and the Rayleigh distance has the form of $\mathscr{D}_{\mathrm{R}}(\omega) = \pi a^2/\lambda = \pi \mathscr{D}_{\mathrm{c}}(\omega)$. Here, the *critical distance* is defined as the point at which the on-axis sound pressure amplitude reaches its ultimate maximum, and it is expressed as

$$\mathscr{D}_{\mathrm{c}}(\omega) = \frac{a^2}{\lambda}. \tag{3.3}$$

DOI: 10.1201/9781003354994-3

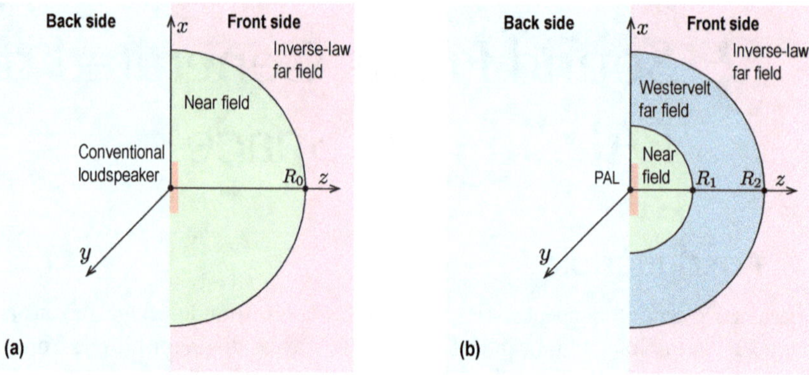

Figure 3.1 Schematic of sound field regions for (a) a conventional loudspeaker and (b) a PAL in free space.

The generation of audio sound by a PAL is a intricate process, resulting in a sound field in front of the PAL that can be effectively categorized into three distinct regions: the *inverse-law far field*, the *Westervelt far field*, and the *near field* [Zhong et al., 2021b]. By segmenting the sound field into these regions, it becomes possible to select appropriate models for each one, facilitating rapid and precise calculations of the sound fields. In Fig. 3.1(b), the transition distance from the near field to the Westervelt far field is denoted by R_1, while that from the Westervelt far field to the inverse-law far field is denoted by R_2. Further details regarding these transitions are presented in Sec. 3.2.

In the inverse-law far field, the sound field can be effectively characterized using the inverse-law equation, similar to Eq. (3.1), making it the simplest region to model. Westervelt derived a closed-form formula for audio sound in the inverse-law far field [Westervelt, 1963]. This derivation was based on the assumption that ultrasound is collimated, and nonlinear ultrasound interactions occur only within a limited aperture area. This closed-form formula is now widely known as the *Westervelt's directivity*, as detailed in Sec. 2.2.4.

Berktay [Berktay, 1965a] and Berktay and Leahy [Berktay and Leahy, 1974] subsequently enhanced the Westervelt's directivity by considering the cylindrical and/or spherical spreading of ultrasonic waves and introducing an aperture factor for the transducer. They also incorporated the product directivity of ultrasonic waves to enhance prediction accuracy. The Berktay solution is expressed in a simple form in the time domain and serves as the foundation for signal modulation techniques in the implementation of PALs. To further enhance prediction accuracy for the sidelobes of PALs, various modifications to the Berktay model were proposed later [Shi and Gan, 2012]. A more precise model involves the convolution of the Westervelt's directivity and the directivities of ultrasonic waves [Mellen and Moffett, 1978; Moffett and Mellen, 1981; Shi and Kajikawa, 2015a; Guasch and Sánchez-Martín, 2018], which is referred to as the convolution directivity model (CDM) in Sec. 2.3. The prediction accuracy of the CDM is further improved in 2023 [Zhong et al., 2023d]. While the audio sound pressure is described simply in the inverse-law far field, it is important to note that the boundary of the inverse-law field is typically located at a significant distance from the transducer, which may render it impractical for certain applications [Zhong et al., 2021b].

The Westervelt far field is defined as the region where the Westervelt equation [Eq. (1.79)] provides accurate predictions, indicating that the local effects characterized by the ultrasonic Lagrangian density are negligible. Consequently, the Westervelt far field is sometimes referred to as the near field when local effects are not taken into account. Under the quasilinear approximation, audio sound can be conceptualized as radiation originating from an infinitely large virtual volume source, with the source density being proportional to the product of the ultrasonic pressure.

In earlier studies, ultrasonic beams were assumed to exhibit spherical spreading with a specific directivity function [Mellen and Moffett, 1978; Moffett and Mellen, 1981]. However, recent research has endeavored to model the nonlinear interactions of actual ultrasonic beams generated by a transducer to enhance prediction accuracy [Zhong et al., 2020b; Zhong et al., 2022b; Zhong et al., 2020c]. Given that the ultrasonic wavelength is typically much smaller than the PAL radius, the paraxial (Fresnel) approximation is commonly applied to ultrasonic waves. This approximation allows for the utilization of a Gaussian beam expansion (GBE) method, as introduced in Sec. 2.4 [Červenka and Bednařík, 2013]. When both ultrasonic and audio waves are assumed to adhere to the paraxial approximation, the Westervelt equation simplifies to the well-known Khokhlov-Zabolotskaya-Kuznetsov (KZK) equation, achieved by approximating a second-order derivative of sound pressure concerning the propagating direction with a first-order derivative. It is important to note that this simplified calculation is accurate only within the paraxial region, which is typically situated within approximately 20 degrees from the transducer axis [Hamilton and Blackstock, 2008]. Recent studies have shown that even within the paraxial region, the predicted results are inaccurate in the near field [Zhuang et al., 2023]. The audio sound field in the Westervelt far field garners more attention than that in the inverse-law far field, and it is within this region that most practical applications are considered.

In the near field of PALs, it becomes imperative not to disregard the ultrasonic Lagrangian density, leading to a more intricate distribution of audio sound pressure. Similar to the near field of conventional loudspeakers, local maxima and minima in sound pressure are observed. While the general second-order nonlinear wave equation [Eq. (1.69)] remains accurate for modeling audio sound in the near field, the Kuznetsov equation [Eq. (1.78)], an equivalent form expressed in terms of the velocity potential, is often preferred to avoid the evaluation of second-order spatial derivatives of the ultrasonic Lagrangian density [Aanonsen et al., 1984; Kagawa et al., 1992; Červenka and Bednařík, 2019]. It is worth noting that calculating the quasilinear solution of the Kuznetsov equation can be quite time-consuming. Consequently, it is relatively uncommon to employ this equation for computing audio sound in the near field of PALs. An alternative approach involves modifying the result obtained using the Westervelt equation with an algebraic correction, as illustrated in

Eq. (1.107) [Červenka and Bednařík, 2022]. Nonetheless, this method necessitates knowledge of the particle velocity field of ultrasound. As a result, it is often more convenient to segment the sound pressure field and apply distinct models to different regions. This division is driven by the intricacies and challenges inherent in calculating the audio sound in the near field generated by PALs.

The audio sound generated behind a conventional loudspeaker is generally less prominent and often of minimal interest compared to the sound in front of it. However, at low frequencies, the diffractive characteristics of these waves can render the sound behind a conventional loudspeaker non-negligible [Aarts and Janssen, 2010; Beranek and Mellow, 2019; Lee and Park, 2023]. In contrast, for a PAL, the sound field behind the device is typically considered to be inaudible due to the inefficient diffraction of ultrasound waves. Recent research has demonstrated that the audio sound pressure level (SPL) behind a PAL is typically at least 10 dB lower than that in front of the transducer [Zhong et al., 2020b]. Further details on this topic are provided in Sec. 3.3.

Sound power is an important metric used to measure the rate at which sound energy is emitted by an acoustic source [Pierce, 2019]. Accurately predicting the audio sound power of a PAL is important for gaining deeper insights into its radiation characteristics, as discussed in Sec. 3.4. The sound power is determined by integrating the sound intensity over an envelope that encloses the sound source. In this context, sound intensity is defined as the real part of the product of sound pressure and the conjugate of particle velocity. Given the complexity of the near field and its challenging calculations, the envelope is typically selected in the far field. In the far field, sound waves can be approximated as spherical spreading with a specific directivity pattern. Consequently, the sound pressure can be described in terms of directivity, while the particle velocity only exhibits a radial component, displaying a straightforward linear relationship with sound pressure. Within this framework, obtaining the directivity of the audio sound represents the initial step in calculating the sound power.

Westervelt initially derived a closed-form expression for sound power utilizing the Westervelt's directivity [Westervelt, 1963]. This expression reveals that audio sound power is directly proportional to the fourth power of the on-surface pressure amplitude, the third power of audio frequency, and the fourth power of aperture size, while inversely proportional to the square of ultrasound frequency. This stands in contrast to sound power for a linear acoustic source, which is proportional to the square of the on-surface pressure amplitude, the square of the frequency, and the 4th power of the aperture size when the wavelength of sound is significantly larger than the aperture size. Conversely, it is nearly independent of frequency and proportional to the square of the aperture size when the wavelength of sound is much smaller than the aperture size. The sound power prediction provided by Westervelt [Westervelt, 1963] is known to be inaccurate due to significant inaccuracies present in the Westervelt's directivity. The predication accuracy can be improved by using advanced directivity calculation method such as the CDM introduced in Sec. 2.3. This model forms the foundation for calculating the far-field directivity of a PAL with an arbitrary profile.

3.2 SOUND FIELD IN FRONT SIDE OF A BAFFLED PAL

The sound field in front of a loudspeaker typically holds greater significance than that behind it. This section commences by conducting a comparative analysis of sound pressure across various regions situated in front of a PAL. Following this, this section delves into an examination of the transition distances separating these regions. Finally, the sound pressure distributions of PALs with typical configurations are presented and compared against those of conventional loudspeakers.

Figure 3.2 shows the 3D physical model for a baffled circular PAL with a radius of a to be investigated in this section. Cartesian (x, y, z), cylindrical (ρ, φ, z), and spherical (r, θ, φ) coordinate systems are established with their origin, O, at the centroid of the PAL and the positive z-axis pointing to the radiation direction, where r, θ, and φ are the radial, zenithal, and azimuthal coordinates, respectively.

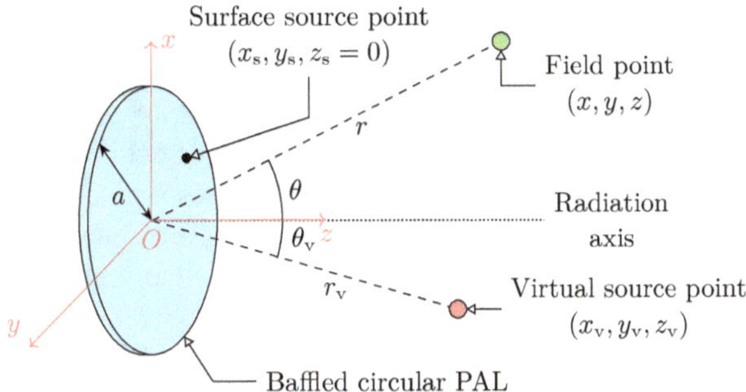

Figure 3.2 Sketch of the 3D physical model for a baffled circular PAL. Extracted from [Zhong et al., 2020c, Fig. 1].

3.2.1 NEAR FIELD, WESTERVELT FAR FIELD, AND INVERSE-LAW FAR FIELD

Figure 3.3 illustrates the audio sound pressure level (SPL) generated by a PAL as a function of the propagation distance at 1 kHz. The curves labeled by "Kuznetsov" and "Westervelt" represent quasilinear solutions derived from the Kuznetsov and Westervelt equations in the frequency domain, as detailed in Sec. 1.4.2. For numerical computation, the spherical wave expansion (SWE) method is used, as detailed in Sec. 2.6. The curve marked by "Inverse-law" is obtained through an extrapolation of the inverse-law formula, following the SWE approach described by Eq. (2.217).

In the near field, where the audio sound is significantly influenced by strong local effects characterized by the ultrasonic Lagrangian density, the use of the Kuznetsov equation is essential for accurate predictions of the audio SPL. However, when examining Figs. 3.3(a) and (b), respectively, it becomes evident that for radial distances exceeding 0.42 m and 0.19 m, respectively, the disparity between the curves labeled "Kuznetsov" and "Westervelt" is less than 0.1 dB. Therefore, in the Westervelt far field, one can effectively employ the Westervelt equation to predict the audio sound.

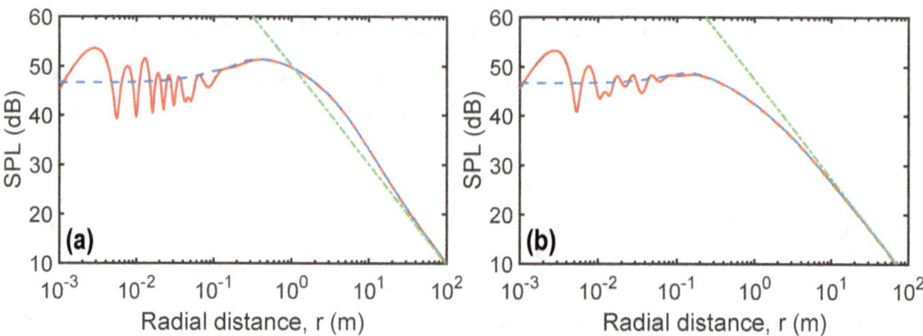

Figure 3.3 Audio SPL generated by a baffled circular PAL with a radius of 5 cm as a function of the propagating (radial) distance at 1 kHz calculated with various methods. (a) On the radiator axis (0°). (b) In the direction of a zenithal angle of 10°. The average ultrasound frequency is 40 kHz and the corresponding sound attenuation coefficients are 2.8×10^{-2} Np/m. ——, Kuznetsov equation; - - - , Westervelt equation; — - —, inverse-law. Extracted from [Zhong et al., 2021b, Fig. 4].

Moving further to the inverse-law far field, this region is defined by radial distances greater than 28.7 m and 7.3 m in Figs. 3.3(a) and (b), respectively. Here, the distinction between the curves marked "Westervelt" and "Inverse-law" is less than 1 dB, indicating that the Westervelt equation can also be suitably utilized to predict the audio sound in this region.

For comparative purposes, Fig. 3.4 displays the audio SPL as a function of the propagation distance at 4 kHz, generated by a circular conventional loudspeaker with a radius of 0.1 m and a uniform profile (referred to as a piston source) defined by Eq. (2.29). The results reveal that, similar to a PAL, the sound pressure in the near field exhibits local maxima and minima, resulting from intricate wave interference within this region. As moving farther away from the loudspeaker, the exact solution converges toward the inward-extrapolated far-field pressure, which follows an inverse-law relationship with distance. The critical distance, $\mathscr{D}_{\mathrm{c}}(\omega_{\mathrm{a}})$, defined by Eq. (3.3), indicating where the SPL reaches its final maximum, is depicted in Fig. 3.4. Here, $\omega_{\mathrm{a}} = 2\pi f_{\mathrm{a}}$ represents the angular frequency of audio sound. Additionally, the Rayleigh distance, $\mathscr{D}_{\mathrm{R}}(\omega_{\mathrm{a}})$, given by Eq. (3.2), is also depicted in Fig. 3.4.

Beyond the critical distance, the discrepancy between the audio SPL obtained from the exact solution and the inward-extrapolated far-field pressure is less than 5.6 dB, considering the parameters employed in Fig. 3.4. As the observation point continues to recede, the sound pressure amplitude consistently diminishes, eventually approaching convergence with the inward-extrapolated far-field pressure. Furthermore, beyond the Rayleigh distance, the variance in audio SPL remains under 0.5 dB. Consequently, both the critical and Rayleigh distances can effectively serve as transition points from the near field to the far field for a conventional loudspeaker [Foote, 2014].

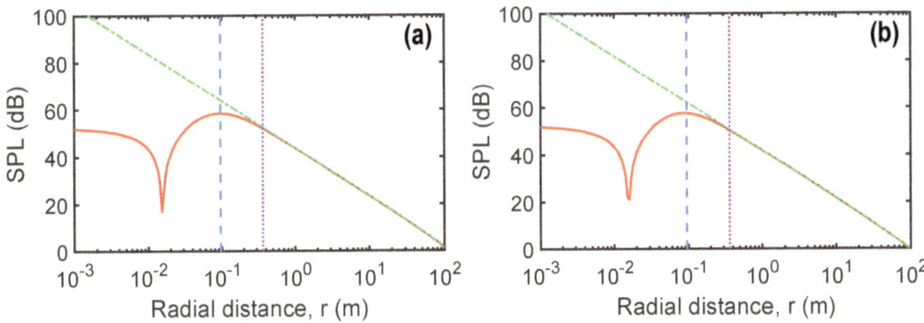

Figure 3.4 Audio SPL generated by a baffled circular conventional loudspeaker with a radius of 10 cm as a function of the propagating (radial) distance at 4 kHz a radius of 10 cm. (a) On the radiator axis (0°). (b) In the direction of a zenithal angle of 10°. The on-surface pressure pressure amplitude is $p_0 =$ 0.1 Pa (71 dB). ——, exact solution; – – –, inward-extrapolated far-field pressure. - - -, critical distance (9.52 cm); ········, Rayleigh distance (36.64 cm).

3.2.2 TRANSITION DISTANCE FROM THE NEAR FIELD TO THE WESTERVELT FAR FIELD

To better understand the sound field in the near field and the Westervelt far field of a PAL, it is important to determine the transition distance between them. Figure 3.5 illustrates the audio SPL at 1 kHz and the ultrasound field at 40 kHz on the radiator axis as a function of the propagating distance for a transducer with a radius of 0.05 m. The ultrasound pressure can be calculated with the closed-form formula as [Pierce, 2019, Eq. (5.7.3)]

$$p_{\mathrm{u}}(\mathbf{r}) = -2\mathrm{i}p_0 \exp\left[\mathrm{i}\frac{k_{\mathrm{u}}a}{2}\left(\sqrt{1+\frac{z^2}{a^2}}+\frac{z}{a}\right)\right] \sin\left[\frac{k_{\mathrm{u}}a}{2}\left(\sqrt{1+\frac{z^2}{a^2}}-\frac{z}{a}\right)\right], \tag{3.4}$$

where $p_0 = \rho_0 c_0 v_0$ is the on-surface pressure amplitude. The ultrasonic Lagrangian density is obtained by

$$\mathcal{L}_{\mathrm{u}}(\mathbf{r}) = \frac{\rho_0}{2} v_{\mathrm{u},z}^2(\mathbf{r}) - \frac{p_{\mathrm{u}}^2(\mathbf{r})}{2\rho_0 c_0^2} \tag{3.5}$$

where the particle velocity components in x- and y- directions are zero, and $v_{\mathrm{u},z}(\mathbf{r})$ is the particle velocity component in the z-direction and can be obtained by the relation $v_{\mathrm{u},z}(\mathbf{r}) = (\mathrm{i}\rho_0\omega_{\mathrm{u}})^{-1}\,\partial p_{\mathrm{u}}(\mathbf{r})/\partial z$ as

$$v_{\mathrm{u},z}(\mathbf{r}) = v_0 \mathrm{e}^{\mathrm{i}k_{\mathrm{u}}z} - v_0\frac{z}{a}\left(1+\frac{z^2}{a^2}\right)^{-1/2} \exp\left(\mathrm{i}k_{\mathrm{u}}a\sqrt{1+\frac{z^2}{a^2}}\right). \tag{3.6}$$

The obtained ultrasound pressure and Lagrangian density on the radiator axis are then normalized by $2\rho_0 c_0 v_0$ and $2\rho_0 v_0$, respectively, and are shown in Fig. 3.5.

The ultrasound pressure amplitude has several local minima and maxima in the near field. From Eq. (3.4), the radial distance at the local minima can be obtained by

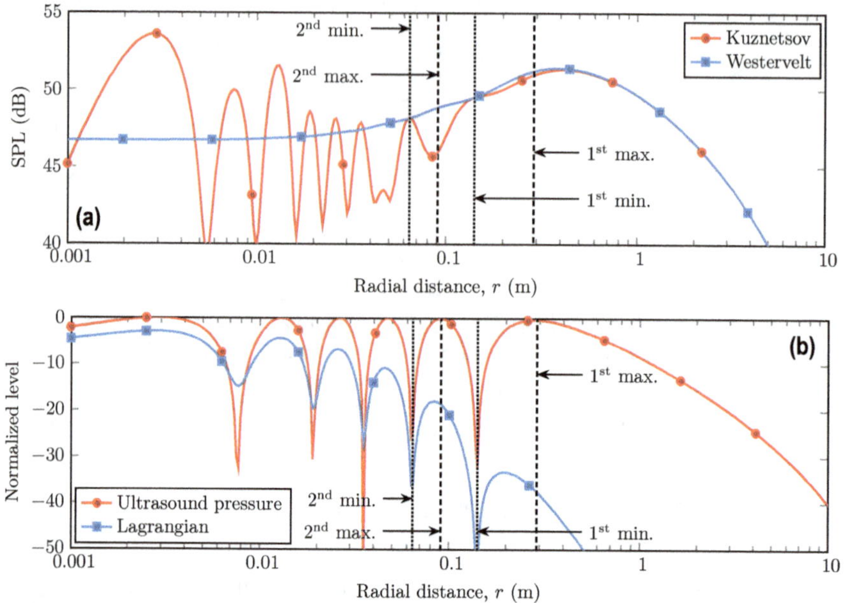

Figure 3.5 (a) The audio SPL at 1 kHz and (b) the level of normalized ultrasound pressure and Lagrangian density at the average ultrasound frequency 40 kHz, as a function of the propagating (radial) distance on the radiator axis. The transducer radius is 0.05 m. Extracted from [Zhong et al., 2021b, Fig. 5].

[Pierce, 2019, Eq. (5.7.4)]

$$r_{\min}(n) = \left[\left(\frac{a}{\lambda_u}\right)^2 - n^2\right]\frac{\lambda_u}{2n}, \quad n = 1, 2, \ldots, \left\lfloor \frac{a}{\lambda_u} \right\rfloor \tag{3.7}$$

where λ_u is the wavelength at the average ultrasound frequency f_u and $\lfloor \cdot \rfloor$ rounds down the quantity inside. Similarly, the radial distance at the local maxima can be obtained by

$$r_{\max}(n) = \left[\left(\frac{a}{\lambda_u}\right)^2 - \left(n - \frac{1}{2}\right)^2\right]\frac{\lambda_u}{2n-1}, \quad n = 1, 2, \ldots, \left\lfloor \frac{a}{\lambda_u} + \frac{1}{2} \right\rfloor \tag{3.8}$$

The first two (n = 1 and 2) local minima and maxima are plotted in Fig. 3.5. The locations of local maxima and minima in the ultrasonic Lagrangian density appear to be closely related to those of the ultrasound pressure.

The non-cumulative nature of the ultrasonic Lagrangian density during the propagation of ultrasound beams has been demonstrated [Aanonsen et al., 1984]. However, in the near field, where the observation point is in close proximity to a PAL, the ultrasonic Lagrangian density displays notable fluctuations, giving rise to a complex audio sound field, as exemplified in Fig. 3.5(a) using the Kuznetsov equation. In this context, the corresponding outcomes derived from the Westervelt equation are notably inaccurate. Conversely, Fig. 3.5(b) illustrates that the ultrasonic Lagrangian density diminishes beyond the initial local maximum (positioned at a radial distance of 0.29 m in this instance), and the results obtained via the Westervelt equation in Fig.3.5(a) align sufficiently with accuracy.

The presence of local maxima and minima in the ultrasound pressure amplitude has a notable impact on the transition from the near field to the Westervelt far field. In Fig. 3.5(b), it is evident that the normalized Lagrangian density at the first two local minima is more than 30 dB lower compared to its proximity to the PAL. Consequently, these minima can be safely disregarded when calculating audio sounds, resulting in nearly identical outcomes from both the Kuznetsov and Westervelt equations. However, at the positions of the first two local maxima, where the Lagrangian density amplitude is close to its peak, their effects become more pronounced, leading to significant discrepancies between the results obtained using the Kuznetsov and Westervelt equations. As one progresses down the sequence of local maxima, this difference diminishes. For instance, at the second local maximum (0.09 m), the difference is 3.0 dB, while at the first local maximum (0.29 m), it reduces to only 0.3 dB.

Figures 3.6 illustrates the difference of the audio SPL using the Kuznetsov and Westervelt equations at different ultrasound frequencies, transducer radii, and distances from the source, all at an audio frequency of 1 kHz. The radial distance to the first maximum of the ultrasound pressure amplitude is obtained from Eq. (3.8) with $n = 1$ and listed in Table 3.1, yielding the expression

$$r_{\max}(1) = \frac{a^2}{\lambda_u} - \frac{\lambda_u}{4}. \tag{3.9}$$

As shown in the figure, the difference between the two equations is negligible at large distances, where the Westervelt equation is sufficiently accurate. Conversely, at small distances, the Kuznetsov equation introduces discrepancies due to the presence of an increasing number of local minima and maxima as the ultrasound frequency and the transducer radius increase, as predicted by Eqs. (3.7) and (3.8).

The distance at the n-th local maximum of the ultrasound pressure amplitude increases as the transducer radius and ultrasound frequency increase, where n is any positive integer number restricted by the condition in Eq. (3.8). The audio SPL difference at the location of the corresponding local maximum decreases as the transducer radius and ultrasound frequency increase. This is because the ultrasound beam is more collimated when the transducer radius and the ultrasound frequency are larger. The ultrasound beams can also be approximated by plane waves when they are highly collimated. In this case, the ultrasonic Lagrangian density $\mathcal{L}$ approaches zero after substituting the plane wave condition $p_u = \rho_0 c_0 |v_{u,z}|$ into Eq. (3.5), which means the local effects are negligible and the Westervelt equation is accurate. Therefore, the magnitude of the audio SPL difference can be determined by how much the ultrasound beams behave like plane waves, which is measured by defining an error

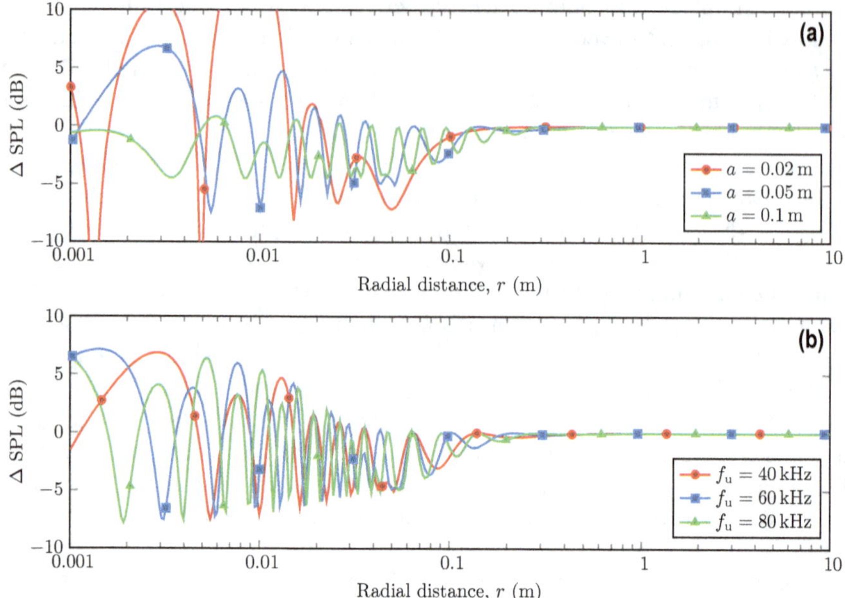

Figure 3.6 The audio SPL difference calculated with the Kuznetsov and Westervelt equations as a function of the propagating (radial) distance on the radiator axis (a) with different transducer radii when the average ultrasound frequency is 40 kHz, and (b) at different average ultrasound frequencies when the transducer radius is 0.05 m, where the audio frequency is 1 kHz. Extracted from [Zhong et al., 2021b, Fig. 6].

function as

$$\varepsilon(z) = \left[1 - \frac{\rho_0 c_0 |v_{u,z}|}{|p_u|}\right] \times 100\% \tag{3.10}$$

If the error function is small, the ultrasound beams are more collimated and the effects of the ultrasonic Lagrangian density would also be small.

Because the audio SPL difference calculated with the two equations is large at points near the local maxima of the ultrasound pressure amplitude, the transition distance from the near field to the Westervelt far field can be defined as the distance at the n_0-th local maximum of the ultrasound pressure amplitude, such that the error function at this point is less than a threshold $\varepsilon_0 > 0$. To obtain n_0, the following condition should be satisfied as

$$\varepsilon[r_{\max}(n_0)] \leq \varepsilon_0 \tag{3.11}$$

After substitution of Eqs. (3.8) and (3.10) into Eq. (3.11), n_0 is obtained by

$$n_0 = \max\left(1, \left\lfloor \frac{a}{\lambda_u} \frac{1}{\sqrt{\varepsilon_0^{-1} - 1}} + \frac{1}{2} \right\rfloor\right) \tag{3.12}$$

Table 3.1

The first maxima of the ultrasound pressure amplitude and the transition distances from the near field to the Westervelt far field for several sets of parameters.

Transducer radius a (m)	Ultrasound frequency f_u (kHz)	First maximum $r_{max}(1)$ (m)	Transition distance from the near field to the Westervelt far field (m)
0.02	40	0.04	0.04 ($n_0 = 1$)
0.05	40	0.29	0.29 ($n_0 = 1$)
0.1	40	1.16	0.38 ($n_0 = 2$)
0.15	40	2.62	0.51 ($n_0 = 3$)
0.2	40	4.67	0.65 ($n_0 = 4$)
0.25	40	7.29	0.79 ($n_0 = 5$)
0.05	60	0.44	0.44 ($n_0 = 1$)
0.05	80	0.58	0.19 ($n_0 = 2$)
0.05	100	0.73	0.24 ($n_0 = 2$)

where $\max(1, \cdot)$ is used to ensure n_0 is at least 1. The choice of a smaller threshold for ε_0 leads to higher precision when using the Westervelt equation to predict audio sounds in the Westervelt far field, region $r \geq r_{max}(n_0)$. The transition distance at several sets of parameters are listed in Table 3.2 when $\varepsilon_0 = 2.5\%$, and the numerical simulations show the error using the Westervelt equation is less than 0.6 dB under this condition.

Figure 3.7 shows the audio SPL difference at different audio frequencies when the transducer radius is 0.05 m, and the average ultrasound frequency is 40 kHz. At high audio frequencies, the audio SPL difference is small at small radial distances. This is because the audio SPL calculated with the Westervelt equation increases by about 12 dB when the audio frequency is doubled, but the amplitude of the ultrasonic Lagrangian density changes little at different audio frequencies, so its effect on the audio SPL is relatively small at high audio frequencies. The locations of local minima and maxima in the ultrasound pressure amplitude do not change with audio frequencies, so the transition distance from the near field to the Westervelt far field does not vary with the audio frequency.

This section derived the formula for the transition distance from the near field to the Westervelt far field for a transducer with a uniform velocity profile. For other velocity profiles, such as parabolic and quartic ones [Červenka and Bednařík, 2019], a similar method can be applied to obtain an appropriate formula. The crucial step in obtaining the formula for a specific velocity profile is to determine the location of the local maxima of the ultrasound pressure amplitude on the transducer axis, which is a linear problem to solve. At these locations, the amplitude of the Lagrangian density is large, and the local effects are significant.

Table 3.2
The transition distance from the Westervelt far field to the inverse-law far field for the parameters.

Transducer radius a (m)	Ultrasound frequency f_u (kHz)	Audio frequency f_a (Hz)	Inverse-law transition distance (m) when ΔSPL < 1 dB
0.02	40	1000	10.6
0.05	40	1000	29.1
0.1	40	1000	31.8
0.15	40	1000	33.0
0.05	60	1000	17.8
0.05	80	1000	12.8
0.05	100	1000	10.2
0.05	40	250	32.3
0.05	40	500	30.6
0.05	40	2000	23.9

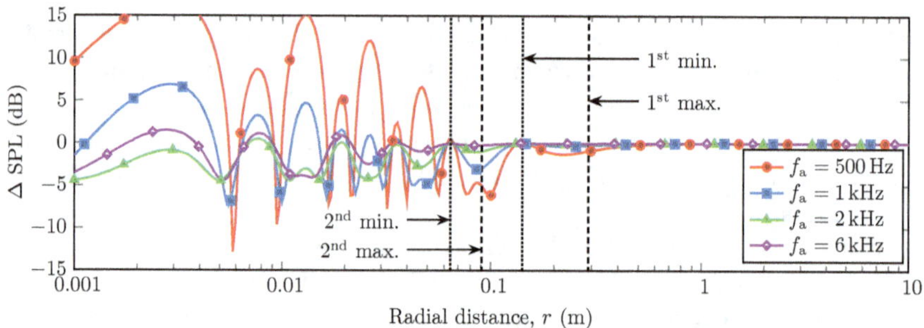

Figure 3.7 The audio SPL difference calculated with the Kuznetsov and Westervelt equations as a function of the propagating (radial) distance on the radiator axis at different audio frequencies (transducer radius is 0.05 m and the average ultrasound frequency is 40 kHz). Extracted from [Zhong et al., 2021b, Fig. 7].

3.2.3 TRANSITION DISTANCE FROM THE WESTERVELT FAR FIELD TO THE INVERSE-LAW FAR FIELD

Figure 3.8 depicts the difference between the audio SPL calculated with the Westervelt equation and the inward-extrapolated inverse-law far-field pressure as a function of the propagating distance on the radiator axis for various transducer radii and ultrasound and audio frequencies. As the radial distance increases, this difference first increases and then decreases, reaching 0 dB at large radial distances where the inverse-law prediction is sufficiently accurate. Defining an error bound as 1 dB, the region where the audio SPL difference is less than 1 dB is identified as the inverse-law far field. The transition distances from the Westervelt far field to the inverse-law far field for the parameters used in Fig. 3.8 are listed in Table 3.2.

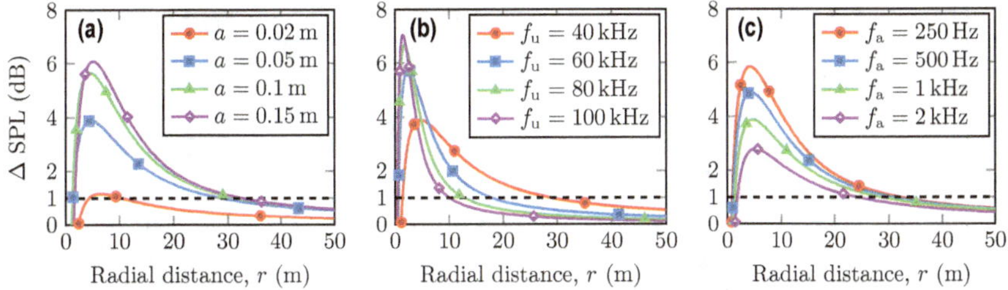

Figure 3.8 The audio SPL difference calculated with the Westervelt equation and the inverse-law property as a function of the propagating (radial) distance on the radiator axis (a) with different transducer radii when the average ultrasound frequency is 40 kHz and the audio frequency is 1 kHz, (b) at different average ultrasound frequencies when the transducer radius is 0.05 m and the audio frequency is 1 kHz, and (c) at different audio frequencies when the transducer radius is 0.05 m and the average ultrasound frequency is 40 kHz, where ΔSPL = 1 dB for dashed lines (- - -). Extracted from [Zhong et al., 2021b, Fig. 8].

In Fig. 3.8(a), it can be observed that the transition distance increases with the transducer radius. For instance, as the transducer radius increases from 0.02 m to 0.05 m, the transition distance increases from 10.6 m to 29.1 m. This is because the effective virtual source that contains most of the ultrasonic energy becomes larger with the increase in transducer radius. On the other hand, Fig. 3.8(b) reveals that the transition distance decreases with the increase in ultrasound frequency. For example, as the ultrasound frequency increases from 40 kHz to 60 kHz, the transition distance decreases from 29.1 m to 17.8 m. This is due to the decrease in size of the effective virtual source as the sound attenuation coefficient of ultrasound beams in air increases. Although there is a small decrease in the transition distance as the audio frequency increases, as shown in Fig. 3.8(c), the effects are relatively minor. For instance, when the audio frequency increases from 500 Hz to 1 kHz, the transition distance reduces by only 1.5 m or 4.9%.

The effects of the transducer radius, ultrasound frequency, and audio frequency on the inverse-law transition distance are more complicated than those on the transition distance from the near field to the Westervelt far field. The ultrasound frequency appears to be the most critical parameter due to significant changes in the ultrasound attenuation coefficient in air with frequency and meteorological conditions. As a result, an empirical formula, such as the one below, can be used to estimate the inverse-law transition distance

$$R_2 = 4\mathscr{D}_{\text{ab}}(\omega_{\text{u}}), \tag{3.13}$$

where $\mathscr{D}_{\text{ab}}$, defined by Eq. (1.114), is the absorption distance for ultrasound. The physical meaning of this formula is that the ultrasound pressure amplitude at this location has been attenuated to 2% ($e^{-4} \approx 0.02$). However, the formula does not hold for the very small sound absorption coefficient.

3.2.4 SOUND FIELD DISTRIBUTIONS

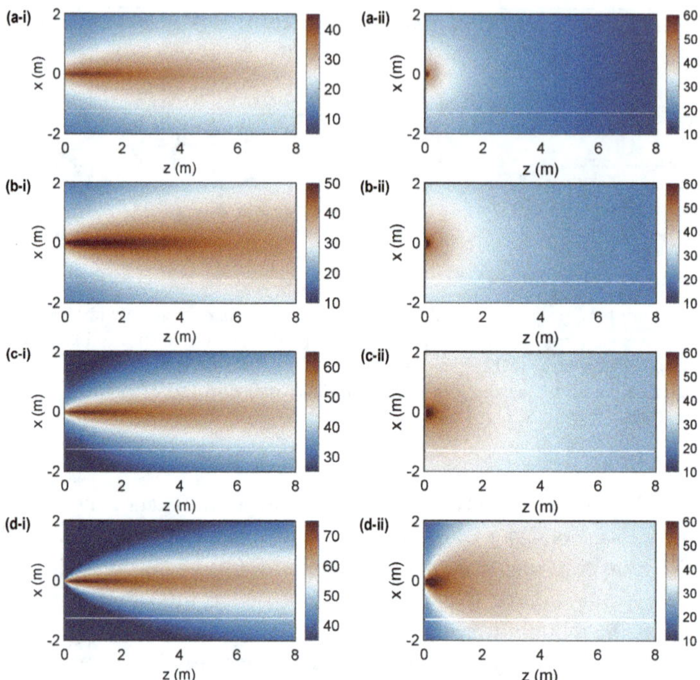

Figure 3.9 Audio SPL distributions generated by a baffled circular (a) PAL and (ii) conventional loudspeaker with a radius of 0.05 m. The velocity profile is uniform in both cases. The audio frequency is (a) 500 Hz, (b) 1 kHz, (c) 2 kHz, and (d) 4 kHz. In (i), the average ultrasound frequency is $f_u = 40\ kHz$. The on-surface pressure pressure amplitude is (i) $p_0 = 50\ Pa\ (125\ dB)$ and (ii) $p_0 = 0.1\ Pa\ (71\ dB)$.

In Figs. 3.9, 3.10, and 3.11, a comparison is depicted between the 2D sound field distributions generated by a PAL and a conventional loudspeaker. Both loudspeakers have same aperture sizes of 0.05 m, 0.1 m, and 0.2 m, respectively. Here, the conventional loudspeaker is modeled by a baffled circular rigid piston. At low frequencies (500 Hz in Fig. 3.9), the audio sound emitted by a conventional loudspeaker exhibits a monopole-like radiation pattern, gradually becoming more directional with increasing frequency. In contrast, the findings reveal that the audio beams generated by a PAL exhibit significantly greater directionality compared to those produced by a conventional loudspeaker across all frequencies and aperture sizes. Notably, the advantage of the PAL is more pronounced at smaller aperture sizes and lower frequencies. At larger aperture sizes and higher frequencies, such as the conditions presented in Fig. 3.11(d) at 0.2 m and 4 kHz, PALs and conventional loudspeakers generate audio beams with comparable directionality.

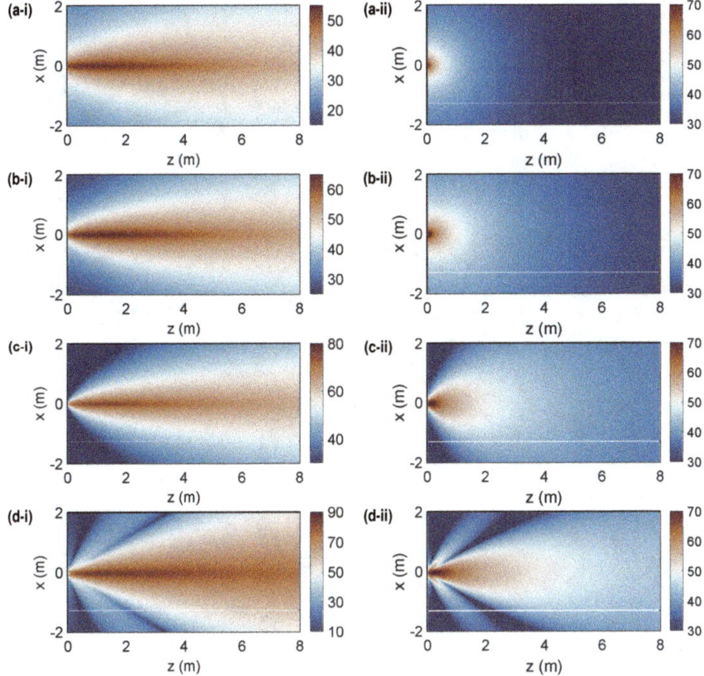

Figure 3.10 Audio SPL distributions generated by a baffled circular (a) PAL and (ii) conventional loudspeaker with a radius of 0.1 m. The velocity profile is uniform in both cases. The audio frequency is (a) 500 Hz, (b) 1 kHz, (c) 2 kHz, and (d) 4 kHz. In (i), the average ultrasound frequency is $f_u = 40$ kHz. The on-surface pressure pressure amplitude is (i) $p_0 = 50$ Pa (125 dB) and (ii) $p_0 = 0.1$ Pa (71 dB).

Figure 3.12 presents a comparison of the far-field directivity for the audio sound generated by a PAL and a conventional loudspeaker. At low frequencies and with small aperture sizes, the conventional loudspeaker exhibits almost omnidirectional characteristics. For instance, at 400 Hz and a source radius of 0.05 m as shown in Fig. 3.12(a-i), the directivity is only −0.2 dB at 90° for the conventional loudspeaker. Under the same parameter configuration, the PAL demonstrates significantly superior directivity compared to the conventional loudspeaker. For example, the directivity drops to −21.2 dB at 90° in Fig. 3.12(a-i), which is 21 dB lower than that of the conventional loudspeaker.

As the audio frequency increases, the difference in directivity around the main lobe between the two loudspeakers becomes smaller. For example, in Fig. 3.12(c-iii), the half power beam width (HPBW) is 8.3° and 12.6° for the PAL and the conventional loudspeaker, respectively. The directivity difference is only 4.3° in this case. Nevertheless, the PAL continues to outperform the conventional loudspeaker in generating directional audio beams. It is worth noting that the HPBW obtained using the Westervelt's direcitivity [Eq. (2.56)] is only 3.6° for the parameters in Fig. 3.12(c-iii), which is significantly smaller than 8.3°, demonstrating the inaccuracy of the Westervelt's directivity.

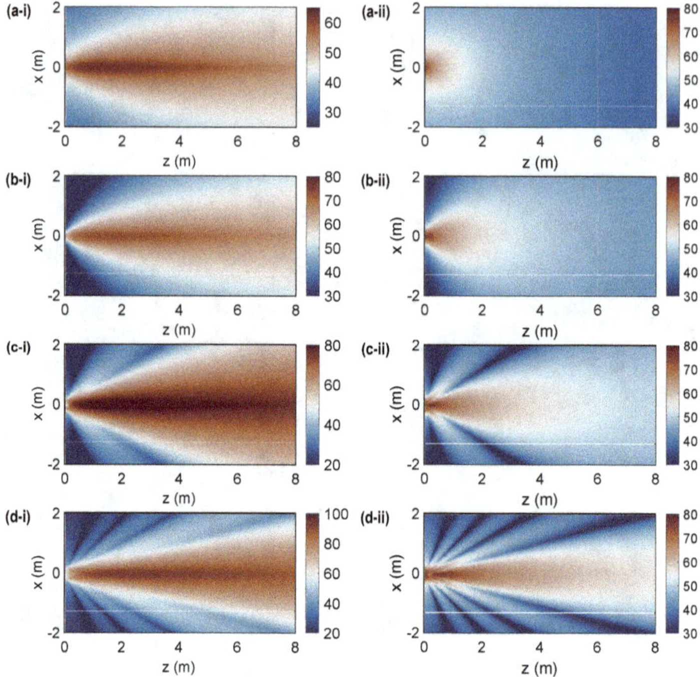

Figure 3.11 Audio SPL distributions generated by a baffled circular (a) PAL and (ii) conventional loudspeaker with a radius of 0.2 m. The velocity profile is uniform in both cases. The audio frequency is (a) 500 Hz, (b) 1 kHz, (c) 2 kHz, and (d) 4 kHz. In (i), the average ultrasound frequency is $f_\mathrm{u} = 40\,\mathrm{kHz}$. The on-surface pressure pressure amplitude is (i) $p_0 = 50\,\mathrm{Pa}$ (125 dB) and (ii) $p_0 = 0.1\,\mathrm{Pa}$ (71 dB).

Figure 3.13 presents the audio SPL at various typical frequencies along the radiator axis as a function of the propagation distance. Meanwhile, Fig. 3.14 displays the audio SPL at several typical distances on the radiator axis as a function of audio frequencies. A noticeable trend emerges where the audio SPL diminishes as the audio frequency decreases, a pattern observed in both PALs and conventional loudspeakers, indicating a poor frequency response at lower frequencies. Specifically, when the radial distance is 1 m, 2 m, 5 m, and 10 m, the audio SPL for PALs decreases by approximately 10.5 dB, 10.7 dB, 10.9 dB, and 11.1 dB per octave, respectively. In the inverse-law far field, the audio sound pressure for PALs, as determined by the CDM and presented in Eq. (2.75), is found to be proportional to the square of the audio frequency, resulting in a 12 dB/octave decrease as the frequency is halved [Bennett and Blackstock, 1975; Yoneyama et al., 1983]. In contrast, for conventional loudspeakers, the directivity can be computed using Eq. (2.72) by inserting the audio frequency. In this case, the audio sound pressure amplitude is shown to be proportional to the audio frequency, resulting in only a 6 dB/octave decrease as the frequency is halved. Consequently, the efficiency of PALs decreases more rapidly at lower frequencies.

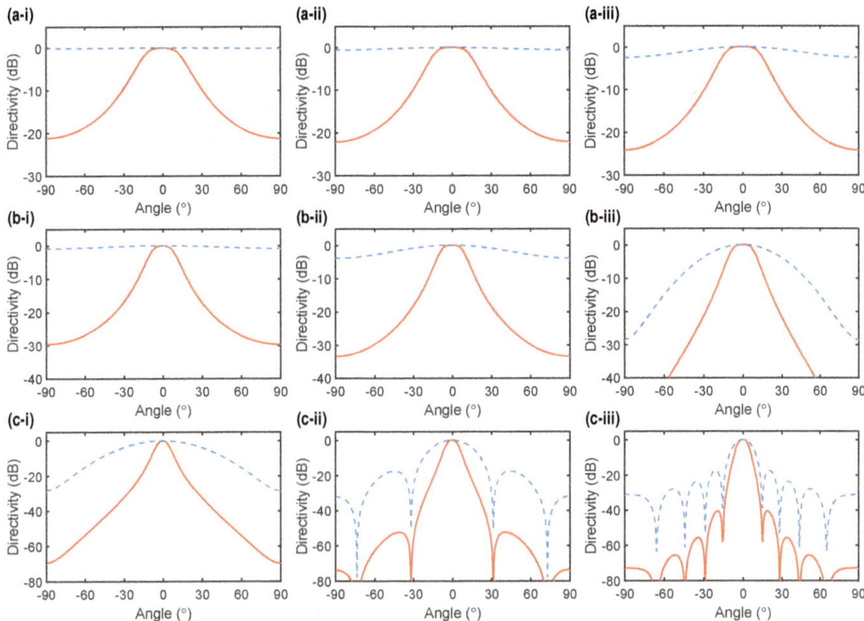

Figure 3.12 Normalized far-field directivity of audio sound generated by (——) a circular PAL or (- - -) a circular conventional loudspeaker with a uniform profile (piston source). The angle is $\theta \cos \varphi$, where $\varphi = 0°$ or $180°$. The radius is (i) $a = 0.05$ m, (ii) $a = 0.1$ m, and (iii) $a = 0.2$ m. The audio frequency is (a) $f_a = 400$ Hz, (b) $f_a = 1$ kHz, and (iii) $f_a = 4$ kHz. The ultrasound frequency is $f_u = 40$ kHz. ——, obtained using the modified CDM Eq. (2.77) - - - , obtained using the SWE Eq. (2.217).

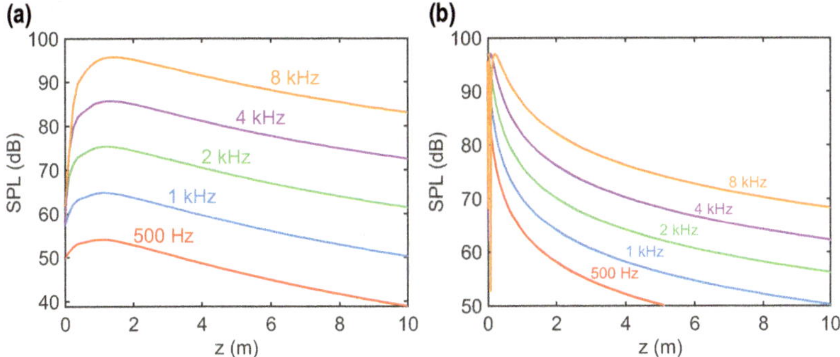

Figure 3.13 On-axis audio SPL of the audio sound at various distances z at 500 Hz, 1 kHz, 2 kHz, 4 kHz, and 8 kHz generated by a circular (a) PAL and (b) conventional loudspeaker with a radius of 0.1 m and a uniform profile. On-surface pressure is (a) $p_0 = 50$ Pa (125 dB) and (b) $p_0 = 1$ Pa (91 dB). Results are obtained using the SWE given by Eq. (2.214) for (a) and the SWE given by Eq. (2.201) for (b). Local nonlinear effects are not included.

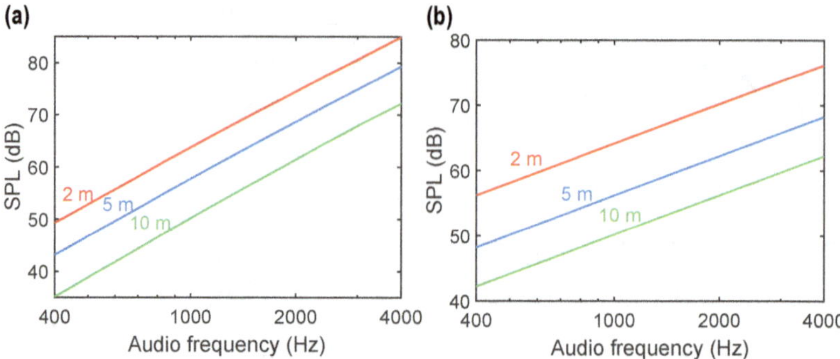

Figure 3.14 On-axis audio SPL at various frequencies at a distance of $z = 2\,\text{m}$, $5\,\text{m}$, and $10\,\text{m}$ to the source. (a) PAL; (b) Conventional loudspeaker. The ultrasound frequency is 40 kHz. On-surface pressure amplitude is (a) $p_0 = 50\,\text{Pa}$ $(125\,\text{dB})$ and (b) $p_0 = 1\,\text{Pa}$ $(91\,\text{dB})$. Results are obtained using the (a) SWE given by Eq. (2.214) and (b) SWE given by Eq. (2.201).

3.2.5 REMARKS

This section investigates the audio sound field in front of a PAL and draws comparisons with that of a conventional loudspeaker. Unlike a conventional loudspeaker, which exhibits a clear distinction between the near and far fields based on the inverse proportional relationship between its sound pressure amplitude and distance, the sound field produced by a PAL is more complex. It is commonly categorized into three regions: the inverse-law far field, the Westervelt far field (also known as the near field without local effects), and the near field (with local effects), as detailed in Sec. 3.2.1.

For a conventional loudspeaker, the transition from the near field to the far field is determined by the critical distance or Rayleigh distance. In contrast, for a PAL, the transition from the near field to the Westervelt far field is influenced by the radial distance at the local maxima of the ultrasound wave and can be predicted using Eq. (3.8) in Sec. 3.2.2. The transition distance from the Westervelt far field to the inverse-law far field is more complicated, as discussed in Sec. 3.2.3. The ultrasound frequency is found to be the most important factor that affects the transition distance, which can be approximately estimated with the empirical formula Eq. (3.13). The sound field distributions in both the near field and the far field are compared in Sec. 3.2.4, revealing that the PAL is more directional than a conventional loudspeaker with the same aperture size. Additionally, the results highlight a weakness of a PAL, namely, it is less efficient at lower frequencies.

3.3 SOUND FIELD BEHIND A NON-BAFFLED PAL

The PAL is commonly simplified in modeling as a vibrating source in an infinitely large baffle for ease of analysis. However, in practical applications, the PAL is typically non-baffled. In Fig. 3.15, the measured directivity of a PAL is compared with that of a conventional loudspeaker, revealing that the audio sound behind the PAL at 500 Hz is more than 30 dB lower than that behind a conventional loudspeaker. Consequently, the audio sound behind the PAL is often overlooked in the literature. Nonetheless, audible sound can be detected behind a non-baffled PAL and may be non-negligible in certain applications. A theoretical model was initially proposed in Zhong et al., 2020b, aiming to predict the audio sound behind the non-baffled PAL. This model is summarized in this section and utilized to examine the sound field properties behind the non-baffled PAL.

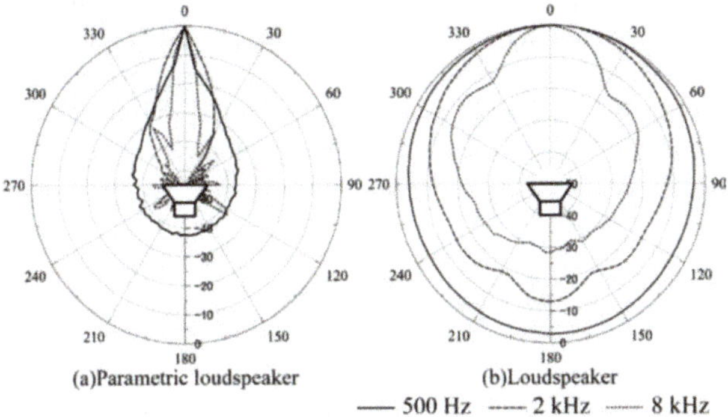

Figure 3.15 Directivity of (a) a PAL and (b) a conventional loudspeaker measured at 2 m. Extracted from [Sugahara et al., 2017, Fig. 3].

3.3.1 PREDICTION MODEL BASED ON THE DISK SCATTERING THEORY

3.3.1.1 Geometry

To investigate the audio sound behind a PAL, a non-baffled PAL is considered here with the same geometry as depicted in Fig. 3.2. For simplicity, the thickness of the PAL is assumed to be significantly smaller than the audio wave length. The vibration velocity on one surface of the PAL is assumed to be uniform with a value of v_0 for simplicity, while the other surface is treated as rigid.

In the case of a linear radiation problem, the sound field behind such a non-baffled source is a combination of the direct and scattered sound fields. The direct sound field can be computed using the Rayleigh integral as given by Eq. (2.38). However, obtaining the scattered sound field is more intricate and depends on the size and shape of the PAL. It is well-established that sound scattering by a disk can be analytically solved under the oblate spheroidal coordinate system [Bowman et al., 1970; Zhong et al., 2018; Zhong et al., 2019a]. Therefore, a disk-shaped PAL with a radius of a is considered in this section.

The oblate spheroidal coordinates (η, ξ, φ) are related to the rectangular coordinates (x, y, z) as follows [Flammer, 2005, Eq. (2.1.2)]

$$\begin{cases} x = a\sqrt{(1-\eta^2)(1+\xi^2)}\cos\varphi, \\ y = a\sqrt{(1-\eta^2)(1+\xi^2)}\sin\varphi, \\ z = a\eta\xi, \end{cases} \tag{3.14}$$

where $2a$ is the focal distance, the coordinate ξ is restricted by $\xi \in [0, \infty)$, η is restricted by $\eta \in [-1, 1]$, and the azimuthal angle φ is restricted by $\varphi \in [0, 2\pi)$.

The isosurfaces of ξ and η at various values are presented in Fig. 3.16 to depict their shapes. When the radial coordinate $\xi = 0$, the oblate shape represents an infinitely thin disk with a radius of a lying on the $z = 0$ plane. This characteristic makes it a suitable tool for modeling the boundary of a non-baffled disk. The boundary of the disk is assumed to be rigid for audio sound, satisfying the condition

$$\left.\frac{\partial p_{\mathrm{a}}(\mathbf{r})}{\partial \xi}\right|_{\xi=\xi_{\mathrm{b}}=0} = 0, \tag{3.15}$$

where $\xi_{\mathrm{b}} = 0$ is the radial coordinate of the disk.

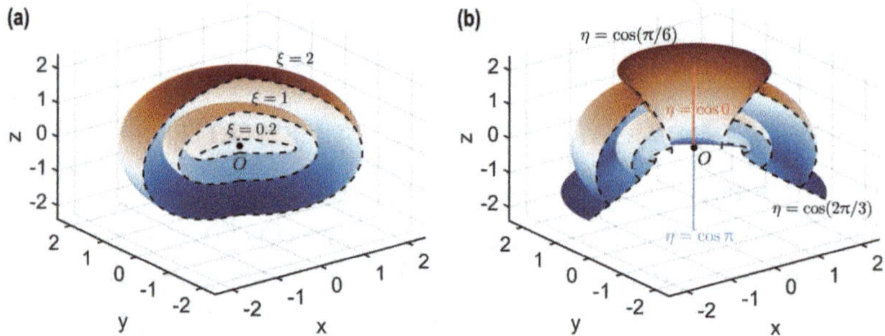

Figure 3.16 Isosurfaces of oblate spheroidal coordinates: (a) $\xi = 0.2, 1, 2$, (b) $\eta = \cos 0, \cos(\pi/6), \cos(2\pi/3), \cos\pi$. Here, the parameter $a = 1$.

The scale factors for oblate spheroidal coordinates are [Flammer, 2005, Eq. (2.2.2b)]

$$\begin{cases} h_\xi = a\sqrt{\dfrac{\xi^2+\eta^2}{1+\xi^2}}, \\ h_\eta = a\sqrt{\dfrac{\xi^2+\eta^2}{1-\eta^2}}, \\ h_\varphi = a\sqrt{(1+\xi^2)(1-\eta^2)}. \end{cases} \tag{3.16}$$

The infinitesimal volume element under the oblate spheroidal coordinate system can be obtained using Eq. (3.16) as

$$dV = h_\xi h_\eta h_\varphi d\xi d\eta d\varphi = a^3(\xi^2 + \eta^2) d\xi d\eta d\varphi. \tag{3.17}$$

3.3.1.2 Audio Sound Field

The total audio sound field can be obtained using Eq. (2.45) as

$$p_a(\mathbf{r}) = -i\rho_0\omega_a a^3 \int_0^{2\pi}\int_0^1\int_0^\infty q_a(\mathbf{r}_v)G(\mathbf{r},\mathbf{r}_v,k_a)(\xi_v^2+\eta_v^2)d\xi_v d\eta_v d\varphi_v, \tag{3.18}$$

where Eq. (3.17) is used, $q_a(\mathbf{r}_v)$ is the source density of virtual audio sound source given by Eq. (1.106), and $\mathbf{r}_v = (\eta_v, \xi_v, \varphi_v)$ is the location of the virtual audio source in the oblate spheroidal coordinate system. $G(\mathbf{r}, \mathbf{r}_v, k_a)$ is the Green's function in the presence of a rigid disk and can be expressed as [Flammer, 2005, Eq. (5.2.5)]

$$\begin{aligned} G(\mathbf{r},\mathbf{r}',k) = \frac{ik}{2\pi}\sum_{m=0}^{\infty}\sum_{n=m}^{\infty} & \varepsilon_m \mathcal{R}_{mn}(-ika, i\xi, i\xi') \\ & \times S_{mn}(-ika,\eta)S_{mn}(-ika,\eta')\cos\left[m(\varphi-\varphi')\right], \end{aligned} \tag{3.19}$$

where ε_m is the Neumann factor, $\varepsilon_m = 1$ when $m = 1$, $\varepsilon_m = 2$ when $m \neq 0$, and the radial component is

$$\begin{aligned} \mathcal{R}_{mn}(-ika, i\xi, i\xi') = & R_{mn}^{(1)}(-ika, i\xi_<)R_{mn}^{(3)}(-ika, i\xi_>) \\ & - \frac{R_{mn}^{(1)\prime}(-ika, i\xi_b)}{R_{mn}^{(3)\prime}(-ika, i\xi_b)} R_{mn}^{(3)}(-ika, i\xi)R_{mn}^{(3)}(-ika, i\xi'), \end{aligned} \tag{3.20}$$

where $\xi_> = \max(\xi, \xi_v)$ and $\xi_< = \min(\xi, \xi_v)$.

The notations of spheroidal wave functions are consistent with those in [Zhang and Jin, 1996, Chap. 15; Van Buren, 2017]. In this context, $S_{mn}(-ik_a a, \eta)$ denotes the normalized angular oblate spheroidal wave function, and $R_{mn}^{(i)}(-ik_a a, i\xi)$ and $R_{mn}^{(i)\prime}(-ik_a a, i\xi)$ represent the i-th kind of the radial oblate spheroidal wave functions and their derivatives with respect to ξ, where $i = 1, 3$. For detailed computations of these special functions, readers are referred to [Zhang and Jin, 1996, Chap. 15; Van Buren, 2017].

Calculating the audio sound can be time-consuming due to the three-fold integral in Eq. (3.18) and the two-fold summation of Green's function in Eq. (3.19). However, when the vibration velocity profile on the transducer's surface is axisymmetric about its axis, and the source density of virtual audio sources ($q_a(\mathbf{r}_v)$) is also axisymmetric, the total audio sound of the non-baffled model can be simplified by integrating over the azimuthal angle φ_v, resulting in [Zhong et al., 2020b, Eq. (10)]

$$p_a(\mathbf{r}) = -i2\pi\rho_0\omega_a a^3 \int_0^1\int_0^\infty q_a(\mathbf{r}_v)G_0(\mathbf{r},\mathbf{r}_v,k_a)(\xi_v^2+\eta_v^2)d\xi_v d\eta_v, \tag{3.21}$$

where the axisymmetric Green's function is simplified to

$$G_0(\mathbf{r},\mathbf{r}_\mathrm{v},k_\mathrm{a}) = \frac{\mathrm{i}k_\mathrm{a}}{2\pi}\sum_{n=0}^{\infty}\mathcal{R}_{0n}(-\mathrm{i}k_\mathrm{a}a,\mathrm{i}\xi,\mathrm{i}\xi_\mathrm{v})S_{0n}(-\mathrm{i}k_\mathrm{a}a,\eta)S_{0n}(-\mathrm{i}k_\mathrm{a}a,\eta_\mathrm{v}). \quad (3.22)$$

3.3.2 NON-BAFFLED MODEL FOR A CONVENTIONAL LOUDSPEAKER

The sound radiation from a baffled conventional loudspeaker is commonly modeled by assuming the loudspeaker cabinet to be a rigid, infinitely large baffle surrounding a vibrating membrane. In most cases, the thickness of a non-baffled conventional loudspeaker cannot be neglected, making it unsuitable for modeling as a disk, as demonstrated for the PAL in Sec. 3.3.1. In contrast, a non-baffled conventional loudspeaker is often represented by a spherical cap source (depicted in pink in Fig. 3.17) set on a rigid sphere [Qiu et al., 2009; Aarts and Janssen, 2010; Liu and Maury, 2017; Beranek and Mellow, 2019]. This modeling approach is based on the assumption that the diffraction effects caused by a sphere and a cube with similar dimensions are comparable if the dimensions of the diffracting bodies are significantly smaller than the wavelength. Although the error resulting from this idealization increases at higher frequencies, it remains useful for illustrating the general directional properties of the non-baffled conventional loudspeaker.

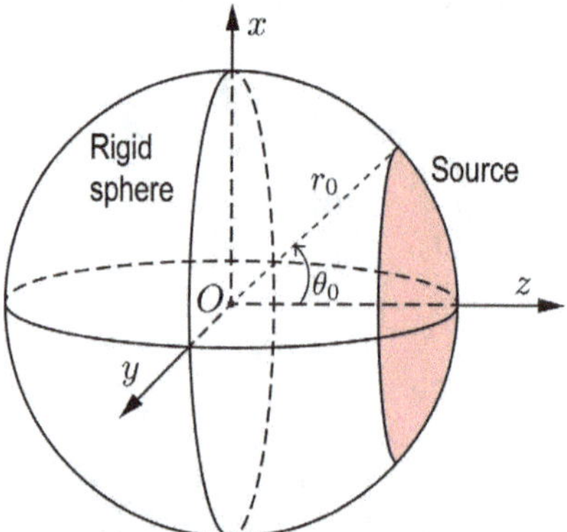

Figure 3.17 Sketch of a non-baffled conventional loudspeaker, where the spherical cap (depicted in pink) represents the source.

The simplest axisymmetric piston source set on a rigid sphere with a radius of r_0 is considered here. As shown in Fig. 3.17, a Cartesian coordinate system $Oxyz$ is established with its origin O at the sphere center. The positive z-axis points to the center of the spherical cap source. The radial velocity distribution of the source is [Skudrzyk, 1971, Eq. (20.32); Beranek and Mellow, 2019, Eq. (12.48)]

$$v_r(r_0,\theta,\varphi) = v_0\begin{cases}\cos\theta_0, & 0\le\theta\le\theta_0,\\ 0, & \theta_0<\theta\le\pi,\end{cases} \quad (3.23)$$

where θ_0 is half of the cap source span angle. The source density in space can be represented by

$$q_a(\mathbf{r}) = \delta(r - r_0) v_r(r_0, \theta, \varphi), \tag{3.24}$$

where $\delta(\cdot)$ is the Dirac delta function.

The sound pressure in the presence of a rigid sphere can be obtained as [Zhong et al., 2022c, Eq. (19)]

$$p_a(\mathbf{r}) = -\mathrm{i}\rho_0\omega_a \iiint_{r_s \geq r_0} q_a(\mathbf{r}_s) G(\mathbf{r}, \mathbf{r}_s, k_a) \mathrm{d}^3\mathbf{r}_s, \quad r \geq r_0, \tag{3.25}$$

where the Green's function for a point source scattered by a rigid sphere is [Zhong et al., 2022c, Eq. (11)]

$$\begin{aligned} G(\mathbf{r}, \mathbf{r}_s, k_a) = \mathrm{i}k_a \sum_{\ell=0}^{\infty} \sum_{m=-\ell}^{\ell} \left[\mathrm{j}_\ell(k_a r_{s,<}) \mathrm{h}_\ell(k_a r_{s,>}) - \frac{\mathrm{j}'_\ell(k_a r_0)}{\mathrm{h}'_\ell(k_a r_0)} \mathrm{h}_\ell(k_a r_s) \mathrm{h}_\ell(k_a r) \right] \\ \times Y_\ell^m(\theta, \varphi) Y_\ell^{m,*}(\theta_s, \varphi_s), \end{aligned} \tag{3.26}$$

where $r_{s,<} = \min(r, r_s), r_{s,>} = \max(r, r_s)$.

Substituting Eqs. (3.24) and (3.26) into Eq. (3.25) and using the Wronskian $W[\mathrm{j}_\ell(k_a r_0), \mathrm{h}_\ell(k_a r_0)] = \mathrm{i}/(k_a r_0)^2$ given by Eq. (B.28), the audio sound pressure is simplified to

$$p_a(\mathbf{r}) = \mathrm{i}\rho_0 c_0 \sum_{\ell=0}^{\infty} \frac{\mathrm{h}_\ell(k_a r)}{\mathrm{h}'_\ell(k_a r_0)} V_{r,\ell}^0(r_0) Y_\ell^0(\theta, \varphi). \tag{3.27}$$

Here, the spherical angular spectrum of the radial particle velocity at $r = r_0$ is [Williams, 1999, Eq. (6.98)]

$$V_{r,\ell}^m(r_0) = \int_0^{2\pi} \int_0^{\pi} v_r(r_0, \theta_s, \varphi_s) Y_\ell^{m,*}(\theta_s, \varphi_s) \sin\theta_s \mathrm{d}\theta_s \mathrm{d}\varphi_s. \tag{3.28}$$

Substituting Eq. (3.23) into Eq. (3.28) and using Eq. (B.2), the spherical angular spectrum is simplified to [Skudrzyk, 1971, Eq.(20.39); Aarts and Janssen, 2010, Eq. (9)]

$$\begin{aligned} V_{r,\ell}^m(r_0) = \delta_{m0} v_0 \sqrt{\frac{\pi}{2\ell+1}} \left\{ \frac{\ell+1}{2\ell+3} [P_\ell(\cos\theta_0) - P_{\ell+2}(\cos\theta_0)] \right. \\ \left. \times + \frac{\ell}{2\ell-1} [P_{\ell-2}(\cos\theta_0) - P_\ell(\cos\theta_0)] \right\}. \end{aligned} \tag{3.29}$$

In Eq. (3.29), the definition $P_{-\ell-1} = P_\ell, \ell = 0, 1, \ldots,$ has been used to deal with the cases $\ell = 0, 1$.

The surface area of the spherical cap is $\mathcal{A}_0 = 4\pi r_0^2 \sin^2(\theta_0/2)$. If this area is chosen to be equal to the area of the circular flat piston in infinite baffle, the relation between the radius of the circular piston and the sphere radius is [Aarts and Janssen, 2010, Eq. (11)]

$$a = 2r_0 \sin(\theta_0/2). \tag{3.30}$$

Figure 3.18 compares the SPL distributions generated by typical non-baffled and baffled sources. These results are compared against those obtained for non-baffled and baffled PALs in the subsequent section.

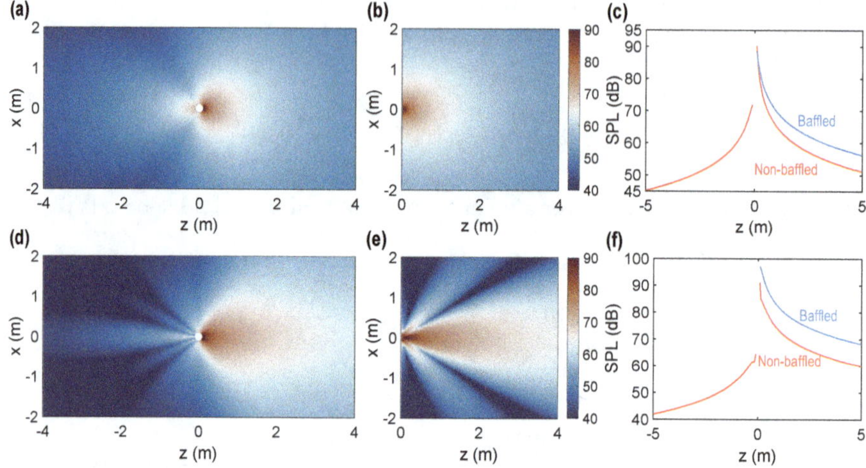

Figure 3.18 SPL distributions generated by (a, d) a spherical cap source set on a rigid sphere (non-baffled source, $r_0 = 0.1$ m, $\theta_0 = \pi/3$) and (b, e) a baffled circular source ($a = 0.1$ m) at (a–b) 1 kHz and (d–e) 4 kHz. Comparison of the on-axis ($x, y = 0$) SPL at (c) 1 kHz and (f) 4 kHz. The on-surface pressure amplitude is $p_0 = 1$ Pa (91 dB).

3.3.3 RESULTS AND DISCUSSIONS

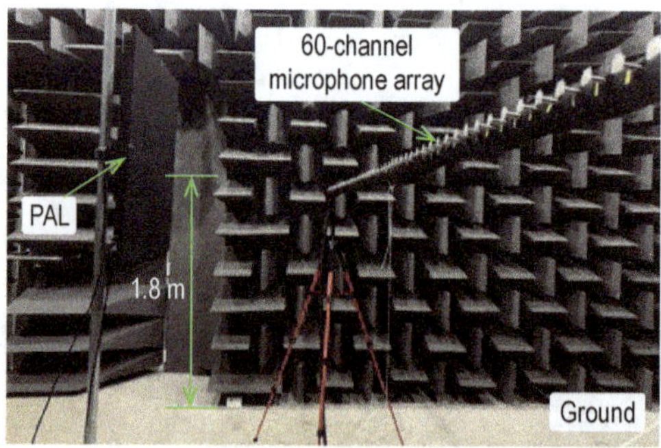

Figure 3.19 Photo of the experimental setup for investigating the audio sound behind a non-baffled PAL. Experiments were conducted in a hemi-anechoic room with dimensions of 7.20 m × 5.19 m × 6.77 m (height). The PAL used in experiments is a Holosonics Audio Spotlight AS-24i [Holosonics, 2019] with a surface size of 0.6 m × 0.6 m. The height of the PAL center is 1.8 m.

In this section, both simulation and experiment results are presented to investigate the properties of the audio sound field behind a non-baffled PAL. The experimental setup is shown in Fig. 3.19. In the simulations, a circular piston was driven with a uniform surface vibration velocity amplitude and the radius was set as 338.5 mm so that its area is the same as that of the rectangular PAL used in the experiments, i.e., $0.6^2 \approx \pi \times 0.3385^2$. The relative humidity and the temperature in the experiments were 60% and 25°C, respectively. The carrier frequency of the PAL is 64 kHz measured by a Brüel & Kjær Type 4939 microphone. All of the aforementioned measured data were set as known parameters in the simulations and the air absorption coefficients were calculated according to Appendix A. The computation of the spheroidal wave functions that were obtained by modifying the codes of [Van Buren, 2017].

The sound field was measured at a rectangular grid comprising $60 \times 61 = 3661$ points in the yOz plane at a height of 1.8 m. In all cases, 60 microphones were positioned in the y-direction to form a microphone array, spanning from $y = -1.45$ m to $y = 1.5$ m, with a uniform spacing of 5 cm. The microphone array was positioned at 61 discrete positions in the z-direction, ranging from $z = -3$ m to $z = 3$ m, with intervals of 0.1 m. All the measurement microphones utilized were Brüel & Kjær Type 4957. The sound pressure at the microphones was sampled using a Brüel & Kjær PULSE system. To mitigate spurious sounds caused by intense ultrasound radiation from the PAL, all microphones were covered with thin plastic films.

In Fig. 3.20, the simulated SPLs of the audio sound along the z-axis are presented, generated by the finite size disk-shaped PAL using the baffled (Sec. 2.6) and non-baffled (Sec. 3.3.1) models, in comparison with experimental results. All SPLs in the simulations are normalized to a consistent reference value, ensuring that the SPL at $z = 2$ m for the baffled PAL matches the corresponding experimentally measured value. For the audio sound on the back side ($z < 0$), it cannot be predicted by the baffled model, resulting in the absence of data for this model.

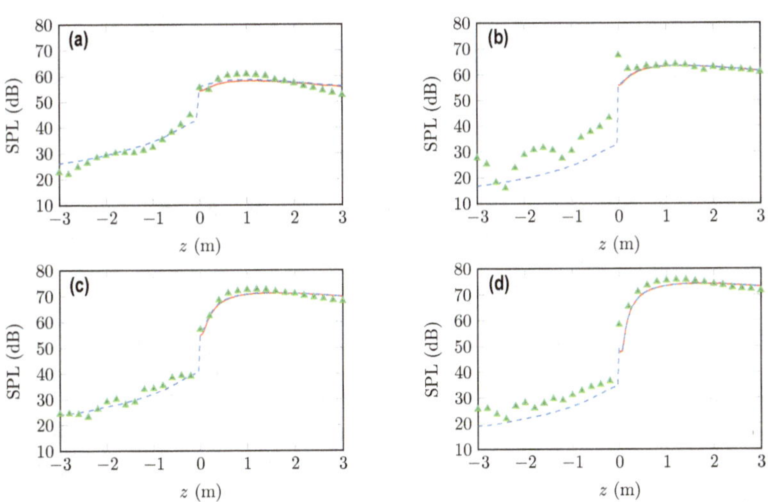

Figure 3.20 Audio SPL along the z-axis at various audio frequencies: (a) 315 Hz, (b) 500 Hz, (c) 800 Hz, and (d) 1 kHz. Solid line, baffled model given by Eq. (2.45); dashed line, non-baffled model given by Eq. (3.21); triangle, measurement. Extracted from [Zhong et al., 2020b, Fig. 2].

The experimental results on both the front and back sides of the PAL generally align with those predicted by the non-baffled model. Substantial errors occur at 500 Hz when $z < 0.2$ m, potentially attributed to measurement inaccuracies, ground reflections, the PAL's shape, and scattering effects from equipment. It is observed that the values obtained from both models exhibit nearly identical trends across all frequencies at locations significantly distant from the PAL, with the maximum difference being less than 0.4 dB when $z > 1$ m. However, as depicted in Fig. 3.18, the difference in SPL in front of non-baffled and baffled conventional loudspeakers is more pronounced. For instance, at a distance of 5 m ($z = 5$ m), the SPL generated by the baffled conventional loudspeaker exceeds that of the non-baffled counterpart by 5.0 dB and 8.2 dB at 1 kHz and 4 kHz, respectively.

In Fig. 3.21, the SPL curves for the non-baffled PAL are presented at various distances $z = -0.1$ m, 0.25 m, -0.5 m, -1.0 m, and -2.0 m across different frequencies. The on-surface pressure amplitude of ultrasound ($p_0 = \rho_0 c_0 v_0$) is consistently set as 125 dB for all curves for better comparison. Both Figs. 3.20 and 3.21 reveal the presence of audible audio sounds on the back side of the PAL, arising from the diffraction of the finite-size disk. For example, at 315 Hz, the SPL is 45 dB at $z = -0.1$ m, while the audio sound at $z = 1.0$ m in front of the PAL is approximately 61.4 dB.

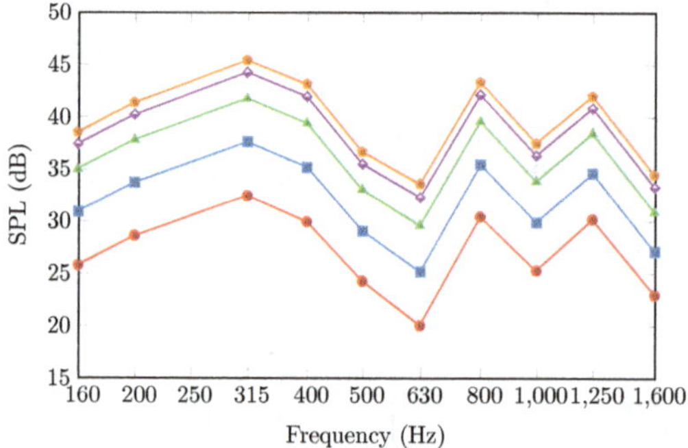

Figure 3.21 Audio SPL along the z-axis obtained using the non-baffled model at 1/3 octave center frequencies from 160 Hz to 1.6 kHz. The on-surface sound pressure amplitude $p_0 = 50$ Pa (125 dB). Red circle, $z = -2$ m; Blue square, $z = -1$ m; green triangle, $z = -0.5$ m; purple diamond, $z = -0.25$ m; orange pentagon, $z = -0.1$ m. Extracted from [Zhong et al., 2020b, Fig. 3].

The SPL attains its peak at 315 Hz (with a corresponding wavelength is 1.09 m) in all cases. This is attributed to the fact that constructive interference of waves is most pronounced for point monopoles when the PAL radius PAL is approximately 0.35 times the wavelength [Zhong et al., 2018]. As the frequency decreases below 315 Hz, the SPL on the back side diminishes due to a substantial reduction (12 dB) in the PAL's frequency response when the audio frequency is halved. On the contrary, as the frequency ascends from 315 Hz, the SPL on the back side initially diminishes before reaching local maxima at 800 Hz and 1250 Hz. This behavior is a consequence of the weakening of diffraction effects as the wavelength decreases, until it aligns with other resonant frequencies where the constructive interference of waves becomes significant.

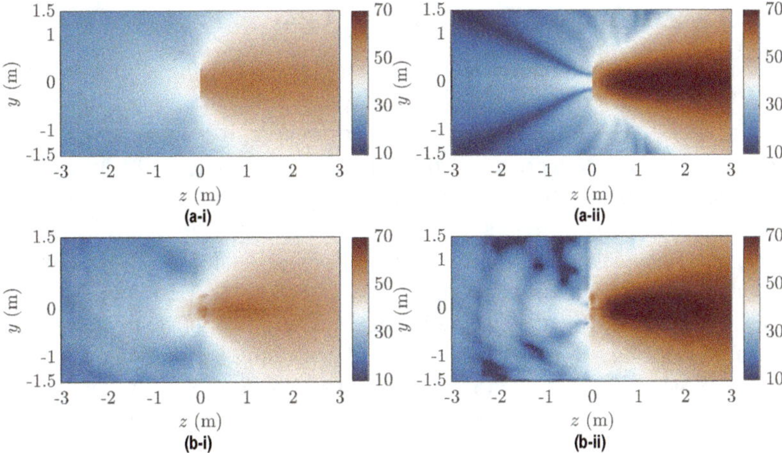

Figure 3.22 Audio SPL distributions at (a) 315 Hz and (b) 800 Hz obtained by (i) the simulations and (ii) the measurements. The PAL is placed at the origin and the radiation surface is on the plane $z = 0$. Extracted from [Zhong et al., 2020b, Fig. 4].

Figure 3.22 illustrates the distribution of SPLs in the yOz plane at 315 Hz and 800 Hz, where a noteworthy agreement between simulation and experimental results is observed. This coherence underscores the efficacy of the non-baffled model in accurately predicting sound fields on the back side of the PAL. The measured sound fields exhibit resonances, attributed to the concrete ground floor in the experimental setup. Moreover, it is evident that the audio sound is audible over a substantial area on the back side of the PAL. For instance, at 315 Hz, there exists an approximately circular region centered around the PAL with a radius of about 1.3 m, where SPLs exceed 35 dB. This emphasizes the necessity to consider the effects of the finite size of the PAL in the analysis.

3.3.4 REMARKS

This section explored the audio sound field behind a non-baffled PAL utilizing a disk scattering model and the quasilinear approximation. The model treats each virtual audio source, generated by the ultrasound, as a point monopole and calculates its scattered sound through a disk of finite size. Both simulations and experiments converge on the finding that audible sound is present on the back side of the PAL, although notably diminished compared to the front side. The audible sound on the back side primarily originates from the scattering of virtual audio sources generated by the ultrasound, rather than the scattering of ultrasound itself. The audio sound behind the PAL becomes more prominent as the audio frequency decreases. This underscores the importance of considering the finite size of the PAL, particularly at lower audio frequencies.

3.4 SOUND POWER GENERATED BY A PAL

The sound power is a fundamental metric for characterizing sound sources. An accurate assessment of the sound power of PALs can enhance our understanding of their energy conversion characteristics. Furthermore, it is imperative to consider the sound power in certain practical applications. For example, PALs have been used in active noise control systems as the control sources to minimize the radiated sound power by noise sources [Tanaka and Tanaka, 2011; Ye et al., 2012]. In such scenarios, accurate analysis of the power outputs of the sources, including PALs, is essential. The sound power calculation as well as the measurement for conventional loudspeakers are well known. While the calculation and measurement of sound power for conventional loudspeakers are well-established, determining the sound power for PALs presents unique challenges due to the involved nonlinear processes. The challenges are explored in Sec. 3.4.1. Section 3.4.2 introduces a model to calculate the sound power based on the CDM introduced in Sec. 2.3. Finally, Sec. 3.4.3 delves into the analysis of power conversion efficiency from ultrasound to audio sound.

3.4.1 CHALLENGES OF THE SOUND POWER MEASUREMENT FOR PALS

The audio sound is demodulated in air due to the nonlinear interactions of intense ultrasonic waves radiated by the PAL. The power conversion from ultrasound to audio sound is widely recognized as highly inefficient for a PAL [Westervelt, 1963]. In linear acoustics, for a conventional acoustic source, electric energy is converted into mechanical vibration to emit sound waves, and the corresponding sound power calculation is well known. However, the power conversion process is different for a PAL. In a PAL, electrical energy is converted into mechanical vibration to emit ultrasound, which is subsequently self-demodulated in air, resulting in the generation of an audio sound wave. The sound power of the audio sound wave is transferred from the ultrasound wave. The audio sound produced by a PAL can be considered as radiation from an infinitely large virtual volume source, where the source density is proportional to the ultrasound field. Consequently, measuring the sound power becomes challenging because it requires enclosing the virtual audio source with an infinitely large surface, which is not feasible in practice.

For instance, ISO 3745 provides a sound-pressure-based method to obtain the sound power of an acoustic source by measuring the sound pressure on a measurement surface (spherical or hemi-spherical envelopes) in the far field in an anechoic or a hemi-anechoic room [ISO 3745, 2012]. The mechanism of this method is that, in free field, the radiated waves are spherically spreading in the far field and the radial sound intensity can then be simply calculated by sound pressure measured at multiple points on the measurement surface. However, the far field of the audio sound generated by a PAL is generally located more than 10 m away from the source as demonstrated in Sec. 3.2, making it challenging to implement this method under practical measurement conditions as it requires a large full anechoic or hemi-anechoic room and accurate positioning of a huge number of measurement microphones.

ISO 3741 offers a precision method based on sound pressure measurement in a reverberation chamber [ISO 3741, 2010]. The mechanism of this method is that, in a reverberation chamber, the sound pressure is nearly uniform except for locations close to the source or chamber walls, and the sound power is proportional to the square of the sound pressure [Kuttruff, 2017]. Therefore, the sound power can be obtained by measuring the sound pressure in the reverberation room. However, the virtual audio sound source produced by a PAL tends to spread throughout the entire space in a reverberation chamber, making it difficult to identify a suitable measurement region far away from the source. Moreover, the direct ultrasound interacts with the reflected ultrasound and produces new virtual audio sound sources, indicating that the audio sound power measured in reverberation room is different from that measured in free field.

ISO 9614 specifies sound-intensity-based methods to obtain the sound power by measuring the sound intensity on a measurement surface that either completely surrounds or hemispherically surrounds the sound source [ISO 9614-1, 1993; ISO 9614-2, 1996]. The sound intensity can be measured at specific discrete points on the surface or by performing a continuous scan over the surface. However, it is challenging to identify a suitable measurement surface as it needs to be large enough to enclose the infinite virtual source of a PAL, so this method is also not suitable for measuring the sound power of PALs.

Vibration-based measurements, such as those outlined in ISO/TS 7849, offer alternative approaches for sound power measurement [ISO/TS 7849-1, 2009; ISO/TS 7849-2, 2009]. This kind of method determines sound power by measuring the vibration on the surface of the sound source. The recently proposed vibration-based radiation mode method claims to accurately measure the sound power from sources with arbitrarily curved surfaces [Bates et al., 2022]. However, these methods are not suitable for measuring the sound power of PALs because the virtual audio source of a PAL does not correspond to an actual vibrating object with a defined surface. Other methods such as the sound intensity scaling method using beamforming can estimate local sound power by converting the beamforming output [Hald, 2005]. However, this method necessitates the array to be positioned at a sufficient distance from the source, making it not practical for PAL sound power measurements. Another method, based on measuring the total energy density in a reverberation chamber [Nutter et al., 2007], also cannot be applied to PAL sound power measurement due to similar reasons as the method in ISO 3741.

3.4.2 SOUND POWER

3.4.2.1 Calculting Sound Power Using the Directivity

The sound power of an acoustic source can be determined by integrating the sound intensity over an enveloping surface that surrounds the source. In the case of a spherical envelope and when the envelope is in the far field, this calculation is expressed as

$$W = \iint_S I_r \mathrm{d}S, \tag{3.31}$$

where S is the envelope surface, I_r represents the radial sound intensity and can be obtained using the equation $I_r = |p|^2/(2\rho_0 c_0)$ in the far field of the sound source, where p is the sound pressure. The sound pressure in the far field can be written as Eq. (3.1). Consequently, the sound power can be calculated as [Pierce, 2019]

$$W = \frac{\rho_0 \omega^2 |Q_0|^2}{32\pi^2 c_0^2} \iint_{4\pi} |\mathcal{D}(\theta, \varphi)|^2 \mathrm{d}\Omega, \tag{3.32}$$

where $\mathcal{D}$ is the directivity function, and the element of the solid angle is $\mathrm{d}\Omega = \sin\theta \mathrm{d}\theta \mathrm{d}\varphi$.

To obtain the sound power of the audio sound generated by a PAL, a baffled circular PAL with a radius of a is considered in this section, which is a typical configuration in literature as well as applications. The sketch of the sound power calculation for a baffled circular PAL is shown in Fig. 3.2. When the PAL radiates two harmonic ultrasound waves with frequencies f_1 and f_2 ($f_1 < f_2$), the boundary condition on the surface is [Eq. (2.28)] $v_z(\mathbf{r}_s) = v_{1,z}(\mathbf{r}_s)\mathrm{e}^{-\mathrm{i}\omega_1 t} + v_{2,z}(\mathbf{r}_s)\mathrm{e}^{-\mathrm{i}\omega_2 t}, r_s \leq a$, where $v_{i,z}(\mathbf{r}_s)$ is an arbitrary complex velocity profile at the source point $\mathbf{r}_s$ at frequency f_i, $i = 1, 2$.

The generated ultrasound pressure level is often limited to ensure safety, so the nonlinearity is weak and the derivation can employ the quasilinear approximation (Sec. 1.4.3). With the successive method, the ultrasound pressure can be expressed as a twofold Rayleigh integral over the surface area of the PAL, and the audio sound pressure can be considered as a superposition of the pressure radiated by infinite virtual audio sources in air with the source density proportional to the ultrasound sound pressure. The audio sound pressure is then obtained by integrating the Green's function over the entire virtual source.

The Rayleigh integral for ultrasound pressure can be approximated in the far field by substituting the equivalent total volume velocity Eq. (2.42) into Eq. (3.1), resulting in

$$p_i(\mathbf{r}) = \frac{p_0 \omega_i a^2}{2\mathrm{i} c_0 r} \mathrm{e}^{\mathrm{i}k_i r} \mathcal{D}_i(\theta, \varphi). \tag{3.33}$$

Here, $i = 1, 2$, the total area of the circular radiation surface $\mathcal{A}_0 = \pi a^2$ is used, and $p_0 = \rho_0 c_0 v_0$ represents the on-surface pressure amplitude. By substituting Eq. (3.33) into Eq. (3.32), the ultrasound power can be obtained as

$$W_i = \frac{p_0^2 \omega_i^2 a^4}{8\rho_0 c_0^3} \int_0^{2\pi} \int_0^{\frac{\pi}{2}} |\mathcal{D}_i(\theta, \varphi)|^2 \sin\theta \mathrm{d}\theta \mathrm{d}\varphi. \tag{3.34}$$

Note that this integration is conducted over a hemisphere in front of the PAL, considering its radiation into half space due to the presence of the baffle.

The CDM is an accurate and computationally efficient model for calculating the far-field sound pressure of PALs, as detailed in Sec. 2.3. The key of this model is to approximate the exact ultrasound pressure by the inward-extrapolated far-field pressure given by Eq. (3.33), so as to obtain the approximated virtual source density of audio sound. Therefore, the directivities of audio sound obtained by both the direct

[Eq. (2.76)] and modified [Eq. (2.77)] CDMs are used here to calculate the sound power generated by a PAL. By substituting Eq. (2.76) or Eq. (2.77) into Eq. (3.32), the audio sound power based on the CDM can be obtained as

$$W_{\mathrm{a}} = \frac{\beta^2 p_0^4 a^8 \omega_1^2 \omega_2^2 \omega_{\mathrm{a}}^4}{512\pi^2 \alpha_{\mathrm{t}}^2 \rho_0^3 c_0^{13}} \int_0^{2\pi} \int_0^{\pi/2} |\mathcal{D}_{\mathrm{a}}(\theta,\varphi)|^2 \sin\theta \mathrm{d}\theta \mathrm{d}\varphi. \tag{3.35}$$

Equation (3.35) shows that the audio sound power, W_{a}, is proportional to the 4th power of the amplitude of the ultrasound, p_0^4. The effects of other physical parameters, such as audio sound frequency, f_{a}, aperture size, a, and ultrasound frequency, f_{u}, cannot be directly observed as the integral term in Eq. (3.35) cannot be simplified to a closed-form expression.

It is worth mentioning that, Westervelt derived the far-field audio sound pressure as Eq. (2.64), and calculated the audio sound power using the method described by Eq. (3.32), providing a closed-form expression as [Westervelt, 1963, Eq. (14)]

$$W_{\mathrm{a,W}} = \frac{\pi^2 \beta^2 p_0^4 a^4 \omega_{\mathrm{a}}^3}{32\alpha_{\mathrm{t}} \rho_0^3 c_0^8}. \tag{3.36}$$

Equation (3.36) shows that the sound power is proportional to the 4th power of the amplitude of ultrasound, p_0^4, the 4th power of the aperture size, a^4, the 3rd power of the audio sound frequency, f_{a}^3, and inversely proportional to the total attenuation coefficient, α_{t}. Since the total attenuation coefficient is approximately proportional to the square of the ultrasound frequency when the relaxation effect is neglected (refer to Appendix A), the sound power is approximately inversely proportional to the square of the ultrasound frequency, f_{u}^2. Additionally, it should be noted that the sound power given by Westervelt is independent of the specific ultrasound velocity profile and the shape of PALs. In contrast, the sound power calculated using Eq. (3.35) remains valid regardless of the profile and shape of the PAL because its directivity can be accurately predicated by the CDM.

3.4.2.2 Simulation Results

Figure 3.23 shows the audio sound power for varying audio frequency, f_{a}, aperture size, a, and average ultrasound frequency, f_{u}. It is observed that the audio sound power increases as the audio frequency and aperture size increase, while it decreases as the ultrasound frequency increases. Compared with the results based on CDMs, the predictions obtained by Westervelt overestimate the sound power for the parameters adopted here. This discrepancy can be attributed to the inaccuracy of audio sound pressure predicted by Westervelt, which is reflected in two aspects. First, the Westervelt's prediction for the on-axis sound pressure is slightly larger. Numerical calculation shows that, the on-axis sound pressure predicted by Westervelt is approximately 1.06 times larger than that obtained from the direct CDM given by Eq. (2.76), and approximately 1.1 times larger than that obtained from the modified CDM given by Eq. (2.77). Second, but more importantly, as demonstrated in Zhong et al., 2023d, the main lobe of the Westervelt's directivity is normally broader, leading to an overestimation of the sound pressure in other directions. Meanwhile, it is observed that the predictions obtained from the direct CDM are also greater than those from the modified CDM. This discrepancy arises because the aperture factor of audio sound is taken into account by the modified CDM to improve the accuracy. As audio frequency f_{a} and the aperture size a increase, the discrepancy in predicted directivity between the two models increases, leading to a greater disparity in integration results in the sound power calculation.

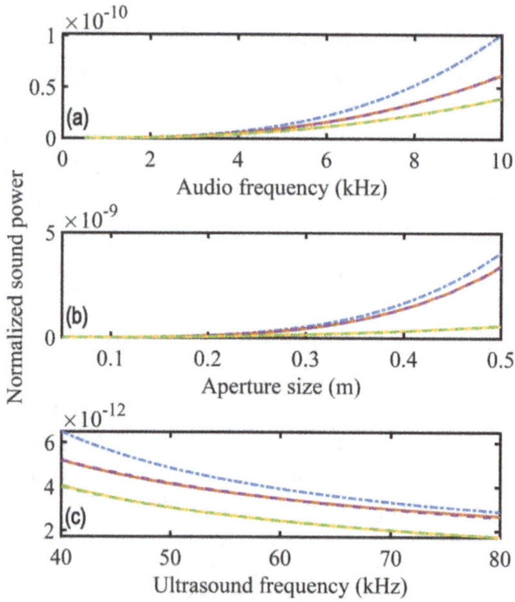

Figure 3.23 The audio sound power generated by a baffled circular PAL with a uniform profile for varying (a) audio frequency, (b) aperture size, and (c) ultrasound frequency. The results are normalized to p_0^4. ---, Westervelt; ——, direct CDM; ——, modified CDM; - - - , - - - , fitting curves. Extracted from [Li et al., 2023, Fig. 2].

As shown by Eq. (3.35), it is challenging to directly determine the relationship between audio sound power and some key physical parameters in the CDM because a closed-form expression for audio sound directivity is unavailable. To address this, the least square fitting method is employed to determine the exponent of a power function in relation to audio frequency, aperture size, and ultrasound frequency. The obtained fitting results for varying audio frequency, aperture size, and ultrasound frequency are presented in Tables 3.3, 3.4, and 3.5, respectively (the coefficients of the fitting results are omitted for simplicity). The corresponding fitting curves are also depicted in Fig. 3.23. The results indicate that for the direct CDM, the audio sound power exhibits proportionality to $f_{\mathrm{a}}^{2.7}$, a^4, and f_{u}^{-1}. In contrast, for the modified CDM, the audio sound power demonstrates proportionality to $f_{\mathrm{a}}^{2.4}$, $a^{2.6}$, and $f_{\mathrm{u}}^{-1.2}$. These relationships differ from the predictions of the Westervelt's sound power equation Eq. (3.36), which states a direct proportionality to f_{a}^3, a^4, and f_{u}^{-2}.

Table 3.3
Fitting results for curves of the audio sound power, W_a, as a function of the audio frequency, f_a, at different aperture sizes, a, and ultrasound frequencies, f_u, based on the direct and modified CDMs. Extracted from [Li et al., 2023, Table I].

	$a = 0.1$ m $f_u = 40$ kHz	$a = 0.2$ m $f_u = 40$ kHz	$a = 0.4$ m $f_u = 40$ kHz	$a = 0.1$ m $f_u = 60$ kHz	$a = 0.1$ m $f_u = 80$ kHz
Direct	$f_a^{2.7}$	$f_a^{2.7}$	$f_a^{2.7}$	$f_a^{2.8}$	$f_a^{2.8}$
Modified	$f_a^{2.4}$	$f_a^{2.2}$	$f_a^{1.9}$	$f_a^{2.5}$	$f_a^{2.4}$

Table 3.4
Fitting results for curves of the audio sound power, W_a, as a function of the aperture size, a, at different audio frequencies, f_a, and ultrasound frequencies, f_u, based on the direct and modified CDMs. Extracted from [Li et al., 2023, Table II].

	$f_a = 1$ kHz $f_u = 40$ kHz	$f_a = 4$ kHz $f_u = 40$ kHz	$f_a = 8$ kHz $f_u = 40$ kHz	$f_a = 4$ kHz $f_u = 60$ kHz	$f_a = 4$ kHz $f_u = 80$ kHz
Direct	a^4	a^4	a^4	a^4	$a^{3.9}$
Modified	$a^{3.2}$	$a^{2.6}$	$a^{2.3}$	$a^{2.4}$	$a^{2.3}$

Table 3.5
Fitting results for curves of the audio sound power, W_a, as a function of the ultrasound frequency, f_u, at different audio frequencies, f_a, and aperture sizes, a, based on the direct and modified CDMs. Extracted from [Li et al., 2023, Table III].

	$f_a = 1$ kHz $a = 0.1$ m	$f_a = 4$ kHz $a = 0.1$ m	$f_a = 8$ kHz $a = 0.1$ m	$f_a = 4$ kHz $a = 0.2$ m	$f_a = 4$ kHz $a = 0.4$ m
Direct	$f_u^{-1.1}$	f_u^{-1}	$f_u^{-0.8}$	f_u^{-1}	f_u^{-1}
Modified	$f_u^{-1.1}$	$f_u^{-1.2}$	$f_u^{-1.2}$	$f_u^{-1.6}$	f_u^{-2}

Tables 3.3 to 3.5 present the fitted expressions at some other typical values. It can be observed that for the direct CDM, the fitting exponent for audio frequency, f_a, the aperture size, a, and the ultrasound frequency, f_u, remains relatively stable

across different parameter values. On the other hand, in the modified CDM, the fitting exponent for audio frequency, f_a, and the ultrasound frequency, f_u, decreases as the aperture size, a, increases. Furthermore, the fitting exponent for the aperture size, a, decreases with an increase in audio frequency, f_a. This can be attributed to the dominance of the effective directivity, $\mathcal{D}_A(\theta, \varphi)$, which is introduced in the modified CDM [Eq. (2.77)] and primarily determined by the product of the audio frequency and the aperture size, $f_a a$. When the audio frequency and/or the aperture size increase, the main lobe of the effective directivity becomes narrower, resulting in a reduced contribution to the overall integral.

3.4.3 CONVERSION EFFICIENCY FROM ULTRASOUND TO AUDIO SOUND

For a PAL, the power conversion from ultrasound to audio sound is widely recognized as highly inefficient due to the intrinsic nonlinear process. To simplify the analysis, a circular piston source is considered when the velocity profile is uniform across the radiation surface, although the method presented in this section is applicable for a PAL with an arbitrary profile. In such a case, the ultrasound directivity is obtained by Eq. (2.83) as $\mathcal{D}_i(\theta, \varphi) = \mathrm{jinc}(k_i a \sin\theta)$, where $\mathrm{jinc}(\cdot)$ is the jinc function defined by Eq. (2.84). Consequently, the effective directivity is $\mathcal{D}_A = \mathrm{jinc}(k_a a \sin\theta)$. By substituting the ultrasound directivity into Eq. (3.34) and using the approximation $\mathrm{jinc}(k_i a)/(k_i a) \approx 1$ as $k_i a \gg 1$, the ultrasound power is obtained as

$$W_i \approx \frac{\pi a^2 p_0^2}{2\rho_0 c_0}. \tag{3.37}$$

The conversion efficiency from ultrasound to audio sound is defined as

$$\eta = \frac{W_a}{W_1 + W_2} \times 100\%, \tag{3.38}$$

where W_a is the audio sound power. Therefore, the conversion efficiency of a circular PAL with a uniform profile can be obtained by substituting Eqs. (3.35) and (3.37) into Eq. (3.38) as

$$\eta = \frac{\beta^2 p_0^2 a^6 \omega_1^2 \omega_2^2 \omega_a^4}{512\pi^3 \alpha_t^2 \rho_0^2 c_0^{12}} \int_0^{2\pi} \int_0^{\pi/2} |\mathcal{D}_a(\theta, \varphi)|^2 \sin\theta \mathrm{d}\theta \mathrm{d}\varphi \times 100\%. \tag{3.39}$$

Equation (3.39) shows that the conversion efficiency is proportional to the square of the amplitude of the on-surface pressure, p_0^2. However, the relationship between the conversion efficiency and other physical parameters cannot be determined directly.

In the Westervelt's model as detailed in Sec. 2.2.4 [Westervelt, 1963], the ultrasound waves are assumed to be collimated planar and confined inside the aperture area $\mathrm{H}(a-\rho)$, where $\mathrm{H}(\cdot)$ is the Heaviside function and $\rho = \sqrt{x^2+y^2}$ is the polar radial coordinate. Consequently, the sound intensity is approximated by $I = |p_0|^2/(2\rho_0 c_0)\mathrm{H}(a-\rho)$, and the sound power is obtained by $W_i = I\pi a^2 = \pi a^2 p_0^2/(2\rho_0 c_0)$. Finally, the expression for the conversion efficiency of the

Westervelt's model is then obtained by using Eq. (3.36) as

$$\eta_{\mathrm{W}} = \frac{\pi\beta^2 p_0^2 a^2 \omega_{\mathrm{a}}^3}{32\alpha_{\mathrm{t}}\rho_0^2 c_0^7} \times 100\%. \tag{3.40}$$

Equation (3.40) illustrates that the power conversion efficiency is proportional to the square of the on-surface ultrasound amplitude, p_0^2, the cube of the audio sound frequency, f_{a}^3, and the square of the aperture size, a^2. Additionally, it is approximately inversely proportional to the square of the ultrasound frequency, f_{u}^2.

Figure 3.24 shows the conversion efficiency from ultrasound to audio sound, as determined by Eqs. (3.39) and (3.40), across varying audio frequencies, f_{a}, aperture sizes, a, and average ultrasound frequencies, f_{u}. The on-surface pressure amplitude p_0 is set to 28.3 Pa (120 dB), which is a typical value for practical applications. The results indicate that the overall conversion efficiency of PALs is remarkably low, with values below 0.15 % for the given parameters. The Westervelt's model overestimates the conversion efficiency, yielding the highest estimation, while the modified CDM gives the lowest but the most accurate prediction, which is below 0.05% for the parameters adopted here.

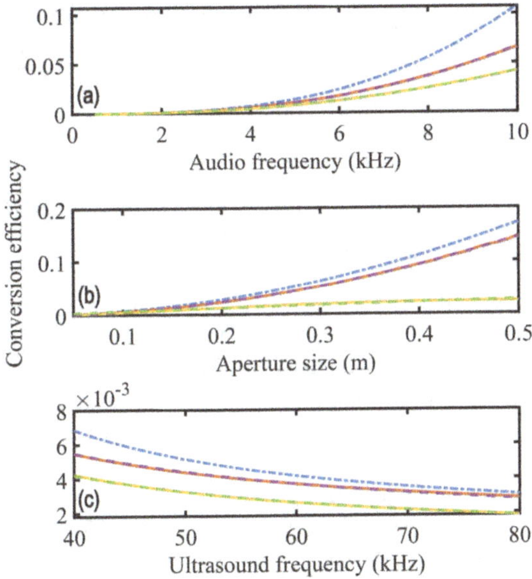

Figure 3.24 The acoustic conversion efficiency, η, of a circular PAL with a uniform profile for varying (a) audio frequency, (b) aperture size, and (c) ultrasound frequency. - - - , Westervelt; —— , direct CDM; —— , modified CDM; - - - , - - - , fitting curves. Extracted from [Li et al., 2023, Fig. 3].

Moreover, the results reveal a consistent trend between conversion efficiency and audio sound power. Specifically, the conversion efficiency increases with as both the audio frequency and aperture size increase, while it decreases as the ultrasound frequency increases. This relationship can be simply deduced by examining the fitting results of the audio sound power and the ultrasound power by Eq. (3.37). Considering that the ultrasound power is independent of the audio and ultrasound frequencies, it follows the same trend of the audio sound power as shown in Tables 3.3 and 3.5 (the coefficients of the fitting results are omitted). Since the ultrasound power is proportional to the square of the aperture size, a^2, the dependence of conversion efficiency on the aperture size can be obtained by dividing the expressions presented in Table 3.4 by a^2, and the results are shown in Table 3.6 (the coefficients of the fitting results are omitted). It shows that the exponent of the fitting results based on the modified CDM is very small, suggesting that the acoustic conversion efficiency is influenced to a slight extent by the aperture size.

Table 3.6

Fitting results for curves of the conversion efficiency, η, as a function of the aperture size, a, at different audio frequencies, f_a, and ultrasound frequencies, f_u, based on the direct and modified CDMs. Extracted from [Li et al., 2023, Table IV].

	$f_u = 40$ kHz $f_a = 1$ kHz	$f_u = 40$ kHz $f_a = 4$ kHz	$f_u = 40$ kHz $f_a = 8$ kHz	$f_u = 60$ kHz $f_a = 4$ kHz	$f_u = 80$ kHz $f_a = 4$ kHz
Direct	a^2	a^2	a^2	a^2	$a^{1.9}$
Modified	$a^{1.2}$	$a^{0.6}$	$a^{0.3}$	$a^{0.4}$	$a^{0.3}$

3.4.4 REMARKS

This section delves into the theoretical analysis and numerical simulations to explore the audio sound power generated by a PAL. The rationale behind this approach stems from the inherent challenges associated with accurately measuring the audio sound power of a PAL using conventional sound power measurement methods. The findings indicate that the audio sound power exhibits an approximate proportionality to the square of the audio frequency and aperture size, while demonstrating an inverse proportionality to the ultrasound frequency. Moreover, the conversion efficiency from ultrasound to audio sound is generally below 0.15% for typical configurations. Despite the complexities in direct measurement, the method presented in this section provides a reliable approach for predicting sound power, offering valuable insights and guidance for the practical applications of PALs.

It is important to note that a currently reliable method for measuring the sound power generated by a PAL is unavailable. Further research is needed to develop appropriate methodologies for accurately measuring the sound power of PALs. The theoretical framework employed in this section relies on the quasilinear approximation assumption. However, in applications characterized by strong nonlinearity, the quasilinear approximation may not be applicable. In such instances, the development of a more comprehensive theory becomes necessary to accurately predict the sound power generated by a PAL.

3.5 SUMMARY

The generation of audio sound by a PAL is a highly intricate process, resulting in a sound field that is complex than that produced by a conventional loudspeaker. The primary focus lies on the sound fields in front of a baffled PAL, which are discussed in detail in Sec. 3.2. In Sec. 3.2.1, the sound field in front of a PAL is divided into three regions based on the properties of audio sound pressure: the near field, the Westervelt far field, and the inverse-law far field. The near field, where local effects cannot be ignored, demands more accurate modeling, such as utilizing the Kuznetsov equation. The Westervelt far field is defined as the region where the Westervelt equation is sufficiently accurate, and the local effects characterized by the ultrasonic Lagrangian density can be neglected. The inverse-law far field is the region where the audio sound pressure is inversely proportional to the propagation distance. The transition distances from the near field to the Westervelt far field and to the inverse-law far field are analyzed in Secs. 3.2.2 and 3.2.3. The audio sound field distribution generated by a PAL is compared to that by a conventional loudspeaker, as presented in Sec. 3.2.4.

The audio beam generated by the PAL is highly directional, resulting in a very small audio sound behind the PAL, even when the PAL is not baffled. Section 3.3 employs the quasilinear approximation and the disk scattering theory to investigate the propagation of audio sound behind a non-baffled PAL. The audio sound power generated by a PAL, as well as the power conversion efficiency, is analyzed in Sec. 3.4. The challenges of sound power measurement for PALs are discussed. The results and conclusions presented in this section provide references and insights into the audio sound field produced by a PAL in free space.

4 Reflection, Transmission, and Scattering of Acoustic Waves Generated by PALs

4.1 INTRODUCTION

In Chap. 3, the reader is introduced to the sound fields generated by a PAL in free space. However, PALs find applications in diverse acoustic environments where sound waves can encounter various interactions such as reflection, transmission, and scattering. Consequently, the wave propagation of sound generated by PALs differs significantly from that in free space. This chapter delves into sound propagation for acoustic waves generated by PALs, covering aspects such as reflection from surfaces, transmission through thin partitions, and scattering by rigid spheres.

When a sound wave encounters a surface, it gives rise to a reflected wave due to changes in the acoustic impedance of the medium. The characteristics of this reflected wave are contingent upon the properties of the surface it interacts with. The reflection of audio sound generated by conventional loudspeakers has been extensively explored [Pierce, 2019, Chap. 3]. In these studies, conventional sound sources are typically represented as omnidirectional point sources at low frequencies. However, at middle and high frequencies, considerations such as radiation directivity and loudspeaker aperture size become important.

In the case of PALs, research has also delved into the reflection phenomena. For instance, PALs have been employed in investigations concerning sound absorption coefficients of materials in air [Castagnède et al., 2008; Romanova et al., 2019; Sugahara et al., 2019], as well as reflection and transmission coefficients of underwater elastomeric materials [Humphrey, 1985; Humphrey et al., 2008]. They have even been used in active noise control applications aimed at mitigating binaural noise at human ears [Tanaka et al., 2017]. In these scenarios, reflections occur at material surfaces, human skin, and hair, leading to complex acoustic interactions.

When a reflecting surface is situated near a PAL, it gives rise to both primary and secondary sound waves generated by the PAL, and these waves undergo reflection off the surface. The reflection phenomenon in the context of underwater applications, particularly involving pressure-release surfaces, has been studied [Muir et al., 1977]. This model operates on the assumption that primary fields initially exhibit plane wave characteristics within the Rayleigh distance, transitioning into spherical waves beyond that point. One key finding of this research is that the difference-frequency wave (DFW) generated by the incident primary waves exhibits an anti-phase relationship with itself after undergoing pressure-release reflection. Conversely, the DFW generated by the reflected primary waves aligns in-phase with the incident DFW.

DOI: 10.1201/9781003354994-4

Consequently, the DFW experiences a phase cancellation effect, a phenomenon that has been observed and validated through experiments [Muir et al., 1977].

These studies have been further expanded to encompass finite-sized planar targets with weak nonlinearity, utilizing a more precise model rooted in the KZK equation [Garrett et al., 1984]. Additionally, researchers have explored the reflection of the water-air (pressure-release) interface at small grazing angles to investigate its implications for acoustic communication in shallow-water channels [Wang et al., 1999]. Two theoretical models have been proposed for this scenario: a simplified Westervelt model where primary waves experience significant attenuation within the collimated zone, and a spherical spreading model where primary wave interactions are substantial in the far field characterized by a spherically spreading beam. Experimental investigations conducted at grazing angles of 5.4° and 7.7° have revealed that only the spherical spreading model aligns well with the observed experimental results.

In addition to the underwater experimental studies discussed in the literature, there have also been experimental investigations into the reflection of audio sound generated by a PAL in an air environment [Pompei, 2002]. These experiments yielded noteworthy findings: sound waves reflected from a rigid wall retain the same directivity as the incident beam, while those reflected from a wall equipped with a diffusive panel lose their directivity entirely. However, it is important to note that the research conducted in this study did not account for the effects of ultrasound reflection [Pompei, 2002]. When a PAL emits sound in the presence of a reflecting surface in an air environment, additional audio sound components are generated due to the reflection of ultrasound waves. While the models proposed in Wang et al., 1999 are applicable solely in the far field, and the model in Garrett et al., 1984 is valid in the near field but confined to the paraxial region, the non-paraxial model introduced in [Červenka and Bednařík, 2013; Zhong et al., 2020c] offers greater accuracy in wide-angle field scenarios but does not address reflections. In a more recent development in 2020, this non-paraxial model was further expanded to account for reflection effects [Zhong et al., 2020e]. This extended model is put to use in Sec. 4.2 to investigate the reflection of audio sound produced by a PAL.

When sound waves impinge on a partition, they give rise to both reflected waves on the incident side and transmitted waves on the transmission side. To quantify the impact of the partition, the concept of *insertion loss (IL)* is commonly employed. The IL refers to the reduction in sound pressure level at a specific location on the transmission side due to the presence of the partition [ISO 10847, 1997]. Understanding the insertion loss of a partition with respect to audio sound generated by a PAL holds significance in various applications. For instance, PALs, known for their ability to produce quasi-plane waves, can be utilized to measure the acoustic properties of materials in situ by gauging the sound pressure on the transmission side of the specimen [Castagnède et al., 2008]. Additionally, the sharp directivity of PALs is of interest to mobile phone designers [Ahn et al., 2019], particularly when seeking to maximize the effective radiating surface to generate substantial sound levels. In scenarios where a PAL is integrated beneath a phone screen, it becomes essential to comprehend the effects of the thin screen on the resulting audio sound. Furthermore, in experimental research, circular PALs may be required for validation purposes, even though commercial PALs are typically square or rectangular in shape. As detailed in Sec. 4.3.3, a circular PAL can be created by enclosing a square PAL with a 6 mm thick perspex panel.

The consideration of a thin partition within this book implies that the partition thickness is significantly smaller than both the audio wavelength and the effective absorption distance [Eq. (1.2)] of the PAL. The transmission of an audio plane wave through such a thin partition is a well-established phenomenon, often predicted using the mass law [Pierce, 2019]. Analytical derivations of transmission loss and insertion loss for spherical waves emitted by a point monopole encountering a thin partition have been conducted, employing the plane wave expansion method [Shi et al., 2008]. Furthermore, extensive studies have explored the transmission of diffuse incident sound through partitions [Pellicier and Trompette, 2007]. The analysis of audio sound transmission through a thin partition generated by a PAL was initially addressed in Zhong et al., 2020d, and comprehensive examination of this analysis is presented in Sec. 4.3.

In certain applications, human presence within the audible range introduces scattering effects on sound generated by a PAL [Nakashima et al., 2005; Ye et al., 2008]. While both the head and torso contribute to scattering, the impact of human heads can be more significant due to the presence of ears and the fact that only a small portion of sound is incident on the torso when highly directional beams are produced by a PAL. This section approximates the human head using a rigid sphere in its physical modeling, even though there is notable sound absorption attributed to the skin and hair at higher audio frequencies [Katz, 2000]. The effects of the rigid sphere on the audio sound generated by conventional loudspeakers have been well studied to reveal the effects of the human head [Zou and Qiu, 2008]. It is worth noting that the effects of a sphere on audio sound generated by a PAL differ from those on conventional loudspeakers because PALs produce sound through nonlinear wave interactions, necessitating the consideration of ultrasound wave scattering.

Utilizing the Westervelt equation and quasilinear approximations, an analytical expression for the sound pressure of the DFW was derived in scenarios where primary plane waves interact with a rigid sphere [Abbasov and Zagrai, 1995]. This study demonstrates that the total sound pressure of the DFW results from a combination of nonlinear interactions originating from the incident primary waves (incident-with-incident), the scattered primary waves (scattered-with-scattered), and interactions between the incident and scattered primary waves (incident-with-scattered). However, this solution exhibits inaccuracies due to the neglect of higher-order spherical harmonics in the Green's function and the assumption that scattered waves adhere to the plane wave model. Additionally, the accuracy of the Westervelt equation remains unexamined, leading to significant prediction errors, particularly in regions with substantial local effects near the sphere and between the sphere and the PAL. Although a study considered full spherical harmonics in Silva and Bandeira, 2013, the calculations were restricted to the far-field solution and involved a small sphere with a radius of only 1.0 mm, exposed to two intersecting plane waves underwater. Therefore, the findings and conclusions may differ significantly for a sphere comparable in size to a human head when exposed to audio sound generated by a piston-like PAL in an air medium. In 2022, a computationally efficient method was developed to calculate quasilinear solutions for audio sound generated by a PAL, incorporating both the Kuznetsov and Westervelt equations [Zhong et al., 2022c]. The methods and results of this approach are presented in Sec. 4.4.

4.2 REFLECTION FROM A SURFACE

4.2.1 PROBLEM DESCRIPTION

As shown in Fig. 4.1, a PAL generates two harmonic ultrasound waves at frequencies of f_1 and f_2 toward an infinitely large reflecting surface with an incident angle θ_0. The audio sound wave is nonlinearly generated at the frequency of $f_a = f_2 - f_1$, where $f_2 > f_1$. The distance between the PAL center and the reflecting surface is D. A Cartesian coordinate system $Oxyz$ is established with its origin, O, at the projection point of the PAL on the reflecting surface and the positive z-axis pointing to the centroid of the PAL. The PAL center is located at $\mathbf{r}_c = (x_c, y_c, z_c)$.

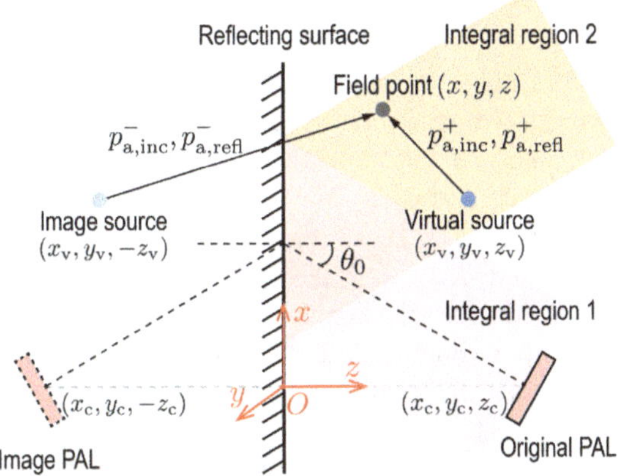

Figure 4.1 A PAL radiating sound toward an infinitely large reflecting surface with an incident angle θ_0.

4.2.2 THEORETICAL MODEL BASED ON THE IMAGE SOURCE METHOD

When the sound beams impinge on the reflecting surface, both ultrasound and audio sound waves are reflected. Following the principles of the image source method (ISM), the reflected sound field can be conceptualized as if it were generated by an image PAL positioned at $(x_v, y_v, -z_v)$, as illustrated in Fig. 4.1. Consequently, the total audio sound mainly consists of four components, and can be expressed as

$$p_{\mathrm{a,tot}}(\mathbf{r}) = p^+_{\mathrm{a,inc}}(\mathbf{r}) + p^-_{\mathrm{a,inc}}(\mathbf{r}) + p^+_{\mathrm{a,refl}}(\mathbf{r}) + p^-_{\mathrm{a,refl}}(\mathbf{r}) \tag{4.1}$$

where $p^+_{\mathrm{a,inc}}$ is generated by the nonlinear interactions of incident ultrasound, $p^-_{\mathrm{a,inc}}$ is the reflection of $p^+_{\mathrm{a,inc}}$ to satisfy the boundary condition on the reflecting surface for audio sound, $p^+_{\mathrm{a,refl}}$ is generated by the nonlinear interactions of reflected ultrasound, and $p^-_{\mathrm{a,refl}}$ is the reflection of $p^+_{\mathrm{a,refl}}$. These four components are analyzed individually in the following paragraphs. The nonlinear interactions of the incident and reflected ultrasound are neglected because of small source density of the virtual source resulted from them.

4.2.2.1 Audio Sound Generated by Incident Ultrasound

Based on Eq. (2.45), the audio sound generated by the incident ultrasound is obtained by

$$p^+_{\mathrm{a,inc}}(\mathbf{r}) = -\mathrm{i}\rho_0\omega_{\mathrm{a}} \iiint_{z_{\mathrm{v}}\geq 0} q_{\mathrm{a,inc}}(\mathbf{r}_{\mathrm{v}}) g_{\mathrm{3D}}(\mathbf{r},\mathbf{r}_{\mathrm{v}},\omega_{\mathrm{a}}) \mathrm{d}^3\mathbf{r}_{\mathrm{v}}, \tag{4.2}$$

where the 3D Green's function in free space $g_{\mathrm{3D}}(\mathbf{r},\mathbf{r}_{\mathrm{v}},\omega_{\mathrm{a}})$ is given by Eq. (2.36) with $|\mathbf{r}-\mathbf{r}_{\mathrm{v}}| = \sqrt{(x-x_{\mathrm{v}})^2+(y-y_{\mathrm{v}})^2+(z-z_{\mathrm{v}})^2}$. As shown by Eq. (2.44), the source density of the virtual source at $\mathbf{r}_{\mathrm{v}}$ is obtained by

$$q_{\mathrm{a,inc}}(\mathbf{r}_{\mathrm{v}}) = \frac{\beta\omega_{\mathrm{a}}}{\mathrm{i}\rho_0^2 c_0^4} p^*_{1,\mathrm{inc}}(\mathbf{r}_{\mathrm{v}}) p_{2,\mathrm{inc}}(\mathbf{r}_{\mathrm{v}}), \tag{4.3}$$

where the incident ultrasound $p_{i,\mathrm{inc}}(\mathbf{r}_{\mathrm{v}})$ is generated by the original PAL in free space at frequencies f_i, $i = 1,2$, which can be calculated using Rayleigh integral Eq. (2.38).

To fulfill the boundary condition on the reflecting surface for the audio sound $p^+_{\mathrm{a,inc}}$, an image for each virtual source located at $\mathbf{r}_{\mathrm{v}} = (x_{\mathrm{v}}, y_{\mathrm{v}}, z_{\mathrm{v}})$ is assumed to be at $\mathbf{r}_{\mathrm{v},-} = (x_{\mathrm{v}}, y_{\mathrm{v}}, -z_{\mathrm{v}})$. These virtual sources possess a source density of $R_{\mathrm{v}}(\omega_{\mathrm{a}})q_{\mathrm{a,inc}}(\mathbf{r}_{\mathrm{v}})$, where $R_{\mathrm{v}}(\omega_{\mathrm{a}})$ is the spherical wave reflection coefficient at frequency f_{a}. When the boundary is completely soft, $R_{\mathrm{v}}(\omega_{\mathrm{a}}) = 0$; when it is rigid, $R_{\mathrm{v}}(\omega_{\mathrm{a}}) = 1$. For an arbitrary impedance boundary, the specific value of $R_{\mathrm{v}}(\omega_{\mathrm{a}})$ varies with frequency, source and field point locations, incident angle, and the boundary impedance [Rudnick, 1947].

The reflection of the audio sound generated by the incident ultrasound ($p^+_{\mathrm{a,inc}}$) is considered as the sound field generated by all image virtual sources. Building upon Eq. (2.45), the sound pressure is expressed as

$$p^-_{\mathrm{a,inc}}(\mathbf{r}) = -\mathrm{i}\rho_0\omega_{\mathrm{a}} \iiint_{z_{\mathrm{v}}\geq 0} R_{\mathrm{v}}(\omega_{\mathrm{a}}) q_{\mathrm{a,inc}}(\mathbf{r}_{\mathrm{v}}) g_{\mathrm{3D}}(\mathbf{r},\mathbf{r}_{\mathrm{v},-},\omega_{\mathrm{a}}) \mathrm{d}^3\mathbf{r}_{\mathrm{v}}. \tag{4.4}$$

In the Green's function $g_{\mathrm{3D}}(\mathbf{r},\mathbf{r}_{\mathrm{v},-},\omega_{\mathrm{a}})$, the distance between the field point at $\mathbf{r}$ and the image virtual source at $\mathbf{r}_{\mathrm{v},-}$ is expressed as $|\mathbf{r}-\mathbf{r}_{\mathrm{v},-}| = \sqrt{(x-x_{\mathrm{v}})^2+(y-y_{\mathrm{v}})^2+(z+z_{\mathrm{v}})^2}$. The spherical wave reflection coefficient is difficult to measure. Because the audio beams generated by the PAL behave like plane waves in most scenarios, the plane wave reflection coefficient can be used for simplicity.

4.2.2.2 Audio Sound Generated by Reflected Ultrasound

The reflected ultrasound can be assumed to be that generated by the same PAL at the position of its image position multiplied by a plane wave reflection coefficient $R(\omega_1)$ and $R(\omega_2)$ at frequencies f_1 and f_2, respectively. The audio sound generated by the reflected ultrasound is then

$$p_{\mathrm{a,refl}}^{+}(\mathbf{r}) = -\mathrm{i}\rho_0\omega_{\mathrm{a}} \iiint_{z_{\mathrm{v}}\geq 0} R^*(\omega_1)R(\omega_2)q_{\mathrm{a,img}}(\mathbf{r}_{\mathrm{v}})g_{\mathrm{3D}}(\mathbf{r},\mathbf{r}_{\mathrm{v}},\omega_{\mathrm{a}})\mathrm{d}^3\mathbf{r}_{\mathrm{v}}, \tag{4.5}$$

where the source density of the image virtual source is

$$q_{\mathrm{a,img}}(\mathbf{r}_{\mathrm{v}}) = \frac{\beta\omega_{\mathrm{a}}}{\mathrm{i}\rho_0^2 c_0^4} p_{1,\mathrm{img}}^*(\mathbf{r}_{\mathrm{v}})p_{2,\mathrm{img}}(\mathbf{r}_{\mathrm{v}}), \tag{4.6}$$

and $p_{1,\mathrm{img}}(\mathbf{r}_{\mathrm{v}})$ and $p_{2,\mathrm{img}}(\mathbf{r}_{\mathrm{v}})$ are the sound pressure for the corresponding ultrasound generated by the image PAL in free space at frequencies f_1 and f_2, respectively. Similarly, to satisfy the boundary condition on the reflecting surface for audio sound, the reflection of $p_{\mathrm{a,refl}}^{+}$ is obtained as

$$p_{\mathrm{a,refl}}^{-}(\mathbf{r}) = -\mathrm{i}\rho_0\omega_{\mathrm{a}} \iiint_{z_{\mathrm{v}}\geq 0} R^*(\omega_1)R(\omega_2)R_{\mathrm{v}}(\omega_{\mathrm{a}})q_{\mathrm{a,img}}(\mathbf{r}_{\mathrm{v}})g_{\mathrm{3D}}(\mathbf{r},\mathbf{r}_{\mathrm{v},-},\omega_{\mathrm{a}})\mathrm{d}^3\mathbf{r}_{\mathrm{v}}. \tag{4.7}$$

4.2.3 REFLECTED SOUND FIELDS

4.2.3.1 Simulation Setups

In Sec. 4.2.2, the reader is introduced to the theoretical model for calculating the reflected sound field when a PAL is placed toward a reflecting surface. The key is to decompose the sound field into four components as shown by Eq. (4.1). The component $p_{\mathrm{a,inc}}^{+}$ represents the audio sound field generated by the incident ultrasound and is calculated by Eq. (4.2). The component $p_{\mathrm{a,inc}}^{-}$ calculated by Eq. (4.4) is the reflection of $p_{\mathrm{a,inc}}^{+}$ to satisfy the boundary condition on the reflecting surface for audio sound. The component $p_{\mathrm{a,refl}}^{+}$ represents the audio sound field generated by the reflected ultrasound and is calculated by Eq. (4.5). The component $p_{\mathrm{a,refl}}^{-}$ calculated by Eq. (4.7) is the reflection of $p_{\mathrm{a,refl}}^{+}$ to satisfy the boundary condition on the reflecting surface for audio sound. The total audio sound field is the superposition of these four components.

In this section, the audio sound field generated by a PAL placed toward a reflecting surface is simulated. The PAL is modeled by a circular piston source with a radius of $a = 0.1$ m. The surface vibration velocity amplitude v_0 is assumed to be $0.12\,\mathrm{m/s}$, so the on-surface ultrasound pressure, $p_0 = \rho_0 c_0 v_0 = 50\,\mathrm{Pa}$, is approximately 125 dB, which is a typical value in applications. The ultrasound frequencies are set as $f_1 = 60\,\mathrm{kHz}$ and $f_2 = 61\,\mathrm{kHz}$, and the audio frequency is then $f_{\mathrm{a}} = 1\,\mathrm{kHz}$. The absorption coefficient of ultrasound in air is 0.23 Np/m, which is calculated according to Appendix A at 20°C with a relative humidity being 50% and the ambient pressure being the standard atmospheric pressure. The Rayleigh distance and the absorption length are 5.5 m and 2.17 m obtained using Eqs. (1.111) and (1.114), respectively.

To simplify the calculation, the triple integral domain in Eq. (4.1) which is infinitely large, is constrained to a specific region encompassing the predominant energy of the ultrasound beams. Specifically, the integral domain is refined into two truncated cylindrical columns. These columns have a radius of 3 m (which is 30 times the PAL radius) and a length of 10 m (exceeding 4 times the effective absorption length). They are symmetrically centered on the axis of the PAL and its corresponding image. The first column, labeled as "Integration region 1" in Fig. 4.1, is dedicated to computing the nonlinear interactions of incident ultrasound, as described in Eqs. (4.2) and (4.4). It begins at the surface of the PAL, extending along the radiator axis, and terminates at the reflecting surface. The second column, labeled by "Integration region 2" in Fig. 4.1, is used for evaluating the nonlinear interactions of reflected ultrasound, as described in Eqs. (4.5) and (4.7). It starts from the end of the first column, proceeding in the direction of the axis corresponding to the image PAL. Only the ultrasound pressure within these two columns is taken into account. The integration of these expressions is performed numerically using the Simpson's 1/3 rule [Chapra and Canale, 2015, Sec. 21.2].

4.2.3.2 Rigid Reflecting Surface

Figure 4.2 shows the audio sound generated by a PAL in free space (at 30° incidence) at its original and image source locations, and the total audio sound field calculated by Eq. (4.1), where the reflecting surface is rigid for both ultrasound and audio sound, and the distance to the PAL is $D = 1$ m. It is clear that the total audio sound in Fig. 4.2(c) is the superposition of the other two in Figs. 4.2(a) and (b). The interference between the reflected and incident waves happens near the reflecting surface like two plane waves because the audio beams generated by the PAL behave like plane waves. The sound pressure on the back side ($z > D = 1$ m) focuses on the reflection axis and is almost equivalent to the sound radiated by the image PAL.

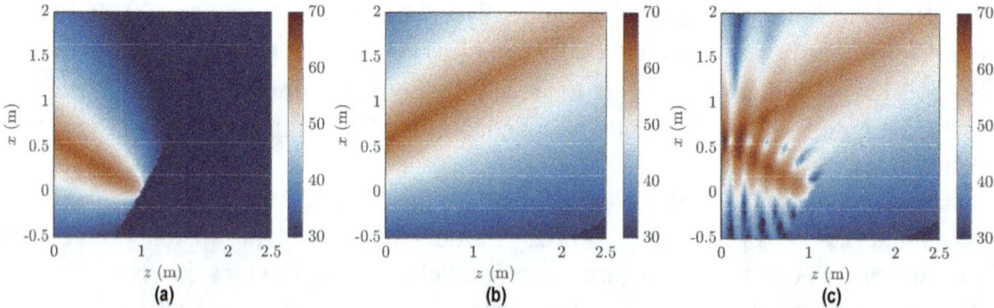

Figure 4.2 Audio SPL at 1 kHz generated by: (a) the original PAL in free space; (b) the image PAL with respect to the reflecting surface; and (c) the PAL near a rigid reflecting surface. Extracted from [Zhong et al., 2020e, Fig. 2].

For comparison with conventional sources, the incident, reflected, and total sound radiated by an audio piston source and a conventional directional sound source are calculated and shown in Fig. 4.3 at 1 kHz. The piston source is the same size as the PAL and mounted on an infinitely large baffle, so the sound radiates only in the forward direction. The directional source is a compact end-fire array consisting of 5 point monopoles with an interval of 45 mm as described in Tu et al., 2016. A comparison between Figs. 4.2 and 4.3 shows that the reflection for the audio sound generated by the PAL is much stronger than that generated by the other two conventional audio sound sources.

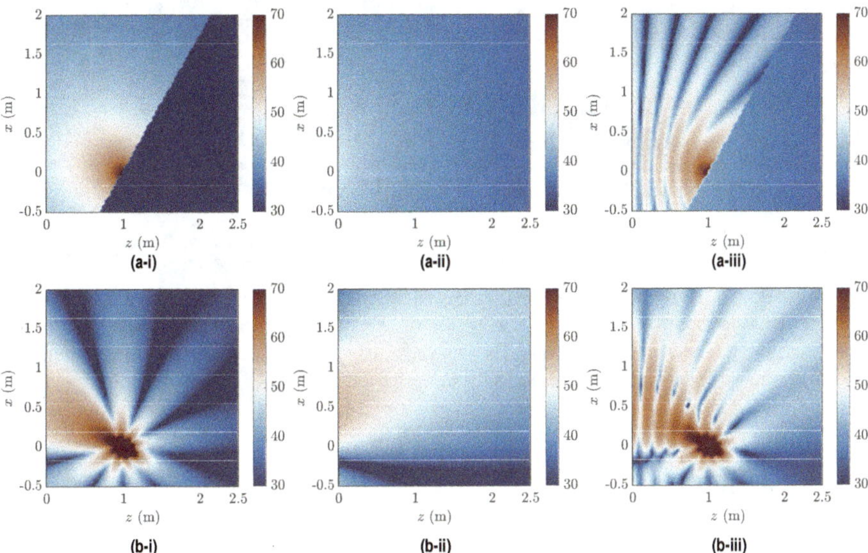

Figure 4.3 Audio SPL at 1 kHz where (i), (ii), and (iii) are the incident, reflected, and total sound radiated by (a) a piston source and a 5-channel end-fire array. Extracted from [Zhong et al., 2020e, Fig. 3].

4.2.3.3 Physical Mechanisms

The mechanism of the reflected audio sound generated by the PAL is different from that generated by conventional audio sources. It can be explained by analyzing the four components in Eq. (4.1) for the PAL. The calculated sound fields are shown in Fig. 4.4 using the same parameters in Fig. 4.2. The total sound pressure shown in Fig. 4.2(c) is the superposition of the audio sound generated by the incident ultrasound $p^+_{\mathrm{a,inc}}$ shown in Fig. 4.4(a), its reflection $p^-_{\mathrm{a,inc}}$ shown in Fig. 4.4(b), the audio sound generated by the reflected ultrasound $p^+_{\mathrm{a,refl}}$ shown in Fig. 4.4(c), and its reflection $p^-_{\mathrm{a,refl}}$ shown in Fig. 4.4(d). The audio sound generated by the original PAL [shown in Fig. 4.2(a)] is the superposition of $p^+_{\mathrm{a,inc}}$ and $p^-_{\mathrm{a,refl}}$, and the one generated by the image PAL [shown in Fig. 4.2(b)] is the superposition of $p^-_{\mathrm{a,inc}}$ and $p^+_{\mathrm{a,refl}}$. It can be found the audio sound generated by the reflected ultrasound ($p^+_{\mathrm{a,refl}}$) is the dominant contributor to the directivity of the reflected audio sound of the PAL. The

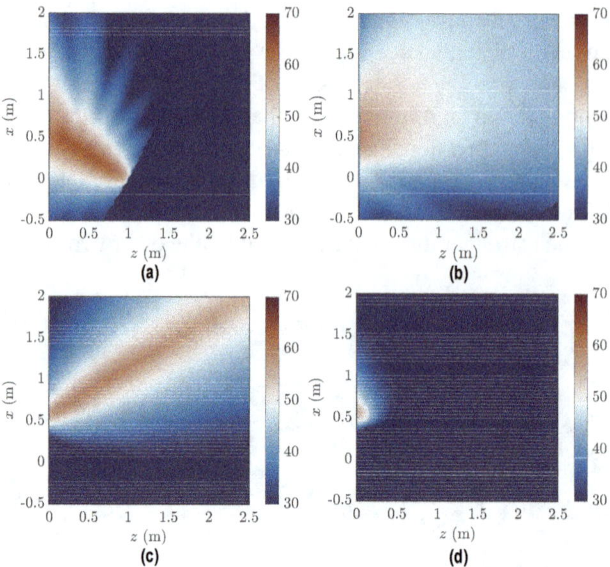

Figure 4.4 Audio SPL of four components generated by the PAL at 30° incidence near a rigid reflecting surface with a distance of 1 m at 1 kHz: (a) $p^+_{\mathrm{a,inc}}$ and (b) $p^-_{\mathrm{a,inc}}$ the audio sound generated by the incident ultrasound and its reflection, respectively; (c) $p^+_{\mathrm{a,refl}}$ and (d) $p^-_{\mathrm{a,refl}}$ the audio sound generated by the reflected ultrasound and its reflection, respectively. Extracted from [Zhong et al., 2020e, Fig. 4].

amplitude of the audio sound generated by reflected sound is affected by the distance between the PAL and the reflecting surface (D).

4.2.3.4 Effects of the Distance to the Reflecting Surface

Figure 4.5 shows the audio sound field generated by the PAL at 30° incidence with reflection surface at $D = 2$ m and 4 m. Compared with Fig. 4.2, the amplitude of the audio sound generated by the reflected sound becomes small as D increases, because the amplitude of the reflected ultrasound becomes smaller when the PAL moves farther away from the reflecting surface, especially when the distance is larger than the effective absorption length (2.17 m in this case).

4.2.3.5 Effects of the Ultrasonic Absorption Coefficients

In some applications, reflecting surfaces such as thin carpets can be highly absorbent for the ultrasound but less absorbent for audio sound. Figure 4.6 shows the audio sound generated by the original PAL and its image, as well as the total sound fields when the sound absorption coefficient of the reflecting surface is 0.5 and 0.9 for ultrasound $(1-|R^*(\omega_1)R(\omega_2)|)$, and 0 for audio sound. Because the reflected ultrasound is small with a large sound absorption coefficient, the total sound pressure mainly consists of the audio sound generated by the incident ultrasound. The directivity of the reflected audio beams becomes worse for a larger ultrasound absorption

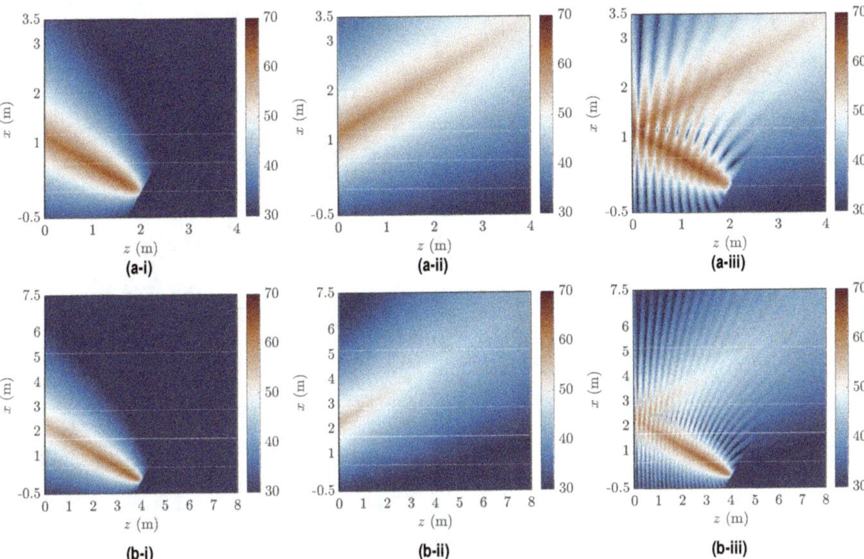

Figure 4.5 Audio SPL generated by the (i) original PAL and the (ii) image PAL, and the (iii) total sound fields with different distance between the PAL and the reflecting surface. The distance is (a) 2 m and (b) 4 m. Extracted from [Zhong et al., 2020e, Fig. 5].

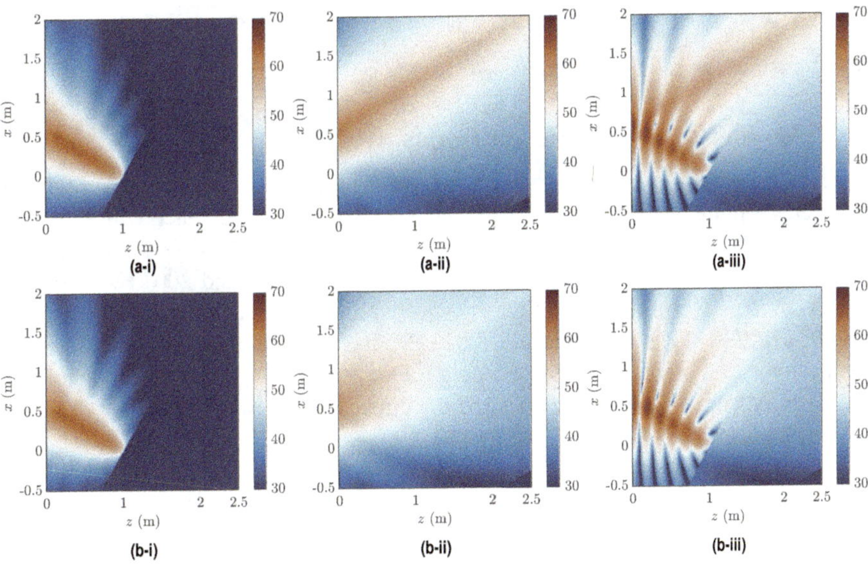

Figure 4.6 Audio SPL generated by the (i) original PAL and its (ii) image, and the (iii) total sound field with different sound absorption coefficient of the reflecting surface. The distance between the PAL and the reflection surface is 1 m. The ultrasound coefficient on the surface $z = 0$ is (a) 0.5 and (b) 0.9. Extracted from [Zhong et al., 2020e, Fig. 6].

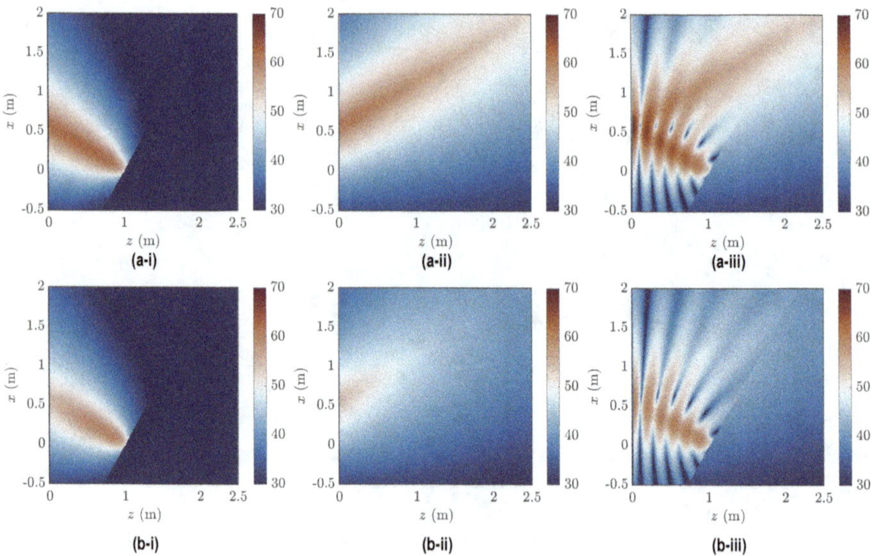

Figure 4.7 Audio SPL generated by the (i) original PAL and its (ii) image, and the (iii) total sound field at 30° incidence near a rigid reflecting surface with $D = 1$ m. (a) are for the ultrasound frequencies of 100 kHz and 101 kHz, (b) are for the ultrasound frequencies of 200 kHz and 201 kHz. The distance between the PAL and the reflection surface is 1 m. Extracted from [Zhong et al., 2020e, Fig. 7].

coefficient of the reflecting surface.

4.2.3.6 Effects of the Ultrasonic Frequencies

Sound absorption in air is different at different frequencies, especially at high frequencies. Figure 4.7 shows the audio sound of PAL at 30° incidence with reflection when $D = 1$ m. The ultrasound frequencies are 100 kHz and 101 kHz, or 200 kHz and 201 kHz. The absorption coefficients at 100 kHz (101 kHz) and 200 kHz (201 kHz) in air are 0.38 Np/m and 0.95 Np/m, respectively. The effective absorption lengths at 100 kHz (101 kHz) and 200 kHz (201 kHz) in free space are 1.32 m and 0.53 m, respectively. It can be found by comparing Fig. 4.7 with Fig. 4.2 that the amplitude of reflected audio beams decreases and the directivity deteriorates as the ultrasound frequency increases.

4.2.3.7 Experimental Observations

This section presents experimental observations of the reflection phenomenon associated with acoustic waves generated by a PAL. A sketch and photos of the experimental setup are shown in Figs. 4.8 and 4.9, respectively. The experiment involves the measurement of sound fields generated by a PAL, a conventional omnidirectional loudspeaker (point monopole), and a horn loudspeaker (directional source), both with and without a cotton sheet placed on the ground. The thickness of the cotton sheet is 250 μm and the surface density is 0.12 kg/m^2. The size of the cotton sheet is 2.8 m × 4 m and placed on the ground so that the projection of the centroid of the loudspeaker is on the bisector with respect to the narrower side (2.8 m), as shown in Fig. 4.8. The impedance tube (Brüel & Kjær Type 4206) was used to measure the absorption coefficient according to the two-microphone method specified in ISO 10523-2 [ISO 10534-2, 2001]. The result measured at 1 kHz is approximately 0.05 demonstrating that a little audio sound energy was absorbed by the cotton sheet. Therefore, it has negligible effects on the audio sound generated by the conventional loudspeakers.

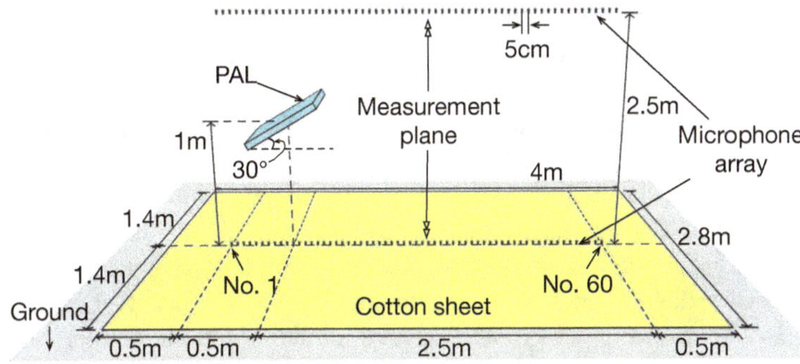

Figure 4.8 Sketch of the experimental setup when a PAL radiates toward ground with and without a cotton sheet. Extracted from [Zhong et al., 2020e, Fig. 8].

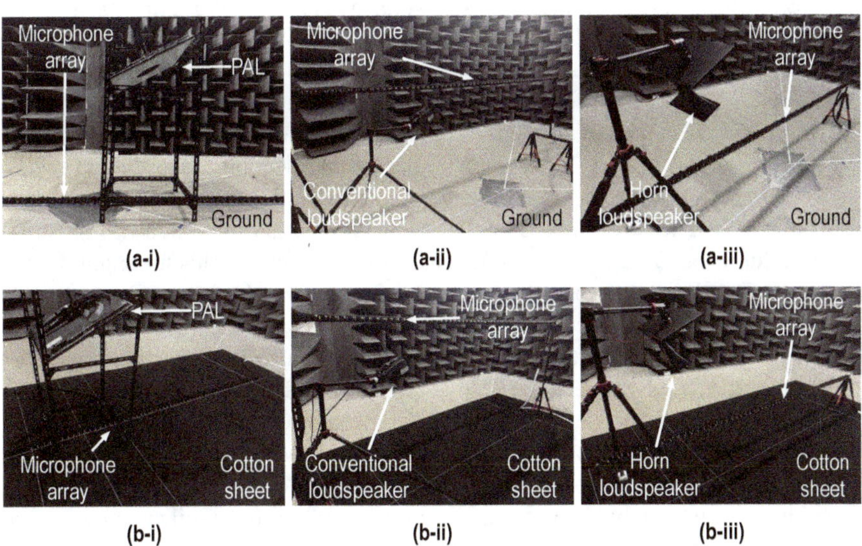

Figure 4.9 Photos of the experimental setups when various loudspeakers radiate toward the ground in the (a) absence and (b) presence of the cotton sheet: (i) the PAL, (ii) the conventional omnidirectional loudspeaker, and (iii) the horn loudspeaker. Extracted from [Zhong et al., 2020e, Fig. 9].

Figure 4.8 shows a sketch of the experimental setup when the PAL radiates toward ground. The sound field was measured at various points distributed on a vertical plane across the centroid of the testing loudspeaker. The length and the height of the measurement plane are 3 m and 2.5 m, respectively. A 60-channel microphone array with the microphone spacing of 5 cm was used to measure the sound pressure. The spacing between measurement points in the vertical direction is 5 cm when the microphone array is close to the loudspeaker, and 10 cm in the other areas. All measurement microphones were Brüel & Kjær Type 4957 microphones. The sound pressure was sampled with a Brüel & Kjær PULSE system (the analyzer 3053-B-120 with the input panel UA-2107-120).

The PAL, point sound source, and conventional directional source used in the experiments are a Holosonics Audio Spotlight AS-24i [Holosonics, 2019] with the surface size of 60 cm × 60 cm, a Genelec 8010A conventional loudspeaker, and a Daichi dome horn loudspeaker with a 24 cm × 8 cm rectangular opening, respectively. The carrier frequency of the PAL is 64 kHz according to measurements with a Brüel & Kjær Type 4939 microphone, and the audio frequency in the experiments was set to 1 kHz. The radiating surface of the PAL is covered by a 6 mm thick perspex panel with a hole of radius 0.1 m at its center to simulate the circular PAL used in simulations, as shown in Fig. 4.9(a). The relative humidity and the temperature in the experiments were 68% and 25.4°C, respectively.

Figure 4.10 shows the measured sound fields at 1 kHz generated by PAL, conventional, and horn loudspeakers at $\theta_0 = 30°$ incidence, with and without the cotton sheet on ground. Due to operation difficulties, the sound fields within the rectangular regions ($0.85\,\mathrm{m} \leq z \leq 2.5\,\mathrm{m}, -0.5\,\mathrm{m} \leq x \leq 0.35\,\mathrm{m}$), ($1\,\mathrm{m} \leq z \leq 1.2\,\mathrm{m}, -0.5\,\mathrm{m} \leq x \leq 0.35\,\mathrm{m}$), and ($0.9\,\mathrm{m} \leq z \leq 1.3\,\mathrm{m}, -0.5\,\mathrm{m} \leq x \leq 0.35\,\mathrm{m}$) were not measured for the 3 configurations, respectively, which are marked as blank regions in the figures.

In Fig. 4.10(a), it is observed that the reflected audio sound remains highly focused on the axis in the direction of reflection, consistent with the numerical simulations shown in Fig. 4.2(c). However, with the cotton sheet placed on the ground, as depicted in Fig. 4.10(d), the audio sound experiences a drop of up to 6 dB along the reflection axis. These results are similar to the simulation presented in Fig. 4.6(f), where an ultrasound absorption coefficient of 0.9 is assumed. This finding illustrates that only a small portion of the ultrasound reflects when the cotton sheet is on the reflective surface, consequently impacting the generation of audio sound by a PAL in close proximity to a reflective surface. This consequently affects the generation of audio sound of a PAL near a reflective surface, but the reflected sound generated by conventional loudspeakers remains nearly unchanged whether the cotton sheet is present or not. These results suggest that the reflected audio sound produced by PALs is influenced not only by the reflection of audio sound generated by incident ultrasound but also by newly generated audio sound resulting from reflected ultrasound. It is this latter component that governs the directivity of the reflected audio sound.

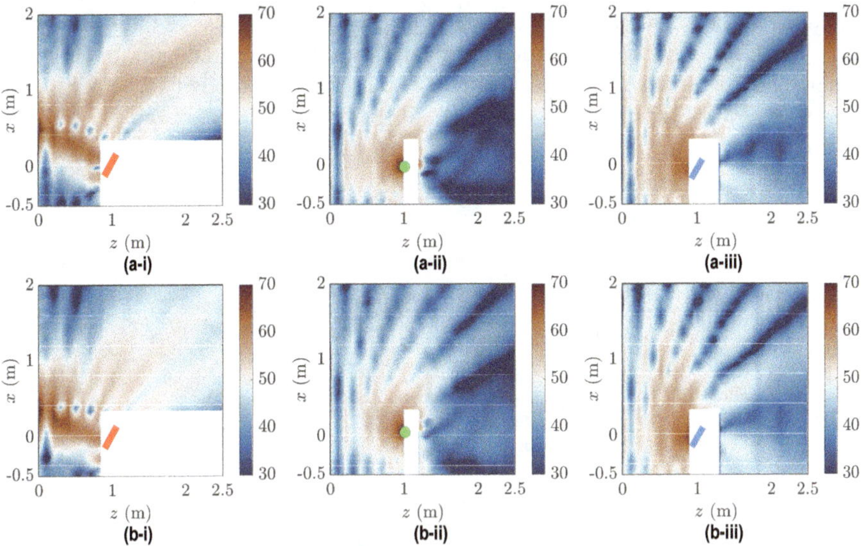

Figure 4.10 Measured audio SPL distributions at 1 kHz generated by a (i) PAL, (ii) conventional omnidirectional loudspeaker, and (iii) horn loudspeaker, in the (a) absence and (b) presence of the cotton sheet at $z = 0$. The incident angle is $\theta_0 = 30°$. Extracted from [Zhong et al., 2020e, Fig. 10].

4.2.4 REMARKS

The reader is introduced to the reflection phenomenon associated with acoustic waves generated by PALs. The simulation results presented in this section demonstrate that when ultrasound waves impinge on a reflecting surface, the generation of audio sound is not solely attributed to the incident ultrasound but also involves contributions from the reflected ultrasound. Consequently, it becomes imperative to consider the reflective properties of ultrasound, leading to distinct characteristics in the reflection of audio sound generated by a PAL compared to a conventional directional source. Notably, the reflected ultrasound preserves the sharp directivity observed in the reflected audio sound. However, in the case of a PAL, if the reflecting surface is highly absorptive for ultrasound, the directivity of the reflected audio sound is attenuated because the reflected ultrasound is of diminished magnitude. The experimental observations are presented to valid the conclusions.

4.3 TRANSMISSION THROUGH A THIN PARTITION

4.3.1 PROBLEM DESCRIPTION

A sketch of a PAL radiating sound through a partition is shown in Fig. 4.11, where the radiator axis of the PAL is perpendicular to the partition surface for simplicity. A Cartesian coordinate system $Oxyz$ is established with its origin O at the projection of the centroid of the PAL on the partition surface and the z-axis is perpendicular to the surface. The centroid of the PAL is at $(0, 0, z_s)$, so that the source point on the radiation surface is denoted by $\mathbf{r}_s = (x_s, y_s, z_s)$, where $z_s < 0$. The thickness of the infinitely large partition is assumed to be small enough compared to the audio wavelength and the effective absorption length of the PAL for simplicity. The partition is placed at $z = 0$ and its area density is denoted by σ.

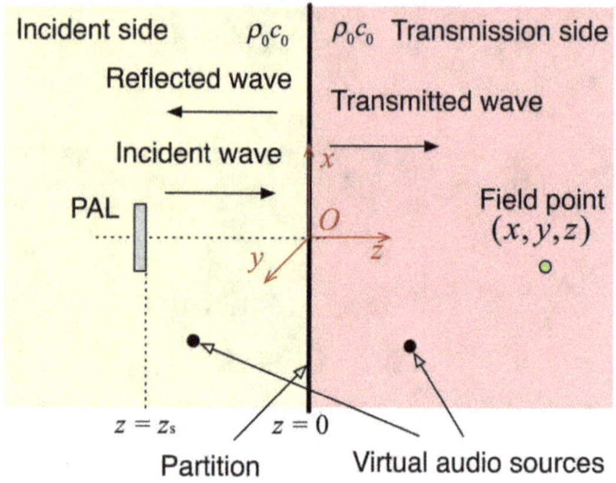

Figure 4.11 Sketch for a PAL radiating toward a thin partition. Extracted from [Zhong et al., 2020d, Fig. 1].

4.3.2 THEORETICAL MODEL

When the PAL generates ultrasound at two different frequencies, f_1 and f_2 (where $f_2 > f_1$), an infinite number of virtual audio sources at frequency $f_a = f_2 - f_1$ emerges throughout the space. Upon the ultrasonic wave encountering a partition, it gives rise to reflected and transmitted waves. These waves generate new virtual audio sources on both the incident side ($z < 0$) and the transmission side ($z > 0$). The audio sound produced within the partition due to ultrasound is minimal, and for simplicity, it is disregarded in the analysis.

The audio sound generated by these three kinds of virtual audio sources—those generated by incident, reflected, and transmitted ultrasound—all propagate through the partition, leading to the emergence of reflected and transmitted audio sound. For the purposes of calculating the insertion loss of the partition in this section, our focus lies solely on the audio sound present on the transmission side ($z > 0$). According to the preceding analysis, four distinct audio sound components exist on the transmission side: the transmitted audio sound resulting from (1) incident and (2) reflected ultrasound, and the audio sound arising from (3) transmitted ultrasound and its (4) reflection on the transmission side.

Because the audio sound generated by the reflected ultrasound radiates in the direction which is away from the partition toward the PAL source direction, the transmitted sound of these audio sound is small and can be neglected. Similarly, the reflection of the audio sound generated by the transmitted ultrasound can be neglected as well. Therefore, two audio sound components dominate the sound field on the transmission side. The total sound pressure of audio sound can be approximately expressed as

$$p_{\mathrm{a,tot}}(\mathbf{r}) = p_{\mathrm{a,inc}}(\mathbf{r}) + p_{\mathrm{a,tr}}(\mathbf{r}), \quad z > 0, \tag{4.8}$$

where $p_{\mathrm{a,inc}}(\mathbf{r})$ and $p_{\mathrm{a,tr}}(\mathbf{r})$ represent the transmitted sound of the audio sound generated by incident ultrasound and the audio sound generated by transmitted ultrasound, respectively.

4.3.2.1 Transmission of Audio Sound Generated by Incident Ultrasound

The incident ultrasound can be calculated by Rayleigh integral Eq. (2.38) as

$$p_i(\mathbf{r}) = -2\mathrm{i}\rho_0\omega_i \iint_{-\infty}^{\infty} v_{i,z}(\mathbf{r}_\mathrm{s}) g_{\mathrm{3D}}(\mathbf{r}, \mathbf{r}_\mathrm{s}, \omega_i) \mathrm{d}^2\mathbf{r}_\mathrm{s}, \quad z_\mathrm{s} < z < 0, \tag{4.9}$$

where $v_{i,z}(\mathbf{r}_\mathrm{s})$ is the velocity profile on the radiation surface at the ultrasonic frequency f_i, and the Green's function in free space $g_{\mathrm{3D}}(\mathbf{r}, \mathbf{r}_\mathrm{s}, \omega_i)$ is given by Eq. (2.36).

The audio sound generated by incident ultrasound at field point $\mathbf{r}$ on the incident side can be expressed as

$$p_{\mathrm{a,inc}}(\mathbf{r}) = -\mathrm{i}\rho_0\omega_\mathrm{a} \int_{z_\mathrm{s}}^{0} \int_{-\infty}^{\infty} \int_{-\infty}^{\infty} q_{\mathrm{a,inc}}(\mathbf{r}_\mathrm{v}) g_{\mathrm{3D}}(\mathbf{r}, \mathbf{r}_\mathrm{v}, \omega_\mathrm{a}) \mathrm{d}^3\mathbf{r}_\mathrm{v}, \quad z_\mathrm{s} < z < 0, \tag{4.10}$$

where the source density function at the virtual source point $\mathbf{r}_\mathrm{v}$ is obtained by Eq. (2.44) as

$$q_{\mathrm{a,inc}}(\mathbf{r}_\mathrm{v}) = \frac{\beta\omega_\mathrm{a}}{\mathrm{i}\rho_0^2 c_0^4} p_{1,\mathrm{inc}}^*(\mathbf{r}_\mathrm{v}) p_{2,\mathrm{inc}}(\mathbf{r}_\mathrm{v}), \quad z_\mathrm{s} < z_\mathrm{v} < 0. \tag{4.11}$$

It is important to note that Eq. (4.10) does not account for local effects. Figure 4.12 illustrates the on-axis audio SPL generated by a circular PAL in free space, both with and without the inclusion of local effects, using the parameters consistent with those employed throughout this section. The results, when considering local effects, exhibit significant fluctuations in the near field, primarily attributed to the Lagrangian density of the ultrasonic waves. Notably, the distance between two

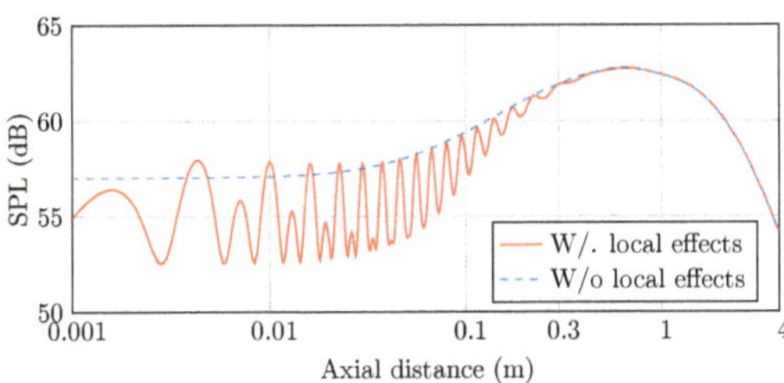

Figure 4.12 On-axis audio SPL at 1 kHz generated by a circular PAL in free space with and without including local effects. The parameters are $a = 0.1$ m, $p_0 = 50$ Pa (125 dB), and $f_\mathrm{u} = 60$ kHz.

peaks of the curve is on the same order of magnitude as the ultrasonic wavelength (5.4 mm). These fluctuation pose practical challenges in experimental measurements presented in the subsequent section, as the distance between neighboring measurement points is only 0.1 m. When the distance between the field point and the PAL exceeds 0.3 m, the difference between the two curves becomes less than 0.2 dB. Therefore, for the specific problem investigated in this section, it is reasonable to neglect the local effects induced by ultrasonic waves without introducing substantial errors.

The transmitted sound of the audio sound generated by incident ultrasound can be calculated by using the plane wave expansion. For a point monopole located at $\mathbf{r}_\mathrm{m} = (x_\mathrm{m}, y_\mathrm{m}, z_\mathrm{m})$ with a source strength of Q_m, the transmitted sound at field point $\mathbf{r}$ on the transmission side of the partition can be expressed as

$$p_{\mathrm{tr,m}}(\mathbf{r}) = -\frac{Q_\mathrm{m}}{4\pi} K(\mathbf{r}, \mathbf{r}_\mathrm{m}), \quad z > 0, \quad z_\mathrm{m} < 0, \tag{4.12}$$

where the Weyl's integral is expressed as [Williams, 1999, Eq. (2.64)]

$$K(\mathbf{r}, \mathbf{r}_\mathrm{m}) = \frac{\mathrm{i}}{2\pi} \int_{-\infty}^{\infty} \int_{-\infty}^{\infty} \frac{T_\mathrm{plane}}{k_{\mathrm{a},z}} \mathrm{e}^{\mathrm{i}[k_x(x-x_\mathrm{m}) + k_y(y-y_\mathrm{m}) + k_{\mathrm{a},z}|z-z_\mathrm{m}|]} \mathrm{d}k_x \mathrm{d}k_y, \tag{4.13}$$

$$k_{\mathrm{a},z} = \sqrt{k_\mathrm{a}^2 - k_\rho^2}, \quad k_\rho^2 = k_x^2 + k_y^2, \tag{4.14}$$

and the unit of $K(\mathbf{r}, \mathbf{r}_\mathrm{m})$ is the same as the wavenumber. T_plane is the sound pressure transmission coefficient for the plane wave with the wave vector $\mathbf{k}_\mathrm{a} = (k_x, k_y, k_{\mathrm{a},z})$ [Pierce, 2019, Sec. 3.8].

$$T_\mathrm{plane} = \frac{2\rho_0 \omega_\mathrm{a} / k_{\mathrm{a},z}}{2\rho_0 \omega_\mathrm{a} / k_{\mathrm{a},z} - \mathrm{i}\omega_\mathrm{a}\sigma}. \tag{4.15}$$

If the thin partition is porous material, the flow resistance is taken into account and the sound pressure transmission coefficient is modified as [Pierce, 2019, Sec. 3.8]

$$T_\mathrm{plane} = \frac{2\rho_0 \omega_\mathrm{a} / k_{\mathrm{a},z}}{2\rho_0 \omega_\mathrm{a} / k_{\mathrm{a},z} + [1/R_\mathrm{f} - 1/(\mathrm{i}\omega_\mathrm{a}\sigma)]^{-1}}, \tag{4.16}$$

provided that the material is homogeneous, where R_f is the specific flow resistance.

With the above expressions, the transmitted sound of the audio sound generated by incident ultrasound on the transmission side in Eq. (4.10) is expressed as

$$p_{\mathrm{a,inc}}(\mathbf{r}) = -\frac{\mathrm{i}\rho_0 \omega_\mathrm{a}}{4\pi} \int_{z_\mathrm{s}}^{0} \int_{-\infty}^{\infty} \int_{-\infty}^{\infty} q_{\mathrm{a,inc}}(\mathbf{r}_\mathrm{v}) K(\mathbf{r}, \mathbf{r}_\mathrm{v}) \mathrm{d}^3\mathbf{r}_\mathrm{v}, \quad z > 0, \tag{4.17}$$

Equation (4.13) is hard to converge because the small sound attenuation coefficient at the audio frequency makes the integrand singular at $k_{\mathrm{a},z} = \sqrt{k_\mathrm{a}^2 - k_\rho^2} \approx 0$ when $k_\rho = \Re k_\mathrm{a}$. By using the variable substitution and the integral representation for the

Bessel function of order zero Eq. (B.21), Eq. (4.13) reduces to

$$
\begin{aligned}
K(\mathbf{r},\mathbf{r}_\mathrm{v}) =& \mathrm{i}\int_0^{\pi/2} T_\mathrm{plane} k_\mathrm{a}^2 J_0(\Re k_\mathrm{a}\rho_\mathrm{v}\cos\gamma)\mathrm{e}^{\mathrm{i}\sqrt{k_\mathrm{a}^2-(\Re k_\mathrm{a})^2\cos^2\gamma}|z-z_\mathrm{v}|} \\
&\times \frac{\sin\gamma\cos\gamma}{\sqrt{k_\mathrm{a}^2-(\Re k_\mathrm{a})^2\cos^2\gamma}}\mathrm{d}\gamma \\
&+\int_0^{\infty} T_\mathrm{plane} k_\mathrm{a}^2 J_0(\Re k_\mathrm{a}\rho_\mathrm{v}\cosh\gamma)\mathrm{e}^{-\sqrt{(\Re k_\mathrm{a})^2\cosh^2\gamma-k_\mathrm{a}^2}|z-z_\mathrm{v}|} \\
&\times \frac{\sinh\gamma\cosh\gamma}{\sqrt{(\Re k_\mathrm{a})^2\cosh^2\gamma-k_\mathrm{a}^2}}\mathrm{d}\gamma,
\end{aligned}
\tag{4.18}
$$

where $\rho_\mathrm{v} = \sqrt{(x-x_\mathrm{v})^2+(y-y_\mathrm{v})^2}$ is the transverse distance between the field point and the virtual source point, both integrals on the right-hand side are numerically calculated by the Simpson's 1/3 rule, and the dummy variable of the second integral, γ, is replaced by $\tan[(\gamma+1)\pi/4]$ to transform the infinitely large interval into a finite one.

4.3.2.2 Audio Sound Generated by Transmitted Ultrasound

Due to the computational inefficiency of applying the plane wave expansion directly to the incident ultrasound, the Gaussian beam expansion (GBE) method, detailed in Sec. 2.4, is employed in this context. The GBE for ultrasound is derived using Eq. (2.125) as

$$
p_{i,\mathrm{inc}}(\mathbf{r}) = p_0\mathrm{e}^{\mathrm{i}k_i(z-z_\mathrm{s})}\sum_{n=1}^{N}\frac{A_n}{1+\mathrm{i}B_n\tilde{z}_i}\exp\left(-\frac{B_n\tilde{\rho}^2}{1+\mathrm{i}B_n\tilde{z}_i}\right), \quad z_\mathrm{s}<z<0. \tag{4.19}
$$

Here, the index $i = 1,2$ represents the ultrasound at frequency f_i.

It should be noted that the GBE assumes the paraxial approximation for ultrasound waves. Figure 4.13 presents a comparison of the audio SPL along a line perpendicular to the radiator axis of the circular PAL in free space. This comparison is based on the Westervelt equation, both with and without the paraxial approximation assumed for ultrasonic waves. The results indicate that the difference is less than 0.4 dB when the axial distance exceeds 1 m, underscoring that the adoption of the GBE does not introduce significant errors.

The ultrasound can be expanded into plane waves using the plane wave expansion method to give

$$
\begin{aligned}
p_{i,\mathrm{inc}}(\mathbf{r}) =& \frac{p_0 a^2}{4\pi}\sum_{n=1}^{N}\frac{A_n}{B_n} \\
&\times \iint_{-\infty}^{\infty}\exp\left\{-\frac{(k_x^2+k_y^2)a^2}{4B_n}+\mathrm{i}[k_x x+k_y y+k_{i,z}(z-z_\mathrm{s})]\right\}\mathrm{d}k_x\mathrm{d}k_y,
\end{aligned}
\tag{4.20}
$$

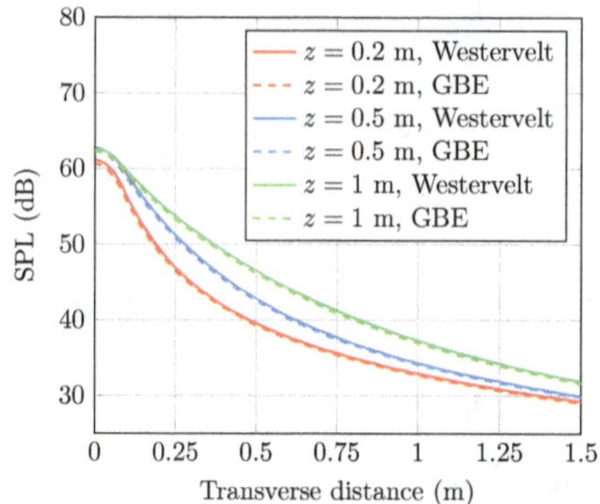

Figure 4.13 The audio SPL at 1 kHz along the line perpendicular to the radiator axis of the circular PAL in free space based on the Westervelt equation with (labeled "GBE") and without (labeled "Westervelt") the paraxial approximation assumed for ultrasonic waves. The parameters are $a = 0.1\,\text{m}, p_0 = 50\,\text{Pa}\ (125\,\text{dB})$, and $f_\text{u} = 60\,\text{kHz}$.

where $k_{i,z} = \sqrt{k_i^2 - k_\rho^2}$. Similar to Eq. (4.13), the transmitted ultrasound can be obtained as

$$p_{i,\text{tr}}(\mathbf{r}) = \frac{p_0 a^2}{2} \sum_{n=1}^{N} \frac{A_n}{B_n} \int_0^\infty T_\text{plane} J_0(k_\rho \rho) k_\rho \exp\left[-\frac{k_\rho^2 a^2}{4B_n} + \mathrm{i}k_{i,z}(z - z_\text{s})\right] \mathrm{d}k_\rho, \quad (4.21)$$

where $z > 0$, T_plane is the sound pressure transmission coefficient for a plane wave with the wave vector $\mathbf{k}_i = (k_x, k_y, k_{i,z})$, $\rho = \sqrt{x^2 + y^2}$, and the integral on the right-hand side is numerically calculated by the Simpson's 1/3 rule here.

The audio sound generated by transmitted ultrasound on the transmission side can be expressed as

$$p_{\text{a,tr}}(\mathbf{r}) = -\mathrm{i}\rho_0 \omega_\text{a} \int_0^\infty \int_{-\infty}^\infty \int_{-\infty}^\infty q_{\text{a,tr}}(\mathbf{r}_\text{v}) g_\text{3D}(\mathbf{r}, \mathbf{r}_\text{v}, \omega_\text{a}) \mathrm{d}^3\mathbf{r}_\text{v}, \quad z > 0, \quad (4.22)$$

where the source density function of the virtual audio source at $\mathbf{r}_\text{v}$ is

$$q_{\text{a,tr}}(\mathbf{r}_\text{v}) = \frac{\beta \omega_\text{a}}{\mathrm{i}\rho_0^2 c_0^4} p_{1,\text{tr}}^*(\mathbf{r}_\text{v}) p_{2,\text{tr}}(\mathbf{r}_\text{v}), \quad z_\text{v} > 0. \quad (4.23)$$

4.3.2.3 Insertion Loss of the Partition

The IL of the partition, concerning the audio sound produced by the PAL, is defined as the difference of the SPL at a designated receiver location $\mathbf{r}$ on the transmission side, in the presence and absence of the partition [ISO 10847, 1997]. It is expressed as

$$\text{IL}_\text{a}(\mathbf{r}) = 20\log_{10}\left(\left|\frac{p_{\text{a},0}(\mathbf{r})}{p_{\text{a,tot}}(\mathbf{r})}\right|\right) \quad (\text{dB}), \quad z > 0, \quad (4.24)$$

where $\log_{10}$ represents the common logarithm with the base of 10. In Eq. (4.24), $p_{a,0}(\mathbf{r})$ is the sound pressure at $\mathbf{r}$ in the absence of the partition. This value can be derived by changing the upper limit of the integral with respect to the coordinate z_v in Eq. (4.10) from 0 into the positive infinity.

For comparison, the insertion loss of the partition for a plane wave with the normal incidence (IL_n), for a spherical wave from a point monopole (IL_m), and for a directional incident sound from a directional end-fire array source consisting of 5 point monopoles used in Tu et al., 2016 (IL_d) are calculated with following equations.

$$IL_n = 20\log_{10}\left(\left|\frac{2\rho_0\omega_a/k_{a,z} - i\omega_a M}{2\rho_0\omega_a/k_{a,z}}\right|\right) \quad \text{(dB)}, \tag{4.25}$$

$$IL_m(\mathbf{r}) = 20\log_{10}\left(\left|\frac{\exp(-\alpha_a|\mathbf{r}-\mathbf{r}_m|)}{|\mathbf{r}-\mathbf{r}_m|K(\mathbf{r},\mathbf{r}_m)}\right|\right) \quad \text{(dB)}, \quad z > 0, \tag{4.26}$$

$$IL_d(\mathbf{r}) = 20\log_{10}\left(\left|\frac{\sum_{n=1}^{5} Q_n|\mathbf{r}-\mathbf{r}_n|^{-1}e^{ik_a|\mathbf{r}-\mathbf{r}_n|}}{\sum_{n=1}^{5} Q_n K(\mathbf{r},\mathbf{r}_n)}\right|\right) \quad \text{(dB)}, \tag{4.27}$$

where the location of the point monopole is at $\mathbf{r}_m = (x_m, y_m, z_m)$, and the n-th point monopole of the end-fire array is at $\mathbf{r}_n = (x_n, y_n, z_n)$ with a volume velocity of Q_n.

4.3.3 TRANSMITTED SOUND FIELDS

4.3.3.1 Simulation Setups

In the following simulations, a circular piston source with a radius of $a = 0.1$ m is considered. The on-surface ultrasound pressure amplitude is set as $p_0 = 50$ Pa (125 dB). The average ultrasound frequency is set as $f_u = 60$ kHz. The ultrasound absorption coefficient in air is determined to be 0.228 Np/m, which is calculated according to Appendix A at a temperature of 20°C and a relative humidity of 50% under standard atmospheric pressure.

Two distinct materials serve as the thin partition in the simulations. The first material is an aluminum foil with a thickness of 50 μm, possessing a bulk density of 2.7×10^3 kg/m^3. The second one is a polyester fiber blanket, which has a thickness of 1 mm, a bulk density of 20 kg/m^3, and a flow resistivity of 2×10^3 Pa·s/m^2 [Garai and Pompoli, 2005, Table 1].

In the calculation, it becomes necessary to confine the infinitely large integral domain in Eqs. (4.10) and (4.22) to a specific region that covers the major energy of ultrasound beams. As a result, the integral domain is constrained to a cylindrical column centered on the axis of the PAL, featuring a radius of 3 m (30 times the PAL radius). The expression for the transmitted audio sound generated by incident ultrasound is presented in Eq. (4.17), involving a fourfold integral that is time-consuming when computing the off-axis field. Because $K(\mathbf{r},\mathbf{r}_v)$ depends only on the transverse distance (ρ_v) and longitudinal distance ($|z-z_v|$) between the two points, it can be pre-calculated at every transverse and longitudinal distance pair. Subsequently, this pre-calculated value can be substituted into Eq. (4.17), simplifying the integral into a threefold one, provided the frequency and thickness of the partition are specified.

4.3.3.2 Effects of a 50 μm Thick Aluminum Partition

In Fig. 4.14, a comparison is presented among the audio sound radiated by a PAL, a point monopole, and a conventional directional source (specifically, a 5-channel end-fire array [Tu et al., 2016]) at $f_a = 1$ kHz, all with and without the presence of an aluminum partition. The source is located at $z_s = -1.0$ m. Figures 4.14(a–c) show the original sound fields of the sources without the partition, while Figs. 4.14(d–f) show the total sound fields of the sources with the partition. The sound fields on the incident side are omitted in Figs. 4.14(d–f) to emphasize sound transmission. Specifically, Figs. 4.14(a) and (d) are for the PAL, Figs. 4.14(b) and (e) are for the point monopole, and Figs. 4.14(c) and (f) are for the end-fire array.

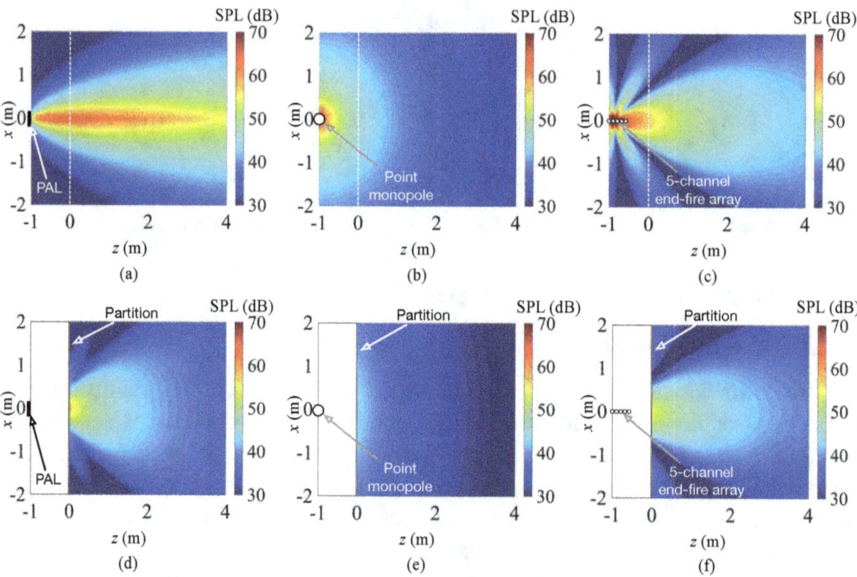

Figure 4.14 Audio SPL at $f_a = 1$ kHz with a 50 μm thick aluminum partition and a source at $z_s = -1$ m without the partition for the: (a) PAL, (b) point monopole, (c) end-fire array, and with the partition for the: (d) PAL, (e) point monopole, (f) end-fire array. The average ultrasound frequency $f_u = 60$ kHz. Extracted from [Zhong et al., 2020d, Fig. 2].

It is evident from Fig. 4.14 that the pattern of the sound field generated by a PAL undergoes significant changes behind the partition. The audio sound becomes less focused on the radiator axis, indicating a diminished directivity. In contrast, the pattern of sound fields radiated by a point monopole or an end-fire array exhibits only slight modifications behind the partition. This discrepancy arises from the fact that the directional audio sound emitted by the PAL is predominantly composed of ultrasound, which is substantially impeded by the partition. On the other hand, the thin aluminum foil is nearly transparent to audio sound radiated by conventional sound sources.

Figure 4.15 depicts the IL along the radiator axis ($\rho = 0$) on the transmission side ($z > 0$) for various source types. The IL for the ultrasound is approximately 35.7 dB, resulting in a considerable reduction of about 71.4 dB in the audio sound generated by the transmitted ultrasound. This reduction is attributed to the approximate proportionality between the magnitudes of the sound pressure for audio sound and the square of the magnitudes of ultrasound. In contrast, the IL for conventional sources such as the plane wave, point monopole, or end-fire array is approximately 3.2 dB. Notably, the IL of the audio sound generated by the PAL behind the partition can exceed 17 dB, indicating a more substantial attenuation compared to conventional sources.

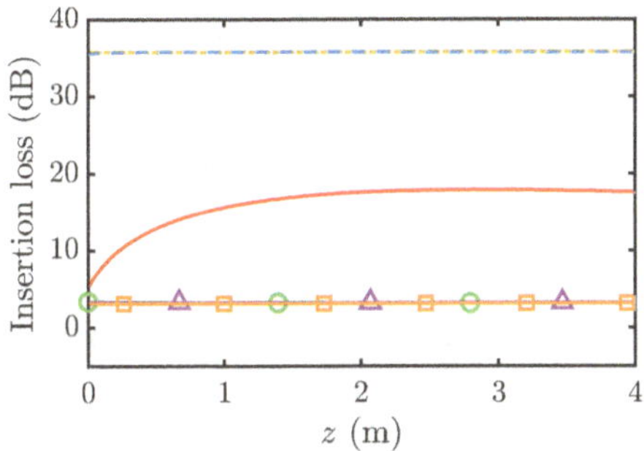

Figure 4.15 Insertion loss of sound radiated by various sources through an aluminum partition with a thickness of 50 μm. ——, PAL, 1 kHz; - - -, PAL, 60 kHz; —o—, point monopole, 1 kHz; —△—, end-fire array, 1 kHz; —□—, plane wave, 1 kHz; - - -, plane wave, 60 kHz. Extracted from [Zhong et al., 2020d, Fig. 3].

The audio sound generating from the PAL behind the partition comprises two components: the transmitted sound of the audio signal produced by incident ultrasound and the audio sound generated by the transmitted ultrasound itself. Given that the partition effectively obstructs most ultrasound, the contribution of audio sound from transmitted ultrasound becomes negligible. Consequently, the audio sound field behind the partition is primarily characterized by the transmitted sound of the audio signal generated by incident ultrasound. This distinction is evident in the significantly higher IL of 17 dB for the PAL compared to the modest IL of 3.2 dB observed for conventional sources. The partition has a substantial impact on the transmission of audio sound for a PAL, emphasizing the necessity to consider the transmission of ultrasound in the context of PAL systems.

4.3.3.3 Effects of a 1 mm Thick Polyester Fiber Blanket

In Fig. 4.16, the audio sound radiated by various sources behind a polyester fiber blanket partition, with a thickness of 1 mm and sources located at $z_s = -1$ m, is presented. For a PAL source, the audio sound behind the partition remains concentrated near the radiator axis, in contrast to a partition made of aluminum. This distinct behavior is attributed to the insufficient IL of the polyester fiber blanket for ultrasound, allowing some ultrasound to transmit through the partition and influence the directivity of the audio sound. This phenomenon is evident in Fig. 4.17, where the IL of the blanket at 60 kHz is only around 10 dB. Further observations from Fig. 4.16 reveal that nearly all sound transmits through the partition for conventional sound sources, because their ILs approach 0 as shown in Fig. 4.17. Conversely, a PAL exhibits a distinct behavior, with a significant portion of its audio sound being impeded by the partition, resulting in an IL of approximately 15 dB.

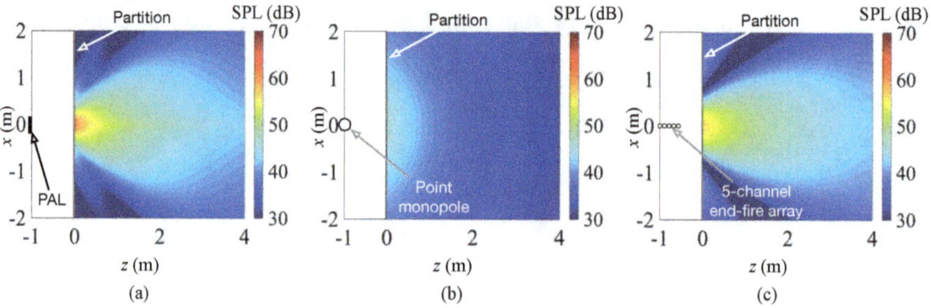

Figure 4.16 Audio SPL at $f_a = 1$ kHz with a 1 mm thick polyester fiber blanket partition and a source at $z_s = -1$ m: (a) for a PAL; (b) for a point monopole; (c) an end-fire array. Extracted from [Zhong et al., 2020d, Fig. 4].

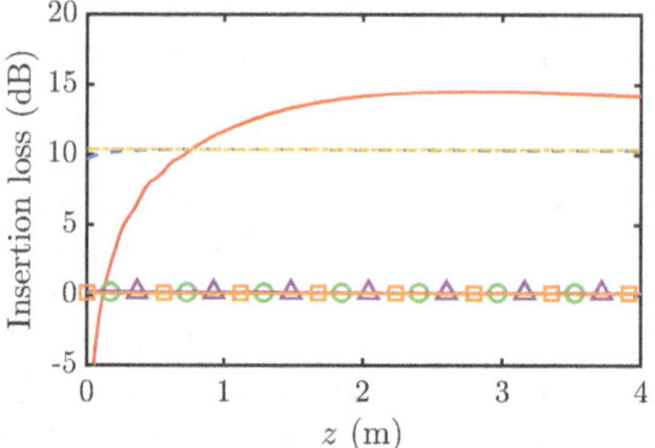

Figure 4.17 Insertion loss of sound radiated by various sources through a polyester fiber blanket partition with a thickness of 1 mm. ——, PAL, 1 kHz; - - -, PAL, 60 kHz; —o—, point monopole, 1 kHz; —△—, end-fire array, 1 kHz; —□—, plane wave, 1 kHz; - - -, plane wave, 60 kHz. Extracted from [Zhong et al., 2020d, Fig. 5].

Based on the above findings, the impact of partition thickness on the amplitude and shape of transmitted audio waves can be discussed. For conventional loudspeakers, it is observed that in the mass law frequency range, doubling either the thickness or the frequency typically results in a 6 dB increase in insertion loss for a thin partition. Consequently, with wideband audio signals, there is a notable alteration in both the amplitude and shape of the transmitted wave. However, when dealing with tonal sound, variations in partition thickness tend to produce only slight changes in the audio beam shape behind the partition. In the case of PALs, if the transmitted audio sound predominantly stems from the transmitted ultrasonic waves (e.g., when the partition thickness is smaller than the ultrasonic wavelength), an increase in thickness substantially elevates the insertion loss for the ultrasonic waves. Consequently, the transmitted audio waves diminish in amplitude and become less focused due to the reduction in ultrasonic wave amplitudes. On the other hand, if the transmitted audio sound is mainly generated by the audio waves resulting from the ultrasonic waves on the source side (e.g., when the thickness is larger than the ultrasonic wavelength), the effects of partition thickness are similar to those for conventional loudspeakers.

4.3.3.4 Experimental Observations

The experiments were conducted in a hemi-anechoic room with dimensions of $7.20\,\mathrm{m} \times 5.19\,\mathrm{m} \times 6.77\,\mathrm{m}$ (height). The sketch and photographs of experimental setups are shown in Fig. 4.18. The sound fields generated by a PAL, a conventional loudspeaker (point monopole), and a horn loudspeaker (directional source) without and with a partition (a sheet of aluminum foil or a cotton sheet) were measured in the experiments.

The sound field was measured at a rectangular grid with $60 \times 41 = 2460$ points and $60 \times 31 = 1860$ points in the xOz plane for the cases without and with the partition, respectively. In all cases, 60 microphones were located in the x-direction from $x = -1.45\,\mathrm{m}$ to $x = 1.5\,\mathrm{m}$ with a spacing of 5 cm and the sound field was measured simultaneously with a 60-channel microphone array shown in Fig. 4.18(b). All measurement microphones were Brüel & Kjær Type 4957. For the cases without the partition, the microphone array was located at 41 different positions in the z direction from $z = -1\,\mathrm{m}$ to $z = 3\,\mathrm{m}$ with the spacing being 10 cm. For the cases with the partition, it was located at 31 positions in the z-direction from $z = 0$ to $z = 3\,\mathrm{m}$ with the same spacing being 10 cm. In all the measurements, the height of the centroid of loudspeakers and the microphones is 1.9 m.

The PAL, the point monopole, and the conventional directional source in the experiments were a Holosonics Audio Spotlight AS-24i [Holosonics, 2019] with the surface size of $60\,\mathrm{cm} \times 60\,\mathrm{cm}$, a Genelec 8010A conventional voice coil loudspeaker, and a Daichi dome horn loudspeaker with a $24\,\mathrm{cm} \times 8\,\mathrm{cm}$ rectangular opening, respectively. The carrier frequency of the PAL is 64 kHz according to the measurements with a Brüel & Kjær Type 4939 microphone and the audio frequency in the experiments was set to 1 kHz. The radiating surface of the PAL was covered by a 6 mm thick perspex panel with a hole of 10 cm radius at the center to construct a circular piston source as shown in Fig. 4.18(c).

The experimental results show that the SPLs on the radiator axis of the PAL decrease by more than 30 dB at 1 kHz when the PAL is covered by a same size square perspex panel. Consequently, the panel can provide sufficient IL for audio sound at 1 kHz, which makes the PAL covered by the panel with a hole a circular piston. To avoid the spurious sound at microphones induced by the intensive ultrasound radiated by the PAL, all the microphones are covered by a piece of small and thin plastic film.

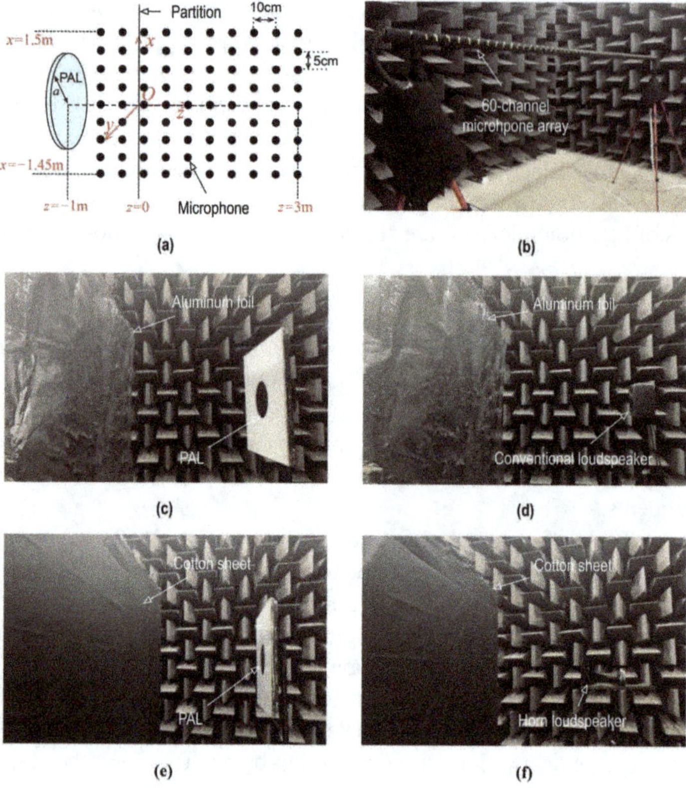

Figure 4.18 (a) Sketch of the experimental setups; a photo of (b) a 60-channel microphone array; (c) the PAL and the aluminum foil; (d) the conventional loudspeaker (point monopole) and the aluminum foil; (e) the PAL and the cotton sheet; (f) the horn loudspeaker (conventional directional source) and the cotton sheet. Extracted from [Zhong et al., 2020d, Fig. 6].

Two partitions were used in the experiments: a sheet of 50 μm thick aluminum foil with a bulk density of 2.7×10^3 kg/m^3 and a 250 μm thick cotton sheet with a surface density of 0.12 kg/m^2. The size of the aluminum foil and the cotton sheet are 3.6 m (x-direction) $\times$ 3 m (y-direction) and 4 m (x-direction) $\times$ 2.9 m (y-direction), respectively. The centroid of the loudspeaker is at the same height as that of the aluminum foil or the cotton sheet shown in Figs. 4.18(c–f). The relative humidity and the temperature in the experiments were 68% and 25.4°C, respectively.

Figure 4.19 shows the simulation and experimental results of audio sound fields at 1 kHz generated by the three different sound sources which are at 1.9 m high above ground without partitions. The experimental results of sound fields generated by the PAL with and without the partition are generally in accordance with the predicted ones shown in Figs. 4.19(a-i) and (b-i). It can be found the measured sound field radiated by the PAL is highly focused on the radiator axis as expected. Some small fluctuations occurred in the z-axis direction in the experimental results might be caused by reflections of the ground. The measured sound fields generated by the conventional loudspeakers shown in Figs. 4.19(a-ii), (a-iii), (b-ii), and (b-iii) also agree well with the simulation ones. Figure 4.19(a-ii) is obtained by using the analytical solution of a point monopole above a rigid ground, and Fig. 4.19(a-iii) is obtained by the boundary element method (BEM) solver of the commercial software Virtual.Lab Acoustics. The mesh size of the horn loudspeaker is set as 13.3 mm which is smaller than 1/10 wavelength at 1 kHz, and the model is shown in Fig. 4.20.

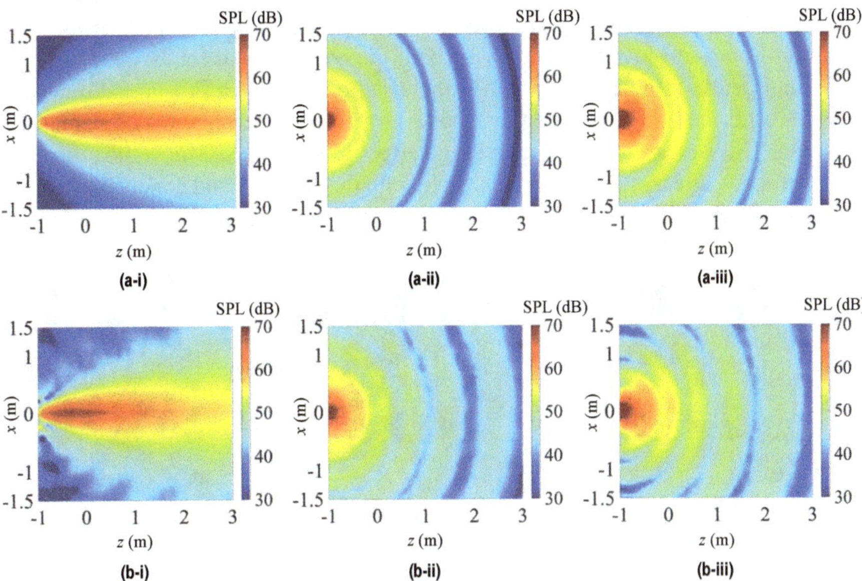

Figure 4.19 Audio SPL at 1 kHz with various sound sources at $z = -1$ m and $x = 0$ above the ground with a height of 1.9 m. (i) PAL; (ii) conventional loudspeaker; and (iii) horn loudspeaker. (a) Simulations; (b) Experiments. Extracted from [Zhong et al., 2020d, Fig. 7].

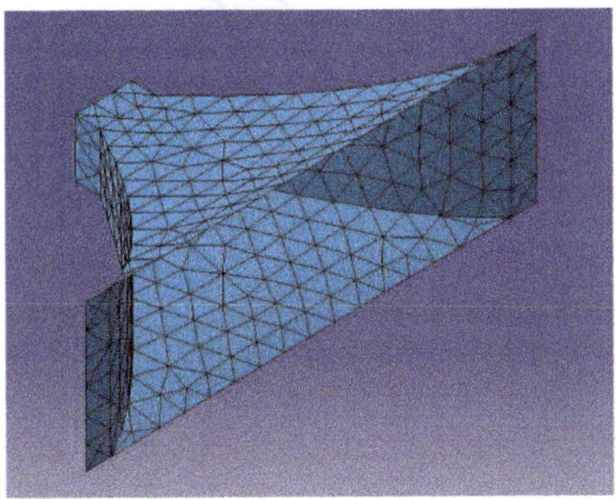

Figure 4.20 The model of the horn loudspeaker used in Virtual.Lab Acoustics.

Figure 4.21 shows the experimental results of audio sound at 1 kHz generated by the three sound sources above the ground with a sheet of aluminum foil and a cotton sheet. The sound fields in the space between the loudspeaker ($z = -1$ m) and the partition ($z = 0$) were not measured, because they are not the primary focus of this study and are also challenging to measure in practice. The theoretical prediction results corresponding to Figs. 4.21(a-i) and (b-i) are presented in Figs. 4.22(a) and (b), respectively, where reasonable agreements between predictions and experiments are observed. The small errors might be caused by the unevenness of the partition surface in the experiments. As depicted in Fig. 4.21, the impact of thin partitions on sound transmission loss is minimal for both the conventional loudspeaker and the horn loudspeaker. However, for a PAL, the partitions lead to a substantial reduction in SPLs, and the audio sound is less concentrated on the radiator axis behind the partition. This outcome is attributed to the effective blockage of most ultrasound by the partition. The experimental results align with the analyses and conclusions drawn from the simulations.

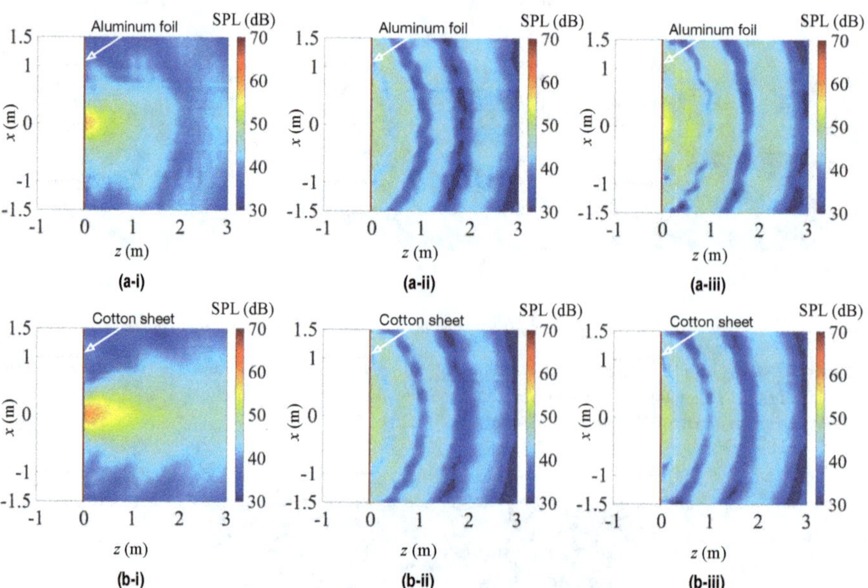

Figure 4.21 Experimental results of the audio SPL at 1 kHz with various sound sources at $z = -1$ m and $x = 0$ above the ground with a height of 1.9 m with a sheet of (a) aluminum foil and (b) cotton sheet. (i) PAL; (ii) conventional loudspeaker; and (iii) horn loudspeaker. Extracted from [Zhong et al., 2020d, Fig. 8].

4.3.4 REMARKS

In this section, the reader is introduced to the transmission properties of acoustic waves generated by PALs through a thin partition. Unlike audio sound produced by

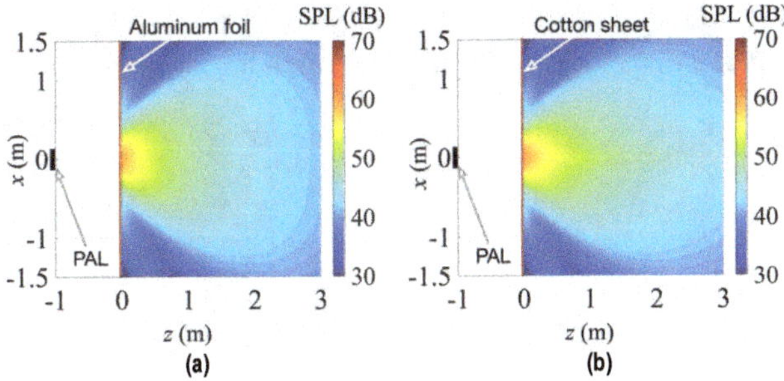

Figure 4.22 Theoretical prediction audio SPL corresponding to Figs. 4.21(a-i) and (b-i), i.e., the sound fields at 1 kHz with the PAL at $z = -1$ m and $x = 0$ with (a) a sheet of aluminum foil and (b) a cotton sheet. Extracted from [Zhong et al., 2020d, Fig. 9].

conventional sources, the directivity of audio generated by a PAL significantly degrades upon passing through a thin partition. To explore this phenomenon, a PAL radiation model utilizing the Westervelt equation and the plane wave expansion method is employed to calculate sound fields behind both aluminum foil and a porous material blanket. These calculations are performed under the quasi-linear assumption, where the paraxial approximation specifically applies to ultrasonic waves. For comparison, the audio sounds generated by a point monopole and a conventional directional (end-fire) source are presented.

Both simulation and experimental results consistently demonstrate that the transmitted sound from a PAL behind a thin partition is reduced and less focused on the radiator axis. This finding suggests the potential to create PALs of various shapes beyond square/rectangular configurations by covering them with thin panels—an option not easily achievable with conventional loudspeakers. Additionally, it highlights the consideration that when implementing PALs in mobile phones, a higher electric power supply may be necessary, especially if the PAL is positioned beneath the screen. Even a thin screen can substantially block audio sounds generated by PALs. It is important to note that this section focuses solely on normal incidence, and future research could explore the transmission properties of acoustic waves under oblique incidence.

4.4 SCATTERING BY A RIGID SPHERE

4.4.1 PROBLEM DESCRIPTION

A circular PAL radiating toward a rigid sphere is shown in Fig. 4.23. Both the PAL and the sphere are assumed to be placed in free space. The radii of the PAL and the sphere are denoted by a and r_0, respectively, and the centroid of the sphere is placed on the radiator axis of the loudspeaker. The distance between their centers is d. A Cartesian coordinate system $Oxyz$ is established with its origin, O, at the centroid of the sphere and the negative z-axis pointing to the centroid of the PAL. The spherical coordinates (r, θ, φ) and the cylindrical coordinates (ρ, φ, z) are established with respect to the Cartesian coordinates (x, y, z) for further calculations, where r, θ, φ, and ρ are the radial distance, zenith angle, azimuth angle, and polar radial distance, respectively.

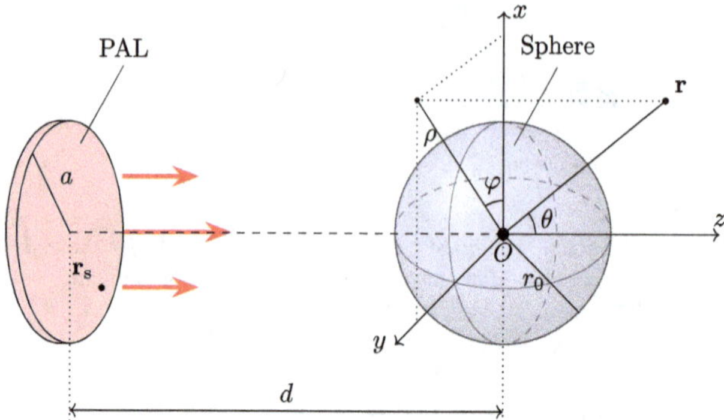

Figure 4.23 Sketch of a circular PAL radiating toward a sphere. The radius of the PAL and the sphere is a and r_0, respectively. Extracted from [Zhong et al., 2022c, Fig. 1].

4.4.2 THEORETICAL MODEL

4.4.2.1 Ultrasound Field

A rigorous approach to modeling the ultrasound field should include scattering from both the non-baffled PAL and the sphere, as well as the interaction between them. It has been demonstrated in Zhong et al., 2020b that the ultrasound generated by a non-baffled PAL can be approximated by the one generated by a baffled PAL, because the size of the PAL is large enough when compared to the wavelength of the ultrasound. The interaction between a PAL and a sphere is known to generate multiple scattering; however, when the separation between them is larger than the ultrasonic wavelength and the dimensions of each component, this scattering is considered to be insignificant [Gaunaurd et al., 1995]. It is therefore neglected in this section for simplicity and to focus on the scattering effects by the sphere.

The ultrasound pressure at a field point $\mathbf{r} = (r, \theta, \varphi)$ in the presence of a rigid sphere shown in Fig. 4.23 is solved using Rayleigh integral Eq. (2.38) as

$$p_i(\mathbf{r}) = -2\mathrm{i}\rho_0\omega_i \iint_{-\infty}^{\infty} v_{i,z}(\mathbf{r}_\mathrm{s}) G(\mathbf{r}, \mathbf{r}_\mathrm{s}, \omega_i) \mathrm{d}^2\mathbf{r}_\mathrm{s}, \quad r \geq r_0 \tag{4.28}$$

where the variables $\mathbf{r}_\mathrm{s} = (x_\mathrm{s}, y_\mathrm{s}, z_\mathrm{s})$ are the coordinates for the source point on the PAL surface, $(\rho_\mathrm{s}, \varphi_\mathrm{s}, z_\mathrm{s})$ are the corresponding cylindrical coordinates, and $z_\mathrm{s} = -d$. The Green's function $G(\mathbf{r}, \mathbf{r}_\mathrm{s}, \omega_i)$ is the velocity potential solution at the field point $\mathbf{r}$ generated by a point monopole at $\mathbf{r}_\mathrm{s}$ with the unit volume velocity [Lin et al., 2004; Zou et al., 2007]

$$G(\mathbf{r}, \mathbf{r}_\mathrm{s}, \omega_i) = \mathrm{i}k_i \sum_{\ell_i=0}^{\infty} R_{\ell_i}(r, r_\mathrm{s}) \sum_{m_i=-\ell_i}^{\ell_i} Y_{\ell_i}^{m_i}(\theta, \varphi) Y_{\ell_i}^{m_i,*}(\theta_\mathrm{s}, \varphi_\mathrm{s}), \tag{4.29}$$

where the radial component is

$$R_{\ell_i}(r,r_s) = j_{\ell_i}(k_i r_{s,<}) h_{\ell_i}(k_i r_{s,>}) - \frac{j'_{\ell_i}(k_i r_0)}{h'_{\ell_i}(k_i r_0)} h_{\ell_i}(k_i r) h_{\ell_i}(k_i r_s). \tag{4.30}$$

Here, $r_{s,<} = \min(r, r_s)$ and $r_{s,>} = \max(r, r_s)$.

By substituting Eq. (4.29) into Eq. (4.28), one obtains

$$\begin{aligned} p_i(\mathbf{r}) = 2p_0 \sum_{\ell_i=0}^{\infty} \sum_{m_i=-\ell_i}^{\ell_i} Y_{\ell_i}^{m_i}(\theta,\varphi) \int_0^a u_i(\rho_s) R_{\ell_i}(r,r_s) \\ \times \left[\int_0^{2\pi} Y_{\ell_i}^{m_i,*}(\theta_s,\varphi_s) \mathrm{d}\varphi_s \right] k_i^2 \rho_s \mathrm{d}\rho_s, \quad r \geq r_0. \end{aligned} \tag{4.31}$$

Here, the source profile is assumed to be axisymmetric, and denoted as $v_{i,z}(\mathbf{r}_s) = v_0 u_i(\rho_s)$. By using Eq. (B.8), Eq. (4.31) reduces to

$$p_i(\mathbf{r}) = 4\pi p_0 \sum_{\ell_i=0}^{\infty} \mathcal{R}_{\ell_i}(r) Y_{\ell_i}^0(\theta, 0) \tag{4.32}$$

where the radial component for ultrasound is

$$\mathcal{R}_{\ell_i}(r) = \int_0^a u_i(\rho_s) R_{\ell_i}(r,r_s) Y_{\ell_i}^0(\theta_s, 0) k_i^2 \rho_s \mathrm{d}\rho_s. \tag{4.33}$$

It is noted that the ultrasound pressure given by Eq. (4.32) is independent of the azimuthal angle, φ, because the velocity profile is assumed to be axisymmetric. According to the definition of the spherical harmonics given by Eq. (B.3), the radial component for ultrasound can be rewritten as

$$\mathcal{R}_{\ell_i}(r) = \sqrt{\frac{2\ell_i+1}{4\pi}} \int_0^a u_i(\rho_s) R_{\ell_i}(r,r_s) P_{\ell_i}(\cos\theta_s) k_i^2 \rho_s \mathrm{d}\rho_s, \tag{4.34}$$

where $P_{\ell_i}(\cdot)$ is the Legendre polynomial.

By using the geometric relations

$$\begin{cases} r_s^2 = \rho_s^2 + d^2 \Longrightarrow r_s \mathrm{d}r_s = \rho_s \mathrm{d}\rho_s, \\ \cos(\pi - \theta_s) = \dfrac{d}{r_s}, \end{cases} \tag{4.35}$$

and the symmetry relation of the Legendre polynomials [Zhang and Jin, 1996, Eq. (4.2.8)]

$$P_{\ell_i}(\cos\theta_s) = P_{\ell_i}\left(-\frac{d}{r_s}\right) = (-1)^{\ell_i} P_{\ell_i}\left(\frac{d}{r_s}\right), \tag{4.36}$$

the radial component for ultrasound can be further represented as

$$\mathcal{R}_{\ell_i}(r) = (-1)^{\ell_i} \sqrt{\frac{2\ell_i+1}{4\pi}} \int_d^{\sqrt{d^2+a^2}} u_i(\rho_s) R_{\ell_i}(r,r_s) P_{\ell_i}\left(\frac{d}{r_s}\right) k_i^2 r_s \mathrm{d}r_s. \tag{4.37}$$

Compared to Rayleigh integral in Eq. (4.28), Eq. (4.32) is more efficient to calculate because the convergence of the series is faster than a double integral and the field coordinates are also decoupled. The components of the particle velocity of the ultrasound under the spherical coordinate system is then obtained by substituting Eq. (4.32) into Eq. (1.103) to give

$$\begin{cases} v_{i,r}(\mathbf{r}) = \dfrac{1}{\mathrm{i}\rho_0\omega_i}\dfrac{\partial p_i(\mathbf{r})}{\partial r} = -4\pi\mathrm{i}v_0\displaystyle\sum_{\ell_i=0}^{\infty}\dfrac{\mathrm{d}\mathcal{R}_{\ell_i}(r)}{\mathrm{d}(k_ir)}Y_{\ell_i}^0(\theta,0), \\ v_{i,\theta}(\mathbf{r}) = \dfrac{1}{\mathrm{i}\rho_0\omega_i r}\dfrac{\partial p_i(\mathbf{r})}{\partial\theta} = -4\pi\mathrm{i}v_0\displaystyle\sum_{\ell_i=0}^{\infty}\dfrac{\mathcal{R}_{\ell_i}(r)}{k_ir}\dfrac{\mathrm{d}Y_{\ell_i}^0(\theta,0)}{\mathrm{d}\theta}, \\ v_{i,\varphi}(\mathbf{r}) = \dfrac{1}{\mathrm{i}\rho_0\omega_i r\sin\theta}\dfrac{\partial p_i(\mathbf{r})}{\partial\varphi} = 0. \end{cases} \tag{4.38}$$

4.4.2.2 Audio Sound Field

The source density for the virtual audio sound source can be obtained by substituting Eq. (4.32) into Eq. (2.44), reading

$$q_\mathrm{a}(\mathbf{r}_\mathrm{v}) = \frac{16\pi^2\beta p_0^2\omega_\mathrm{a}}{\mathrm{i}\rho_0^2c_0^4}\sum_{\ell_1,\ell_2=0}^{\infty}\mathcal{R}_{\ell_1}^*(r_\mathrm{v})\mathcal{R}_{\ell_2}(r_\mathrm{v})Y_{\ell_1}^0(\theta_\mathrm{v},0)Y_{\ell_2}^0(\theta_\mathrm{v},0). \tag{4.39}$$

The audio sound pressure can be solved by integrating the source density using the Green's function with spherical scattering,

$$p_\mathrm{a}(\mathbf{r}) = -\mathrm{i}\rho_0\omega_\mathrm{a}\iiint_{r_\mathrm{v}\geq r_0} q_\mathrm{a}(\mathbf{r}_\mathrm{v})G(\mathbf{r},\mathbf{r}_\mathrm{v},\omega_\mathrm{a})\mathrm{d}^3\mathbf{r}_\mathrm{v}, \quad r\geq r_0. \tag{4.40}$$

The substitution of the spherical expansion of Green's function given by Eq. (4.29) into Eq. (4.40) yields

$$\begin{aligned} p_\mathrm{a}(\mathbf{r}) = \rho_0c_0k_\mathrm{a}^2\sum_{\ell_\mathrm{a}=0}^{\infty}\sum_{m_\mathrm{a}=-\ell_\mathrm{a}}^{\ell_\mathrm{a}}\int_0^{2\pi}\int_0^{\pi}\int_{r_0}^{\infty} q_\mathrm{a}(\mathbf{r}_\mathrm{v})\mathrm{j}_{\ell_\mathrm{a}}(k_\mathrm{a}r_{\mathrm{v},<})\mathrm{h}_{\ell_\mathrm{a}}(k_\mathrm{a}r_{\mathrm{v},>}) \\ \times Y_{\ell_\mathrm{a}}^{m_\mathrm{a}}(\theta,\varphi)Y_{\ell_\mathrm{a}}^{m_\mathrm{a},*}(\theta_\mathrm{v},\varphi_\mathrm{v})r_\mathrm{v}^2\sin\theta_\mathrm{v}\mathrm{d}r_\mathrm{v}\mathrm{d}\theta_\mathrm{v}\mathrm{d}\varphi_\mathrm{v}, \end{aligned} \tag{4.41}$$

where $r_{\mathrm{v},<} = \min(r,r_\mathrm{v}), r_{\mathrm{v},>} = \min(r,r_\mathrm{v})$. After substituting the source density Eq. (4.39) into Eq. (4.41), it is written as

$$\begin{aligned} p_\mathrm{a}(\mathbf{r}) = \frac{16\pi^2\beta p_0^2}{\mathrm{i}\rho_0c_0^2}\sum_{\ell_1,\ell_2=0}^{\infty}\sum_{\ell_\mathrm{a}=0}^{\infty}\sum_{m_\mathrm{a}=-\ell_\mathrm{a}}^{\ell_\mathrm{a}} Y_{\ell_\mathrm{a}}^{m_\mathrm{a}}(\theta,\varphi) \\ \times\int_{r_0}^{\infty}\mathrm{j}_{\ell_\mathrm{a}}(k_\mathrm{a}r_{\mathrm{v},<})\mathrm{h}_{\ell_\mathrm{a}}(k_{\mathrm{v},>})\mathcal{R}_{\ell_1}^*(r_\mathrm{v})\mathcal{R}_{\ell_2}(r_\mathrm{v})k_\mathrm{a}^3r_\mathrm{v}^2\mathrm{d}r_\mathrm{v} \\ \times\int_0^{2\pi}\int_0^{\pi}Y_{\ell_1}^0(\theta_\mathrm{v},0)Y_{\ell_2}^0(\theta_\mathrm{v},0)Y_{\ell_\mathrm{a}}^{m_\mathrm{a},*}(\theta_\mathrm{v},\varphi_\mathrm{v})\sin\theta_\mathrm{v}\mathrm{d}\theta_\mathrm{v}\mathrm{d}\varphi_\mathrm{v}. \end{aligned} \tag{4.42}$$

By using Eqs. (B.5) and (B.9), the triple integral with respect to the spherical harmonics in Eq. (4.42) can be simplified to

$$\begin{aligned}&\int_0^{2\pi}\int_0^{\pi} Y_{\ell_1}^0(\theta_v,0)Y_{\ell_2}^0(\theta_v,0)Y_{\ell_a}^{m_a,*}(\theta_v,\varphi_v)\sin\theta_v d\theta_v d\varphi_v \\ &\quad = \sqrt{\frac{(2\ell_1+1)(2\ell_2+1)(2\ell_a+1)}{4\pi}}\begin{pmatrix}\ell_1 & \ell_2 & \ell_a\\ 0 & 0 & 0\end{pmatrix}^2 \delta_{m_a,0}.\end{aligned} \tag{4.43}$$

Accordingly, the SWE of audio sound pressure is obtained as

$$p_a(\mathbf{r}) = \frac{16\pi^2\beta p_0^2}{i\rho_0 c_0^2}\sum_{\ell_a=0}^{\infty}\mathcal{R}_{\ell_a}(r)Y_{\ell_a}^0(\theta,0), \tag{4.44}$$

where the radial component for audio sound is

$$\mathcal{R}_{\ell_a}(r) = \sum_{\ell_1,\ell_2=0}^{\infty} A_\ell \int_{r_0}^{\infty} \mathrm{j}_{\ell_a}(k_a r_{v,<})\mathrm{h}_{\ell_a}(k_a r_{v,>})\mathcal{R}_{\ell_1}^*(r_v)\mathcal{R}_{\ell_2}(r_v)k_a^3 r_v^2 dr_v, \tag{4.45}$$

and the coefficient is

$$A_\ell = \sqrt{\frac{(2\ell_1+1)(2\ell_2+1)(2\ell_a+1)}{4\pi}}\begin{pmatrix}\ell_1 & \ell_2 & \ell_a\\ 0 & 0 & 0\end{pmatrix}^2. \tag{4.46}$$

The boundary condition on the surface of the sphere is assumed to be rigid in this section. When the boundary becomes non-rigid, the radial component of the Green's function given by Eq. (4.30) should be modified to satisfy the specific boundary condition. The scattered field would then be more complicated, especially for the field between the PAL and the sphere.

4.4.3 SCATTERED SOUND FIELDS

4.4.3.1 Accuracy of Using the Westervelt Equation

Equation (4.44) represents the audio sound pressure obtained based on the Westervelt equation. As shown in Sec. 1.3.3, the Westervelt equation is unable to account for the local effects which are significant near the source. As shown by Eq. (1.107), the local effects can be included by adding a correction to the audio sound pressure obtained using the Westervelt equation. The scattered sound field can be seen as a "secondary" source generated by the sphere. Therefore, it is necessary to verify the accuracy of using the Westervelt equation at the locations close to the sphere.

In the simulations, the center frequency of the ultrasound is set as $f_u = 64$ kHz, which is a common value used in commercial PALs. The radii of the PAL and the sphere are both set as 0.1 m. The distance between the PAL and the sphere, d, is set as 1.0 m. The audio sound fields generated by a PAL without the sphere are obtained using the SWE method in Sec. 2.6. The audio sound fields generated by a conventional loudspeaker with and without the sphere are calculated by setting k_a in Eq. (4.32) and using the method described in Zhong and Qiu, 2020, respectively. A uniform vibration velocity profile is assumed for all cases and an amplitude of 0.12 m/s and 2×10^{-4} m/s are set for the ultrasound generated by the PAL and the audio sound generated by the conventional loudspeaker, respectively.

Figure 4.24 presents a comparison of SPLs for audio sound generated by a PAL using the Westervelt equation with and without local effects at various zenithal angles, θ, both at distances of 0.1 m and 1.0 m from the centroid of the sphere, at frequencies of 500 Hz, 1 kHz, and 2 kHz. It is clear that the SPLs with local effects exhibit variations from those obtained using the Westervelt equation without local effects, a phenomenon consistent with observations in the absence of the sphere [Červenka and Bednařík, 2019; Zhong et al., 2021b]. The disparity between SPLs with and without local effects becomes pronounced when the observation point is in close proximity to the sphere or the PAL. For example, at the angle $\theta = 170°$, when the distance from the field point to the centroid of the sphere is 0.1 m, the SPL difference is 16.5 dB, 2.6 dB, and 0.5 dB, at 500 Hz, 1 kHz, and 2 kHz, respectively. As the distance increases to 1.0 m, fluctuations emerge, particularly as the zenith angle approaches 180°, comparable to the ultrasonic wavelength of 5.4 mm at 64 kHz. This behavior stems from the intricate nature of the ultrasonic field, where pressure is large near the sphere, leading to strong local effects that the Westervelt equation is unable to accurately model.

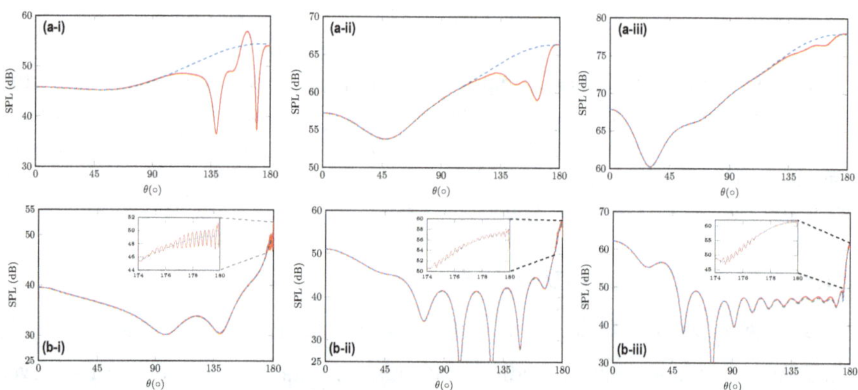

Figure 4.24 Audio SPL at various zenith angles with the distance of d to the centroid of the sphere generated by a PAL. The distance is (a) $d = 0.1$ m and (b) $d = 1.0$ m. The audio frequency is (i) 500 Hz, (ii) 1 kHz, and (iii) 2 kHz. ——, with local effects; - - -, without local effects. Extracted from [Zhong et al., 2022c, Fig. 2].

As the frequency increases, the difference between SPLs calculated with and without local effects diminishes. This phenomenon arises from the fact that while the audio SPL, calculated using the Westervelt equation, exhibits an increase of approximately 12 dB when the audio frequency doubles, corresponding variations in the amplitude of the ultrasonic Lagrangian density remain relatively small. Consequently, the impact of the ultrasonic Lagrangian density on the audio SPL becomes less pronounced at higher audio frequencies. This observation leads to the conclusion that including local effects in the calculation is crucial, particularly for observation points near the sphere at lower frequencies. Consequently, all subsequent simulations take into account these local effects.

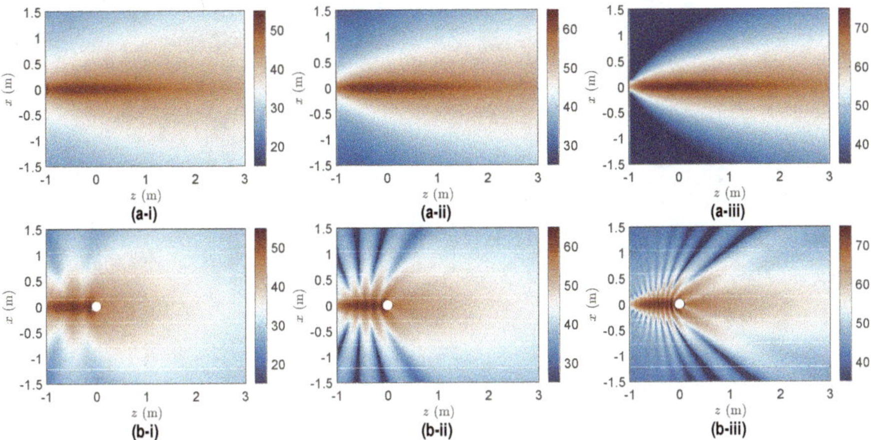

Figure 4.25 Audio SPL generated by a PAL in the (a) absence and (b) presence of the sphere at (i) 500 Hz, (ii) 1 kHz, and (iii) 2 kHz. The on-surface pressure pressure amplitude is $p_0 = 50\,\text{Pa}$ (125 dB). Extracted from [Zhong et al., 2022c, Fig. 3].

4.4.3.2 Simulated Scattered Sound Fields

The audio sound fields generated by a PAL with and without the sphere at 500 Hz, 1 kHz, and 2 kHz are shown in Fig. 4.25. For comparison, the sound fields generated by a conventional loudspeaker with the same size as that of the PAL are presented in Fig. 4.26. Because the generated SPL is increased by respectively 12 dB and 6 dB for the PAL and conventional loudspeaker when the frequency is doubled, the range of the color bar in Figs. 4.25 and 4.26 is increased by 10 dB and 5 dB as frequency is doubled for better comparison.

Compared to the conventional loudspeaker, the directivity of the audio sound generated by the PAL is more focused. After introducing the sphere, the scattering effects become progressively more significant as the frequency increases for both the PAL and the conventional loudspeaker. This is because the audio sound wavelength becomes smaller compared to the sphere and so more audio sound is reflected at higher frequencies. For the conventional loudspeaker, the effects of the sphere on the back side ($z > 0$) are negligible at low frequencies because the audio wavelength is much larger than the size of the sphere [Zou and Qiu, 2008]. However, the SPL on the back side of a sphere generated by a PAL decreases significantly. This phenomenon is attributed to the ultrasound wavelength, measuring 5.36 mm at 64 kHz, which is considerably smaller than the dimensions of the sphere. As a result, the ultrasound beam is nearly obstructed by the sphere, resulting in minimal energy transmission behind it, as depicted in Fig. 4.27. Consequently, the audio beam behind the sphere is no longer directional.

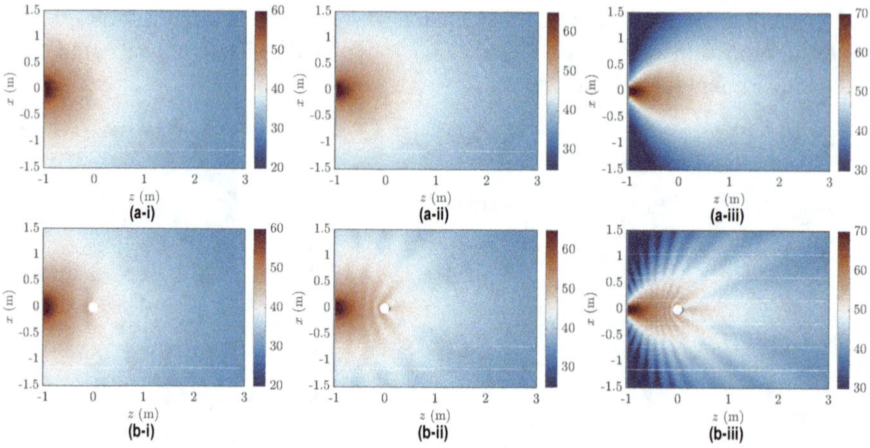

Figure 4.26 Audio SPL generated by a conventional loudspeaker in the (a) absence and (b) presence of the sphere at 500 Hz, (ii) 1 kHz, and (iii), and 2 kHz. The on-surface pressure pressure amplitude is $p_0 = 0.08$ Pa (69 dB). Extracted from [Zhong et al., 2022c, Fig. 4].

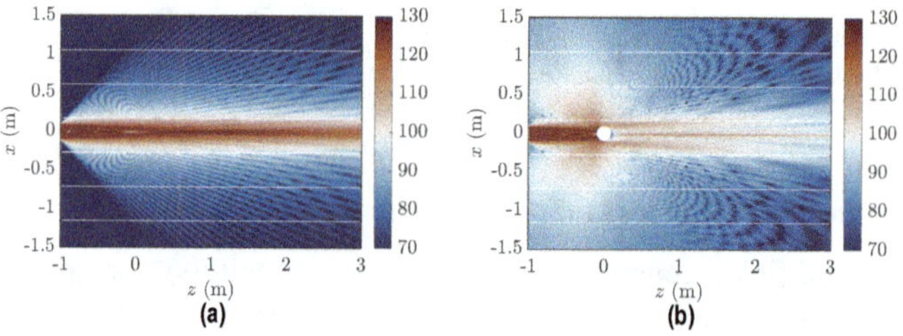

Figure 4.27 Ultrasound SPL generated by a PAL in the (a) absence and (b) presence of the sphere at 64 kHz. The on-surface pressure pressure amplitude is $p_0 = 50$ Pa (125 dB).

Figure 4.28 shows the SPL at different zenith angles, θ, with a distance of 1.0 m to the centroid of the sphere generated by a PAL or a conventional loudspeaker with and without the sphere. After introducing the sphere, the half sound pressure (−6 dB) angles for the audio sound generated by the PAL are increased from 15.9°, 13.1°, and 10.8° to 76.4°, 50.2°, and 21.8°, at 500 Hz, 1 kHz, and 2 kHz, respectively, while there is little change observed for the conventional loudspeaker. This demonstrates that the sphere severely deteriorates the directivity of the audio sound generated by a PAL, because the ultrasound maintaining the directivity of the audio sound are almost completely blocked by the sphere.

It is interesting to note that the audio sound generated by the PAL are augmented at some angles and frequencies on the front side ($z < 0$) after introducing the sphere, while the SPL changes a little for the case with the conventional loudspeaker. For example, the SPL at the zenith angle of 135° increases from 36.7 dB and 26.3 dB to 41.7 dB and 47.2 dB at 1 kHz and 2 kHz, respectively. Figure 4.29 compares the SPL at the zenith angle of 135° with and without the sphere, for both the PAL and the

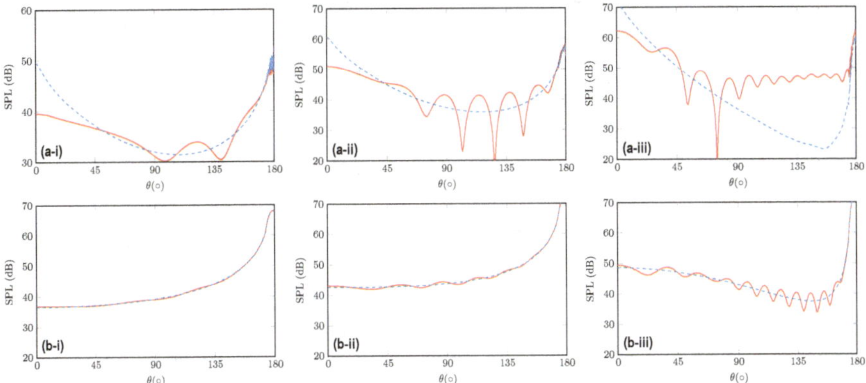

Figure 4.28 Audio SPL at various zenith angles with a distance of 1.0 m to the centroid of the sphere generated by (a) a PAL and (b) a conventional loudspeaker. The audio frequency is (i) 500 Hz, (ii) 1 kHz, and (iii) 2 kHz. ——, with sphere; - - - , without sphere. Extracted from [Zhong et al., 2022c, Fig. 5].

conventional loudspeaker, at different frequencies. The difference of the SPL with and without the sphere fluctuates for the conventional loudspeaker and the fluctuation becomes larger as the frequency increases. The increment of the SPL is no larger than 6 dB below 4 kHz for the conventional loudspeaker. However, the SPL is generally enlarged after introducing the sphere, as the frequency increases for the audio sound generated by the PAL. The increment of the SPL is more than 6 dB at frequencies larger than 1620 Hz. This means that a listener outside the audible region of the PAL can still hear a significant augmentation of the audio sound when a human head moves across the radiation direction of the PAL. The reason is that the ultrasound are almost completely reflected from the sphere because their wavelength is much smaller than the sphere radius. The reflected ultrasound form another virtual array which augments the audio sound on the front side of the sphere.

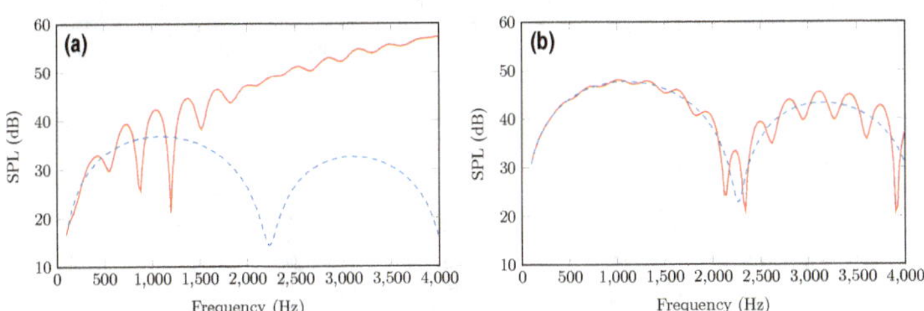

Figure 4.29 Audio SPL at the zenith angle $\theta = 135^\circ$ and the radius of 1.0 m generated by (a) a PAL and (b) a conventional loudspeaker from 100 Hz to 4 kHz. ——, with sphere; - - - , without sphere. Extracted from [Zhong et al., 2022c, Fig. 6].

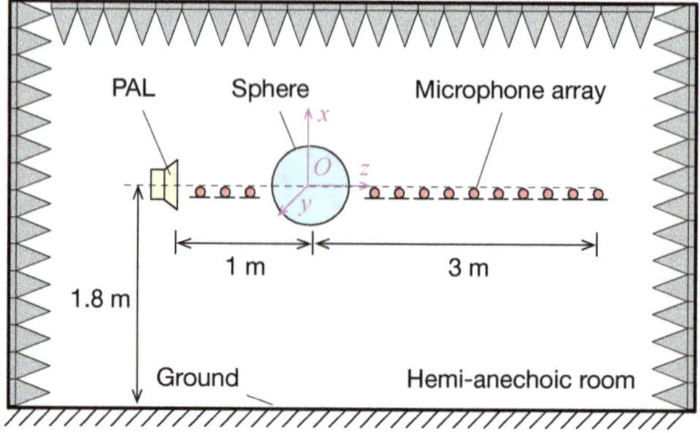

Figure 4.30 Sketch of the experiment setup. Extracted from [Zhong et al., 2022c, Fig. 7].

4.4.3.3 Experimental Observation

This section presents the experimental observation results on the sphere scattering by a rigid sphere. The experiments were conducted in the hemi-anechoic room with dimensions of $7.2\,\mathrm{m} \times 5.19\,\mathrm{m} \times 6.77\,\mathrm{m}$ (height). The sketch and photos of the experimental setup are shown in Figs. 4.30 and 4.31, respectively. The sound field generated by a PAL with and without the sphere was measured in the experiments. A solid wooden sphere with a radius of 0.1 m and density of $1.2 \times 10^3\,\mathrm{kg/m^3}$ was used, and

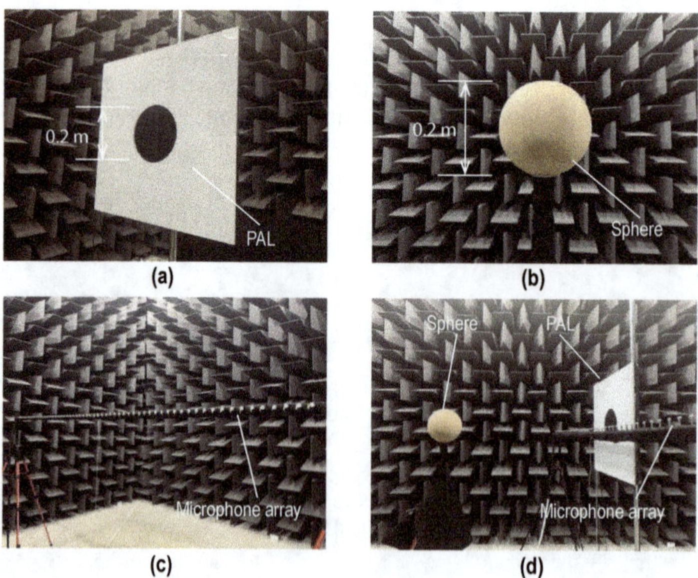

Figure 4.31 Photos of the experiment setup: (a) a PAL; (b) a solid wooden sphere; (c) a 60-channel microphone array; (d) the measurement system. Extracted from [Zhong et al., 2022c, Fig. 8].

height of both the loudspeaker and sphere are 1.8 m. The wooden sphere was supported by a tripod, which was covered by absorption materials to avoid additional scattering effects from the tripod.

The PAL in the experiments is a Holosonics Audio Spotlight AS-24i [Holosonics, 2019] with a surface size of 60 cm × 60 cm. The carrier frequency of the PAL is 64 kHz. The radiating surface of the PAL is covered by a 6 mm thick perspex panel with a hole of 10 cm radius at its center, as shown in Fig. 4.31(a). It has been demonstrated in Sec. 4.3 that the sound pressure levels on the radiator axis of the PAL decrease by more than 30 dB at 1 kHz when the PAL is covered by this perspex panel, which indicates that the panel successfully reproduces a circular piston source with a hole at the center.

The sound field was measured in a 2.95 m × 4 m rectangular plane at the same height as the loudspeaker ($x = 0, -1.45\,\text{m} \le y \le 1.5\,\text{m}, -1\,\text{m} \le z \le 3\,\text{m}$). Sixty measurement positions were taken in the y-direction from $y = -1.45$ m to $y = 1.5$ m with a spacing of 5 cm. The SPLs were measured simultaneously using a 60-channel microphone array, as shown in Fig. 4.31(c). The microphone array was moved along the z-axis with a step of 10 cm to obtain the sound fields in the measurement plane. All the measurement microphones are Brüel & Kjær Type 4957. To avoid spurious sound at the microphones induced by the intensive ultrasound radiated by the PAL, all microphones were covered by a piece of small and thin plastic film. The relative humidity and temperature were 70% and 25.4°C, respectively.

Figure 4.32 shows the measured results of the audio sound generated by a PAL with and without the sphere. Values close to the sphere cannot be measured, so they are left blank in Fig. 4.32(b). The measured results are well in accordance with the simulations shown in Fig. 4.25. Some small fluctuations occur in the z-axis direction, and this is caused by reflections from the ground floor. Other measurement errors may arise from the imperfect positioning of the centroid of the sphere on the radiator axis of the PAL, as well as the location error of the array microphone and scattering from the microphones and other measurement equipment. It can be observed that the directivity of the audio sound generated by the PAL is severely deteriorated on the back side of the sphere, which is consistent with the simulation results.

Figure 4.33(a) compares the measured audio SPL with and without the sphere for the microphone located at $x = 0.7$ m and $z = -0.7$ m which is approximately at the azimuthal angle $\theta = 135°$ and the radius of 1.0 m to the centroid of the sphere. Only the results at the center frequencies in the 1/3 octave band from 315 Hz to 4 kHz were measured and plotted. Figure 4.33(b) compares the SPL increment for the simulation and experiment results after introducing the sphere. It can be found the measured results are generally in accordance with the numerical ones. The large mismatches between the simulation and experimental results are observed at high frequencies, which might be caused by the facts that the positioning of the equipment is more sensitive at small wavelengths and the scattering effects of the microphone array become more prominent. It is clear the audio SPL is enlarged after introducing the sphere. Informal hearing tests also showed that a listener in front of the PAL can hear the audio sound scattered back after introducing the sphere as if they are reflected by

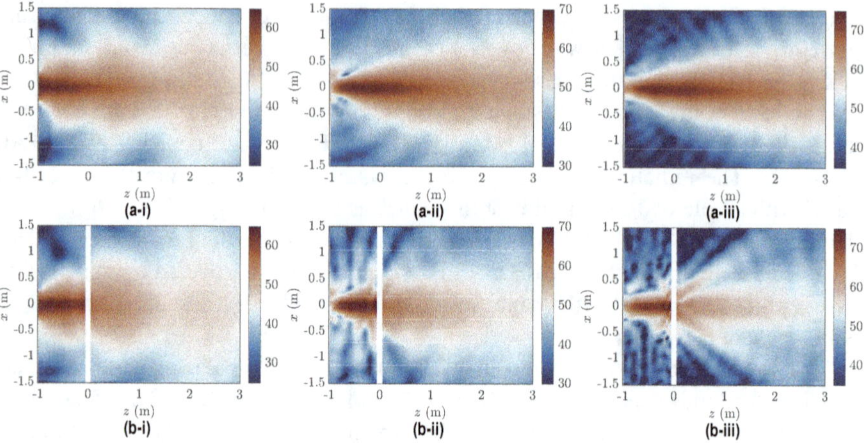

Figure 4.32 Measured audio SPLs generated by a PAL in the (a) absence and (b) presence of the sphere at (i) 500 Hz, (ii) 1 kHz, and (iii) 2 kHz. Extracted from [Zhong et al., 2022b, Fig. 9].

the sphere. This is different to the perception of the scattered audio sound generated by a conventional loudspeaker.

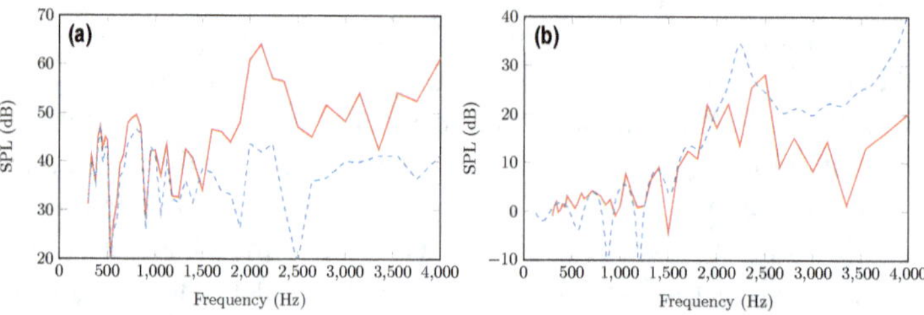

Figure 4.33 Measured audio SPL at the microphone located at $x = 0.7$ m and $z = -0.7$ m generated by a PAL at the frequencies from 315 Hz to 4 kHz: (a) the SPL with and without the sphere; ——, with sphere; - - -, without sphere; (b) the SPL increment by simulations and experiments; ——, experiment; - - -, simulation; Extracted from [Zhong et al., 2022c, Fig. 10].

4.4.4 REMARKS

In this section, both simulation and experimental results reveal a pronounced degradation in the directivity of audio sound generated by a PAL following the introduction of the sphere. Consequently, a listener positioned behind another listener whose head is in the radiation direction of the PAL cannot anticipate a highly collimated audio beam, as would typically be expected. The primary reason for this deviation is that the ultrasound responsible for forming the directivity of the audio sound are nearly entirely obstructed by the sphere.

An interesting observation is that a listener situated on the front side of the sphere, between the PAL and the sphere, perceives some audio sound as if it is emanating from the sphere itself. This occurrence is attributed to the reflected ultrasound forming another virtual array, enhancing the audio sound on the front side of the sphere. In contrast, the impact of a sphere on listeners positioned on the front side is negligible for conventional loudspeakers. This difference can be attributed to the substantial size of the sphere relative to the ultrasound wavelength, significantly influencing the propagation of ultrasound waves and thereby shaping the behavior of the demodulated audio sound.

In this section, a rigid boundary condition is assumed on the surface of the sphere. However, when the boundary becomes non-rigid, as is the case when considering the absorption effects of skin and hair, adjustments to the radial component of the Green's function [Eq. (4.29)] become necessary to satisfy the specific boundary condition. This modification introduces a higher degree of complexity into the scattered field, particularly in the region between the PAL and the sphere, thereby necessitating further investigation into this area. Moreover, an interesting avenue for future research involves exploring the behavior of audio sound in scenarios where multiple spheres are present, simulating conditions with multiple listeners. Investigating the implications of such configurations could provide valuable insights into the interactive effects of multiple listeners on the audio sound generated by a PAL.

4.5 SUMMARY

Physical generation of audio sound by PALs involves intricate nonlinear processes, leading to distinct wave properties compared to conventional loudspeakers. This chapter conducts a comprehensive examination of the reflection, transmission, and scattering of acoustic waves generated by PALs.

Section 4.2 delves into the reflection from a surface. A theoretical model based on the image source method is introduced in Sec. 4.2.2 to simulate the reflected sound fields. The simulation and experimental findings, presented in Sec. 4.2.3, highlight that the reflected ultrasound retains the sharp directivity observed in the reflected audio sound. However, if the reflecting surface strongly absorbs ultrasound, the directivity of the reflected audio sound diminishes due to the reduced magnitude of reflected ultrasound.

The performance of sound transmission through a thin partition is analyzed in Sec. 4.3. A theoretical model, as described in Sec. 4.3.2, is developed based on the GBE and the plane wave expansion is developed in for modeling the transmitted sound fields. Both the simulation and experimental outcomes in Sec. 4.3.3 reveal that audio sound transmitted from a PAL positioned behind a thin partition is weaker and less focused on the radiator axis. This phenomenon primarily arises from the blockage of most ultrasonic waves by the thin partition. This effect is minimal in the case of conventional audio sources.

In Sec. 4.4, the scattering by a rigid sphere of audio sound generated by a PAL is examined. The spherical wave expansion is developed in Sec. 4.4.2 to model this sphere scattering problem. The simulation and experimental observations in Sec. 4.4.3 show that, in contrast to conventional loudspeakers, the directivity of audio sound generated by a PAL significantly deteriorates behind a rigid sphere. This is because the ultrasounds responsible for maintaining the audio sound directivity are blocked by the sphere. Instead, these ultrasounds are reflected and produce audio sound on the front side of the sphere.

In summary, the directivity of audio sound generated by PALs severely deteriorates when sound waves are reflected from a non-rigid surface, obstructed by a thin partition, or scattered by a rigid sphere. This occurs because the directivity of audio sound relies heavily on ultrasounds, which are highly sensitive to the acoustic environment compared to audio sound. Consequently, the expected sharp directivity of PALs cannot be guaranteed in complex acoustic environments, and their directional performance often deteriorates in audio applications. The methods and results presented in this chapter provide some guidance for analyzing the effects of reflection, transmission, and scattering on audio systems employing PALs.

5 Sound Fields Generated by PALs in Rooms

5.1 INTRODUCTION

Chapter 3 investigates sound fields generated by a PAL in free space, while Chap. 4 explores reflection, transmission, and scattering of acoustic waves emitting from a PAL. In these scenarios, the acoustic environment is unbounded, extending infinitely. Given that practical applications often involve sound sources in reverberant environments like rooms, museums, art galleries, and automobile cabins, this chapter explores the sound fields produced by PALs in such enclosed spaces.

While irregularly shaped rooms are prevalent in practical applications, rectangular rooms are often the preferred choice for simulating acoustic environments. This preference stems from their simplicity, which makes it easier to study and understand the properties of sound fields in enclosed space [Samarasinghe et al., 2018; Nolan and Davy, 2019]. The rectangular room also serves as a building block for solving problems involving more complex structures. For instance, solutions of the reverberant sound fields in rectangular rooms are utilized to calculate the sound field in coupled rectangular rooms [Poblet-Puig and Rodríguez-Ferran, 2013], door slits on ground [Wang et al., 2023a], baffled openings in walls [Wang et al., 2015], and rectangular-like enclosures with leaning walls [Li and Cheng, 2004]. Therefore, only rectangular rooms are considered in this chapter.

There are many methods for simulating the reverberant sound fields in rectangular rooms [Savioja and Svensson, 2015; Kuttruff, 2017]. Among them the image source method (ISM) is one of the most popular methods due to its simplicity. Early studies focus on modeling omnidirectional sources in a room [Allen and Berkley, 1979], as well as the fast computation algorithms [McGovern, 2009]. In 2012, the ISM was generalized for modeling directional sources placed in two-dimensional (2D) rectangular rooms [Betlehem and Poletti, 2012], and it was then extended for three-dimensional (3D) rectangular rooms [Samarasinghe et al., 2018].

The ISM provides an exact solution to the wave equation for a rectangular room with rigid walls. It is inherently inaccurate for the room with finite impedance walls especially when source and/or receiver is close to the wall due to the assumption of angle independent wall reflection coefficients [Allen and Berkley, 1979]. Although the modeling accuracy can be improved by using angle dependent wall reflection coefficients, the computation is complicated because these coefficients depend on multiple parameters and involve numerical integrations over angles in the complex plane [Lehmann and Johansson, 2008; Aretz et al., 2014]. In addition, the required number of image sources to obtain a satisfactory converged solution scales with the reverberation time [Betlehem and Poletti, 2012]. The computation becomes more time-consuming for a lightly damped room which has a long reverberation time.

DOI: 10.1201/9781003354994-5

Moreover, the computational cost becomes heavier if there are many receivers in the enclosure [Duraiswami et al., 2007]. Although the convergence is fast when the reverberation time is short, the prediction accuracy is deteriorated if the angle independent wall reflection coefficients are used.

An alternative to the ISM is the *modal expansion method (MEM)* which expresses the sound field as the superposition of a complete set of modal functions (eigenfunctions). The MEM was firstly used to solve the sound field in a rectangular enclosure with rigid or lightly damped walls [Morse and Bolt, 1944], and then extended to accommodate the finite impedance walls [Du et al., 2011; Nolan and Davy, 2019]. A hybrid MEM was proposed to combine the free space Green's function and a modal expansion to improve the convergence speed [Xu and Sommerfeldt, 2010]. Although most of MEMs are used in frequency domain, it can be extended to simulate the transient response [Meissner, 2013]. The major advantage of the MEM is that it provides exact solutions to the wave equation with appropriate boundary conditions, so it is in principle more accurate than the ISM when a room has finite impedance walls.

For an arbitrary source placed in a room, a key of using MEMs is to obtain the modal source density of this source, which is the inner product of the source density and the modal function [Meissner, 2013]. However, the source density for a directional source described by cylindrical or spherical harmonics is usually unknown. Another challenge of using MEMs is that the number of required room modes to obtain converged results scales with the frequency and/or the room size. This limits the applications of MEMs to low-frequency responses in small rooms. Fortunately, the computational speed can be improved utilizing the fast Fourier transform (FFT) along the dimensions of rooms where walls are lightly damped or rigid [Zhong et al., 2023c].

In this chapter, the acoustic waves generated by conventional linear acoustic sources are discussed in Sec. 5.2, where the MEM is used for solving sound fields generated by 2D and 3D directional sources. The MEM is then extended to calculate the audio sound field generated by PALs in a rectangular room in Sec. 5.3, where the properties of the audio sound fields generated by PALs are discussed and compared to conventional loudspeakers.

5.2 CONVENTIONAL LINEAR SOURCES IN A RECTANGULAR ROOM

Figure 5.1 shows the physical model to be investigated. Both 2D [Betlehem and Poletti, 2012] and 3D [Samarasinghe et al., 2018] rooms are considered. It is noted that the 2D room is an ideal simplified model and can be used to approximate the case when one dimension of the source is sufficiently large. The 2D and 3D (global) Cartesian coordinate systems *Oxy* and *Oxyz* are established with their origin at one corner of the 2D and 3D rooms, respectively. The dimensions of the 2D and 3D rooms are $V = L_x \times L_y$ and $V = L_x \times L_y \times L_z$, respectively, and the walls are denoted by ∂V. An arbitrary linear sound source is placed in the room with its acoustic center located at $\boldsymbol{\rho}_\mathrm{c} = (x_\mathrm{c}, y_\mathrm{c})$ or $\mathbf{r}_\mathrm{c} = (x_\mathrm{c}, y_\mathrm{c}, z_\mathrm{c})$. The sound source is assumed to be spatially confined and is depicted as a finite extent distributed source inscribed within a circle of radius $\rho_{\mathrm{s}0}$ and a sphere of radius $r_{\mathrm{s}0}$ as shown in Figs. 5.1(a) and (b), respectively.

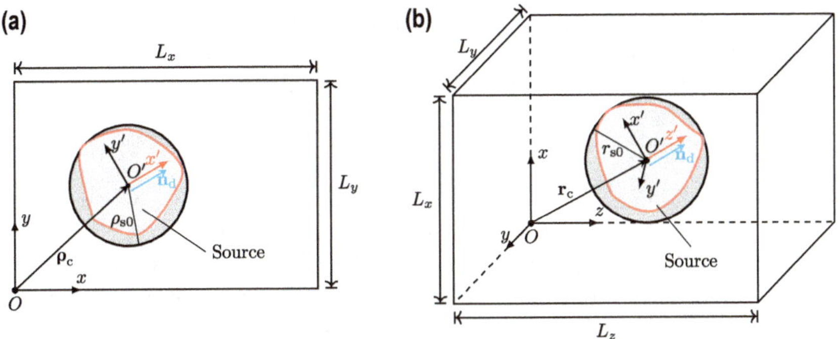

Figure 5.1 Sketch of an acoustic source in (a) 2D and (b) 3D rectangular rooms. The source is represented by a finite extent distributed source inscribed within (a) a circle with a radius ρ_{s0} and (b) a sphere with a radius r_{s0}. Extracted from [Zhong et al., 2023c, Fig. 1].

5.2.1 MODAL EXPANSION METHOD

This section focuses on deriving the MEM in the 3D case. The extension to the 2D case is straightforward. For an arbitrary source with a source density of $q(\mathbf{r})$, the wave propagation in the frequency domain is governed by the Helmholtz equation

$$(\nabla^2 + k^2)p(\mathbf{r}) = \mathrm{i}\rho_0\omega q(\mathbf{r}), \quad \mathbf{r} \in V, \tag{5.1}$$

where i is the imaginary unit, ρ_0 is the air density, $\omega = 2\pi f$ is the angular frequency at f, the wavenumber $k = \omega/c_0 + \mathrm{i}\alpha$, c_0 is the sound speed in air, and α is the attenuation coefficient at frequency f due to atmospheric absorption (Appendix A). Note that when the source is just a point monopole source with a volume velocity of Q_0, the right-hand side of Eq. (5.1) reduces to $\mathrm{i}\rho_0\omega Q_0\delta(\mathbf{r}-\mathbf{r}_\mathrm{c})$, where $\delta(\cdot)$ is the Dirac delta function.

To solve Eq. (5.1), the wall boundary conditions need to be defined. Suppose the walls have a finite specific acoustic impedance of $Z(\mathbf{r})$, and define the normalized specific acoustic admittance as

$$\eta(\mathbf{r}) \equiv \frac{\rho_0 c_0}{Z(\mathbf{r})}. \tag{5.2}$$

The choice of this definition is because it is convenient to represent a commonly occurred case when the wall is rigid, i.e., $\eta = 0$. The boundary conditions for the pressure then read

$$[\mathbf{n}\cdot\nabla - \mathrm{i}\eta(\mathbf{r})k]p(\mathbf{r}) = 0, \quad \mathbf{r} \in \partial V, \tag{5.3}$$

where the vector $\mathbf{n}$ is the unit normal vector of ∂V, which points outward of the room walls. Note that Eq. (5.3) differs by a factor of -1 from that used in some literatures, because this book adopts the time-harmonic convention of $\exp(-\mathrm{i}\omega t)$ instead of $\exp(\mathrm{i}\omega t)$. The reason of this preference can be found in Shin, 2019.

The MEM expresses the solution of the inhomogeneous wave equation (5.1) with the boundary condition Eq. (5.3) as the superposition of a set of *modal functions* (also known as *eigenfunctions*), which describe the characteristic spatial distribution of sound pressure in a room. For an arbitrary source with a source density of $q(\mathbf{r})$, the sound pressure in an arbitrary room can be calculated with

$$p(\mathbf{r}) = \mathrm{i}\rho_0\omega \sum_{n=0}^{\infty} \frac{Q(\mathbf{k}_n)\psi(\mathbf{k}_n,\mathbf{r})}{\Lambda_n^2(k^2-k_n^2)}, \tag{5.4}$$

where k_n is the norm of the vector $\mathbf{k}_n$ representing eigenvalues.

The modal functions $\psi(\mathbf{k}_n,\mathbf{r})$ shown in Eq. (5.4) depend on the shape of the room and its wall impedance. They are determined by the wave equation and the boundary as

$$(\nabla^2 + k_n^2)\psi(\mathbf{k}_n,\mathbf{r}) = 0, \quad \mathbf{r} \in V, \tag{5.5}$$

$$[\mathbf{n}\cdot\nabla - \mathrm{i}\eta(\mathbf{r})k]\psi(\mathbf{k}_n,\mathbf{r}) = 0, \quad \mathbf{r} \in \partial V, \tag{5.6}$$

where the index n denotes the *modal number*, and k_n is the wavenumber (eigenvalue) at this mode.

The modal function is orthogonal over the room, i.e.,

$$\iiint_V \psi(\mathbf{k}_n,\mathbf{r})\psi(\mathbf{k}_{n'},\mathbf{r})\mathrm{d}^3\mathbf{r} = \Lambda_n^2\delta_{nn'}, \tag{5.7}$$

where $\delta_{nn'}$ is the Kronecker delta function, which equals to 1 when $n = n'$ and 0 otherwise. The normalization factor is then obtained as

$$\Lambda_n^2 \equiv \iiint_V \psi^2(\mathbf{k}_n,\mathbf{r})\mathrm{d}^3\mathbf{r}. \tag{5.8}$$

The integration on the right-hand side of Eq. (5.8) is called the inner product of modal functions. Note that there is another definition for the inner product which takes the complex conjugate of one of the modal functions before multiplying the modes together, i.e., $\psi\psi^*$. However, it has been demonstrated the definition given by Eq. (5.8) is more suitable in calculating sound fields [Naka et al., 2005; Nolan and Davy, 2019].

The modal source density in Eq. (5.4) has the form of

$$Q(\mathbf{k}_n) \equiv \iiint_V q(\mathbf{r})\psi(\mathbf{k}_n,\mathbf{r})\mathrm{d}^3\mathbf{r}, \tag{5.9}$$

which can be considered as the inner product of the source density $q(\mathbf{r}_\mathrm{s})$ and the modal function $\psi(\mathbf{k}_n,\mathbf{r})$.

For a special case when a point monopole source is placed at $\mathbf{r}_\mathrm{s}$ with a volume velocity of Q_0, Eq. (5.9) is simplified to

$$Q(\mathbf{k}_n) = Q_0\psi(\mathbf{k}_n,\mathbf{r}_\mathrm{s}). \tag{5.10}$$

Substituting Eq. (5.10) into Eq. (5.4) yields the sound pressure at $\mathbf{r}$ as

$$p(\mathbf{r}) = \mathrm{i}\rho_0\omega Q_0 \sum_{n=0}^{\infty} \frac{\psi(\mathbf{k}_n,\mathbf{r}_\mathrm{s})\psi(\mathbf{k}_n,\mathbf{r})}{\Lambda_n^2(k^2-k_n^2)}. \tag{5.11}$$

5.2.1.1 Modal Functions in Rectangular Rooms

In a rectangular room shown in Fig. 5.1, the wave equation for the modal function can be written with Cartesian coordinates x, y, z as

$$\left(\frac{\partial^2}{\partial x^2}+\frac{\partial^2}{\partial y^2}+\frac{\partial^2}{\partial z^2}+k_n^2\right)\psi(\mathbf{k}_n,\mathbf{r})=0. \tag{5.12}$$

The modal function can then be separated into three functions that

$$\psi(k_{n_x},k_{n_y},k_{n_z},x,y,z)=\psi(k_{n_x},x)\,\psi(k_{n_y},y)\,\psi(k_{n_z},z). \tag{5.13}$$

Substituting Eq. (5.13) into Eq. (5.12) yields that

$$\begin{aligned}\frac{1}{\psi(k_{n_x},x)}\frac{\mathrm{d}^2\psi(k_{n_x},x)}{\mathrm{d}x^2}+\frac{1}{\psi(k_{n_y},y)}\frac{\mathrm{d}^2\psi(k_{n_y},y)}{\mathrm{d}y^2}+\frac{1}{\psi(k_{n_z},z)}\frac{\mathrm{d}^2\psi(k_{n_z}z)}{\mathrm{d}z^2}\\=-\left(k_{n_x}^2+k_{n_y}^2+k_{n_z}^2\right),\end{aligned} \tag{5.14}$$

where the modal number n denotes the index set of n_x, n_y, and n_z, and the modal wavenumber vector consisting of the corresponding eigenvalues is expressed as $\mathbf{k}_n = (k_{n_x}, k_{n_y}, k_{n_z})$ with $k_n^2 = k_{n_x}^2 + k_{n_y}^2 + k_{n_z}^2$.

The terms on the left-hand side of Eq. (5.14) depend only on one coordinate variable while the equation has to be valid for any values of the variables, so this equation can be split into three single-variable equations

$$\left(\frac{\mathrm{d}^2}{\mathrm{d}\xi^2}+k_{n_\xi}^2\right)\psi(k_{n_\xi},\xi)=0,\quad \xi=x,y,z. \tag{5.15}$$

The solution of Eq. (5.15) is

$$\psi(k_{n_\xi},\xi)=A_\xi \mathrm{e}^{\mathrm{i}k_{n_\xi}\xi}+B_\xi \mathrm{e}^{-\mathrm{i}k_{n_\xi}\xi}. \tag{5.16}$$

The coefficients A_ξ and B_ξ in Eq. (5.16) are determined by the boundary conditions in the ξ-direction, and will be discussed in details in Secs. 5.2.1.2 and 5.2.1.3, where $\xi = x, y, z$. Based on Eq. (5.4), the sound pressure generated by a source with a source density of $q(\mathbf{r})$ is then obtained as

$$p(\mathbf{r})=\mathrm{i}\rho_0\omega\sum_{n_x,n_y,n_z=0}^{\infty}\frac{Q(\mathbf{k}_n)\,\psi(\mathbf{k}_n,\mathbf{r})}{\Lambda_n^2(k^2-k_{n_x}^2-k_{n_y}^2-k_{n_z}^2)}. \tag{5.17}$$

5.2.1.2 Finite Impedance Walls

The boundary conditions for eigenfunctions with finite impedance walls are given by Eq. (5.3). For simplicity, the normalized specific acoustic admittance is assumed to be uniform on each wall in this chapter, and $\eta_{\xi,0}$ and η_{ξ,L_ξ} denote the value on the wall $\xi = 0$ and $\xi = L_\xi$, respectively, where $\xi = x, y, z$.

In x-direction, the boundary condition on two ends are

$$\begin{cases} \left.\dfrac{\partial \psi(k_{n_x},x)}{\partial x}\right|_{x=0} + \mathrm{i}k_{n_x}\eta_{x,0}\psi(k_{n_x},0) = 0, \\ \left.\dfrac{\partial \psi(k_{n_x},x)}{\partial x}\right|_{x=L_x} - \mathrm{i}k_{n_x}\eta_{x,L_x}\psi(k_{n_x},L_x) = 0. \end{cases} \tag{5.18}$$

Substituting Eq. (5.16) into Eq. (5.18) yields

$$\begin{cases} (k_{n_x} - \eta_{x,0}k)A_x - (k_{n_x} + \eta_{x,0}k)B_x = 0, \\ (k_{n_x} - \eta_{x,L_x}k)A_x \mathrm{e}^{\mathrm{i}k_{n_x}L_x} - (k_{n_x} + \eta_{x,L_x}k)B_x \mathrm{e}^{-\mathrm{i}k_{n_x}L_x} = 0. \end{cases} \tag{5.19}$$

The analysis on y- and z-directions is the same, so it is not repeated for simplicity. By solving Eq. (5.19), the modal functions can be described as

$$\psi(\mathbf{k}_n,\mathbf{r}) = \cos(k_{n_x}x + \gamma_{n_x})\cos\left(k_{n_y}y + \gamma_{n_y}\right)\cos\left(k_{n_z}z + \gamma_{n_z}\right), \tag{5.20}$$

where the modal phases $\gamma_n = \left(\gamma_{n_x}, \gamma_{n_y}, \gamma_{n_z}\right)$ are determined by the equation

$$k_{n_\xi}\tan\gamma_{n_\xi} = \mathrm{i}k\eta_{\xi,0}, \tag{5.21}$$

and the eigenvalues k_{n_ξ} are obtained by solving the transcendental equation [Nolan and Davy, 2019]

$$(k^2\eta_{\xi,0}\eta_{\xi,L_\xi} + k_{n_\xi}^2)\tan\left(k_{n_\xi}L_\xi\right) = -\mathrm{i}kk_{n_\xi}\left(\eta_{\xi,0} + \eta_{\xi,L_\xi}\right), \tag{5.22}$$

which are complex numbers in general. The normalization factor given by Eq. (5.8) is then obtained by directly evaluating the integral as

$$\Lambda_n^2 = \frac{V}{8}\prod_{\xi=x,y,z}\left[1 + \mathrm{sinc}(k_{n_\xi}L_\xi)\cos\left(k_{n_\xi}L_\xi + 2\gamma_{n_\xi}\right)\right], \tag{5.23}$$

where the volume $V = L_xL_yL_z$ and $\mathrm{sinc}\,x \equiv (\sin x)/x$ is the sinc function.

5.2.1.3 Lightly Damped or Rigid Walls

For lightly damped or rigid walls, the formulation can be simplified. The boundary conditions for rigid walls are obtained by setting $\eta = 0$ in Eq. (5.6), so it has

$$\mathbf{n}\cdot\nabla\psi(\mathbf{k}_n,\mathbf{r}) = 0, \quad \mathbf{r}\in\partial V. \tag{5.24}$$

From Eq. (5.21), the modal phases are zero. The eigenvalues are then simplified to be

$$k_{m_\xi} = \frac{m_\xi\pi}{L_\xi}, \tag{5.25}$$

and the corresponding modal functions are

$$\psi(\mathbf{k}_n,\mathbf{r}) = \cos(k_{n_x}x)\cos\left(k_{n_y}y\right)\cos\left(k_{n_z}z\right), \tag{5.26}$$

which can be considered as a simplified form of Eq. (5.20) by setting $\gamma_{n_\xi} = 0$. The normalization factor given by Eq. (5.8) reduces to

$$\Lambda_n^2 = \begin{cases} \dfrac{S}{\varepsilon_n}, & \text{2D}, \\ \dfrac{V}{\varepsilon_n}, & \text{3D}, \end{cases} \tag{5.27}$$

where $\varepsilon_n = \varepsilon_{n_x}\varepsilon_{n_y}\varepsilon_{n_z}$ and the Neumann factor $\varepsilon_{n_\xi} = 1$ when $n_\xi = 0$, $\varepsilon_{n_\xi} = 2$ when $n_\xi \neq 0$.

For lightly damped walls with $0 < |\eta| \ll 1$, the eigenvalues and eigenfunctions remain the same as given by Eqs. (5.25) and (5.26). However, the wavenumber $k = \omega/c_0 + \mathrm{i}\alpha$ is modified as $k = \omega/c_0 + \mathrm{i}(\alpha + D_n/2)$ to include the damping effects, where D_n is the damping term of each mode [Meissner, 2013; Zhong et al., 2022a]

$$D_n \equiv \frac{1}{\Lambda_n^2} \iint_{\partial V} \eta(\mathbf{r}) \psi^2(\mathbf{k}_n, \mathbf{r}) \mathrm{d}^2\mathbf{r}. \tag{5.28}$$

Based on the assumption that $\eta(\mathbf{r})$ is uniform on each wall, it is obtained that

$$D_n = \frac{1}{V}\left[\varepsilon_{n_x} S_x(\eta_{x,0} + \eta_{x,L_x}) + \varepsilon_{n_y} S_y(\eta_{y,0} + \eta_{y,L_y}) + \varepsilon_{n_z} S_z(\eta_{z,0} + \eta_{z,L_z})\right], \tag{5.29}$$

where $S_x = L_y L_z$, $S_y = L_x L_z$, and $S_z = L_x L_y$ are the areas of the wall perpendicular to x, y, and z axes, respectively. For a 2D room, Eq. (5.29) reduces to

$$D_n = \frac{1}{S}\left[\varepsilon_{n_x} L_y(\eta_{x,0} + \eta_{x,L_x}) + \varepsilon_{n_y} L_x(\eta_{y,0} + \eta_{y,L_y})\right]. \tag{5.30}$$

Furthermore, if $\eta(\mathbf{r}) = \eta_0$ over all walls, then Eqs. (5.29) and (5.30) are simplified respectively to

$$D_n = \frac{2\eta_0}{V}\left(\varepsilon_{n_x} S_x + \varepsilon_{n_y} S_y + \varepsilon_{n_z} S_z\right), \tag{5.31}$$

and

$$D_n = \frac{2\eta_0}{S}(\varepsilon_{n_x} L_y + \varepsilon_{n_y} L_x). \tag{5.32}$$

5.2.1.4 Fast Computation Using Fast Fourier Transform

In numerical computations, the modal expansion given by Eq. (5.17) needs to be truncated. Assume that the truncation terms are $N_x/2, N_y/2$, and $N_z/2$ for n_x, n_y, and n_z, respectively. Meanwhile, by using the relation $\cos\left(k_{n_\xi}\xi\right) = \left(\mathrm{e}^{\mathrm{i}k_{n_\xi}\xi} + \mathrm{e}^{-\mathrm{i}k_{n_\xi}\xi}\right)/2$, Eq. (5.4) can be rewritten as

$$p(\mathbf{r}) \approx \sum_{n_x=-N_x/2}^{N_x/2-1} \sum_{n_y=-N_y/2}^{N_y/2-1} \sum_{n_z=-N_z/2}^{N_z/2-1} P(\mathbf{k}_n) \mathrm{e}^{\mathrm{i}\mathbf{k}_n \cdot \mathbf{r}}, \tag{5.33}$$

for lightly damped rooms, where

$$P(\mathbf{k}_n) \equiv \frac{\mathrm{i}\rho_0\omega Q(\mathbf{k}_n)}{V(k^2 - k_n^2)}. \tag{5.34}$$

The calculation of Eq. (5.33) is time-consuming especially when the sound pressure at many observation points is required and at high frequencies and/or in large rooms. To reduce the calculation time, it can be transformed to a form of the inverse discrete Fourier transform (DFT) for efficient computation with the aid of the FFT [Williams, 1999, Sec. 1.8.2]. The direct calculation of Eq. (5.33) requires the order of $N_x^2 N_y^2 N_z^2$ multiplications and additions to obtain the sound field. However, the complexity of the computation with FFT reduces to only $N_x N_y N_z (\log N_x)(\log N_y)(\log N_z)$ without loss of accuracy.

By defining the interval between modal wavenumber vector,

$$\Delta\mathbf{k} \equiv (\Delta k_x, \Delta k_y, \Delta k_z), \quad \Delta k_\xi = \frac{\pi}{L_\xi}, \quad \xi = x, y, z, \tag{5.35}$$

the sound pressure given by Eq. (5.33) is rewritten as

$$p(\mathbf{r}) \approx \sum_{n_x=-N_x/2}^{N_x/2-1} \sum_{n_y=-N_y/2}^{N_y/2-1} \sum_{n_z=-N_z/2}^{N_z/2-1} P(\mathbf{k}_n) \mathrm{e}^{\mathrm{i}\left(n_x \Delta k_x x + n_y \Delta k_y y + n_z \Delta k_z z\right)}. \tag{5.36}$$

In the spatial domain, the corresponding discretization of coordinates is adopted as

$$\xi_{m_\xi} = m_\xi \Delta\xi = m_\xi \frac{2L_\xi}{N_\xi}, \quad \xi = x, y, z, \tag{5.37}$$

where the spatial domain index $m_\xi = -N_\xi/2, -N_\xi/2+1, \cdots, N_\xi/2-1$. It is then obtained that

$$\mathbf{k}_n \cdot \mathbf{r}_m = (m_x n_x \Delta k_x \Delta x, m_y n_y \Delta k_y \Delta y, m_z n_z \Delta k_z \Delta z) = 2\pi\left(\frac{m_x n_x}{N_x}, \frac{m_y n_y}{N_y}, \frac{m_z n_z}{N_z}\right). \tag{5.38}$$

The values at two end points and the center point are then

$$\begin{cases} \xi(-N_\xi/2) = \xi_{-N_\xi/2} = -L_\xi, \\ \xi(0) = \xi_0 = 0, \\ \xi(N_\xi/2 - 1) = \xi_{N_\xi - 1} = L_\xi - \dfrac{2L_\xi}{M_\xi}. \end{cases} \tag{5.39}$$

Therefore, the values at spatial point $\mathbf{r}_m \equiv (x_{m_x}, y_{m_y}, z_{m_z})$ can be obtained by

$$p(\mathbf{r}_m) \approx \sum_{n_x=-N_x/2}^{N_x/2-1} \sum_{n_y=-N_y/2}^{N_y/2-1} \sum_{n_z=-N_z/2}^{N_z/2-1} P(\mathbf{k}_n) \exp\left[\mathrm{i}2\pi\left(\frac{m_x n_x}{N_x} + \frac{m_y n_y}{N_y} + \frac{m_z n_z}{N_z}\right)\right]. \tag{5.40}$$

By defining the following change of indices

$$m'_\xi = m_\xi + N_\xi/2, \quad n'_\xi = n_\xi + N_\xi/2, \quad \xi = x,y,z, \tag{5.41}$$

$m'_\xi, n'_\xi = 0, 1, \ldots, N_\xi/2 - 1$, Eq. (5.40) can be transformed to the standard form of the inverse DFT as

$$p'(m') \approx \frac{1}{N_x N_y N_z} \sum_{n'_x=0}^{N_x-1} \sum_{n'_y=-0}^{N_y-1} \sum_{n'_z=-0}^{N_z-1} P'(n') \exp\left[\mathrm{i}2\pi\left(\frac{m'_x n'_x}{N_x} + \frac{m'_y n'_y}{N_y} + \frac{m'_z n'_z}{N_z}\right)\right]. \tag{5.42}$$

where the following functions are defined

$$p'(m') = \frac{(-1)^{m'_x+m'_y+m'_z}}{N_x N_y N_z} p((m'_x - N_x/2)\Delta x, (m'_y - N_y/2)\Delta y, (m'_z - N_z/2)\Delta z), \tag{5.43}$$

and

$$P'(n') = (-1)^{n'_x+n'_y+n'_z} P((n'_x - N_x/2)\Delta k_x, (n'_y - N_y/2)\Delta k_y, (n'_z - N_z/2)\Delta k_z). \tag{5.44}$$

It is noted that the computation of Eq. (5.33) using FFT is valid only for lightly damped walls when $\mathbf{k}_n$ are real valued as given by Eq. (5.25). Nevertheless, the computation using FFT still significantly reduces the computation load unless walls have finite impedance in all directions. For example, when the room walls are nonrigid only in the z-direction [Nolan and Davy, 2019], the computation efficiency can still be improved by using the FFT in both x- and y-directions.

5.2.1.5 Remarks

In this section, the reader is introduced to the MEM, a wave-based simulation method to calculate the non-diffuse reverberant sound field generated by a directional source in both 2D and 3D rectangular rooms. The modal functions in rectangular rooms are presented in Sec. 5.2.1.1. The boundary conditions of finite impedance and lightly damped walls are discussed in Secs. 5.2.1.2 and 5.2.1.3, respectively. To improve the computation efficiency, the summation of rooms modes can be written in the form of discrete Fourier transform so that the FFT can be used for lightly damped or rigid walls, as illustrated in Sec. 5.2.1.4. This enables the accurate calculation of sound fields in rooms even at high frequencies, extending up to 40 kHz, all while maintaining a relatively low computational burden.

5.2.2 DIRECTIONAL SOURCES IN A 2D RECTANGULAR ROOM

Since a PAL is one kind of directional acoustic sources, the sound fields generated by an arbitrary linear directional source are analyzed at first. As shown in Fig. 5.1(a), a directional sound source is placed in the room with its acoustic center at $\boldsymbol{\rho}_\mathrm{c} = (x_\mathrm{c}, y_\mathrm{c})$. The sound source is assumed to be spatially confined, so it is represented by a finite extent distributed source inscribed within a circle of radius ρ_{s0}. For the method used for this problem, it is convenient to define a primed (local) coordinate system $O'x'y'$ in the 2D room with the positive x' axis being the direction of the main lobe and the origin O' at the acoustic center of the source. The radiation directions of the source are denoted by unit vectors $\mathbf{n}_\mathrm{d} = (\cos\varphi_\mathrm{d}, \sin\varphi_\mathrm{d})$. The angle φ_d is the azimuthal angle in the polar coordinate system (ρ, φ). The objective in this section is to calculate the sound field generated by an arbitrary directional source in a 2D rectangular room.

5.2.2.1 Cylindrical Harmonic Representation of a Directional Source in a 2D Space

There are generally two ways to represent a directional source: collections of point monopole sources and multipoles. Representation using a large collection of point monopole sources is based on the fact that any sound field can be approximated by that radiated by a set of point sources. By assuming an array of point sources, the source strength (also known as the driving function) for each source is obtained via fitting against a predefined or measured directivity [Escolano et al., 2007; Murphy et al., 2014]. The multipole-based methods express the sound field radiated by a directional source using the cylindrical [Betlehem and Poletti, 2012] harmonic representations in 2D models. The concept is that the cylindrical harmonics form an orthogonal and complete basis for any well behaved functions defined on a circle, so that the directivity can be represented using this basis. This allows for a sparse representation of the source directivity than the point monopole sources, and is therefore prevailing in wave-based simulations [Poletti, 2005]. This section aims to introduce the cylindrical harmonic representation of a 2D directional source.

The sound pressure at an exterior field point $\boldsymbol{\rho}'$ generated by an arbitrary directional source in a 2D space can be represented by the cylindrical harmonics as [Betlehem and Poletti, 2012]

$$p(\boldsymbol{\rho}') = \frac{\rho_0 \omega}{4} \sum_{m=-\infty}^{\infty} A_m(k) H_m(k\rho') \mathrm{e}^{\mathrm{i}m\varphi'}, \quad \rho' > \rho_{\mathrm{s}0}, \tag{5.45}$$

where ρ' and φ' are the polar and azimuthal coordinates of the point $\boldsymbol{\rho}'$, the normalization factor $\rho_0\omega/4$ represents the self-radiation resistance of a point monopole source with the wavenumber k in 2D free space, $\rho_{\mathrm{s}0}$ is the radius of the circle just enclosing the source as shown in Fig. 5.1(a), $H_m(\cdot)$ is the Hankel function of the first kind, and $A_m(k)$ are expansion coefficients which can be obtained by analytical derivations and/or measured results [Betlehem and Abhayapala, 2005]. This directional source can be considered as the radiation from a finite extent distributed source with a source density of $q'(\boldsymbol{\rho}'_\mathrm{s})$, so that $q'(\boldsymbol{\rho}'_\mathrm{s}) = 0$ for $\rho'_\mathrm{s} > \rho_{\mathrm{s}0}$. Note that the field (or observation) point $\boldsymbol{\rho}'$ is represented in the primed coordinate system.

In free space, the sound pressure generated by such a distributed source can be obtained by [Zhong et al., 2021a]

$$p(\boldsymbol{\rho}') = -\mathrm{i}\rho_0\omega \iint_{\rho'_\mathrm{s} \le \rho_{\mathrm{s}0}} q'(\boldsymbol{\rho}'_\mathrm{s}) g_{\mathrm{2D}}(\boldsymbol{\rho}' - \boldsymbol{\rho}'_\mathrm{s}) \mathrm{d}^2 \boldsymbol{\rho}'_\mathrm{s}, \tag{5.46}$$

where the 2D Green's function is expressed as

$$g_{2D}(\boldsymbol{\rho}' - \boldsymbol{\rho}'_s) = \frac{i}{4} H_0(k|\boldsymbol{\rho}' - \boldsymbol{\rho}'_s|). \tag{5.47}$$

It can be represented as the superposition of cylindrical waves by [Poletti, 2019; Zhong et al., 2021a]

$$g_{2D}(\boldsymbol{\rho}' - \boldsymbol{\rho}'_s) = \frac{i}{4} \sum_{m=-\infty}^{\infty} J_m(k\rho'_<) H_m(k\rho'_>) e^{im(\varphi' - \varphi'_s)} \tag{5.48}$$

where $J_m(\cdot)$ is the Bessel function, $\rho'_< = \min(\rho', \rho'_s)$, and $\rho'_> = \max(\rho', \rho'_s)$. Substituting Eq. (5.48) into Eq. (5.46) yields

$$p(\boldsymbol{\rho}') = \frac{\rho_0 \omega}{4} \sum_{m=-\infty}^{\infty} \left[\iint_{\rho'_s \le \rho_{s0}} q'(\boldsymbol{\rho}'_s) J_m(k\rho'_s) e^{-im\varphi'_s} d^2 \boldsymbol{\rho}'_s \right] H_m(k\rho') e^{im\varphi'}. \tag{5.49}$$

By comparing Eqs. (5.49) and (5.45), it can be found that the cylindrical harmonic expansion coefficients $A_m(k)$ and the source density $q'(\boldsymbol{\rho}'_s)$ are related by

$$A_m(k) = \iint_{\rho'_s \le \rho_{s0}} q'(\boldsymbol{\rho}'_s) J_m(k\rho'_s) e^{-im\varphi'_s} d^2 \boldsymbol{\rho}'_s. \tag{5.50}$$

Equation (5.50) shows that an arbitrary directional source described by Eq. (5.45) can be considered as the radiation from an equivalent source with a source density of $q'(\boldsymbol{\rho}'_s)$ inside the circle $\rho' \le \rho_{s0}$.

5.2.2.2 Modal Source Density

It is shown in Sec. 5.2.1 that the sound field in a rectangular room can be obtained using the modal expansion given by Eq. (5.17) once the modal source density given by Eq. (5.9) is known. Although the cylindrical expansion coefficients of a directional source shown in Eq. (5.45) can be measured, the source density $q(\mathbf{r}_s)$ is generally unknown . To obtain the modal source density for a 2D directional source, an auxiliary integral is introduced as

$$I(\mathbf{k}_n) = \iint_S q(\boldsymbol{\rho}_s) e^{-i\mathbf{k}_n \cdot \boldsymbol{\rho}_s} d^2 \boldsymbol{\rho}_s. \tag{5.51}$$

It is observed that the modal source density given by Eq. (5.9) can be expressed using the auxiliary integral as

$$Q(\mathbf{k}_n) = \frac{1}{4} \sum_{\pm} e^{-i(\gamma_{n_x} + \gamma_{n_y})} I(\pm k_{n_x}, \pm k_{n_y}), \tag{5.52}$$

where the symbol $\sum_{\pm}$ denotes the summation of four possible combinations of positive and negative signs. Consequently, once the integral given by Eq. (5.51) is determined, the modal source density is obtained according to Eq. (5.52). The approach presented here is to solve the integral by using the cylindrical harmonic expansion coefficients represented by Eq. (5.50).

The source density $q(\boldsymbol{\rho}_\mathrm{s})$ in Eq. (5.51) is given in the unprimed coordinate system Oxy, the origin of which is coincident with a corner of the room. In most cases, the source density is convenient to be represented in a primed coordinate system $O'x'y'$ with the origin at the acoustic center of the source and the positive x' axis in the direction of $\mathbf{n}_\mathrm{d} = (\cos\varphi_\mathrm{d}, \sin\varphi_\mathrm{d})$, which is the center of the main lobe of the source. The primed axes $x'y'$ can be considered as rotating the unprimed ones xy through an angle φ_d followed by a translation of $\boldsymbol{\rho}_\mathrm{c}$. The coordinates of the source point in the primed system $\boldsymbol{\rho}'_\mathrm{s}$ and the coordinates in the unprimed one $\boldsymbol{\rho}_\mathrm{s}$ are then related by

$$\boldsymbol{\rho}'_\mathrm{s} = \mathbf{R}(\boldsymbol{\rho}_\mathrm{s} - \boldsymbol{\rho}_\mathrm{c}), \tag{5.53}$$

where the rotation matrix is

$$\mathbf{R} = \begin{bmatrix} \cos\varphi_\mathrm{d} & \sin\varphi_\mathrm{d} \\ -\sin\varphi_\mathrm{d} & \cos\varphi_\mathrm{d} \end{bmatrix}. \tag{5.54}$$

By changing the integration region in Eq. (5.51) from S to $\rho'_\mathrm{s} \le \rho_{\mathrm{s}0}$, it has

$$I(\mathbf{k}_n) = \iint_{\rho'_\mathrm{s} \le \rho_{\mathrm{s}0}} q'(\boldsymbol{\rho}'_\mathrm{s}) \mathrm{e}^{-\mathrm{i}\mathbf{k}_n \cdot \boldsymbol{\rho}_\mathrm{s}} \mathrm{d}^2 \boldsymbol{\rho}'_\mathrm{s}, \tag{5.55}$$

where the relation $\mathrm{d}^2\boldsymbol{\rho}'_\mathrm{s} = \mathrm{d}^2\boldsymbol{\rho}_\mathrm{s}$ is used because the determinant of the rotation matrix is unit. Note that the source point $\boldsymbol{\rho}_\mathrm{s}$ in the exponential function is expressed in the unprimed coordinate system. Because $\mathbf{R}$ is orthonormal, $\mathbf{R}^{-1} = \mathbf{R}^\mathrm{T}$, where $\mathbf{R}^{-1}$ is the inverse matrix of $\mathbf{R}$, and the superscript "T" denotes the transpose. The inverse transformation from $\boldsymbol{\rho}'_\mathrm{s}$ to $\boldsymbol{\rho}_\mathrm{s}$ is obtained by $\boldsymbol{\rho}_\mathrm{s} = \mathbf{R}^{-1}\boldsymbol{\rho}'_\mathrm{s} + \boldsymbol{\rho}_\mathrm{c} = \mathbf{R}^\mathrm{T}\boldsymbol{\rho}'_\mathrm{s} + \boldsymbol{\rho}_\mathrm{c}$. The phase term in Eq. (5.55), i.e., $\mathbf{k}_n \cdot \boldsymbol{\rho}_\mathrm{s}$, can be written as $\mathbf{k}_n^\mathrm{T}\boldsymbol{\rho}_\mathrm{s}$ when $\mathbf{k}_n$ and $\boldsymbol{\rho}_\mathrm{s}$ are represented as column vectors. Accordingly, $\mathbf{k}_n^\mathrm{T}\boldsymbol{\rho}_\mathrm{s} = \mathbf{k}_n^\mathrm{T}(\mathbf{R}^\mathrm{T}\boldsymbol{\rho}'_\mathrm{s} + \boldsymbol{\rho}_\mathrm{c}) = (\mathbf{R}\mathbf{k}_n)^\mathrm{T}\boldsymbol{\rho}'_\mathrm{s} + \mathbf{k}_n^\mathrm{T}\boldsymbol{\rho}_\mathrm{c}$, which can also be written as $(\mathbf{R}\mathbf{k}_n) \cdot \boldsymbol{\rho}'_\mathrm{s} + \mathbf{k}_n \cdot \boldsymbol{\rho}_\mathrm{c}$ using the inner product notation. Then

$$I(\mathbf{k}_n) = \mathrm{e}^{-\mathrm{i}\mathbf{k}_n \cdot \boldsymbol{\rho}_\mathrm{c}} \iint_{\rho'_\mathrm{s} \le \rho_{\mathrm{s}0}} q'(\boldsymbol{\rho}'_\mathrm{s}) \mathrm{e}^{-\mathrm{i}\mathbf{k}'_n \cdot \boldsymbol{\rho}'_\mathrm{s}} \mathrm{d}^2 \boldsymbol{\rho}'_\mathrm{s}, \tag{5.56}$$

where a primed modal wavenumber vector is defined as $\mathbf{k}'_n = \mathbf{R}\mathbf{k}_n$. It is clear that $k'_n \equiv |\mathbf{k}'_n| = |\mathbf{k}_n| = k_n$.

It is known that the cylindrical expansion holds that

$$\mathrm{e}^{-\mathrm{i}\mathbf{k}'_n \cdot \boldsymbol{\rho}'_\mathrm{s}} = \sum_{m=-\infty}^{\infty} \mathrm{i}^{-m} J_m(k_n \rho'_\mathrm{s}) \mathrm{e}^{\mathrm{i}m(\varphi_{k'_n} - \varphi'_\mathrm{s})}, \tag{5.57}$$

where $(k_n, \varphi_{k'_n})$ are the polar coordinates of $\mathbf{k}'_n$ such that $\varphi_{k'_n} = \tan^{-1}(k'_{ny}/k'_{nx})$. Equation (5.57) represents the cylindrical expansion of a plane wave arriving from the direction $-\mathbf{k}'_n$ when $\mathbf{k}'_n$ are real valued. Substituting Eq. (5.57) into Eq. (5.56) yields

$$I(\mathbf{k}_n) = \mathrm{e}^{-\mathrm{i}\mathbf{k}_n \cdot \boldsymbol{\rho}_\mathrm{c}} \sum_{m=-\infty}^{\infty} \mathrm{i}^{-m} \mathrm{e}^{\mathrm{i}m\varphi_{k'_n}} \iint_{\rho'_\mathrm{s} \le \rho_{\mathrm{s}0}} q'(\boldsymbol{\rho}'_\mathrm{s}) J_m(k_n \rho'_\mathrm{s}) \mathrm{e}^{-\mathrm{i}m\varphi'_\mathrm{s}} \mathrm{d}^2 \boldsymbol{\rho}'_\mathrm{s}. \tag{5.58}$$

By comparing Eqs. (5.58) and (5.50), it has

$$I(\mathbf{k}_n) = \mathrm{e}^{-\mathrm{i}\mathbf{k}_n\cdot\boldsymbol{\rho}_\mathrm{c}} \sum_{m=-\infty}^{\infty} \mathrm{i}^{-m} A_m(k_n) \mathrm{e}^{\mathrm{i}m\varphi_{k'_n}}. \tag{5.59}$$

Equation (5.59) is one of main results of this section. For an arbitrary directional source, the sound field generated by it can be represented using the cylindrical harmonics as shown by Eq. (5.45). The radiation from such as a source can be considered as that from an equivalent distributed source with a source density of $q'(\boldsymbol{\rho}'_\mathrm{s})$ spatially confined in a circle with a radius of $\rho_{\mathrm{s}0}$. Once the expansion coefficients $A_m(k)$ are determined by analytical and/or experimental methods, the auxiliary integral $I(\mathbf{k}_n)$ can be obtained using Eq. (5.59), where the modal eigenvalues $\mathbf{k}_n$ are obtained using Eqs. (5.22) or (5.25) and the rotated modal eigenvalues $\mathbf{k}'_n = \mathbf{R}\mathbf{k}_n$ with the rotation matrix given by Eq. (5.54). Then the modal source density for this directional source is obtained by Eq. (5.52), where the phase term (if needed) γ_{n_ξ} is obtained using Eq. (5.21). Finally, the sound field in a room is calculated using Eq. (5.4) with the eigenfunctions given by Eqs. (5.20) or (5.26). The modal source density Q_n is indirectly obtained using the cylindrical harmonic expansion coefficients $A_m(k)$.

It is noted that the integral in Eq. (5.56) represents a 2D spatial Fourier transform of the primed source density $q'(\boldsymbol{\rho}'_\mathrm{s})$, which reads

$$\tilde{q}'(\mathbf{k}'_n) \equiv \iint_{\rho'_\mathrm{s} \leq \rho_{\mathrm{s}0}} q'(\boldsymbol{\rho}'_\mathrm{s}) \mathrm{e}^{-\mathrm{i}\mathbf{k}'_n\cdot\boldsymbol{\rho}'_\mathrm{s}} \mathrm{d}^2\boldsymbol{\rho}'_\mathrm{s}. \tag{5.60}$$

This observation allows for a faster computation of the auxiliary integral than that given by Eq. (5.59), i.e., $I(\mathbf{k}_n) = \mathrm{e}^{-\mathrm{i}\mathbf{k}_n\cdot\boldsymbol{\rho}_\mathrm{c}}\tilde{q}'(\mathbf{k}'_n)$, when $\tilde{q}'(\mathbf{k}'_n)$ is known for a directional source.

5.2.2.3 Low-Frequency Sound Fields

In the following simulations, the dimensions of the 2D room are set as $L_x \times L_y = 1.414 \times 1.156\ \mathrm{m}^2$. The normalized acoustic specific admittance of the finite impedance wall considered in the simulations is assumed to be $\eta = 0.5$, which corresponds to an absorption coefficient of 0.9 [Nolan and Davy, 2019]. In all cases, three configurations of the room are considered: (i) all walls are rigid, i.e., $\eta = 0$; (ii) the wall on the plane $x = L_x$ is absorptive with a normalized acoustic specific admittance of $\eta = 0.5$ while other walls are rigid; (iii) the wall on the plane $y = 0$ is absorptive with $\eta = 0$ while other walls are rigid. The eigenvalues determined by Eq. (5.22) are numerically evaluated using the built-in function "FindRoot" in Wolfram Mathematica 13.0, even though there are more efficient ways [Naka et al., 2005; Nolan and Davy, 2019]. The total surface/volume velocity of the source is set as $Q_0 = 2\times10^{-5}\ \mathrm{m}^2/\mathrm{s}$. The sound pressure level (SPL) is presented in all following figures with a reference pressure of 2×10^{-5} Pa. All calculations with FEM were performed using the commercial software COMSOL Multiphysics 6.0. To ensure converged results in the FEM, the maximal mesh size is set to be one tenth of the wavelength.

The sound fields generated by a point monopole array are presented as the first example. Supposing an array consisting of N_s point monopole sources with the i-th source located at $\boldsymbol{\rho}'_{\mathrm{s},i} = (\rho'_{\mathrm{s},i}, \varphi'_{\mathrm{s},i})$ described under the primed coordinates system, the source density can be described as

$$q'(\boldsymbol{\rho}'_\mathrm{s}) = \sum_{i=1}^{N_\mathrm{s}} Q_{\mathrm{s},i}\delta(\boldsymbol{\rho}'_\mathrm{s} - \boldsymbol{\rho}'_{\mathrm{s},i}), \tag{5.61}$$

where $Q_{\mathrm{s},i}$ is the source strength of the i-th source. The cylindrical harmonic expansion coefficients can be obtained by using Eqs. (5.45) and (5.48) as

$$A_m(k) = \sum_{i=1}^{N_\mathrm{s}} Q_{\mathrm{s},i} J_m(k\rho'_{\mathrm{s},i})\mathrm{e}^{-\mathrm{i}m\varphi'_{\mathrm{s},i}}. \tag{5.62}$$

The 2D spatial Fourier transform of the source density is

$$\tilde{q}'(\mathbf{k}'_n) = \sum_{i=1}^{N_\mathrm{s}} Q_{\mathrm{s},i}\mathrm{e}^{-\mathrm{i}\mathbf{k}'_n\cdot\boldsymbol{\rho}'_{\mathrm{s},i}}. \tag{5.63}$$

A simple directional source that generates a dipole outgoing sound field is considered at first. The dipole is approximated by 2 point monopole sources located at $(L_x/2 - 0.05\,\mathrm{m}, L_y/2)$ and $(L_x/2 + 0.05\,\mathrm{m}, L_y/2)$ with source strengths of $-Q_0$ and Q_0. Figure 5.2 illustrates the sound fields obtained using the normal mode expansion method at 191 Hz, which is the (1,1) mode defined for a room with rigid walls. For comparison, the sound field in free space is given as Fig. 5.2(a), where a clear dipole radiation pattern can be observed. When this dipole is placed in the room with rigid walls, Fig. 5.2(b) demonstrates that the radiate field undergoes a noticeable change. To further examine the effects of the room's boundaries, the walls are set to be absorptive on the boundaries $x = L_x$ and $y = 0$ in Figs. 5.2(c) and (d), respectively. It can be observed that these absorptive walls significantly mitigate the reflections of waves in their corresponding directions. The sound fields obtained using the FEM software are almost the same as those obtained using the modal expansion method; therefore, they are not presented to save the space. The spatially averaged relative error, defined as $|\mathbf{p} - \mathbf{p}_\mathrm{FEM}|/|\mathbf{p}|$ with $\mathbf{p}_\mathrm{FEM}$ the sound pressure vector obtained using the FEM, is less than 0.05% in all cases.

The second example is a line source with a length of $2a$ placed on the y' axis, which is usually used to simulate a baffled line piston source in 2D models [Poletti, 2019; Zhong et al., 2021a]. It has a directional beam pattern at large ka due to the aperture effects. The source density can be written as [Zhong et al., 2022a]

$$q'(\boldsymbol{\rho}'_\mathrm{s}) = \frac{Q_0}{2a}\Pi\left(\frac{y'_\mathrm{s}}{2a}\right)\delta(x'_\mathrm{s}), \tag{5.64}$$

where the denominator $2a$ is to ensure that the line source has a total surface velocity of Q_0 as evident by Eq. (2.20), and the rectangle function $\Pi(\zeta) = 1$ when $-1/2 <$

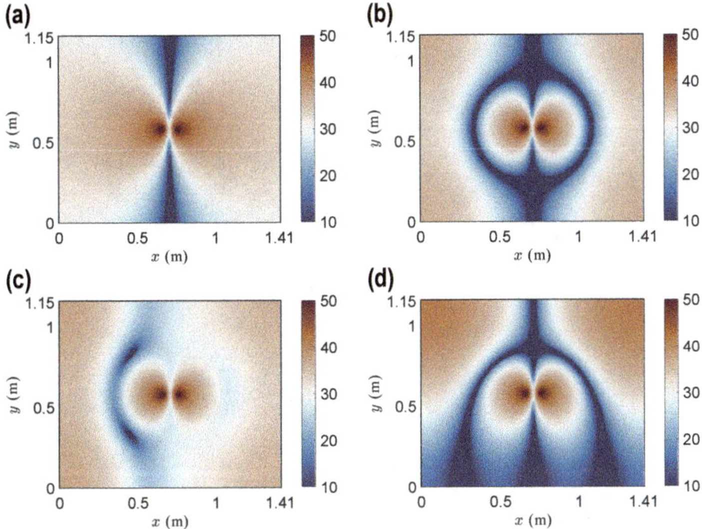

Figure 5.2 Sound fields at 191 Hz (1,1 mode) generated by a dipole consisting of two point sources located at $(L_x/2 - 0.05\,\text{m}, L_y/2)$ and $(L_x/2 + 0.05\,\text{m}, L_y/2)$ in (a) free space and (b–d) a 2D rectangular room. (b) All walls are rigid; (c) $x = L_x$ is covered with absorptive material with an absorption coefficient of 0.9; (d) $y = 0$ is covered with absorptive material with an absorption coefficient of 0.9. Extracted from [Zhong et al., 2023c, Fig. 2].

$\zeta < 1/2$, and $\Pi(\zeta) = 0$ otherwise. The cylindrical expansion coefficients are [Poletti, 2019; Zhong et al., 2021a]

$$A_{2m}(k) = \frac{Q_0}{a} \int_0^a J_{2m}(k\rho_s')\mathrm{d}\rho_s', \tag{5.65}$$

and $A_{2m+1}(k) = 0$, where the integral with respect to J_{2m} can be efficiently evaluated using the Gauss-Legendre quadrature [Zhong et al., 2021a], hypergeometric functions [Poletti, 2019], and Struve functions [Rosenheinrich, 2017]. The 2D spatial Fourier transform of the source density is

$$\tilde{q}'(\mathbf{k}_n') = Q_0 \operatorname{sinc}\left(k_{n_y}' a\right). \tag{5.66}$$

Figure 5.3 shows the sound fields obtained using the MEM at 191 Hz (1,1 mode). The centroid of the line source is located at $\boldsymbol{\rho}_c = (0.2\,\text{m}, 0.2\,\text{m})$, and the direction angle is $\varphi_d = 15°$. For comparison, the sound field in free space is given as Fig. 5.3(a), where an approximately omni-directional radiation pattern can be observed because $ka = 0.17$ is small. When the source is placed in the rectangular room with rigid walls, Fig. 5.3(b) illustrates a clear modal pattern at the 1,1 mode, primarily because the line source is situated near a corner of the room. The mode in the x and y directions tends to diminish when the wall is absorptive on $x = L_x$ and $y = 0$, as shown in Figs. 5.3(c) and (d), respectively. The spatially averaged relative error against the FEM is less than 0.05% which validates the accuracy of the MEM.

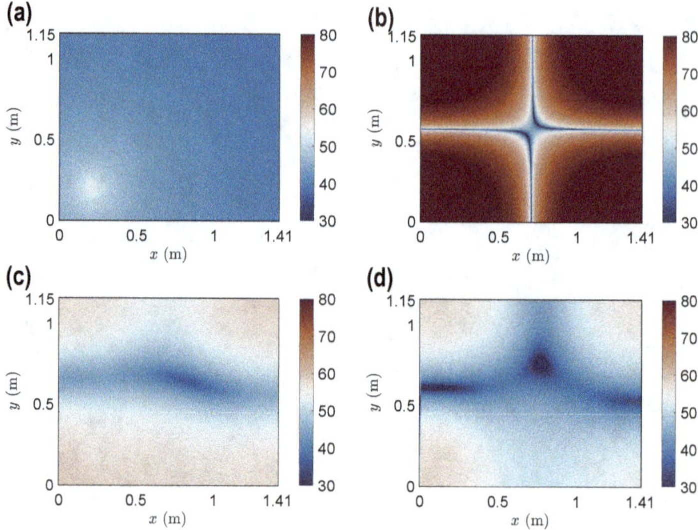

Figure 5.3 Sound fields at 191 Hz (1,1 mode) generated by a line source in (a) free space and (b–d) a 2D rectangular room. (b) All walls are rigid; (c) $x = L_x$ is covered with absorptive material; (d) $y = 0$ is covered with absorptive material. The source centroid is located at $\rho_c = (0.2\,\text{m}, 0.2\,\text{m})$ and the radiation angle is $\varphi_d = 15°$. Extracted from [Zhong et al., 2023c, Fig. 3].

5.2.2.4 High-Frequency Sound Fields

Here, the fields at an extremely high frequency (40 kHz) are presented to investigate the wave propagation of a highly directional source in a room. The accurate ultrasound field around 40 kHz is required to obtain the audio sound field generated by a PAL. Although the sound field at such a high frequency is able to be approximately predicted using various kinds of geometric acoustics models [Savioja and Svensson, 2015], the phase information is neglected or inaccurate so that they cannot be used for the calculations involved in the PAL. The attenuation coefficient due to atmospheric absorption is set as $\alpha = 0.15\,\text{Np/m}$ at 40 kHz, which is evaluated according to ISO 9613 with a relative humidity of 70% and temperature 20°C (Appendix A).

To show the sound fields generated by a highly directional source, the SPL at 40 kHz is given in Fig. 5.4 with other parameters being the same as those used in Fig. 5.3. This frequency is a typical carrier frequency for the PAL, and there is a need for accurately calculating the sound fields at this frequency in a room [Zhong et al., 2022a]. For comparison, the sound field in free space is given as Fig. 5.4(a), showcasing a clear directional beam due to a large value of $ka = 36.6$. When the source is placed in the rectangular room with rigid walls, Fig. 5.4(b) illustrates that the highly directional beam propagates along the radiator axis and experience multiple reflections between walls. The truncation of the directional beam after the incidence on the absorptive wall can be clearly observed in Figs. 5.4(c–d). In such a case, the spatially averaged relative error against the FEM is found to be less than 0.8% which validates the accuracy of the modal expansion method even at a very high frequency.

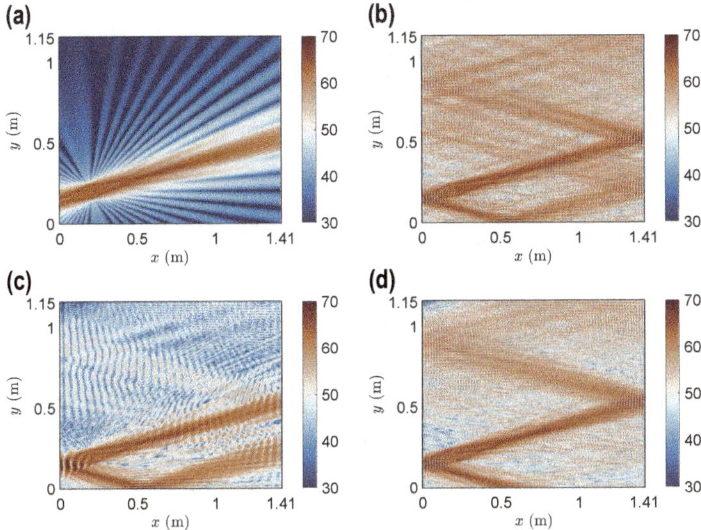

Figure 5.4 The simulated sound fields at 40 kHz generated by a line source with a half-width of 0.05 m in (a) free space and (b–d) a 2D rectangular room. (b) All walls are rigid; (c) $x = L_x$ is covered with absorptive material; (d) $y = 0$ is covered with absorptive material. The source centroid is located at $\rho_c = (0.2\,\text{m}, 0.2\,\text{m})$ and the radiation angle is $\varphi_d = 15°$. Extracted from [Zhong et al., 2023c, Fig. 4].

Figure 5.4 illustrates the propagation of a directional beam in both the front and back sides of the source. However, in practical applications, ultrasound sources exhibit high directionality, and radiation toward the back side is minimal. Consequently, the line source model depicted in Fig. 5.4 is unsuitable for accurately representing a real ultrasound source. A more appropriate approach is to directly eliminate the component propagating in the backward direction. In this case, the source density, as defined by Eq. (5.66), is revised as

$$\tilde{q}'(\mathbf{k}'_n) = Q_0 \operatorname{sinc}\left(k'_{n_y} a\right) \mathrm{H}(k'_{n_y}), \tag{5.67}$$

where $\mathrm{H}(\cdot)$ is the Heaviside function. Figure 5.5 displays the sound fields at 40 kHz after truncating the directivity on the back side of the source. It is evident that the beam is focused exclusively in front of the source, providing a more realistic approximation of ultrasound source behavior in practical applications.

5.2.2.5 Remarks

In this section, the reader is introduced to the sound fields radiated by directional linear sources in 2D rectangular rooms. The cylindrical harmonic representation is employed to represent the directional source. However, it is worth noting that directional sources can also be characterized by either a set of discrete point sources [Escolano et al., 2007] or continuously distributed sources [Bilbao and Ahrens, 2020]. In both cases, the source can be represented by a source density function. By applying the

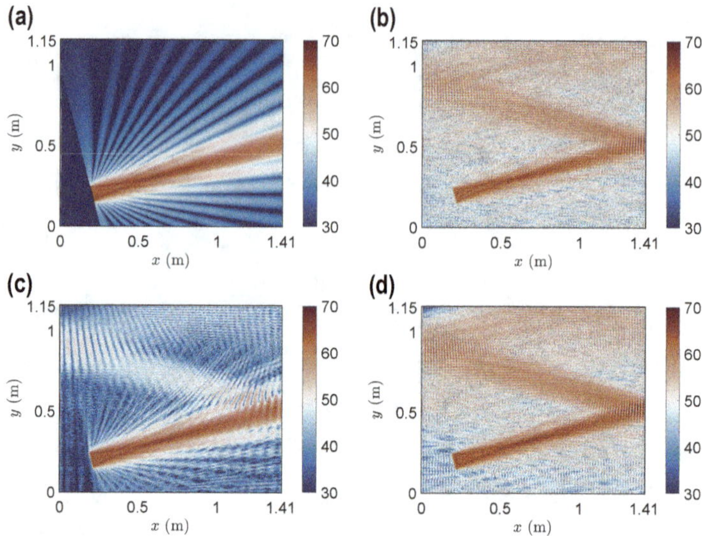

Figure 5.5 The simulated sound fields at 40 kHz generated by a line source with a half-width of 0.05 m in (a) free space and (b–d) a 2D rectangular room. The radiation on the back side of the source is truncated. (b) All walls are rigid; (c) $x = L_x$ is covered with absorptive material; (d) $y = 0$ is covered with absorptive material. The source centroid is located at $\boldsymbol{\rho}_\mathrm{c} = (0.2\,\mathrm{m}, 0.2\,\mathrm{m})$ and the radiation angle is $\varphi_\mathrm{d} = 15°$.

methodology described in Sec. 5.2.2.2, the modal source density can be obtained, which enable solving the sound field in the 2D rectangular room in a similar manner. Typical sound fields at low and high frequencies are presented in Secs. 5.2.2.3 and 5.2.2.4, respectively. Low-frequency sound fields are commonly used to model the sound produced by conventional loudspeakers, while high-frequency sound fields serve as the foundation for solving the audio sound generated by PALs.

5.2.3 DIRECTIONAL SOURCES IN A 3D RECTANGULAR ROOM

As shown in Fig. 5.1(b), a directional sound source is placed in a 3D room with its acoustic center at $\mathbf{r}_\mathrm{c} = (x_\mathrm{c}, y_\mathrm{c}, z_\mathrm{c})$. The sound source is assumed to be spatially confined, so it is represented by a finite extent distributed source inscribed within a sphere of radius $r_{\mathrm{s}0}$. For the method used for this physical problem, it is more convenient to define a primed (local) coordinate system $O'x'y'z'$ with the positive z' axis being the direction of the main lobe and the origin O' at the acoustic center of the source. The radiation directions of the source are denoted by the unit vector $\mathbf{n}_\mathrm{d} = (\sin\theta_\mathrm{d}\cos\varphi_\mathrm{d}, \sin\theta_\mathrm{d}\sin\varphi_\mathrm{d}, \cos\theta_\mathrm{d})$. The angles θ_d and φ_d correspond to the zenithal and azimuthal angles in the spherical coordinate system (r, θ, φ), respectively. The objective is to calculate the sound fields in rooms generated by an arbitrary directional source in a 3D rectangular room.

5.2.3.1 Spherical Harmonic Representation of a Directional Source in a 3D Space

The sound pressure at an exterior field point $\mathbf{r}'$ generated by a directional source in a 3D space can be represented by the spherical harmonics as [Williams, 1999; Samarasinghe et al., 2018; Ben Hagai et al., 2011]

$$p(\mathbf{r}') = \frac{\rho_0 \omega k}{4\pi} \sum_{\ell=0}^{\infty} \sum_{m=-\ell}^{\ell} A_\ell^m(k) \mathrm{h}_\ell(kr') Y_\ell^m(\theta', \varphi'), \tag{5.68}$$

where r', θ', and φ' are respectively the radial, zenithal, and azimuthal coordinates of the point $\mathbf{r}'$, the normalization factor $\rho_0 \omega k/(4\pi)$ represents the self-radiation resistance of a point monopole source with the wavenumber k in 3D free space, $\mathrm{h}_\ell(\cdot)$ is the spherical Hankel function of the first kind, $Y_\ell^m(\cdot,\cdot)$ is the spherical harmonics of degree ℓ and order m, and $A_\ell^m(k)$ are expansion coefficients. Similar to the 2D model, this directional source can be considered as the radiation from a finite extent distributed source with a source density of $q'(\mathbf{r}_\mathrm{s}')$, so that $q'(\mathbf{r}_\mathrm{s}') = 0$ for $r_\mathrm{s}' > r_{\mathrm{s}0}$. Note that the field (or observation) point $\mathbf{r}'$ is represented in the primed coordinate system.

In free space, the sound pressure generated by such a distributed source can be obtained by [Eq. (2.7)]

$$p(\mathbf{r}') = -\mathrm{i}\rho_0 \omega \iiint_{r_\mathrm{s}' \le r_{\mathrm{s}0}} q'(\mathbf{r}_\mathrm{s}') g_{\mathrm{3D}}(\mathbf{r}' - \mathbf{r}_\mathrm{s}') \mathrm{d}^3 \mathbf{r}_\mathrm{s}', \tag{5.69}$$

where the 3D Green's function is expressed as

$$g_{\mathrm{3D}}(\mathbf{r}' - \mathbf{r}_\mathrm{s}') = \frac{\mathrm{e}^{\mathrm{i}k|\mathbf{r}' - \mathbf{r}_\mathrm{s}'|}}{4\pi|\mathbf{r}' - \mathbf{r}_\mathrm{s}'|}. \tag{5.70}$$

It is known that g_{3D} can be represented as the superposition of spherical waves by [Zhong et al., 2020c]

$$g_{\mathrm{3D}}(\mathbf{r}' - \mathbf{r}_\mathrm{s}') = \mathrm{i}k \sum_{\ell=0}^{\infty} \mathrm{j}_\ell(kr'_<) \mathrm{h}_\ell(kr'_>) \sum_{m=-\ell}^{\ell} Y_\ell^m(\theta', \varphi') Y_\ell^{m,*}(\theta_\mathrm{s}', \varphi_\mathrm{s}'), \tag{5.71}$$

where the superscript "*" represents the complex conjugation, $\mathrm{j}_\ell(\cdot)$ is the spherical Bessel function of the first kind, $r'_< = \min(r', r_\mathrm{s}')$, and $r'_> = \max(r', r_\mathrm{s}')$. Substituting Eq. (5.71) into Eq. (5.69) yields

$$p(\mathbf{r}') = \rho_0 \omega k \sum_{\ell=0}^{\infty} \sum_{m=-\ell}^{\ell} \left[\iiint_{r_\mathrm{s}' \le r_{\mathrm{s}0}} q'(\mathbf{r}_\mathrm{s}') \mathrm{j}_\ell(kr_\mathrm{s}') Y_\ell^{m,*}(\theta_\mathrm{s}', \varphi_\mathrm{s}') \mathrm{d}^3 \mathbf{r}_\mathrm{s}' \right] \mathrm{h}_\ell(kr') Y_\ell^m(\theta', \varphi'). \tag{5.72}$$

By comparing Eqs. (5.72) and (5.68), it is observed that the spherical expansion coefficients $A_\ell^m(k)$ and the source density $q'(\mathbf{r}_\mathrm{s}')$ are related by

$$A_\ell^m(k) = 4\pi \iiint_{r_\mathrm{s}' \le r_{\mathrm{s}0}} q(\mathbf{r}_\mathrm{s}') \mathrm{j}_\ell(kr_\mathrm{s}') Y_\ell^{m,*}(\theta_\mathrm{s}', \varphi_\mathrm{s}') \mathrm{d}^3 r_\mathrm{s}'. \tag{5.73}$$

Equation (5.73) shows that an arbitrary directional source described by Eq. (5.68) can be considered as the radiation from an equivalent source with a source density of $q'(\mathbf{r}_s')$ inside the sphere $r' \leq r_{s0}$.

5.2.3.2 Modal Source Density

Similar to Eq. (5.51), an auxiliary integral is defined as

$$I(\mathbf{k}_n) = \iiint_V q(\mathbf{r}_s)\mathrm{e}^{-\mathrm{i}\mathbf{k}_n\cdot\mathbf{r}_s}\mathrm{d}^3\mathbf{r}_s. \tag{5.74}$$

Then the modal source density given by Eq. (5.9) is

$$Q(\mathbf{k}_n) = \frac{1}{8}\sum_{\pm}\mathrm{e}^{-\mathrm{i}(\gamma_{nx}+\gamma_{ny}+\gamma_{nz})}I(\pm k_{nx}, \pm k_{ny}, \pm k_{nz}; \pm\gamma_{nx}, \pm\gamma_{ny}, \pm\gamma_{nz}). \tag{5.75}$$

where the symbol $\sum_{\pm}$ denotes the summation of 8 possible combinations of positive and negative signs.

The primed axes $x'y'z'$ in Fig. 5.1(b) can be considered as rotating the unprimed axes xyz through an angle pair $(\theta_\mathrm{d}, \varphi_\mathrm{d})$ followed by a translation of $\mathbf{r}_\mathrm{c}$. The coordinates of a point in the primed system $\mathbf{r}' = (x', y', z')$ and the coordinates in the unprimed one $\mathbf{r} = (x, y, z)$ are then related by

$$\mathbf{r}' = \mathbf{R}(\mathbf{r} - \mathbf{r}_\mathrm{c}), \tag{5.76}$$

where the rotation matrix is obtained by two elementary rotations $\mathbf{R}_y$ and $\mathbf{R}_z$ as

$$\mathbf{R} = \mathbf{R}_y(\theta_\mathrm{d})\mathbf{R}_z(\varphi_\mathrm{d}) = \begin{bmatrix}\cos\theta_\mathrm{d} & 0 & -\sin\theta_\mathrm{d}\\ 0 & 1 & 0\\ \sin\theta_\mathrm{d} & 0 & \cos\theta_\mathrm{d}\end{bmatrix}\begin{bmatrix}\cos\varphi_\mathrm{d} & \sin\varphi_\mathrm{d} & 0\\ -\sin\varphi_\mathrm{d} & \cos\varphi_\mathrm{d} & 0\\ 0 & 0 & 1\end{bmatrix}. \tag{5.77}$$

By using Eq. (5.76), it is obtained that $\mathbf{k}_n\cdot\mathbf{r}_s = \mathbf{k}_n\cdot(\mathbf{R}^{-1}\mathbf{r}_s' + \mathbf{r}_\mathrm{c}) = (\mathbf{R}\mathbf{k}_n)\cdot\mathbf{r}_s' + \mathbf{k}_n\cdot\mathbf{r}_\mathrm{c}$. By defining the primed wavenumber vector as the rotation of the unprimed one, i.e., $\mathbf{k}_n' = \mathbf{R}\mathbf{k}_n$, the auxiliary integral can be rewritten as

$$I(\mathbf{k}_n) = \mathrm{e}^{-\mathrm{i}\mathbf{k}_n\cdot\mathbf{r}_\mathrm{c}}\iiint_{r_s'\leq r_{s0}} q'(\mathbf{r}_s')\mathrm{e}^{-\mathrm{i}\mathbf{k}_n'\cdot\mathbf{r}_s'}\mathrm{d}^3\mathbf{r}_s'. \tag{5.78}$$

For the spherical harmonic expansion, it has [Poletti, 2005, Eq. (9)]

$$\mathrm{e}^{-\mathrm{i}\mathbf{k}_n'\cdot\mathbf{r}_s'} = 4\pi\sum_{\ell=0}^{\infty}\mathrm{i}^{-\ell}\mathrm{j}_\ell(k_n r_s')\sum_{m=-\ell}^{\ell}Y_\ell^m(\theta_{k_n'}, \varphi_{k_n'})Y_\ell^{m,*}(\theta_s', \varphi_s'), \tag{5.79}$$

where $(k_n, \theta_{k_n'}, \varphi_{k_n'})$ are the spherical coordinates of $\mathbf{k}_n'$. Equation (5.79) represents the spherical harmonic expansion of a plane wave arriving from direction $-\mathbf{k}_n'$ when $\mathbf{k}_n'$ are real valued. Substituting Eq. (5.79) into Eq. (5.78) yields

$$I(\mathbf{k}_n) = 4\pi\mathrm{e}^{-\mathrm{i}\mathbf{k}_n\cdot\mathbf{r}_\mathrm{c}}\sum_{\ell=0}^{\infty}\sum_{m=-\ell}^{\ell}\mathrm{i}^{-\ell}Y_\ell^m(\theta_{k_n'}, \varphi_{k_n'})\iiint_{r_s'\leq r_{s0}} q'(\mathbf{r}_s')\mathrm{j}_\ell(k_n r_s')Y_\ell^{m,*}(\theta_s', \varphi_s')\mathrm{d}^3\mathbf{r}_s'. \tag{5.80}$$

By comparing Eqs. (5.80) and (5.73), it has

$$I(\mathbf{k}_n) = \mathrm{e}^{-\mathrm{i}\mathbf{k}_n\cdot\mathbf{r}_\mathrm{c}} \sum_{\ell=0}^{\infty}\sum_{m=-\ell}^{\ell} \mathrm{i}^{-\ell} A_\ell^m(k_n) Y_\ell^m(\theta_{k'_n}, \varphi_{k'_n}). \tag{5.81}$$

Equation (5.81) is the 3D version of Eq. (5.59). For an arbitrary directional source, the sound field generated by it can be represented using the spherical harmonics as shown by Eq. (5.68). The radiation from such as a source can be considered as that from an equivalent distributed source with a source density of $q'(\mathbf{r}'_\mathrm{s})$ spatially confined in a sphere with a radius of $r_{\mathrm{s}0}$. The calculation process of the modal source density is similar to that for the 2D model as described below Eq. (5.59).

The integral in Eq. (5.78) represents a 3D spatial Fourier transform of the primed source density $q'(\mathbf{r}'_\mathrm{s})$, which reads

$$\tilde{q}'(\mathbf{k}'_n) \equiv \iiint_{r'_\mathrm{s}\le r_{\mathrm{s}0}} q'(\mathbf{r}'_\mathrm{s}) \mathrm{e}^{-\mathrm{i}\mathbf{k}'_n\cdot\mathbf{r}'_\mathrm{s}} \mathrm{d}^3\mathbf{r}'_\mathrm{s}. \tag{5.82}$$

This allows for a faster computation of the auxiliary integral than that given by Eq. (5.81), i.e., $I(\mathbf{k}_n) = \mathrm{e}^{-\mathrm{i}\mathbf{k}_n\cdot\mathbf{r}_\mathrm{c}}\tilde{q}'(\mathbf{k}'_n)$, when $\tilde{q}'(\mathbf{k}'_n)$ is known for a directional source.

5.2.3.3 Low-Frequency Sound Fields

In a 3D space, supposing an array consisting of N_s point monopole sources with the i-th source located at $\mathbf{r}'_{\mathrm{s},i} = (r'_{\mathrm{s},i}, \theta'_{\mathrm{s},i}, \varphi'_{\mathrm{s},i})$ described under the primed coordinate system, the source density can be described as $q'(\mathbf{r}'_\mathrm{s}) = \sum_{i=1}^{N_\mathrm{s}} Q_{\mathrm{s},i}\delta(\mathbf{r}'_\mathrm{s} - \mathbf{r}'_{\mathrm{s},i})$. The spherical harmonic expansion coefficients can be obtained by using Eqs. (5.68) and (5.71) as [Samarasinghe et al., 2018]

$$A_\ell^m(k) = \sum_{i=1}^{N_\mathrm{s}} Q_{\mathrm{s},i} \mathrm{j}_\ell(kr'_{\mathrm{s},i}) Y_\ell^{m,*}(\theta'_{\mathrm{s},i}, \varphi'_{\mathrm{s},i}). \tag{5.83}$$

The 3D spatial Fourier transform of the source density is

$$\tilde{q}'(\mathbf{k}'_n) = \sum_{i=1}^{N_\mathrm{s}} Q_{\mathrm{s},i} \mathrm{e}^{-\mathrm{i}\mathbf{k}'_n\cdot\mathbf{r}'_{\mathrm{s},i}}. \tag{5.84}$$

A simple directional source that generates a dipole outgoing sound field is considered at first. The dipole is approximated by 2 point monopole sources located at $(L_x/2 - 0.05\,\mathrm{m}, L_y/2 - 0.05\,\mathrm{m}, L_z/2 - 0.05\,\mathrm{m})$ and $(L_x/2 + 0.05\,\mathrm{m}, L_y/2 + 0.05\,\mathrm{m}, L_z/2 + 0.05\,\mathrm{m})$ with source strengths of $-Q_0$ and Q_0. Figure 5.6 presents the SPL in the orthogonal slice planes in the x-, y-, and z-directions with the cross point at the source centroid, obtained using the MEM at 223 Hz, which is (1,1,1) mode of the room with rigid walls. For comparison, the sound field in free space is given as Fig. 5.6(a), where a clear dipole radiation pattern can be observed. The spatially averaged relative error against the FEM is found to be less than 0.1%. Similar to the

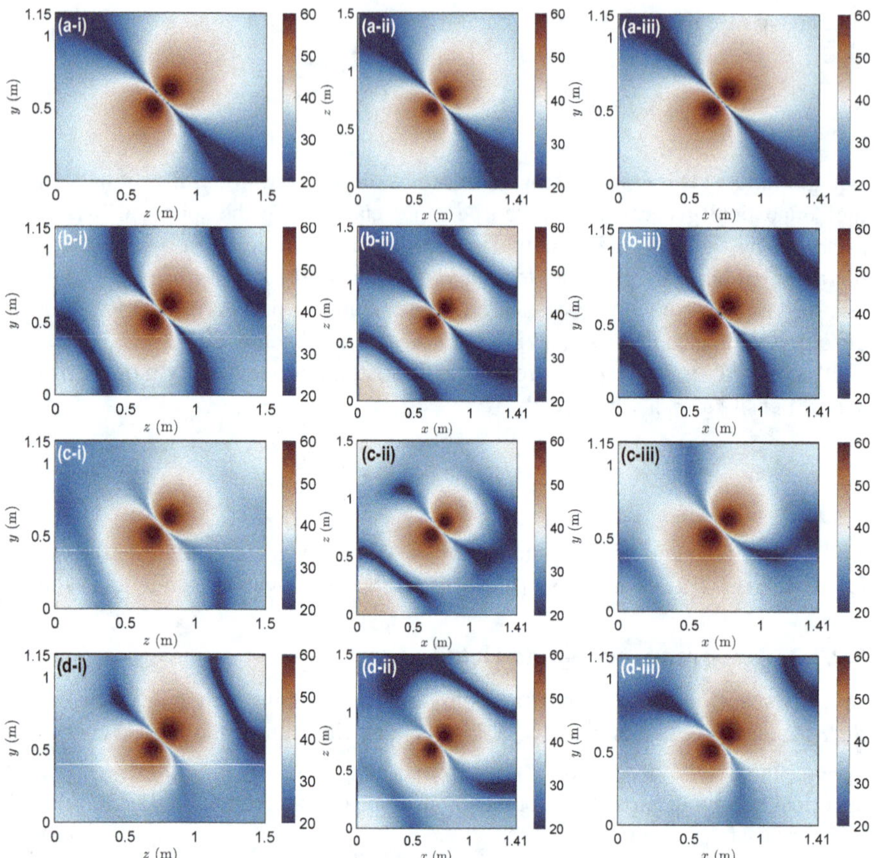

Figure 5.6 Sound fields at 223 Hz (1,1,1 mode) generated by a dipole in (a) free space and (b–d) a 3D rectangular room. (b) All walls are rigid; (c) $x = L_x$ is covered with absorptive material; (d) $y = 0$ is covered with absorptive material. The source centroid is located at $\mathbf{r}_c = (L_x/2, L_y/2, L_z/2)$ and the coordinates of two point sources are $\mathbf{r}'_{s,1} = (-0.05\,\text{m}, -0.05\,\text{m}, -0.05\,\text{m})$ and $\mathbf{r}'_{s,2} = (0.05\,\text{m}, 0.05\,\text{m}, 0.05\,\text{m})$. Results are obtained using the modal expansion method in the plane (i) $x = x_c$, (ii) $y = y_c$, or (iii) $z = z_c$. Extracted from [Zhong et al., 2023c, Fig. 5].

dipole in a 2D rectangular room as shown in Fig. 5.2, the dipole radiation pattern as well as the effects of the absorptive wall can be clearly observed in Fig. 5.6.

The second example is a circular source with a radius of a placed on the plane $z' = 0$, which is usually used to simulate a baffled circular piston source [Poletti, 2018; Zhong et al., 2020c]. It has a directional beam pattern at large ka. The source density can be written as [Zhong et al., 2023c]

$$q'(\mathbf{r}'_s) = \frac{Q_0}{\pi a^2}\mathrm{H}(a - \rho'_s)\delta(z'_s), \tag{5.85}$$

where the denominator πa^2 is to ensure that the source has a total volume velocity of Q_0 as evident by Eq. (2.41), and $\mathrm{H}(\zeta)$ is the Heaviside function which is 1 when

$\zeta \geq 0$ and 0 when $\zeta < 0$. The spherical expansion coefficients are [Poletti, 2018; Zhong et al., 2020c]

$$A_{2\ell}^{m}(k) = \frac{Q_0}{a^2}\delta_{m,0}Y_{2\ell}^{0}(\pi/2,0)\int_0^a \mathrm{j}_{2\ell}(kr_\mathrm{s})r_\mathrm{s}\mathrm{d}r_\mathrm{s}, \tag{5.86}$$

and $A_{2\ell+1}^{m}(k) = 0$, where the integral involving $\mathrm{j}_{2\ell}$ can be efficiently evaluated using the Gauss-Legendre quadrature [Zhong et al., 2020c], hypergeometric functions [Poletti, 2018], and closed-form solutions [Zhong and Qiu, 2020]. The 3D spatial Fourier transform of the source density is

$$\tilde{q}'(\mathbf{k}_n') = Q_0 \mathrm{jinc}\left(\sqrt{k_{n_x}'^2 + k_{n_y}'^2}a\right), \tag{5.87}$$

where the jinc function is defined by Eq. (2.84).

Figure 5.7 shows the sound fields in the rectangular room obtained using the MEM at 223 Hz (1,1,1 mode). The centroid of the circular source is located at $\mathbf{r}_\mathrm{c} = (0.2\,\mathrm{m}, 0.2\,\mathrm{m}, 0.2\,\mathrm{m})$, and the radiation angles are $(\theta_\mathrm{d}, \varphi_\mathrm{d}) = (90^\circ, 15^\circ)$. For comparison, the sound field in free space is given as Fig. 5.7(a), where an approximately omni-directional radiation pattern can be observed because $ka = 0.41$ is small. When this circular source is placed in the room with rigid walls, Fig. 5.7(b) shows a clear modal pattern at the (1,1,1) mode. This pattern is observed due to the source's proximity to the room's corner, facilitating the excitation of this mode. The mode in the x- and y- directions tends to diminish when the wall is absorptive on $x = L_x$ and $y = 0$, as evident by Figs. 5.7(c) and (d), respectively. This is similar to that presented for a 2D room as shown in Fig. 5.3. The spatially averaged relative error against the FEM is found to be less than 0.04% in this example.

5.2.3.4 High-Frequency Sound Fields

Figure 5.8 displays the sound fields in the rectangular room at 40 kHz, with all other parameters matching those used in Fig. 5.7. For reference, Fig. 5.8(a) shows the sound field in free space, revealing a prominent directional ultrasound beam, primarily concentrated in the $z = z_\mathrm{c}$ plane [Fig. 5.8(a-iii)]. When this source is placed in the rectangular room with rigid walls, Fig. 5.8(b) still exhibits a highly directional beam, with the major energy concentrated in the $z = z_\mathrm{c}$ plane [Fig. 5.8(b-iii)]. However, the directional beam undergoes multiple reflections due to the room's walls, resulting in a distinct pattern from that observed in free space [Fig. 5.8(a-iii)]. This outcome underscores that the directional source maintains its directivity within the room with rigid walls, thanks to the small wavelength relative to the room's dimensions. When the room's walls are absorptive, Figs. 5.8(c–d) clearly illustrate the significant impact of absorption, reducing wall reflections.

Figure 5.8 illustrates the propagation of a directional beam in both the front and back sides of the source. However, in practical applications, ultrasound sources exhibit high directionality, and radiation toward the back side is minimal. Consequently, the circular source model depicted in Fig. 5.8 is unsuitable for accurately

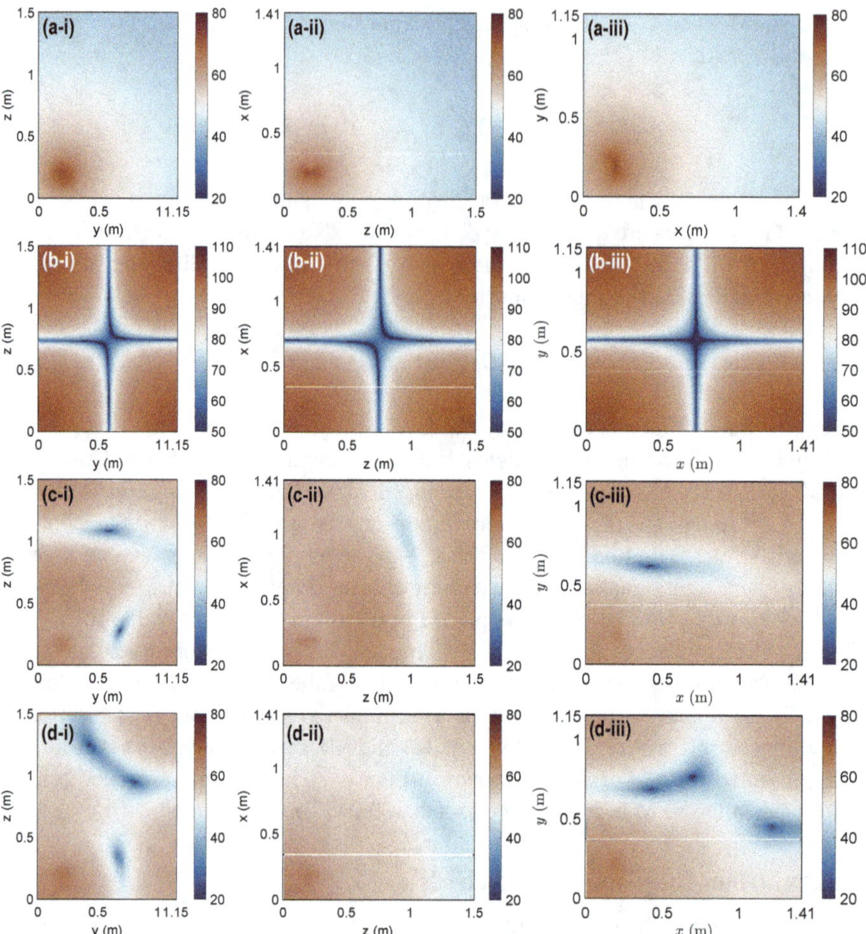

Figure 5.7 Sound fields at 223 Hz (1,1,1 mode) generated by a circular source with a radius of 0.1 m in (a) free space and (b–d) a 3D rectangular room. (b) All walls are rigid; (c) $x = L_x$ is covered with absorptive material with an absorption coefficient of 0.9; (d) $y = 0$ is covered with absorptive material with an absorption coefficient of 0.9. The source centroid is located at $\mathbf{r}_c = (0.2\,\text{m}, 0.2\,\text{m}, 0.2\,\text{m})$ and the radiation angles are $(\theta_d, \varphi_d) = (90°, 15°)$. The total volume velocity is $Q_0 = 2\times10^{-5}\,\text{m}^3/\text{s}$. Results are obtained in the plane (i) $x = x_c$, (ii) $y = y_c$, or (iii) $z = z_c$. Extracted from [Zhong et al., 2023c, Fig. 6].

representing a real ultrasound source. A more appropriate approach is to directly remove the component propagating in the backward direction. Similar to Eq. (5.67), the source density, as defined by Eq. (5.87) is revised as

$$\tilde{q}'(\mathbf{k}'_n) = Q_0 \operatorname{jinc}\left(\sqrt{k'^2_{n_x} + k'^2_{n_y}}\,a\right)\mathrm{H}(k'_{n_z}). \tag{5.88}$$

Figure 5.9 displays the sound fields at 40 kHz after truncating the radiation toward the back side of the source. It is evident that the beam is focused exclusively in front

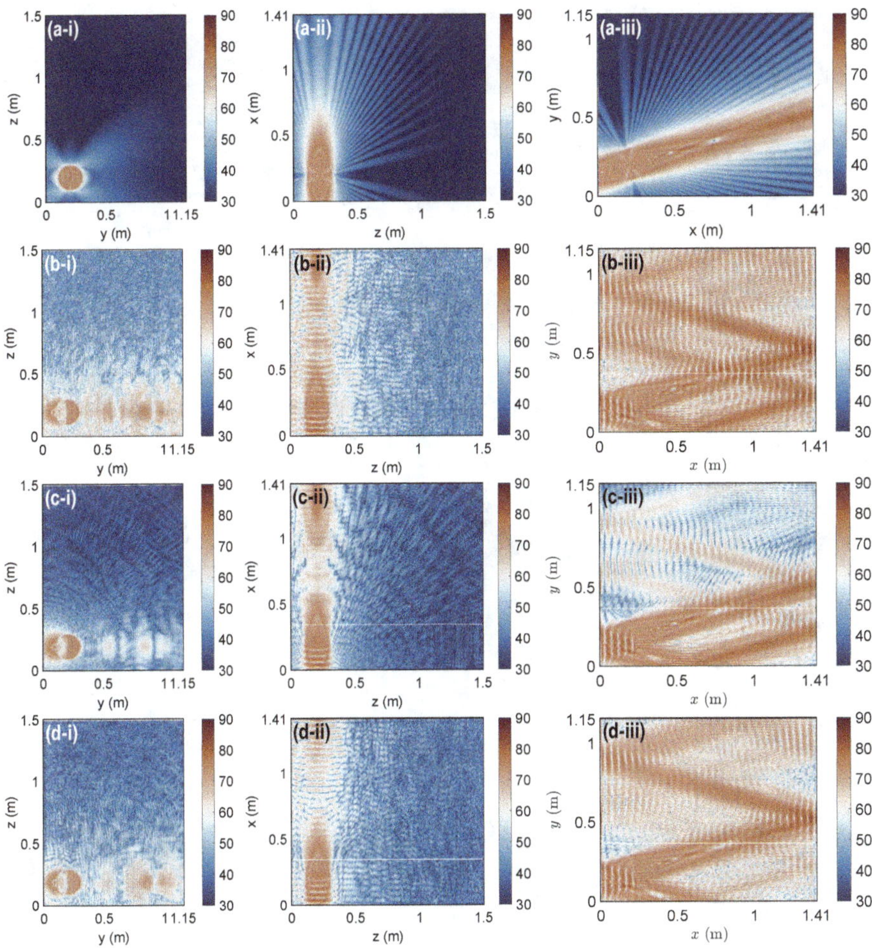

Figure 5.8 The simulated sound fields at 40 kHz generated by a circular source with a radius of 0.1 m in (a) free space and (b–d) a 3D rectangular room. (b) All walls are rigid; (c) $x = L_x$ is covered with absorptive material with an absorption coefficient of 0.9; (d) $y = 0$ is covered with absorptive material with an absorption coefficient of 0.9. The source centroid is located at $\mathbf{r}_c = (0.2\,\text{m}, 0.2\,\text{m}, 0.2\,\text{m})$ and the radiation angles are $(\theta_d, \varphi_d) = (90°, 15°)$. The total volume velocity is $Q_0 = 2 \times 10^{-5}\,\text{m}^3/\text{s}$. Results are obtained in the plane (i) $x = x_c$, (ii) $y = y_c$, or (iii) $z = z_c$. Extracted from [Zhong et al., 2023c, Fig. 7].

of the source, providing a more realistic approximation of ultrasound source behavior in practical applications.

5.2.3.5 Computational Efficacy

Because computation of high-frequency sound fields is essential for PALs, the computational efficiency of the modal expansion method is analyzed based on the numerical results obtained using MATLAB 2022a on a personal computer with an AMD

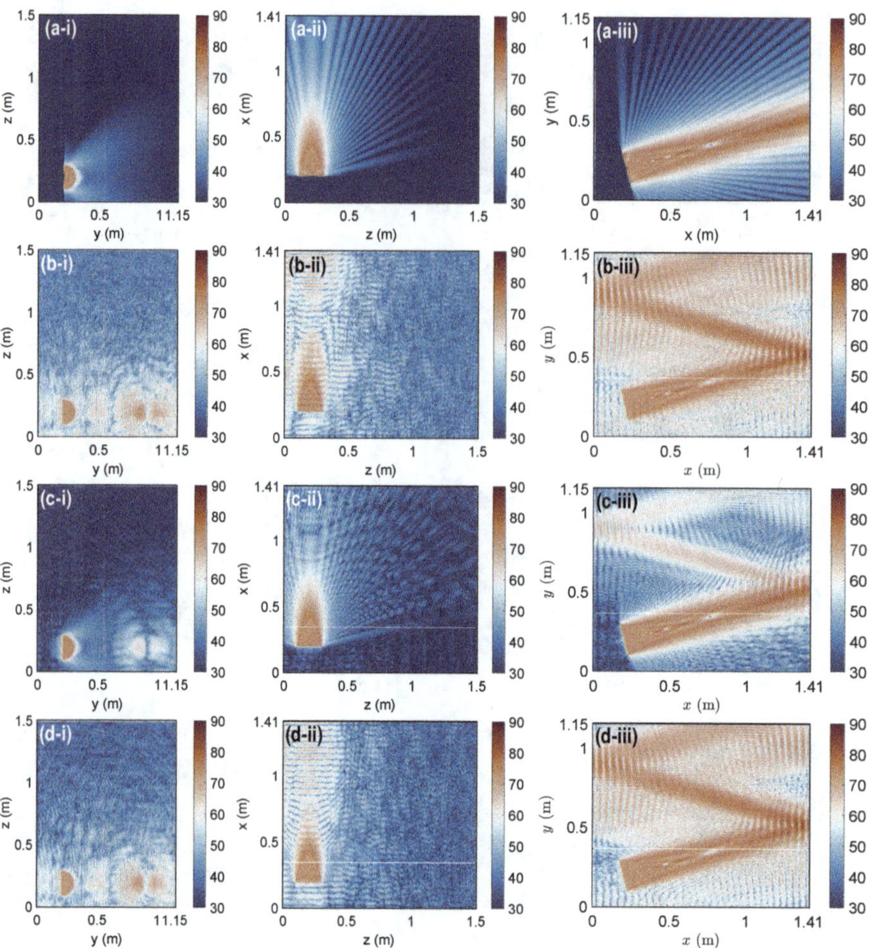

Figure 5.9 The simulated sound fields at 40 kHz generated by a circular source with a radius of 0.1 m in (a) free space and (b–d) a 3D rectangular room. (b) All walls are rigid; (c) $x = L_x$ is covered with absorptive material with an absorption coefficient of 0.9; (d) $y = 0$ is covered with absorptive material with an absorption coefficient of 0.9. The radiation on the back side of the source is truncated. The source centroid is located at $\mathbf{r}_c = (0.2\,\text{m}, 0.2\,\text{m}, 0.2\,\text{m})$ and the radiation angles are $(\theta_d, \varphi_d) = (90°, 15°)$. The total volume velocity is $Q_0 = 2 \times 10^{-5}\,\text{m}^3/\text{s}$. Results are obtained in the plane (i) $x = x_c$, (ii) $y = y_c$, or (iii) $z = z_c$.

Ryzen™ Threadripper™ 3960X central processing unit (CPU) with 256 GB of random access memory (RAM). Table 5.1 compares the calculation time to obtain the sound fields in rectangular rooms using the MEM and the FEM. The calculation time is measured for evaluating the sound pressure at all grid points in the room. The grid points in x-,y-, and z- directions are 35, 30, and 35, respectively, at 191 Hz and 223 Hz. At 40 kHz, the grid points in x-,y-, and z-directions are 350, 300, and 350, respectively. In Table 5.1, the exact calculation time is not presented for the FEM at

40 kHz in the 3D room because the calculation is time-consuming. Simulation results show that the elapsed time for this case is at least more than 2 hours (720 s). It can be found that the FEM can obtain converged results with an acceptable calculation time at low frequencies, but it requires huge long time at high frequencies. When all walls are rigid, the modal expansion method can be more than 100 times faster than the FEM. The improvement of the computation efficiency of the modal expansion method deteriorates when one wall is absorptive because the FFT is unable to be used in the direction perpendicular to this wall as discussed in Sec. 5.2.1.4.

Table 5.1
Calculation time of the modal expansion method and the FEM.

	Frequency	MEM with FFT	FEM
2D room with rigid walls	191 Hz	0.05 s	2 s
	40 kHz	0.07 s	244 s
2D room with one absorptive wall	191 Hz	0.25 s	3 s
	40 kHz	0.36 s	271 s
3D room with rigid walls	223 Hz	0.13 s	22 s
	40 kHz	81 s	> 7200 s
3D room with one absorptive wall	223 Hz	0.75 s	34 s
	40 kHz	243 s	> 7200 s

5.2.3.6 Remarks

In this section, the reader is introduced to the sound fields radiated by directional linear sources in 3D rectangular rooms. As discussed in Sec. 5.2.3.1, the directional source is modeled by the spherical harmonic representation, which has garnered increased interest in audio applications [Klippel and Bellmann, 2016]. The sound fields in the rectangular room are solved using the MEM, where the key is to obtain the modal source density (Sec. 5.2.3.2). Typical sound fields at low and high frequencies are presented in Secs. 5.2.3.3 and 5.2.3.4, respectively. Similar to the 2D problems in Sec. 5.2.2, low-frequency sound fields are commonly used to model the sound produced by conventional loudspeakers, while high-frequency sound fields serve as the foundation for solving the audio sound generated by PALs.

5.3 PALS IN A RECTANGULAR ROOM

The sound fields generated by a PAL in a reverberant environment like a rectangular room are more complicated than that by a conventional loudspeaker. According to the quasilinear solution of the Westervelt equation as discussed in Sec. 1.4, the audio sound can be considered as the radiation from a virtual source with a source density proportional to the ultrasound pressure product. An algebraic term given by

Eq. (1.107) can be used to include the local effects. Therefore, the first step for solving the audio sound field generated by a PAL in a room is to calculate the linear ultrasound field.

The ultrasound with a source density of $q_i(\mathbf{r})$ is determined by the Helmholtz equation

$$(\nabla^2 + k_i^2) p_i(\mathbf{r}) = \mathrm{i}\rho_0 \omega_i q_i(\mathbf{r}), \quad \mathbf{r} \in V, \tag{5.89}$$

which is similar to Eq. (5.1), where k_i and p_i are the complex wavenumber and pressure at frequency f_i, and the subscript $i = 1, 2$ indicates the ultrasound at frequency f_i. The boundary condition is

$$[\mathbf{n} \cdot \boldsymbol{\nabla} - \mathrm{i}\eta_i(\mathbf{r}) k_i] p_i(\mathbf{r}) = 0, \quad \mathbf{r} \in \partial V, \tag{5.90}$$

which is similar to Eq. (5.3), where $\eta_i(\mathbf{r})$ is the normalized specific acoustic admittance for the ultrasound f_i.

For the audio sound, the simplified governing equation reads

$$(\nabla^2 + k_\mathrm{a}^2) p_\mathrm{a}(\mathbf{r}) = \mathrm{i}\rho_0 \omega_\mathrm{a} q_\mathrm{a}(\mathbf{r}), \quad \mathbf{r} \in V, \tag{5.91}$$

and the boundary condition is

$$[\mathbf{n} \cdot \boldsymbol{\nabla} - \mathrm{i}\eta_\mathrm{a}(\mathbf{r}) k_\mathrm{a}] p_\mathrm{a}(\mathbf{r}) = 0, \quad \mathbf{r} \in \partial V, \tag{5.92}$$

where $\eta_\mathrm{a}(\mathbf{r})$ is the normalized specific acoustic admittance for the audio sound, and the virtual source density is [Eq. (2.44)]

$$q_\mathrm{a}(\mathbf{r}) = \frac{\beta \omega_\mathrm{a}}{\mathrm{i}\rho_0^2 c_0^4} p_1^*(\mathbf{r}) p_2(\mathbf{r}). \tag{5.93}$$

Solving Eqs. (5.89) to (5.92) is same as solving Eqs. (5.1) and (5.3). Therefore, the MEM adopted in Sec. 5.2.1 can be employed here.

5.3.1 MODAL EXPANSION METHOD

5.3.1.1 Ultrasound Field

The objective is to solve the problem determined by Eqs. (5.89) and (5.90). According to Eq. (5.17), the ultrasound pressure at frequency of f_i is obtained by

$$p_i(\mathbf{r}) = \mathrm{i}\rho_0 \omega_i \sum_{n=0}^{\infty} \frac{Q_i(\mathbf{k}_n) \psi(\mathbf{k}_n, \mathbf{r})}{\Lambda_n^2 (k_i^2 - k_n^2)}, \tag{5.94}$$

where the modal source density for ultrasound is obtained by Eq. (5.9) as

$$Q_i(\mathbf{k}_n) \equiv \iiint_V q_i(\mathbf{r}) \psi(\mathbf{k}_n, \mathbf{r}) \mathrm{d}^3\mathbf{r}. \tag{5.95}$$

For simplicity, all walls in the room are assumed to be lightly damped or rigid, so $\psi(\mathbf{k}_n, \mathbf{r})$ takes the form of Eq. (5.26).

According to the FFT employed in Sec. 5.2.1.4, the ultrasound pressure at grids points $\mathbf{r}_m$ can be obtained by Eq. (5.40) as

$$p_i(\mathbf{r}_m) \approx \sum_{n_x=-N_x/2}^{N_x/2-1} \sum_{n_y=-N_y/2}^{N_y/2-1} \sum_{n_z=-N_z/2}^{N_z/2-1} P_i(\mathbf{k}_n) \exp\left[\mathrm{i}2\pi\left(\frac{m_x n_x}{N_x} + \frac{m_y n_y}{N_y} + \frac{m_z n_z}{N_z}\right)\right], \tag{5.96}$$

where the propagating factor is

$$P_i(\mathbf{k}_n) = \frac{\mathrm{i}\rho_0 \omega_i Q_i(\mathbf{k}_n)}{V(k_i^2 - k_n^2)}. \tag{5.97}$$

The particle velocity field for the ultrasound is required to calculate the local effects. It can be obtained by substituting Eq. (5.33) into $\mathbf{v}_i(\mathbf{r}) = \nabla p_i(\mathbf{r})/(\mathrm{i}\rho_0\omega_i)$,

$$\mathbf{v}_i(\mathbf{r}) \approx \sum_{n_x=-N_x/2}^{N_x/2-1} \sum_{n_y=-N_y/2}^{N_y/2-1} \sum_{n_z=-N_z/2}^{N_z/2-1} \mathbf{V}(\mathbf{k}_n) \mathrm{e}^{\mathrm{i}\mathbf{k}_n \cdot \mathbf{r}}, \tag{5.98}$$

where the propagating factor is modified as

$$\mathbf{V}_i(\mathbf{k}_n) = \frac{\mathrm{i}\mathbf{k}_n Q_i(\mathbf{k}_n)}{V(k_i^2 - k_n^2)}. \tag{5.99}$$

Similar to Eq. (5.96), the particle velocity field at grid points $\mathbf{r}_m$ can be obtained by

$$\mathbf{v}_i(\mathbf{r}_m) \approx \sum_{n_x=-N_x/2}^{N_x/2-1} \sum_{n_y=-N_y/2}^{N_y/2-1} \sum_{n_z=-N_z/2}^{N_z/2-1} \mathbf{V}_i(\mathbf{k}_n) \exp\left[\mathrm{i}2\pi\left(\frac{m_x n_x}{N_x} + \frac{m_y n_y}{N_y} + \frac{m_z n_z}{N_z}\right)\right]. \tag{5.100}$$

The ultrasound pressure given by Eq. (5.96) can be efficiently calculated using the FFT as demonstrated in Sec. 5.2.1.4. The ultrasound velocity field given by Eq. (5.100) can be similarly calculated using the FFT and the details are not presented here for simplicity.

5.3.1.2 Audio Sound Field

The audio sound pressure without including local effects can be obtained using the MEM given by Eq. (5.17) as

$$p_\mathrm{a}(\mathbf{r}) = \mathrm{i}\rho_0\omega_\mathrm{a} \sum_{n=0}^{\infty} \frac{Q_\mathrm{a}(\mathbf{k}_n)\psi(\mathbf{k}_n, \mathbf{r})}{\Lambda_n^2(k_\mathrm{a}^2 - k_n^2)}, \tag{5.101}$$

where the modal source density for the audio sound is obtained by Eq. (5.9) as

$$Q_\mathrm{a}(\mathbf{k}_n) = \iiint_V q_\mathrm{a}(\mathbf{r})\psi(\mathbf{k}_n, \mathbf{r})\mathrm{d}^3\mathbf{r}. \tag{5.102}$$

By substituting the ultrasound pressure Eq. (5.96) into Eq. (5.93), the source density at grid points $\mathbf{r}_m = (m_x\Delta x, m_y\Delta y, m_z\Delta z)$ can be obtained as

$$q_\mathrm{a}(\mathbf{r}_m) \approx \frac{\beta\omega_\mathrm{a}}{\mathrm{i}\rho_0^2 c_0^4} p_1^*(\mathbf{r}_m) p_2(\mathbf{r}_m). \tag{5.103}$$

By using the explicit form of the modal functions given by Eq. (5.26), Eq. (5.102) can be rewritten as

$$Q_\mathrm{a}(\mathbf{k}_n) = \int_{-L_z}^{L_z}\int_{-L_y}^{L_y}\int_{-L_x}^{L_x} q_\mathrm{a}(\mathbf{r})\mathrm{e}^{-\mathrm{i}\mathbf{k}_n\cdot\mathbf{r}}\mathrm{d}^3\mathbf{r}. \tag{5.104}$$

Equation (5.104) shows that $Q_\mathrm{a}(\mathbf{k}_n)$ is the spatial Fourier transform of $q_\mathrm{a}(\mathbf{r})$ at discrete spectral $\mathbf{k}_n$. Discretizing $\mathbf{r}$ into $\mathbf{r}_m$, the above integral can be approximated by

$$\begin{aligned} Q_\mathrm{a}(\mathbf{k}_n) \approx \frac{8V}{N_xN_yN_z} & \sum_{m_x=-N_x/2}^{N_x/2-1}\sum_{m_y=-N_y/2}^{N_y/2-1}\sum_{m_z=-N_z/2}^{N_z/2-1} q_\mathrm{a}(\mathbf{r}_m) \\ & \times \exp\left[-\mathrm{i}2\pi\left(\frac{m_xn_x}{N_x}+\frac{m_yn_y}{N_y}+\frac{m_zn_z}{N_z}\right)\right]. \end{aligned} \tag{5.105}$$

Defining the primed change of indices m' and n' as given by Eq. (5.41), Eq. (5.105) can be transformed to the standard form of the DFT as

$$Q'_\mathrm{a}(n') = \sum_{m'_x=0}^{N_x-1}\sum_{m'_y=0}^{N_y-1}\sum_{m'_z=0}^{N_z-1} q'_\mathrm{a}(m')\exp\left[-\mathrm{i}2\pi\left(\frac{m'_xn'_x}{N_x}+\frac{m'_yn'_y}{N_y}+\frac{m'_zn'_z}{N_z}\right)\right]. \tag{5.106}$$

where the auxiliary functions are defined as

$$q'_\mathrm{a}(m') = (-1)^{m'_x+m'_y+m'_z} q_\mathrm{a}((m'_x-N_x/2)\Delta x,(m'_y-N_y/2)\Delta y,(m'_z-N_z/2)\Delta z), \tag{5.107}$$

and

$$Q'_\mathrm{a}(n') \equiv \frac{N_xN_yN_z}{8V} Q_\mathrm{a}((n'_x-N_x/2)\Delta k_x,(n'_y-N_y/2)\Delta k_y,(n'_z-N_z/2)\Delta k_z). \tag{5.108}$$

Equation (5.106) can be efficiently calculated using the FFT.

The audio sound pressure at grid points $\mathbf{r}_m$ is expressed as

$$p_\mathrm{a}(\mathbf{r}_m) \approx \sum_{n_x=-N_x/2}^{N_x/2-1}\sum_{n_y=-N_y/2}^{N_y/2-1}\sum_{n_z=-N_z/2}^{N_z/2-1} P_\mathrm{a}(\mathbf{k}_n)\exp\left[\mathrm{i}2\pi\left(\frac{m_xn_x}{N_x}+\frac{m_yn_y}{N_y}+\frac{m_zn_z}{N_z}\right)\right], \tag{5.109}$$

which can be efficiently calculated using the inverse FFT, where the propagating factor

$$P_\mathrm{a}(\mathbf{k}_n) = \frac{\mathrm{i}\rho_0\omega_\mathrm{a}Q_\mathrm{a}(\mathbf{k}_n)}{V(k_\mathrm{a}^2-k_n^2)}. \tag{5.110}$$

The procedure for the calculation using the FFT is summarized in Fig. 5.10. The calculation for in a 2D rectangular room is the same and not presented here for simplicity.

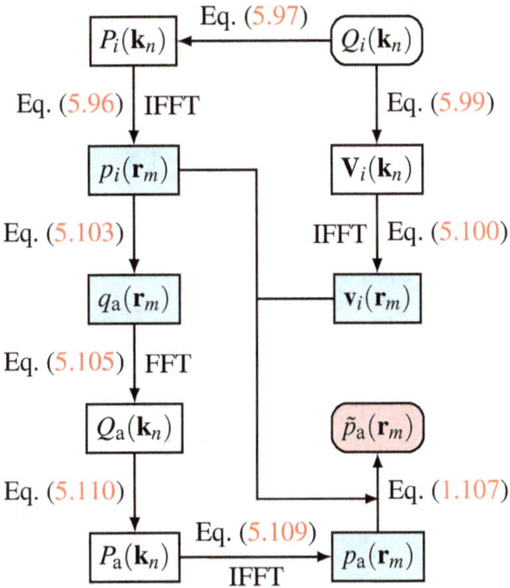

Figure 5.10 Procedure for calculating the audio sound fields generated by a PAL in a rectangular room. Shaded variables are in the spatial domain, and those unshaded are in the spatial frequency domain.

5.3.2 2D SOUND FIELDS

The sound fields generated by a PAL in a 2D rectangular room are considered in this section. In the following simulations, the average ultrasound frequency $f_u = (f_1 + f_2)/2$ is set to 40 kHz. The audio sound frequency is set to $f_a = 1$ kHz. The sound attenuation coefficients due to the atmospheric absorption are calculated according to ISO 9613 with a temperature of 20°C and a relative humidity of 70% (Appendix A). The half-width of the source is set to $a = 5$ cm. The surface velocity amplitude is set to $Q_0 = 0.01\,\text{m}^2/\text{s}$ and $Q_0 = 2\times10^{-5}\,\text{m}^2/\text{s}$ for the PAL and the conventional loudspeaker, respectively. The dimensions of the room are $L_x \times L_y = \mathrm{e}/\pi \times 1\,\text{m}^2$. The sound field generated by a PAL in free field is obtained using the cylindrical wave expansion method as introduced in Sec. 2.5. The normalized specific acoustic admittance is set to a constant $\eta(\rho) = 0.01$ at all frequencies.

Figure 5.11 compares the linear sound field generated by a line source in free field and in the 2D rectangular room, where Fig. 5.11(c) represents the audio sound field generated by a conventional loudspeaker in a room. It is clear that the ultrasound waves at 40 kHz are highly directional in free field since the aperture size (5 cm) is much larger than the ultrasound wavelength (8.6 mm). In the rectangular room, Fig. 5.11(b) shows that the major energy of the ultrasound waves is still focused around the radiator axis $x = L_x/2$. This is because the side length in the x-direction, L_x, is much larger than the ultrasound wavelength so that the walls at $x = 0$ and $x = L_x$ have little effects. However, many local peaks and valleys are presented in the SPL in the y-direction, which is resulted from the multiple reflections of the walls at

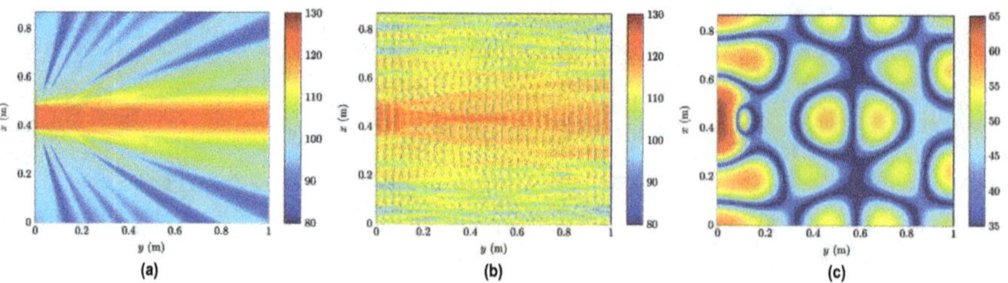

Figure 5.11 (a) Sound field at 40 kHz in free field with the source centroid at $\boldsymbol{\rho}_s = (L_x/2, 0)$. Sound field at (b) 40 kHz and (c) 1 kHz in the rectangular room with the source centroid at $\boldsymbol{\rho}_s = (L_x/2, 0.1\,\text{m})$. Extracted from [Zhong et al., 2022a, Fig. 2].

$y = 0$ and $y = L_y$. At the audio frequency of 1 kHz, Fig. 5.11(c) shows that there are fluctuations and no directional beams are presented.

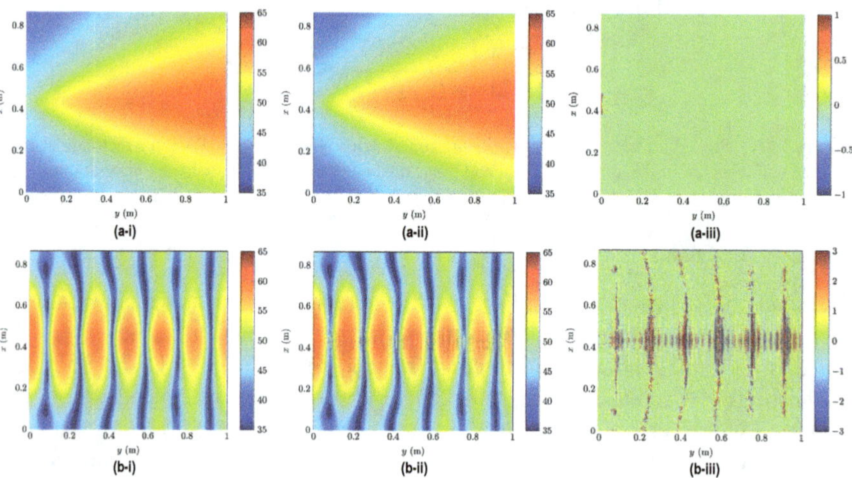

Figure 5.12 Audio SPL generated by a PAL at 1 kHz in (a) free space and $\boldsymbol{\rho}_s = (L_x/2, 0)$; and (b) a 2D rectangular room and $\boldsymbol{\rho}_s = (L_x/2, 0.1\,\text{m})$. (i) Without local effects; (ii) with local effects; and (iii) SPL difference with and without local effects. Extracted from [Zhong et al., 2022a, Fig. 3].

Figure 5.12 compares the audio sound field at 1 kHz generated by a PAL in free field and in the rectangular room. As expected, the audio sound field is highly directional in free field even the aperture size (5 cm) is much smaller than the audio wavelength (34.3 cm). In the rectangular room, Fig. 5.12(b-ii) shows that the audio sound appears to be directional despite some fluctuations in the y-axis direction. The distance between adjacent peaks (or valleys) is found to be around half wavelength of the audio sound (17.15 cm). By comparing to the sound of a conventional loudspeaker in a room as shown in Fig. 5.11(c), it is clear that the advantage of generating directional audio beams is retained for a PAL in a room. It is also found in the top row of Fig. 5.12 that the SPL difference with and without local effects is negligible except at some locations close to the PAL. For the audio sound in a rectangular room, the bottom row of Fig. 5.12 shows that the local effects are also negligible in most locations.

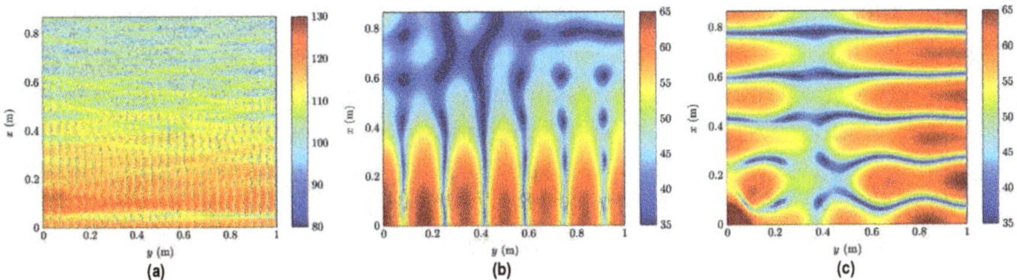

Figure 5.13 (a) Ultrasound field at 40 kHz and (b) audio sound field at 1 kHz generated by a PAL in a room. (c) Audio sound field at 1 kHz generated by a conventional loudspeaker in a room. The line source centroid is located at $\boldsymbol{\rho}_s = (0.1\,\text{m}, 0.1\,\text{m})$ and the radiation direction $\mathbf{n}_d = (0,1)$. Extracted from [Zhong et al., 2022a, Fig. 4].

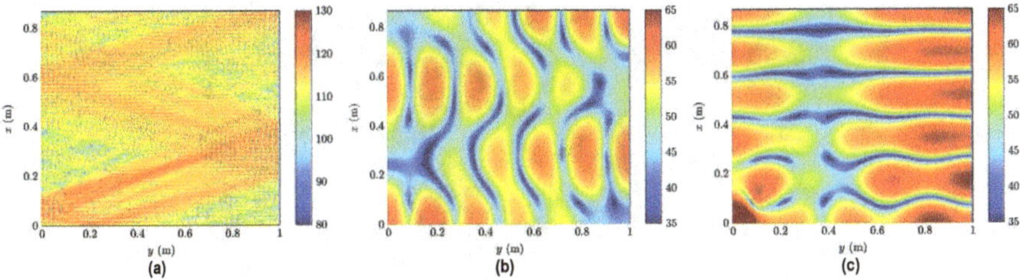

Figure 5.14 (a) Ultrasound field at 40 kHz and (b) audio sound field at 1 kHz generated by a PAL in a room. (c) Audio sound field at 1 kHz generated by a conventional loudspeaker in a room. The line source centroid is located at $\boldsymbol{\rho}_s = (0.1\,\text{m}, 0.1\,\text{m})$ and the radiation direction is $\mathbf{n}_d = (0.3,1)$. Extracted from [Zhong et al., 2022a, Fig. 5].

Figure 5.13 shows the sound field generated by a PAL and a conventional loudspeaker when the source centroid is moved close to the left corner of the room. It is observed in Fig. 5.13(a) that the major part of ultrasound waves is located below $x = L_x/2$. By comparing Fig. 5.13(b) to Fig. 5.12(e), it is clear that the audio beam is focused on the radiation direction and the SPL is relatively low above $x = L_x/2$. However, the audio sound field generated by a conventional loudspeaker as shown in Fig. 5.13(c) does not present any directional beams in the radiation direction.

Figure 5.14 shows the sound field generated by a PAL and a conventional loudspeaker when the line source is not perpendicular to the walls. It can be observed in Fig. 5.14(a) that the ultrasound waves propagate roughly along the radiator axis and reflect after impinging the walls. Consequently, the audio sound field presented in Fig. 5.14(b) appears to be directional in the radiation direction, but with some fluctuations in space. However, the audio sound generated by a conventional loudspeaker shown in Fig. 5.14(c) is similar to that shown in Fig. 5.13(c) indicating that the sound field is insensitive to the radiation direction. The room has significant effects on the radiation of PALs, but in a different way with that on conventional loudspeakers.

5.3.3 3D SOUND FIELDS

In this section, the 3D sound fields generated by a circular PAL in a rectangular room are investigated both numerically and experimentally. The experimental setup is shown in Fig. 5.15. The dimensions of the small room are $L_x \times L_y \times L_z = 1.414 \times 1.156 \times 1.5\,\text{m}^3$. All room walls are assumed to be have a small admittance of $\eta(\mathbf{r}) = 0.01$. The circular PAL used in both simulations and experiments has a radius of $a = 0.1\,\text{m}$. The PAL is the same as that used in Fig. 6.41. The centroid of the PAL is located at $\mathbf{r}_\text{c} = (0.15\,\text{m}, 0.578\,\text{m}, 0.75\,\text{m})$. The sound field at 68×56 grid points are measured with a spacing of 2 cm. A condenser microphone is affixed to a PC-controlled scanning stage, enabling the scanning of the field at grid points.

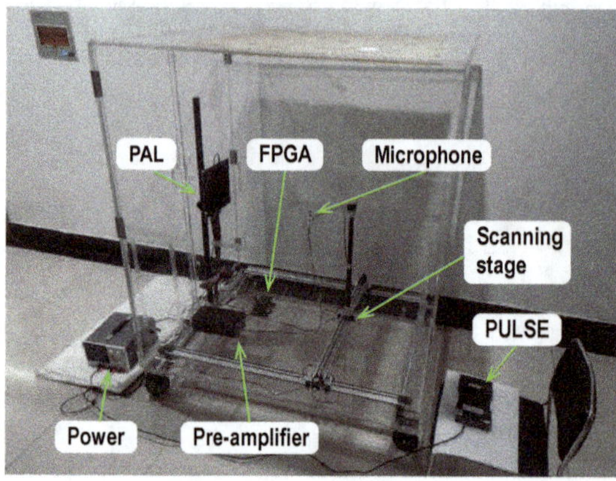

Figure 5.15 A photo of the experimental setup for measuring audio sound fields generated by a PAL in a rectangular room.

5.3.3.1 Ultrasound Field

The average ultrasound frequency is set as 40 kHz in both simulations and experiments. Figures 5.16 and 5.17 show the simulated ultrasound field when the radiation angles are $(\theta_\text{d}, \varphi_\text{d}) = (90^\circ, 0^\circ)$ and $(\theta_\text{d}, \varphi_\text{d}) = (90^\circ, 15^\circ)$, respectively. It is clear that the directional ultrasound beam can be observed due to the small ultrasound wavelength (8.6 mm) compared to the room dimension. The experimental results in the plane $z = z_\text{c} = 0.75\,\text{m}$ are presented in Fig. 5.18 for comparison. It is shown that simulations can predict the main lobe of the ultrasound beam. Nevertheless, the measurements exhibit fewer reflections from the room walls, possibly due to the nonrigidity of the actual environmental walls. Additionally, given that the size of the microphone is comparable to the ultrasound wavelength, the scattering effects of the microphone can introduce measurement errors as well.

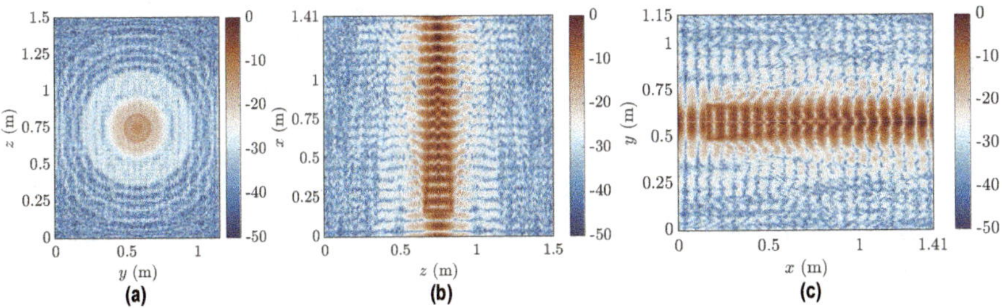

Figure 5.16 The simulated sound fields at 40 kHz generated by a circular source in a 3D rectangular room with lightly damped walls. The source centroid is located at $\mathbf{r}_\mathrm{c} = (0.15\,\mathrm{m}, 0.578\,\mathrm{m}, 0.75\,\mathrm{m})$. The radiation direction is $(\theta_\mathrm{d}, \varphi_\mathrm{d}) = (90^\circ, 0^\circ)$. Results are obtained using the MEM in the plane (a) $x = x_\mathrm{c}$, (b) $y = y_\mathrm{c}$, or (c) $z = z_\mathrm{c}$.

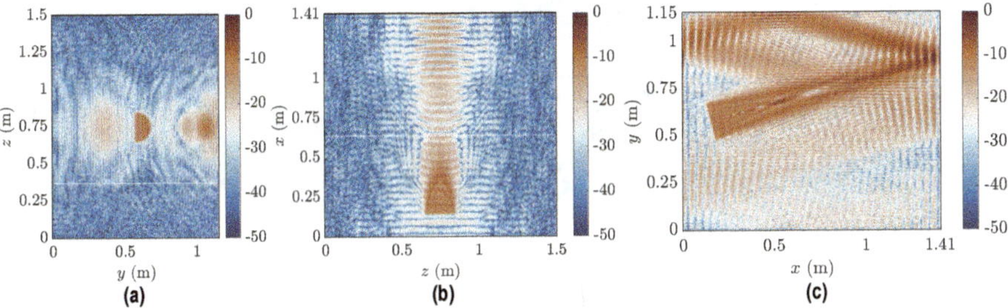

Figure 5.17 The simulated sound fields at 40 kHz generated by a circular source in a 3D rectangular room with lightly damped walls. The source centroid is located at $\mathbf{r}_\mathrm{c} = (0.15\,\mathrm{m}, 0.578\,\mathrm{m}, 0.75\,\mathrm{m})$. The radiation direction is $(\theta_\mathrm{d}, \varphi_\mathrm{d}) = (90^\circ, 15^\circ)$. Results are obtained using the MEM in the plane (a) $x = x_\mathrm{c}$, (b) $y = y_\mathrm{c}$, or (c) $z = z_\mathrm{c}$.

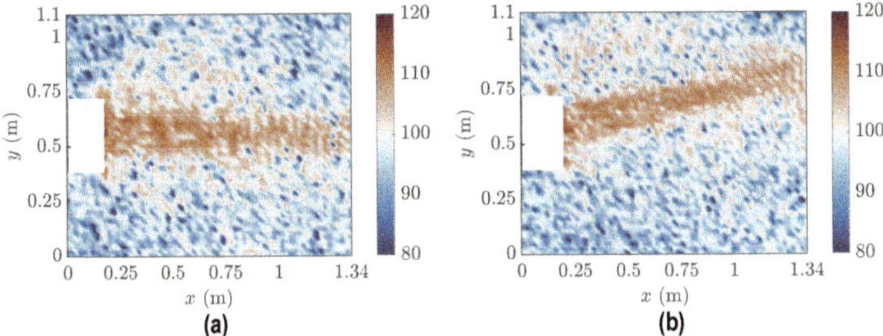

Figure 5.18 Measured 2D ultrasound fields in the plane $z = 0.75\,\mathrm{m}$ at 40 kHz in a rectangular room. The radiation angle is (a) $(\theta_\mathrm{d}, \varphi_\mathrm{d}) = (90^\circ, 0^\circ)$ and (b) $(\theta_\mathrm{d}, \varphi_\mathrm{d}) = (90^\circ, 15^\circ)$.

5.3.3.2 Audio Sound Field

Figures 5.19 to 5.22 present the simulation results of the sound fields at audio frequencies of 512 Hz, 1024 Hz, 2049 Hz, and 3977 Hz. Figure 5.23 presents the corresponding measurement results. It can be observed that the audio sound fields generated by PALs exhibit a clear modal distribution at low frequencies. For example,

4 and 8 pressure nodes can be observed along the x-direction at 512 Hz as shown in Figs. 5.19 and 5.20. At high audio frequencies, the beam is focused more on its radiator axis as evident by Figs. 5.21 and 5.22 at 2049 Hz and 3977 Hz, respectively.

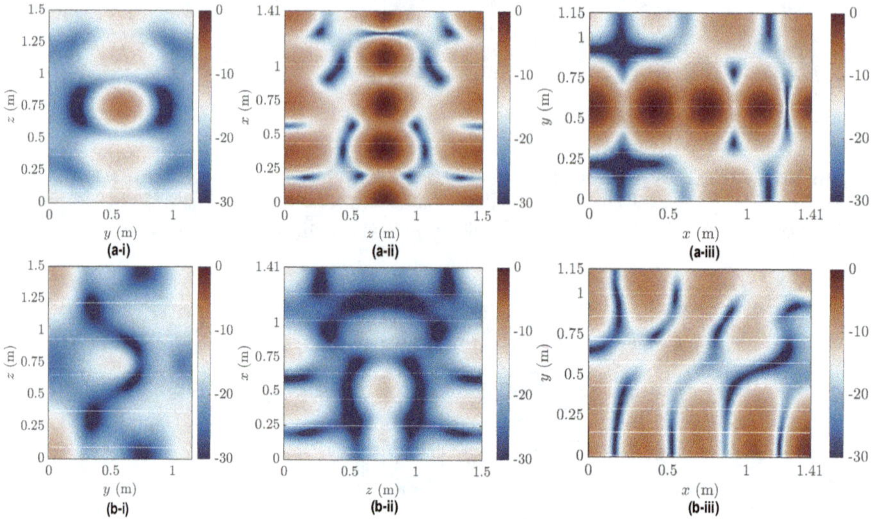

Figure 5.19 Sound fields at 512 Hz generated by a circular source in a 3D rectangular room with lightly damped walls. The source centroid is located at $\mathbf{r}_c = (0.15\,\text{m}, 0.578\,\text{m}, 0.75\,\text{m})$. The radiation angles are (a) $(\theta_d, \varphi_d) = (90°, 0°)$ and (b) $(\theta_d, \varphi_d) = (90°, 15°)$. Results are obtained using the MEM in the plane (i) $x = x_c$, (ii) $y = y_c$, or (iii) $z = z_c$.

The measured results, as shown in Fig. 5.23, closely resemble the results obtained through simulations, confirming the validity of the MEM method. Directional beams can be observed at 2049 Hz and 3977 Hz in Figs. 5.23(c) and 5.23(d), respectively. It is important to take into account that the finite size of the PAL's baffle and the presence of the scanning stage within the room, which can influence the generation of ultrasound and audio sound fields, potentially contributing to measurement errors.

5.3.4 REMARKS

In this section, the reader is introduced to the sound fields generated by a PAL in both 2D and 3D rectangular rooms. As discussed in Sec. 5.3.1, the MEM employed for linear sources is extended to calculate both ultrasound and audio sound fields based on the quasilinear solution. Both 2D and 3D sound fields in rectangular rooms are presented in Secs. 5.3.2 and 5.3.3, respectively. Furthermore, the measurement results are presented for the 3D case, demonstrating a good agreement between the measurements and simulations.

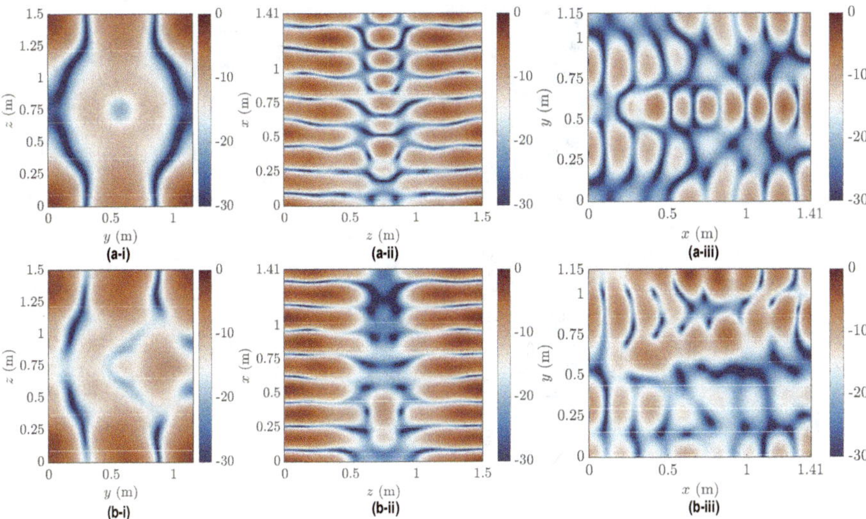

Figure 5.20 Sound fields at 1024 Hz generated by a circular source in a 3D rectangular room with lightly damped walls. The source centroid is located at $\mathbf{r}_c = (0.15\,\text{m}, 0.578\,\text{m}, 0.75\,\text{m})$. The radiation angles are (a) $(\theta_d, \varphi_d) = (90°, 0°)$ and (b) $(\theta_d, \varphi_d) = (90°, 15°)$. Results are obtained using the MEM in the plane (i) $x = x_c$, (ii) $y = y_c$, or (iii) $z = z_c$.

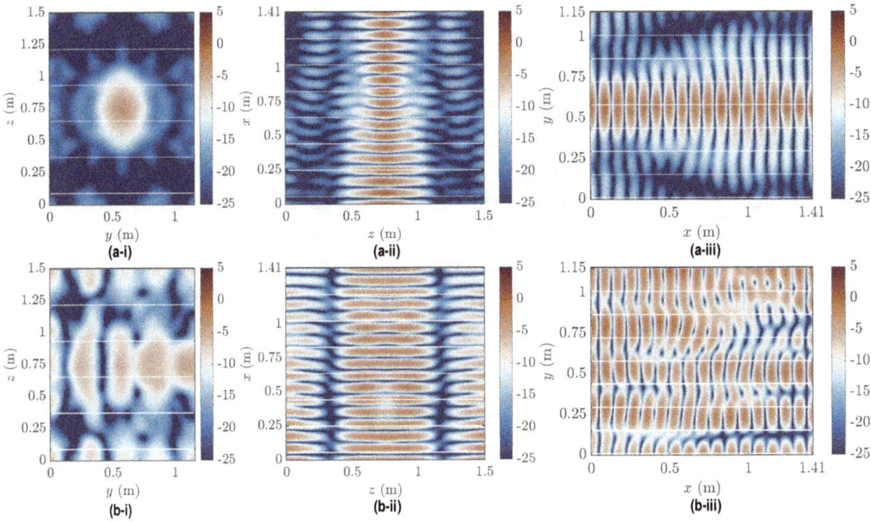

Figure 5.21 Sound fields at 2049 Hz generated by a circular source in a 3D rectangular room with lightly damped walls. The source centroid is located at $\mathbf{r}_c = (0.15\,\text{m}, 0.578\,\text{m}, 0.75\,\text{m})$. The radiation angles are (a) $(\theta_d, \varphi_d) = (90°, 0°)$ and (b) $(\theta_d, \varphi_d) = (90°, 15°)$. Results are obtained using the MEM in the plane (i) $x = x_c$, (ii) $y = y_c$, or (iii) $z = z_c$.

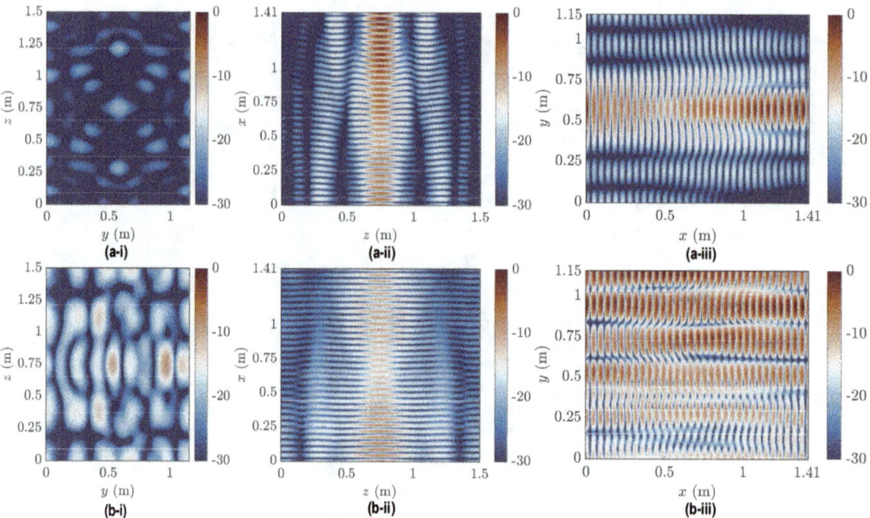

Figure 5.22 Sound fields at 3,977 Hz generated by a circular source in a 3D rectangular room with lightly damped walls. The source centroid is located at $\mathbf{r}_c = (0.15\,\text{m}, 0.578\,\text{m}, 0.75\,\text{m})$. The radiation angles are (a) $(\theta_d, \varphi_d) = (90°, 0°)$ and (b) $(\theta_d, \varphi_d) = (90°, 15°)$. Results are obtained using the MEM in the plane (i) $x = x_c$, (ii) $y = y_c$, or (iii) $z = z_c$.

5.4 SUMMARY

In this chapter, the readers are introduced to the audio sound fields generated by both conventional loudspeakers and PALs in rectangular rooms. The reverberant sound fields generated by linear sources in both 2D and 3D rectangular rooms are discussed in Sec. 5.2 at first. Since the PAL is one kind of directional sources, the sound fields generated by an arbitrary directional source in linear acoustics are analyzed using the cylindrical and spherical harmonic representations in Secs. 5.2.2 and 5.2.3, respectively. The MEM is introduced to calculate the sound fields, which is found to be more computationally efficient than the FEM. This is beneficial to the modeling of PALs in rectangular rooms as it requires the ultrasound field at large number of observation points, as demonstrated in Sec. 5.3.1. Both the simulation and experimental results of audio sound fields generated by PALs are presented in Sec. 5.3 and their properties in the rectangular rooms are discussed.

The sound fields in a rectangular room generated by linear acoustic sources have been investigated extensively in literature, such as the classical textbooks [Kleiner and Tichy, 2014; Kuttruff, 2017]. However, the investigation of the sound fields generated by a PAL in a room is still at its infancy, even though there is potential for applications of PALs in reverberant environments, such as rooms, museums, art galleries, and automobile cabins. This chapter presents only a preliminary exploration

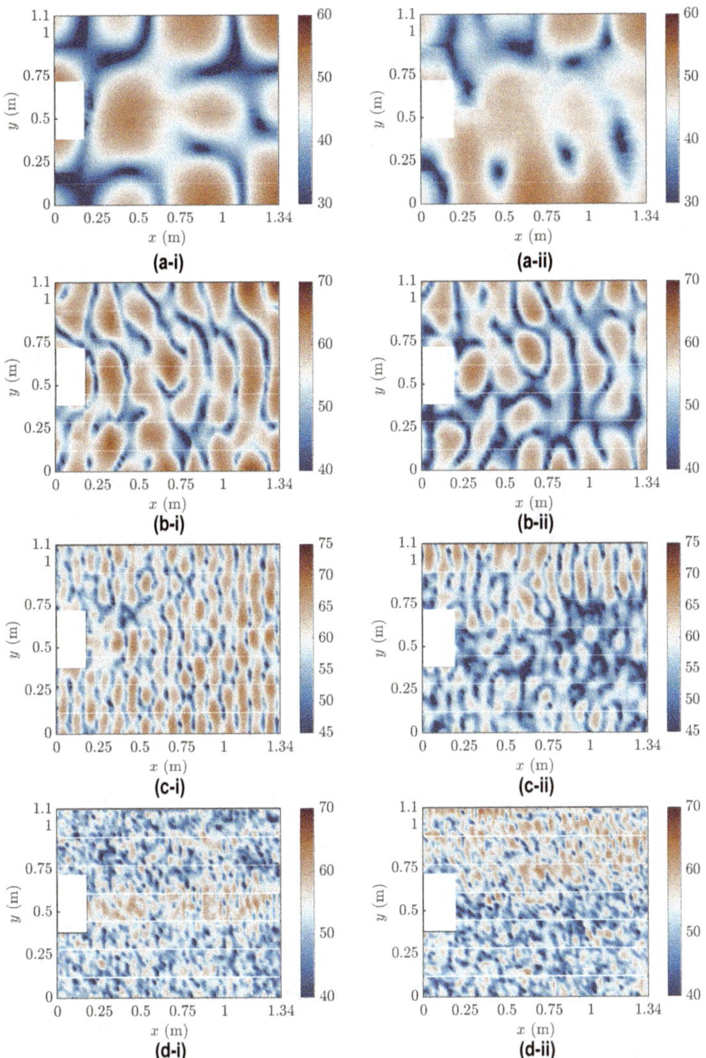

Figure 5.23 Measured 2D ultrasound fields in the plane $z = 0.75$ m in a rectangular room. The radiation angle is (i) $(\theta_d, \varphi_d) = (90°, 0°)$ and (ii) $(\theta_d, \varphi_d) = (90°, 15°)$. The audio frequency is (a) 512 Hz, (b) 1024 Hz, (c) 2049 Hz, and (d) 3977 Hz.

on this topic, and further research using advanced theories and experimental measurements is necessary to gain a deeper understanding. For instance, this includes the examination of broadband audio sound propagation within enclosed spaces and the development of models for audio sound propagation in rooms with irregular shapes.

6 Steerable and Focusing PALs

6.1 INTRODUCTION

6.1.1 STEERABLE PALS

The audio beam generated by parametric array loudspeakers (PALs) is highly directional compared to that generated by conventional loudspeakers. In certain applications, a PAL is anticipated to offer the flexibility of delivering audio beams to specific directions, such as tracking human ears [Okano and Kajikawa, 2022]. The *steerable PAL*, as it is commonly referred to, endeavors to achieve beam steering without the need for mechanical rotation of the PAL. Pompei considered the design of a steerable PAL in 2002 [Pompei, 2002], then Olszewski et al. realized a prototype using the phased array technique [Olszewski et al., 2005; Olszewski, 2009]. They proposed a steerable PAL system that was installed on the ceiling of the room to deliver multiple beams to different directions simultaneously [Olszewski and Linhard, 2006]. In a digital system, the time delay is related to the sampling interval of the digital signal processor (DSP). In most cases, the sampling frequency is not high enough for steering a PAL with enough precision, so the smallest steering angle available is large, which limits the applications of a steerable PAL. However, it is shown that this can be resolved using the fractional delay filter [Wu et al., 2012]. Some efficient digital implementations of the beam steering algorithms were reported [Tan et al., 2004; Gan et al., 2006].

Due to the flexibility of steerable PALs, they have been used in many audio applications. A steerable PAL was used as the secondary source in an ANC system to control the noise at the target point without adversely affecting other areas [Tanaka and Tanaka, 2010]. Compared to the ANC system using a conventional PAL [Zhong et al., 2020f], the advantage of using a steerable PAL is that it can track a moving target in the space. A steerable PAL has also been used in a multilingual teleconferencing system to deliver the directional audio contents to a specified meeting participant with the aid of a head tracking system [Gan et al., 2011]. More applications of steerable PALs in sound reproduction systems can be found in Shi et al., 2014.

Calculation of the audio sound generated by a steerable PAL is similar to that by a conventional PAL, while the only difference is that the steerable velocity profile should be used. Therefore, all numerical models presented in Chap. 2 can be used for modeling steerable PALs. However, these models have more limitations than the calculation for a conventional PAL without beam steering. The direct integration method (DIM) presented in Sec. 2.2 requires the calculation of the ultrasound at more locations to include the major energy of the ultrasonic beam especially when the steering angle is large. This makes the calculation much more time-consuming

DOI: 10.1201/9781003354994-6

and demands more memory. The accuracy of the predication results obtained using the Gaussian beam expansion (GBE) method presented in Sec. 2.2 deteriorates significantly as the steering angle increases because the paraxial approximation is assumed. The calculation of the audio sound generated by a rectangular steerable PAL can be more time-consuming than a circular one because it cannot be efficiently evaluated using the spherical wave expansion (SWE) method as presented in Sec. 2.6.

6.1.2 FOCUSING PALS

Although a PAL can generate highly directional audio beams, one major weakness is its poor low-frequency response. Specifically, the audio sound pressure level (SPL) received at an observation point generated by a conventional PAL decreases by about 12 dB when the audio frequency is halved [Yoneyama et al., 1983]. The reason is that the audio sound pressure radiated by the virtual source is proportional to the square of the audio frequency, see Eq. (2.25) as an example. In general, the audio sound below 500 Hz cannot be generated by a conventional PAL with an acceptable SPL. Clearly, this limitation restricts broad applicability of PALs. One possible way to overcome this problem is to focus audio beams at a target region, which is termed as the *focusing PAL* in this book.

The concept and the realization of focusing parametric acoustic arrays (PAAs) were explored at first by Lucas et al. in 1983 [Lucas et al., 1983]. A focusing PAA was fabricated using a spherical concave source and underwater experiments were conducted to verify their findings. The difference-frequency wave was found to be more effectively focused than a conventional source. However, the paraxial approximation of the Westervelt equation was used in the mathematical modeling leading to limited accuracy. Jing et al. conducted an examination of the accuracy of the Westervelt equation in modeling a focusing transducer's generation of the second-harmonic field in water [Jing et al., 2011]. It was found the error using the Westervelt equation becomes progressively larger in the near field when ka becomes smaller, where k is the acoustic wavenumber and a is the aperture size. Although ka is usually large enough in medical and underwater applications (e.g., $ka = 28$ for the difference-frequency wave in Lucas et al., 1983), it is not the case for a PAL as the audio wavelength can be comparable or much larger than the aperture size (e.g., $ka = 0.9$ at 500 Hz with an aperture size of 0.1 m). Besides, the Lagrangian density is of order $1/\sqrt{kF}$ compared to the nonlinear term, where F is the focal length [Kamakura et al., 2000], so the error using the Westervelt equation increases as the focal point moves close to the radiation surface. It is clear that the reported results including measurements in Lucas et al., 1983 cannot be used to measure the performance of a focusing PAL in air. In 2022, a systematical work carried out by Zhong et al. shows the performance of a focusing PAL [Zhong et al., 2022d]. It is found that the local effects play an important role to improve the low-frequency response.

Apart from improving low-frequency response, focusing PALs have also been explored for many potential applications. For example, it was used in an ANC system to effectively reduce the global noise radiation in 2011 [Tanaka and Tanaka, 2011]. It was demonstrated that a virtual sound source can be remotely constructed by focusing the sound beams generated by multiple PALs [Ogami et al., 2019]. A so-called holographic whisper was designed to render the localized audio based on the mechanism of focusing PALs [Ochiai et al., 2017]. As demonstrated in Sec. 4.2, it is known that the reflection of the audio sound generated by a conventional PAL is prominent which is annoying in some applications. A focusing PAL might be able to address this issue as most of the energy of the audio sound is transferred from the far field to the near field.

Calculation of the audio sound generated by a focusing PAL is similar to that generated by a conventional PAL, and the difference is to include a focusing velocity profile for ultrasound. It is noted that the local effects dominate the wave behavior around the focal point. Accordingly, the predictions using the Westervelt equation or its paraxial form, i.e., KZK equation, are inaccurate because of neglecting the Lagrangian density. Instead, the Kuznetsov equation or the general second-order nonlinear equation must be used (Sec. 1.3).

In this chapter, the performance of both 2D and 3D steerable PALs, along with the generated sound fields, are examined in Secs. 6.2 and 6.3, respectively. The generation of grating lobes for steerable PALs implemented through arrays of ultrasonic emitters, accompanied by discussions of relevant suppression techniques, is discussed in Sec. 6.4. Finally, the performance of focusing PALs and the generated sound fields are illustrated in Sec. 6.5.

6.2 TWO-DIMENSIONAL BEAM STEERING

6.2.1 PROBLEM DESCRIPTION

When one dimension of the radiation surface of the PAL significantly exceeds the wavelength, the radiated sound field remains approximately invariant in that dimension. In such cases, a 2D model, as previously described in Sec. 2.2, can be employed for simplicity. This simplified model is usually used for modeling the radiation from a linear phased array PAL [Shi and Kajikawa, 2015a; Zhong et al., 2021a]. The sketch of a steerable PAL in the 2D model is shown in Fig. 6.1. Cartesian (x,y) and polar (ρ,φ) coordinate systems are established with their origin, O, at the centroid of the PAL, and the positive x-axis pointing to the radiator axis, where ρ and φ are the radial and azimuthal coordinates, respectively. The half width of the PAL along the y-axis is denoted by a.

An ideal 2D steerable PAL adopts a continuous velocity profile for ultrasound as [Eq. (2.13)]

$$v_{i,x}(\boldsymbol{\rho}_\mathrm{s}) = v_0 \mathrm{e}^{\mathrm{i}\Re(k_i)\boldsymbol{\rho}_\mathrm{s}\cdot\hat{\mathbf{s}}}\Pi\left(\frac{y_\mathrm{s}}{2a}\right) = v_0 \mathrm{e}^{\mathrm{i}\Re(k_i)y_\mathrm{s}\sin\varphi_\mathrm{d}}\Pi\left(\frac{y_\mathrm{s}}{2a}\right), \tag{6.1}$$

where $\Pi(\cdot)$ is the rectangle function defined by Eq. (2.11), v_0 is the velocity amplitude constant, $\Re(\cdot)$ takes the real part of the argument, $\hat{\mathbf{s}} = (\cos\varphi_\mathrm{d}, \sin\varphi_\mathrm{d})$ is the unit vector indicating the beam direction, $\boldsymbol{\rho}_\mathrm{s} = (x_\mathrm{s}, y_\mathrm{s})$ with $x_\mathrm{s} = 0$ is denoted as the source point, and φ_d is the steering angle. To generate multiple beams, the velocity

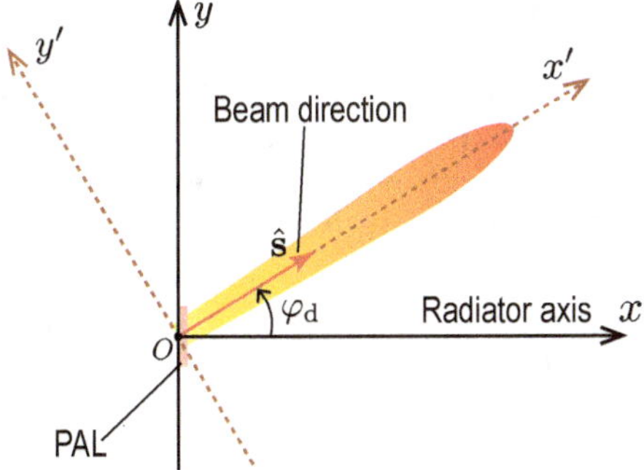

Figure 6.1 Sketch of a steerable PAL in a 2D model with a steering angle of φ_d. The PAL lies in $x = 0$.

profile Eq. (6.1) is revised as

$$v_{i,x}(\boldsymbol{\rho}_\mathrm{s}) = v_0 \Pi\left(\frac{y_\mathrm{s}}{2a}\right) \sum_{n=1}^{N} \mathrm{e}^{\mathrm{i}\Re(k_i)\boldsymbol{\rho}_\mathrm{s}\cdot\hat{\mathbf{s}}_n}, \tag{6.2}$$

where N is the total number of beams, and $\hat{\mathbf{s}}_n = (\cos\varphi_{\mathrm{d},n}, \sin\varphi_{\mathrm{d},n})$ is the steering vector of the n-th beam.

6.2.2 ULTRASOUND FIELD

Figure 6.2 presents an example of ultrasound fields for generating audio sound at 2 kHz with a steering angle of 15°. The steerable profile Eq. (6.1) is used in the simulations. The cylindrical wave expansion (CWE) method presented in Sec. 2.5.2 is employed for calculations. It is clear that the main lobe of both ultrasound beams are steered to the specified steering angle. It is therefore anticipated that the main lobe of the demodulated audio beam would also be steered to this angle.

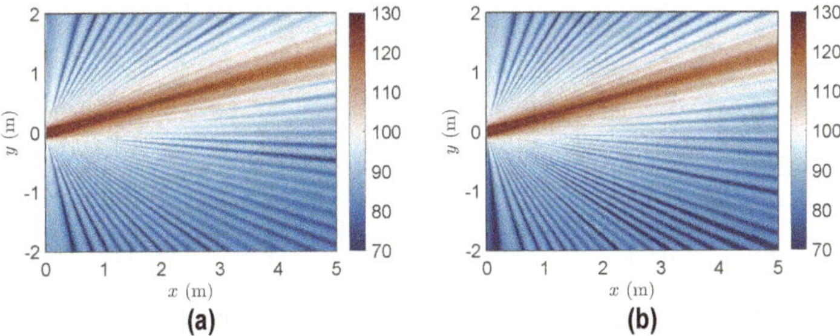

Figure 6.2 Ultrasound pressure level (dB) distributions generated by a baffled line source with a half width of $a = 10\,\mathrm{mm}$ at (a) 39 kHz and (b) 41 kHz for generating audio sound at 2 kHz. The steering angle is $\varphi_\mathrm{d} = 15°$. The on-surface sound pressure amplitude $p_0 = 50\,\mathrm{Pa}$ (125 dB).

6.2.3 AUDIO SOUND IN THE FAR FIELD

In the far field, the directivity can be efficiently calculated using the convolution directivity model (CDM). The results presented in this section is obtained using the 2D modified CDM, as illustrated by Eq. (2.71). The readers are referred to Sec. 2.3.1 for more details on the CDM. In this method, the ultrasound directivity is required to obtain the audio sound directivity.

For a steerable profile given by Eq. (6.1), the far-field directivity of ultrasound is [Schmerr Jr, 2014, Eq. (3.22)]

$$\mathcal{D}_i(\varphi) = \operatorname{sinc}\left[\Re(k_i)a(\sin\varphi - \sin\varphi_{\mathrm{d}})\right], \tag{6.3}$$

where the sinc function is defined as $\operatorname{sinc} x \equiv (\sin x)/x$. For a special case without beam steering, Eq. (6.3) reduces to

$$\mathcal{D}_i(\varphi) = \operatorname{sinc}\left[\Re(k_i)a\sin\varphi\right], \tag{6.4}$$

which is the directivity of a rigid piston source.

The aperture factor of audio sound is determined by an audio source with a same aperture size and a velocity profile of $\exp(ik_{\mathrm{a}}y_{\mathrm{s}}\sin\varphi_{\mathrm{d}})\Pi[y_{\mathrm{s}}/(2a)]$. The effective directivity is then

$$\mathcal{D}_{\mathrm{A}}(\varphi) = \operatorname{sinc}[k_{\mathrm{a}}a(\sin\varphi - \sin\varphi_{\mathrm{d}})]. \tag{6.5}$$

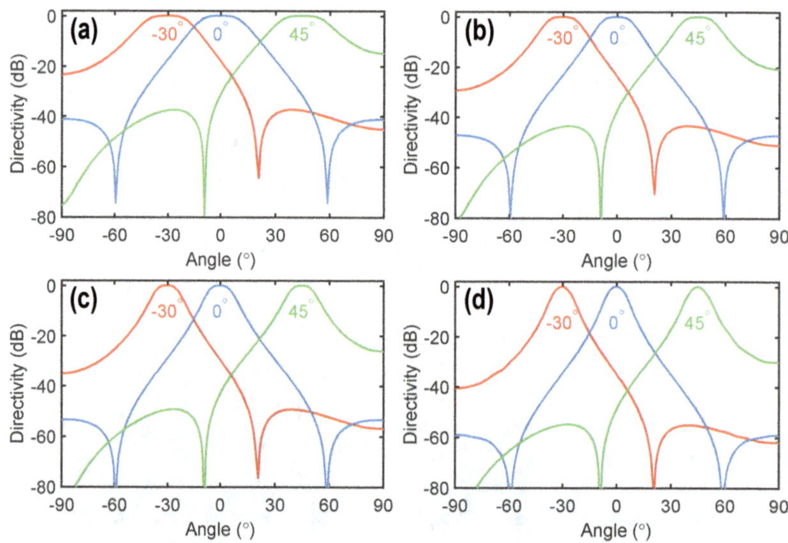

Figure 6.3 Normalized audio sound directivity generated by a 2D steerable PAL at different steering angles $-30°$, $0°$, and $45°$. The ultrasound frequency $f_{\mathrm{u}} = 40$ kHz. The audio frequency is (a) 500 Hz, (b) 1 kHz, (c) 2 kHz, and (d) 4 kHz. The source size is (a) 0.4 m, (b) 0.2 m, (c) 0.1 m, and (d) 0.05 m. In all cases, the Helmholtz number for audio sound is constant, i.e., $k_{\mathrm{a}}a = 3.66$.

Figure 6.3 shows the audio sound directivity generated by a steerable PAL at different steering angles. It is observed that the audio beam can be steered to a preset steering angle. For a same steering angle, it is interesting to note that the directivity shows difference at different configurations even though the Helmholtz number for audio sound is constant, i.e., $k_a a = 3.66$ in Fig. 6.3. However, for a linear source such as conventional loudspeakers, the directivity remains identical if the Helmholtz number is constant, as observed by Eq. (6.3). In contrast, the audio beam generated by a PAL is more directional at higher audio frequencies.

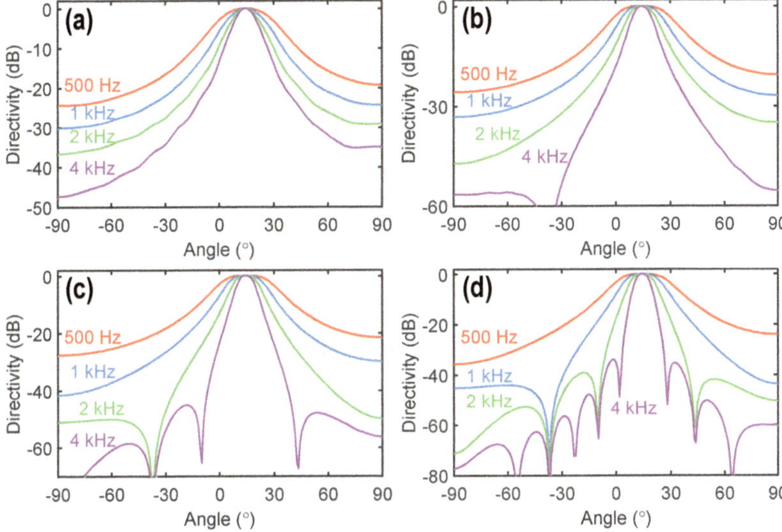

Figure 6.4 Normalized audio sound directivity generated by a 2D steerable PAL at different audio frequencies 500 Hz, 1 kHz, 2 kHz, and 4 kHz. The steering angle is $\varphi_d = 15°$. The ultrasound frequency $f_u = 40$ kHz. The half-width of the source is (a) $a = 0.02$ m, (b) $a = 0.05$ m, (c) $a = 0.1$ m, and (d) $a = 0.2$ m.

Figures 6.4 and 6.5 compare the audio sound directivity at different audio frequencies and source sizes, respectively, when the steering angle is 15°. Compared to the case without beam steering as presented in Sec. 3.2.4, similar trends can be observed. For example, the audio beam is more directional at higher frequencies and larger aperture sizes.

6.2.4 AUDIO SOUND IN THE NEAR FIELD

Section 6.2.3 has demonstrated the far field directivity of the audio sound generated by a steerable PAL. As demonstrated in Chap. 3, the inverse-law far field is usually more than 10 m away from the source. Therefore, the results presented in Sec. 6.2.3 have limitations in practical applications when the observation point is less than 10 m. It is imperative to investigate the audio sound field generated by a steerable PAL in the near field.

Figure 6.6 presents the audio sound field distributions generated by a 2D steerable PAL in the near field at different audio frequencies and source aperture sizes.

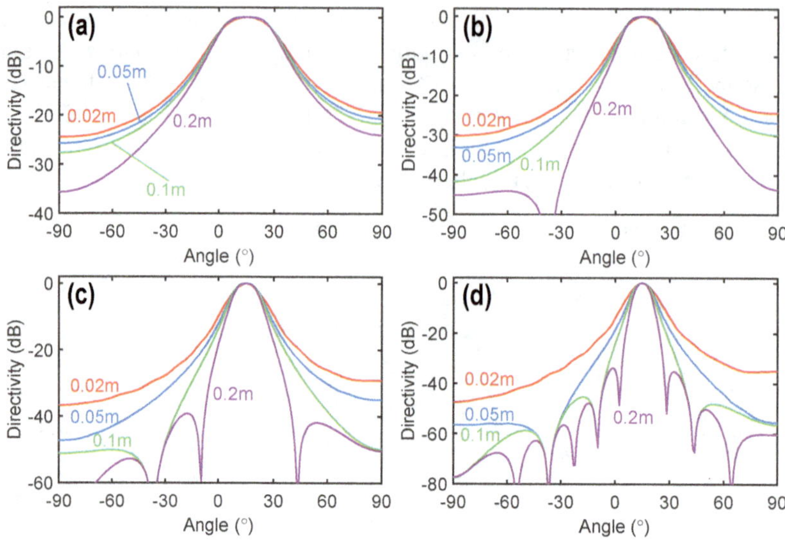

Figure 6.5 Normalized audio sound directivity generated by a 2D steerable PAL at different source sizes 0.02 m, 0.05 m, 0.1 m, and 0.2 m. The steering angle is $\varphi_d = 15°$. The ultrasound frequency $f_u = 40$ kHz. The audio frequency is (a) 500 Hz, (b) 1 kHz, (c) 2 kHz, and (d) 4 kHz.

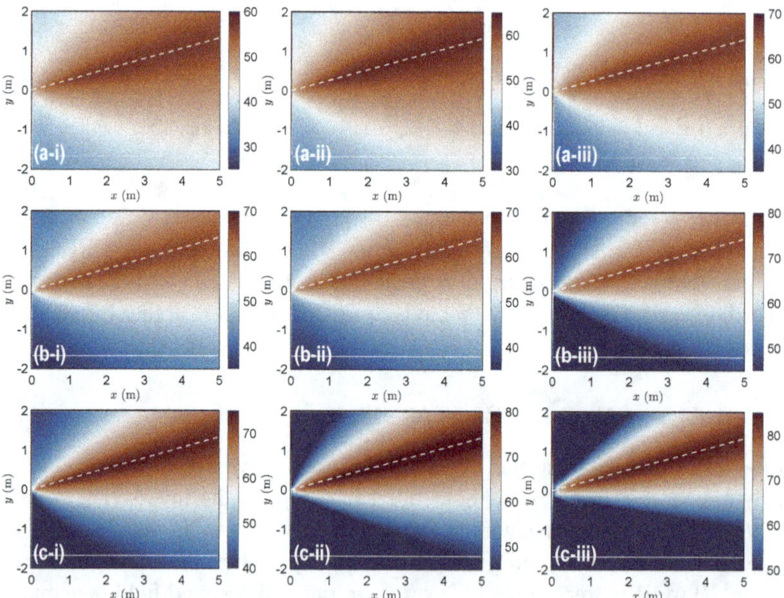

Figure 6.6 Audio sound field distributions generated by a 2D steerable PAL. The steering angle is $\varphi_d = 15°$, and the dashed lines denote the steering direction. The ultrasound frequency is $f_u = 40$ kHz. The on-surface pressure is $p_0 = 50$ Pa (125 dB). The audio frequency is (a) 500 Hz, (b) 1 kHz, and (c) 2 kHz. The source half-width is (i) 50 mm, (ii) 100 mm, and (iii) 150 mm.

Here, the steering angle is set to $\varphi_d = 15°$. The CWE presented in Sec. 2.5 was used to obtain the results. It is clear that the main lobe of audio beams can be steered to the desired direction in all cases. The directionality of the audio beam is more pronounced at higher audio frequencies and larger source aperture sizes, which is similar to the observation of that generated by a conventional PAL without beam steering.

6.2.5 REMARKS

This section presents the simplest physical model to investigate the properties of sound fields generated by a 2D steerable PAL. The results show that the highly directional audio beam can be electronically steered to the desired direction after applying the steerable source profile for ultrasound as given by Eq. (6.1). However, it is noted that this ideal steerable source profile is a continuous function, which is commonly unattainable in practical applications because a steerable PAL is implemented by an array of transducers and the practical source profile is instead a discrete function [Zhong et al., 2021a].

6.3 THREE-DIMENSIONAL BEAM STEERING

6.3.1 PROBLEM DESCRIPTION

A 3D steerable PAL is able to electronically steer the directional audio beam in both azimuthal and zenithal directions, which is more flexible than a 2D steerable PAL as introduced in Sec. 6.2. Figure 6.7 shows the sketch of a baffled circular steerable PAL radiating ultrasound in 3D free space. In the 3D model, Cartesian (x, y, z) and spherical (r, θ, φ) coordinate systems are established with their origin, O, at the centroid of the PAL, and the positive z-axis pointing to the radiation direction, where r, θ, and φ, are the radial, zenithal, and azimuthal coordinates, respectively. The beam direction can be described by a unit vector $\hat{\mathbf{s}} = (\sin\theta_d \cos\varphi_d, \sin\theta_d \sin\varphi_d, \cos\theta_d)$. The source

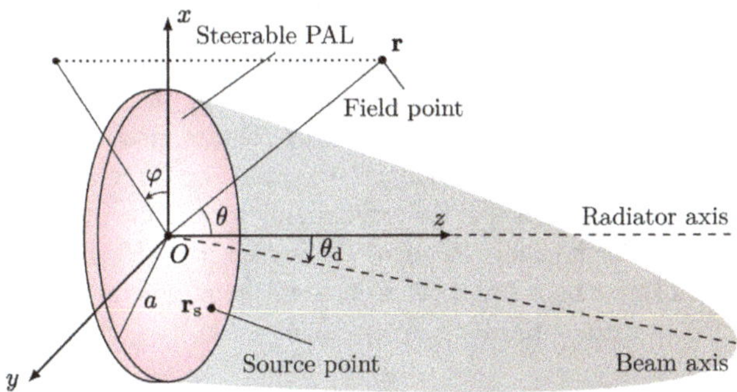

Figure 6.7 Sketch of a circular steerable PAL in the 3D model. The PAL lies in the plane $z = 0$. Extracted from [Zhong et al., 2022b, Fig. 1].

profile of ultrasound for a steerable PAL is [Eq. (2.34)]

$$v_{i,z}(\mathbf{r}_s) = \exp[\mathrm{i}\Re(k_i)\mathbf{r}_s \cdot \hat{\mathbf{s}}]\mathcal{A}(\mathbf{r}_s). \tag{6.6}$$

The source point $\mathbf{r}_s = (x_s, y_s, z_s)$. Without loss of generality, the source is placed in the plane $z = 0$ so that $z_s = 0$. Here, $\mathcal{A}(\mathbf{r}_s)$ represents the aperture area. For a circular or rectangular PAL, the aperture area is given by Eqs. (2.32) and (2.35), respectively.

6.3.2 AUDIO SOUND IN THE FAR FIELD

In this section, the results are obtained using the 3D modified CDM, as illustrated by Eq. (2.77). For a steerable source in a 3D model, the steering direction is assumed to be determined by a unit vector of $\hat{\mathbf{s}}$ and the source profile of ultrasound is given by Eq. (6.6). When the steerable PAL is circular with a radius of a, the directivity of ultrasound is then obtained as

$$\mathcal{D}_i(\hat{\mathbf{r}}) = \mathrm{jinc}\left[\Re(k_i)a\Phi(\hat{\mathbf{r}},\hat{\mathbf{s}})\right]. \tag{6.7}$$

Here, $\hat{\mathbf{r}} = (\theta, \varphi)$ denotes both zenithal and azimuthal angles in spherical coordinates, and the jinc function is defined by Eq. (2.84). An auxiliary function is introduced as

$$\Phi(\hat{\mathbf{r}},\hat{\mathbf{s}}) \equiv \sqrt{(\cos\varphi\sin\theta - \cos\varphi_\mathrm{d}\sin\theta_\mathrm{d})^2 + (\sin\varphi\sin\theta - \sin\varphi_\mathrm{d}\sin\theta_\mathrm{d})^2}. \tag{6.8}$$

The aperture factor of audio sound is determined by an audio source with a same aperture size and a velocity profile of $v_{\mathrm{a},z}(\mathbf{r}_s) = \exp(\mathrm{i}k_\mathrm{a}\mathbf{r}_s \cdot \hat{\mathbf{s}})\mathcal{A}(\mathbf{r}_s)$ [similar to Eq. (6.6)]. The effective directivity is then $\mathcal{D}_\mathrm{A}(\hat{\mathbf{r}}) = \mathrm{jinc}[k_\mathrm{a}a\Phi(\hat{\mathbf{r}},\hat{\mathbf{s}})]$.

When the steerable PAL is rectangular with side lengths of $2a_x$ and $2a_y$ in x- and y-directions, respectively, the ultrasound directivity is obtained as

$$\begin{aligned}\mathcal{D}_i(\hat{\mathbf{r}}) = {} & \mathrm{sinc}\left[\Re(k_i)a_x(\sin\theta\cos\varphi - \sin\theta_\mathrm{d}\cos\varphi_\mathrm{d})\right] \\ & \times \mathrm{sinc}\left[\Re(k_i)a_y(\sin\theta\sin\varphi - \sin\theta_\mathrm{d}\sin\varphi_\mathrm{d})\right].\end{aligned} \tag{6.9}$$

The effective directivity is then $\mathcal{D}_\mathrm{A}(\hat{\mathbf{r}}) = \mathrm{sinc}[k_\mathrm{a}a_x(\sin\theta\cos\varphi - \sin\theta_\mathrm{d}\cos\varphi_\mathrm{d})] \cdot \mathrm{sinc}[k_\mathrm{a}a_y(\sin\theta\sin\varphi - \sin\theta_\mathrm{d}\sin\varphi_\mathrm{d})]$.

Figures 6.8 and 6.9 show the audio directivity respectively generated by a circular and a rectangular steerable PAL in both zenithal and azimuthal directions at different audio frequencies and steering angles. The main lobe of the audio beam can be steered to the desired directional in all cases. The directivity pattern for a circular steerable PAL resembles that for a rectangular one at the lower audio frequency (1 kHz in this case). As the audio frequency increases to 4 kHz, nulls are presented in the radiation pattern for both PALs. The shape of the null is circular and rectangular for the circular and rectangular PAL, respectively.

6.3.3 AUDIO SOUND IN THE NEAR FIELD

6.3.3.1 Zernike Expansion of the Steerable Profile

In this section, the audio sound in the near field generated by a circular steerable PAL is modeled using the SWE with a nonaxisymmetric profile, which is introduced in

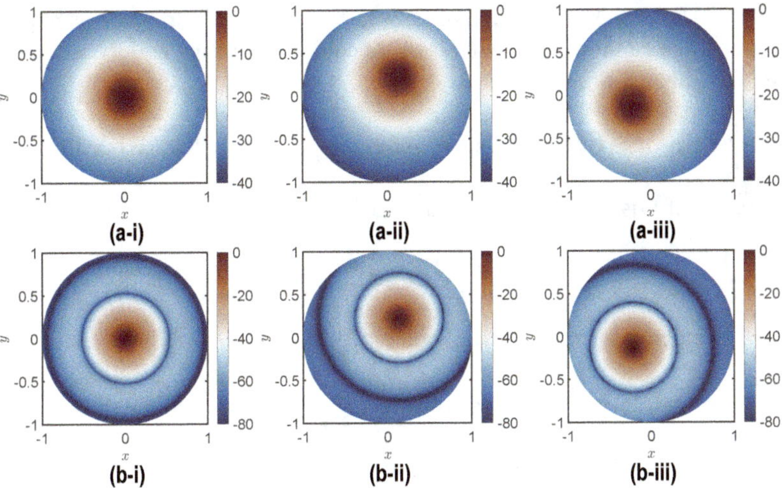

Figure 6.8 Normalized audio sound directivity (dB) generated by a steerable PAL at different steering angles. The radius of the circular size is $a = 0.1$ m. The ultrasound frequency $f_u = 40$ kHz. The audio frequency is (a) 1 kHz and (b) 4 kHz. The steering angles are (i) $(\theta_d, \varphi_d) = (0°, 0°)$, (ii) $(\theta_d, \varphi_d) = (60°, 15°)$, and (iii) $(\theta_d, \varphi_d) = (210°, 15°)$. Horizontal and vertical coordinates are $x = \sin\theta\cos\varphi$ and $y = \sin\theta\sin\varphi$, respectively.

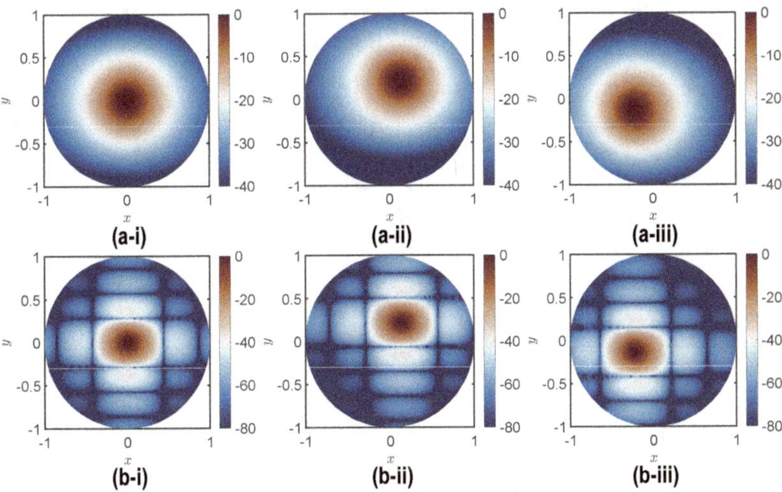

Figure 6.9 Normalized audio sound directivity (dB) generated by a steerable PAL at various steering angles. The sizes of the rectangular source are $a_x = 0.1$ m and $a_y = 0.15$ m. The ultrasound frequency $f_u = 40$ kHz. The audio frequency is (a) 1 kHz and (b) 4 kHz. The steering angles are (i) $(\theta_d, \varphi_d) = (0°, 0°)$, (ii) $(\theta_d, \varphi_d) = (60°, 15°)$, and (iii) $(\theta_d, \varphi_d) = (210°, 15°)$. Horizontal and vertical coordinates are $x = \sin\theta\cos\varphi$ and $y = \sin\theta\sin\varphi$, respectively.

Sec. 2.6.3. The Zernike expansion of the velocity profile of ultrasound is required to obtained at first as shown by Eq. (2.218). According to the integral representation of Bessel functions, it is obtained that

$$\frac{1}{2\pi}\int_0^{2\pi} \exp\left[\mathrm{i}(\Re(k_i)a\sigma\cos\varphi\sin\theta_\mathrm{d} - m_i\varphi)\right]\mathrm{d}\varphi = \mathrm{i}^{m_i}J_{m_i}(\Re(k_i)a\sigma\sin\theta_\mathrm{d}). \quad (6.10)$$

Substituting of Eqs. (6.6) and (6.10) into Eq. (2.219) yields the coefficients

$$M_{n_i}^{m_i} = 2(n_i+1)\mathrm{i}^{m_i}\int_0^1 R_{n_i}^{m_i}(\sigma)J_{m_i}(\Re(k_i)a\sigma\sin\theta_\mathrm{d})\sigma\mathrm{d}\sigma, \quad (6.11)$$

where the integral can be analytically obtained, so Eq. (6.11) reduces to

$$M_{n_i}^{m_i} = 2(n_i+1)\mathrm{i}^{n_i}\frac{J_{n_i+1}(\Re k_i a\sin\theta_\mathrm{d})}{\Re k_i a\sin\theta_\mathrm{d}}. \quad (6.12)$$

Note that Eq. (6.12) is independent of m_i.

Figure 6.10 gives an example of approximating the steerable profile at $y=0$ using the first 15 terms of Zernike circular polynomials when the ultrasound at 40 kHz is steered to the angle of $\theta_\mathrm{d}=15°$. The exact result is directly obtained using the closed-form expression given by Eq. (6.6) and presented for comparison. It is observed in Fig. 6.10(a) that an approximation of the velocity profile obtained using Eq. (2.219) agrees well with the exact result. Figure 6.10(b) presents the relative percentage error defined as $(\tilde{u}-u)/u\times 100\%$, where u and $\tilde{u}$ are the real or imaginary parts of the exact result and the approximation of the velocity profile, respectively. The relative error is less than 0.5% in this case so that the steerable profile is well approximated by the Zernike polynomials.

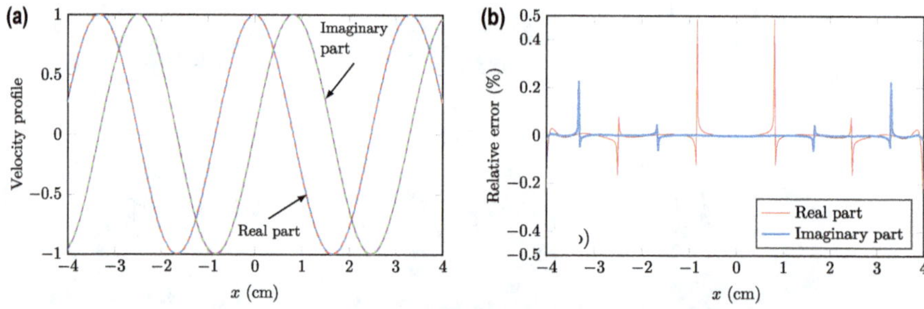

Figure 6.10 Comparison of the steerable profile at $y=0$ and the approximation using the first 15 terms of Zernike circular polynomials, where the ultrasound at 40 kHz is steered to the angle of $\theta_\mathrm{d}=15°$. The results are normalized to v_0. (a) The velocity profile; solid line, the exact result obtained using Eq. (6.6); dashed line, the approximation result obtained using Eq. (2.219). (b) The relative error of the approximation. Extracted from [Zhong et al., 2022b, Fig. 3].

6.3.3.2 Steerable Ultrasound Field

Figure 6.11 shows the ultrasound field at 40 kHz steered to the angle of 15°. For comparison, the ultrasound field steered to the angle of 0°, i.e., without beam steering, is also presented. Figure 6.12 presents the radial SPL along the beam direction and the angular SPL at a radial distance of 1 m for these two cases. It is noted that the azimuthal angle φ can be either 0° or 180° in the plane xOz, so the angular PAL as a function of $\theta \cos\varphi$ is presented in Fig. 6.12(b). When the ultrasonic beam is not steered, i.e., $\theta_d = 0°$, the energy is focused in the radiator direction. When the ultrasonic beam is steered to the angle of 15°, it is clear that the main lobe lies in the beam direction. It is noted in Fig. 6.12(a) that the SPL difference of these two cases is negligible when the radial distance is larger than the location of the last minimal SPL, i.e., 8.8 cm. Before this location in the near field, the fluctuations of the SPL for the steered beam are smaller compared to that without beam steering. This implies that the major part of the energy of the steered beam remains at the same level when compared to that without beam steering, so the generated audio sound in the main lobe would be similar.

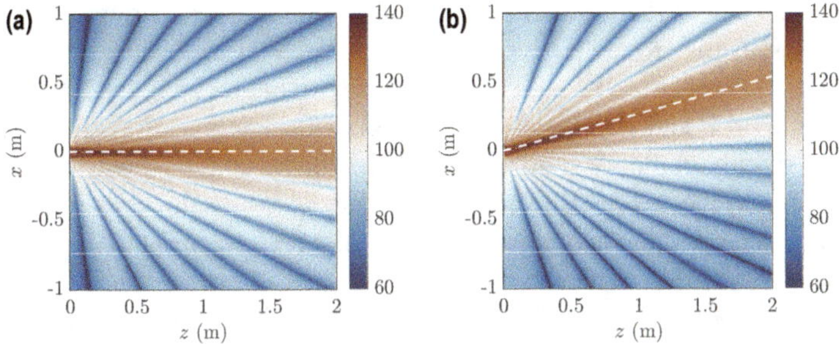

Figure 6.11 The ultrasound field at 40 kHz with the steering angle at (a) $\theta_d = 0°$ and (b) $\theta_d = 15°$, where the truncated degree of Zernike polynomials is 15. The dashed lines denote the beam direction. Extracted from [Zhong et al., 2022b, Fig. 5].

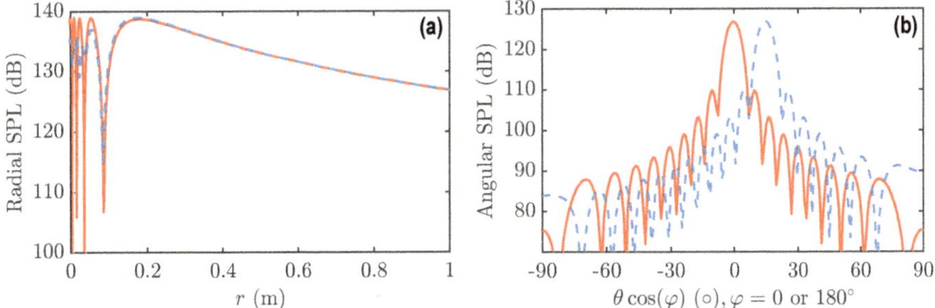

Figure 6.12 The ultrasound field at 40 kHz with the steering angle at 0° and 15°: (a) the radial SPL at the beam direction; (b) the angular SPL at a radial distance of 1 m. ——, $\theta_d = 0°$; - - -, $\theta_d = 15°$. Extracted from [Zhong et al., 2022b, Fig. 6].

6.3.3.3 Steerable Source Generating One Beam

To examine the accuracy of the audio SPL when using the Zernike polynomials to approximate the steering profile of the ultrasound, the audio SPL at several typical field points as a function of the truncated terms of the Zernike degree is presented in Fig. 6.13. It can be found that the audio SPL at all field points converges at large Zernike degrees n. The numerical results show the error is less than 0.1 dB for all points when the truncation terms are larger than 8.

Figure 6.14 shows the 2D audio sound field at 1 kHz with and without the local effects, as well as the difference obtained with and without local effects. The radial audio SPL at the beam direction as a function of the radial distance is shown in Fig. 6.15. It is observed that the audio beam can be steered to the desired direction without side lobes. The local effects make the SPL fluctuate in the near field for both cases, while the effects are negligible at large radial distances. Figure 6.15 shows that the transition distance can be used to determine if the local effects are significant. As the radial distance increases, starting from the transition distance at 0.19 m, the difference between the SPL obtained with and without the local effects decreases and finally converges to zero.

Figure 6.16 shows the 2D audio sound field with the local effects at 250 Hz for a steering angle of 0° and 15°. Figure 6.17 compares the radial SPL at the beam direction with and without the local effects. It can be seen that a low audio frequency can also be steered to the desired direction without generating side lobes. The difference of the SPL obtained with and without the local effects is larger in the near field when compared to the case at 1 kHz, as shown in Fig. 6.15. For example, the SPL difference increases from 0.6 dB to 8.5 dB at the transition distance 0.19 m, with a steering angle of 15° when the frequency decreases from 1 kHz to 250 Hz. The reason is that the audio sound pressure determined by cumulative effects drops about 12 dB as the frequency is halved, while the Lagrangian density of ultrasound changes only a little, which results in significant local effects in the near field. However, it still holds that

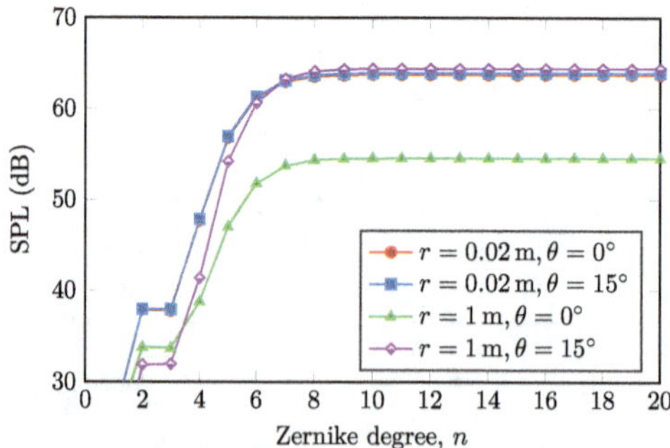

Figure 6.13 The audio SPL at several typical field points as a function of the truncated terms of Zernike degree n, where the audio sound at 1 kHz is steered to the angle of $\theta_d = 15°$. Extracted from [Zhong et al., 2022b, Fig. 7].

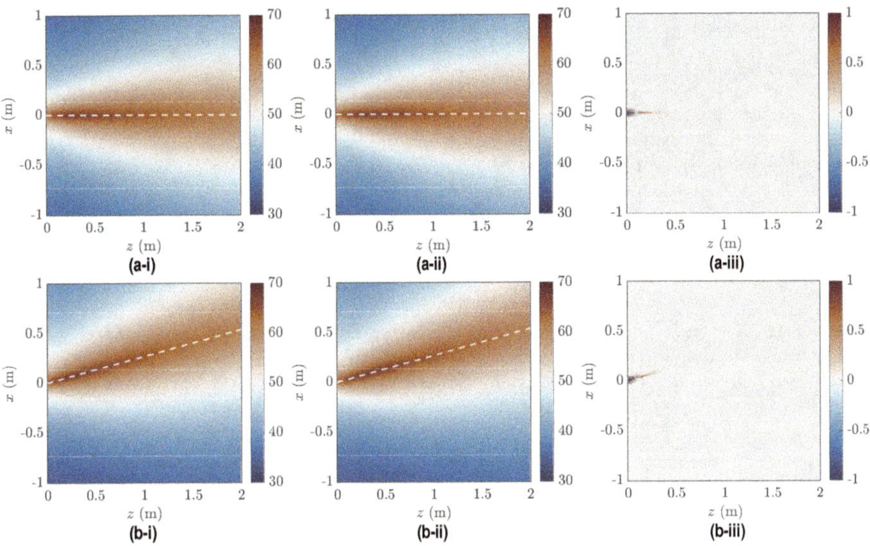

Figure 6.14 The audio sound field at 1 kHz (i) with and (ii) without the local effects, and (iii) the SPL difference. The steering angle is (a) $\theta_d = 0°$ and (b) $\theta_d = 15°$. The dashed lines denote the beam direction. Extracted from [Zhong et al., 2022b, Fig. 8].

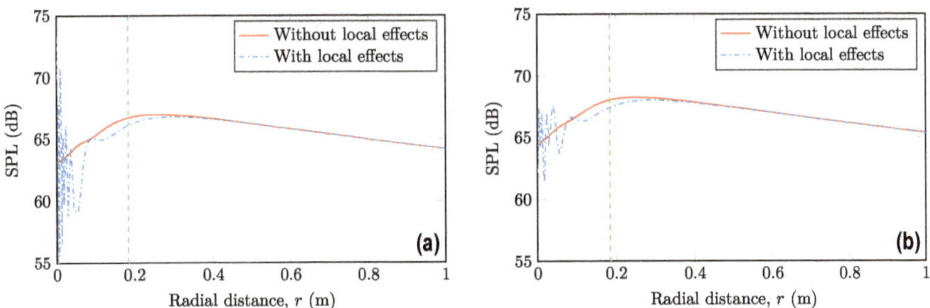

Figure 6.15 The audio SPL at the beam direction as a function of the radial distance with and without the local effects at 1 kHz. The steering angle is (a) $\theta_d = 0°$ and (b) $\theta_d = 15°$. The dashed lines (- - -) denote the transition distance at 0.19 m. Extracted from [Zhong et al., 2022b, Fig. 9].

the difference decreases beyond the transition distance and eventually converges to zero.

6.3.3.4 Steerable Source Generating Dual Beams

Figure 6.18(a) shows the 2D audio sound field when a steerable PAL generates dual beams at angles of 15° and −20°. Figure 6.18(b) presents the angular SPL at various radial distances as a function of the angle $\theta \cos\varphi$, where $\varphi = 0°$ or 180°. All the results are given with the local effects. It can be found that the audio beams are successfully steered to dual directions and the main lobes are exactly at the angles

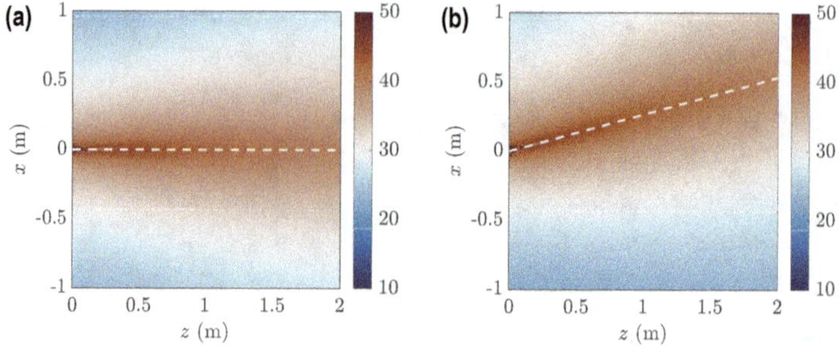

Figure 6.16 The audio sound field with the local effects at 250 Hz. The steering angle is (a) $\theta_d = 0°$ and (b) $\theta_d = 15°$. The dashed lines denote the beam direction. Extracted from [Zhong et al., 2022b, Fig. 10].

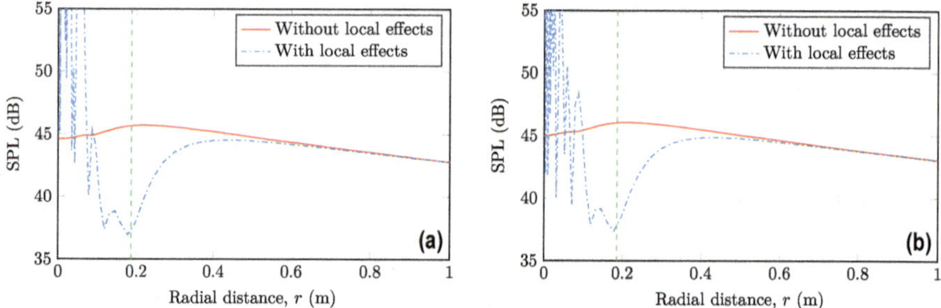

Figure 6.17 The audio sound field along the beam direction with and without the local effects at 250 Hz. The steering angle is (a) $\theta_d = 0°$ and (b) $\theta_d = 15°$. The dashed lines (- - -) denote the transition distance at 0.19 m. Extracted from [Zhong et al., 2022b, Fig. 11].

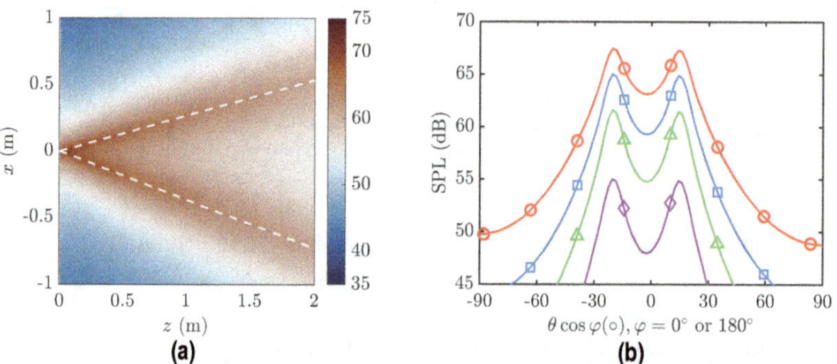

Figure 6.18 The audio sound field generated by a steerable PAL at 1 kHz generating dual beams at the angles of 15° and −20°. (a) 2D sound fields. The dashed lines denote the beam directions. (b) Angular sound field at various radial distances, r. —○—, $r = 0.5$ m; —□—, $r = 1$ m; —△—, $r = 2$ m; —◇—, $r = 5$ m. Extracted from [Zhong et al., 2022b, Figs. 12–13].

of 15° and −20°. The results also show that the maxima obtained at two peaks at the beam directions are almost the same. It is noted in Fig. 6.18(b) that there is a local minimum near 0°. The differences between the maxima at two beam directions and the local minimum become smaller at shorter radial distances. For example, the difference between the local maximum and minimum decreases from 7 dB to 4.4 dB as the radial distance decreases from 5 m to 0.5 m. This implies that the two beams are not well separated at small radial distances in the near field.

6.3.4 REMARKS

The reader is introduced to the concept of the 3D steerable PAL in this section. Compared to its 2D counterpart, as discussed in Sec. 6.2, the 3D steerable PAL offers increased flexibility since it facilitates beam steering in both zenithal and azimuthal directions. Section 6.3.2 examines the far field directivities of audio sound generated by a 3D steerable PAL using the modified CDM. Furthermore, Sec. 6.3.3 explores the audio sound in the near field by employing the SWE. The results demonstrate that by applying the appropriate velocity profile for ultrasound, such as Eq. (6.7), highly directional audio beams can be electronically steered to the desired direction.

6.4 GRATING LOBE AND SUPPRESSION TECHNIQUES

6.4.1 PROBLEM DESCRIPTION

The results presented in previous sections demonstrate that the main lobe of audio beams can be steered to the desired direction by applying steerable profiles for ultrasound. However, these ideal profiles are continuous function which are commonly unattainable in practical applications. Instead, the beam steering is implemented by an array consisting of discrete number of ultrasonic transducers. Consequently, the source profiles are commonly discrete functions. Furthermore, elements of conventional transducer arrays are typically uniformly distributed.

Figure 6.19(a) presents an example of the uniform array configuration consisting of $N = 6$ elements, where the *interelement spacing*, also referred to as the *array pitch*, is consistent and represented as s. A rectangular coordinate system, denoted as $Oxyz$, is established with its origin situated at the centroid of the array, and the x-axis is oriented perpendicular to the radiation surface. For simplicity, all array elements have an identical size in the y-direction, denoted as $2a$. It is assumed that the size of all elements in the z-direction greatly exceeds the wavelength, allowing for the modeling of the radiation solely within the 2D plane Oxy. The *kerf width*, the blank region between adjacent array elements, is denoted as κ, and the total kerf width is determined as $\kappa_{\text{tot}} = (N-1)\kappa$. If the array is labeled by 1, 2, ..., N in order, the y-coordinate of the centroid of the n-th element is determined as

$$y_n = \left(n - \frac{N+1}{2}\right)s, \quad n = 1, 2, ..., N. \tag{6.13}$$

The total size of the array is denoted as $2A$, where

$$A = Ns/2. \tag{6.14}$$

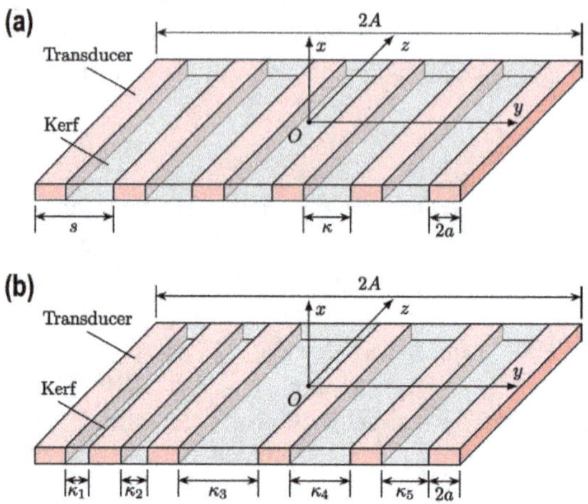

Figure 6.19 Sketch of the steerable PAL consisting of N (6 in this figure) ultrasonic transducer elements with a width of $2a$: (a) uniform array with the equal interelement spacing; (b) random array with the unequal interelement spacing. Extracted from [Zhong et al., 2023b, Fig. 1].

In practical applications, the amplitude and phase distributions on the radiation surface of each element are uniform. Therefore, the total discrete source profile can be given by a relation to the continuous profile Eq. (6.2) as

$$v_{i,x}^{\text{dist}}(\boldsymbol{\rho}_\text{s}) = W(y_\text{s}) \begin{cases} v_{i,x}\left(x_\text{s}, \left(\left\lfloor \dfrac{y_\text{s}}{s} \right\rfloor + \dfrac{1}{2}\right)s\right), & N \bmod 2 = 0, \\ v_{i,x}\left(x_\text{s}, \left(\left\lfloor \dfrac{y_\text{s}}{s} + \dfrac{1}{2} \right\rfloor\right)s\right), & N \bmod 2 = 1. \end{cases} \tag{6.15}$$

Here, the weight function is unit over all radiation surfaces, expressed as

$$W(y_\text{s}) = \sum_{n=1}^{N} \Pi\left(\frac{y_\text{s} - y_n}{2a}\right) = \begin{cases} 1, & -a \le y_\text{s} - y_n \le a, \\ 0, & \text{otherwise}. \end{cases} \tag{6.16}$$

Here, $\Pi(\cdot)$ is the rectangle function defined by Eq. (2.11). Two examples of comparison of continuous and discrete source profiles are shown in Fig. 6.20.

In practical applications, achieving the beam steering without the undesirable occurrence of grating lobes necessitates the use of an ultrasonic array with an interelement spacing smaller than half the wavelength of the carrier ultrasound (typically around 4.3 mm at 40 kHz), a condition referred to as the spatial *Nyquist criterion*. However, ultrasonic wavelengths are typically shorter than the size of an ultrasonic transducer, typically around 10 mm as illustrated in Sec. 8.2.1. This results in the

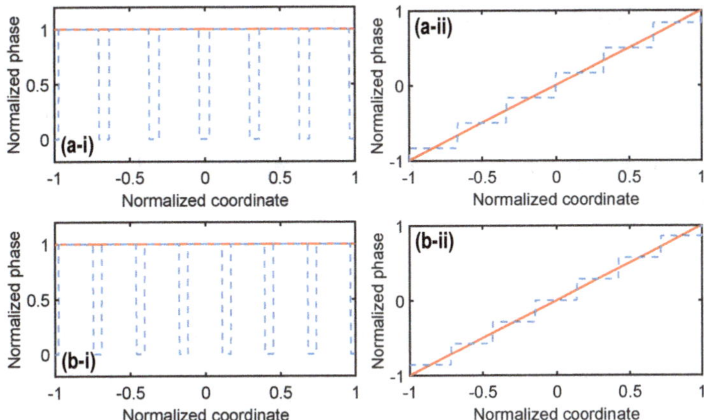

Figure 6.20 Comparison of the continuous and discrete source profiles for the ultrasound. The normalized (i) amplitude, $\left|v_{i,x}^{\text{dist}}(\rho_s)/v_0\right|$, and (ii) phase, $\arg[v_{i,x}^{\text{dist}}(\rho_s)/v_0]/(\Re(k_i)A\sin\varphi_d)$, distributions are presented as a function of the normalized coordinate y_s/A. The array element width is $2a = 10$ mm. The element separation is $s = 12.5$ mm. The number of array elements is (a) $N = 6$ and (b) $N = 7$. ——, continuous profile; - - -, discrete profile.

generation of grating lobes in the ultrasound radiation pattern, subsequently leading to grating lobes in the demodulated audio beams, known as the *spatial aliasing effect*. In contrast, when generating steerable audio beams using a conventional loudspeaker array, spatial aliasing typically occurs only at high audio frequencies because it is easier to satisfy the Nyquist criterion. For example, at 1 kHz, the wavelength is 0.34 m, indicating that the interelement spacing of the array can be up to 0.17 m. Consequently, appropriate techniques must be employed to eliminate grating lobes in practical applications of steerable PALs.

The sparse random array technique is a promising approach to suppressing grating lobes, where "sparse" means the average interelement spacing is larger than half wavelength and "random" means the elements are unequally spaced so that the sound waves emitted from the array elements arrive randomly at the grating lobe, although they do arrive in-phase at the main lobe [Goss et al., 1996]. Figure 8.13(b) presents a sketch of a sparse random array configuration, where the interelement spacing is chosen at random, and the kerf width between the n-th and $(n+1)$-th elements is denoted as κ_n, where $n = 1, 2, \ldots, N-1$. This technique has been successfully and widely used in designing antenna [Kurup et al., 2003] and medical ultrasound [Goss et al., 1996] arrays. Recently, it receives the increasing interests in designing PAL arrays [Zhong et al., 2023b; Tang et al., 2023]. In this section, the simulated annealing (SA) algorithm is used to optimize the beam pattern of a 2D steerable PAL [Zhong et al., 2023b]. Throughout this section, the directivity of the audio sound generated by the PAL is calculated using the 2D direct convolution directivity model (CDM), as demonstrated by Eq. (2.70).

The ultrasound directivity at the frequency f_i generated by the array is given as [Schmerr Jr, 2014, Eq. (4.26)]

$$\mathcal{D}_i(\varphi) = \mathcal{D}_a(\varphi, k_i)\mathcal{D}_s(\varphi, k_i), \quad i = 1,2. \tag{6.17}$$

Here, the element directivity $\mathcal{D}_a(\varphi, k_i)$ and the discrete point source directivity $\mathcal{D}_s(\varphi, k_i)$ are, respectively,

$$\mathcal{D}_a(\varphi, k_i) = \mathrm{sinc}(\Re(k_i) a \sin\varphi), \tag{6.18}$$

$$\text{and } \mathcal{D}_s(\varphi, k_i) = \sum_{n=1}^{N} w_n \exp\left[-\mathrm{i}\Re(k_i) y_n(\sin\varphi - \sin\varphi_{\mathrm{d}})\right]. \tag{6.19}$$

In (6.19), w_n and y_n represent the weight coefficient and the y-coordinate of the centroid of the n-th element, respectively, and φ_{d} is the steering angle. The special case when $\varphi_{\mathrm{d}} = 0$ (0°) represents a conventional PAL without beam steering.

6.4.2 UNIFORM ARRAY GENERATING GRATING LOBES

The uniform array configuration illustrated in Fig. 6.19(a) serves as the foundational setup for performance comparison. By employing (6.13), the discrete point source directivity described by Eq. (6.19) can be further simplified to [Schmerr Jr, 2014, Eq. (4.28)]

$$\mathcal{D}_s(\varphi, k_i) = \frac{\sin\left[N\Re(k_i)s(\sin\varphi - \sin\varphi_{\mathrm{d}})/2\right]}{N\sin\left[\Re(k_i)s(\sin\varphi - \sin\varphi_{\mathrm{d}})/2\right]}. \tag{6.20}$$

Grating lobes manifest when the following condition is met [Schmerr Jr, 2014, Eq. (4.29)]

$$\Re(k_i)s(\sin\varphi_m - \sin\varphi_{\mathrm{d}}) = 2m\pi, \quad m = \pm 1, \pm 2, \pm 3, \ldots, \tag{6.21}$$

where the angle of the grating lobe, φ_m, is determined as

$$\varphi_m = \arcsin\left(\sin\varphi_{\mathrm{d}} + \frac{m\lambda_i}{s}\right). \tag{6.22}$$

Here, λ_i represents the wavelength at frequency f_i. It is noteworthy that if the interelement spacing s exceeds $\lambda_i/2$, the array generates at least one grating lobe at an angle φ_m. However, it is important to note that the audio sound may not fully inherit the grating lobes from the ultrasound calculated by Eq. (6.22), due to the phenomenon known as the grating lobe elimination, which has been investigated in Shi and Gan, 2011. This is because the audio sound directivity is influenced by the product of the directivity of two ultrasonic waves, as presented in Eq. (2.70). When the grating lobes of two ultrasonic waves are separated in location, their product can be small enough to be negligible.

For example, using the parameters from Fig. 6(b) of [Shi and Gan, 2011], when the sound beam is steered to the angle of 15° with the ultrasound frequencies of 40 kHz and 50 kHz, and an interelement spacing of 17.15 mm, Eq. (6.22) predicts

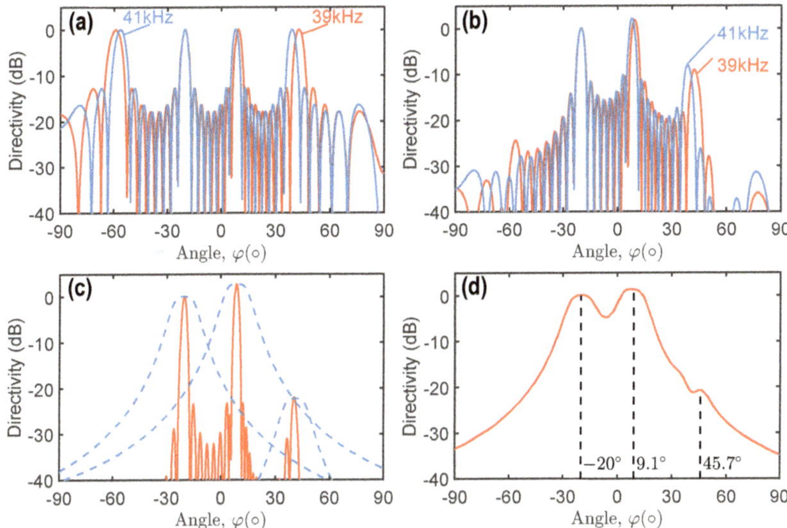

Figure 6.21 The directivities generated by a uniform array where the element number $N = 8$, the audio frequency $f_a = 2$ kHz, and the average ultrasound frequency $f_u = 40$ kHz. (a) the discrete point source directivity and (b) the total array directivity for ultrasound; ——, $f_1 = 39$ kHz; ——, $f_2 = 41$ kHz. (c) ——, the product directivity of ultrasound; - - -, Westervelt's directivity. (d) The audio sound directivity. Extracted from [Zhong et al., 2023b, Fig. 3].

the first grating lobe ($m = 1$) to be located at $-14.0°$ and $-8.1°$. Due to a separation of 5.9° between these grating lobes, the product of their directivity is insignificant in this direction. However, it is important to note that the audio sound directivity is also influenced by the Westervelt's directivity, as demonstrated in Eq. (2.70), even though this factor was not considered in Shi and Gan, 2011. The Westervelt's directivity characterizes the effective length and the diffraction capabilities of the virtual audio source, and its impact on the radiation pattern of the steerable PAL will be elucidated in Sec. 6.4.3.

Figure 6.21 presents both the ultrasound and audio sound directivities generated by a uniform array comprising $N = 8$ elements at an audio frequency of $f_a = 2$ kHz. The interelement spacing is set at $s = 17.15$ mm, and the steering angle is $\varphi_d = -20°$. Due to the interelement spacing being greater than half a wavelength, three grating lobes are present, as depicted in Figs. 6.21(a–b), and their locations can be predicted using Eq. (6.22). Figure 6.21(b) highlights the significance of taking into account the element directivity of ultrasound, as described by Eq. (6.18). The element directivity is governed by the ultrasound Helmholtz number, $k_i a$. The radiation pattern appears nearly uniform when $k_i a$ is small enough to disregard its effects. However, the typical size of ultrasonic transducers (10 mm) used in a steerable PAL is usually larger than the wavelength (8.7 mm at 40 kHz), resulting in highly directional element directivity [Shi et al., 2015]. Consequently, the grating lobe observed at $-57.4°$ is sufficiently small to be disregarded. Figure 6.21(c) presents the product directivity of the ultrasound and the envelopes of the Westervelt's directivity at the main lobe, as well as two grating lobes at 9.1° and 40.8°. Following the convolution process in Eq. (2.70), the audio sound directivity is obtained, as shown in Fig. 6.21(d). It is evident that, apart from the main lobe at −20°, two additional peaks appear at around 9.1° and 45.7°, representing two grating lobes with levels of 1.3 dB and −20.8 dB, respectively.

6.4.3 OPTIMAL ARRAY OBTAINED USING SIMULATED ANNEALING

Without loss of generality, the total kerf width in Fig. 6.19(b) is constrained to $(N-1)\kappa$, ensuring that the average kerf and interelement spacing equal κ and s, respectively. The term "sparse" in this context implies that the average interelement spacing exceeds half the wavelength of the carrier ultrasound, thereby failing to meet the Nyquist criterion and resulting in the presence of grating lobes in the radiation pattern. The SA is a stochastic-based algorithm used for addressing global optimization problems, offering a reliable approximation of the global optimum for objective functions defined across extensive parameter spaces [Kirkpatrick et al., 1983]. In the quest to minimize both grating lobes and sidelobe levels, the objective function is set as the maximal magnitude of the audio sound directivity for the angles outside the main lobe. This objective function depends on element positions and weight coefficients, expressed as [Bevelacqua and Balanis, 2007]

$$\mathcal{J}(\mathbf{y},\mathbf{w}) = \max_{\varphi,\text{s.t.}|\varphi-\varphi_0|\geq\Delta\varphi} |\mathcal{D}_\text{a}(\varphi)|, \tag{6.23}$$

where $\Delta\varphi > 0$ is set to encompass the main lobe, the element position vector is denoted as $\mathbf{y} = [y_1, y_2, \cdots, y_N]^\text{T}$, the weight coefficient vector is represented by $\mathbf{w} = [w_1, w_2, \cdots, w_N]^\text{T}$, and the superscript "T" denotes matrix transposition. The level of Eq. (6.23), i.e., $20\log_{10}(J)$, is recognized as the peak sidelobe level (PSL). In this section, PSL serves as a pivotal metric for evaluating array performance. The optimization challenge lies in identifying the global minimum of the objective function through the optimization of $\mathbf{y}$ and $\mathbf{w}$.

During the optimization process, it is imperative to impose appropriate constraints on the positions and weight coefficients of array elements. To this end, the weight coefficients are confined within the interval $[w_\text{min}, w_\text{max}]$. The overall array size, denoted as $2A$, is predetermined, which implies that $y_1 = a - A$, $y_N = A - a$. Consequently, the total kerf width, κ_tot, becomes fixed as $2A - 2Na$. As illustrated in Fig. 6.19(b), it is essential to ensure that, for all kerf widths, $\kappa_n \geq 0$, ensuring that the spacing between adjacent elements remains no smaller than the element size, $2a$. To facilitate this constraint, a partition of the interval $[0, \kappa_\text{tot}]$ is introduced as

$$\mathcal{P} = \{0 = d_0 \leq d_1 \leq \cdots \leq d_{N-1} = \kappa_\text{tot}\}, \tag{6.24}$$

which divides the permissible total kerf width into $N-1$ subintervals. The n-th subinterval is represented as $[d_{n-1}, d_n]$, with its length equal to the kerf width between the n-th and $(n+1)$-th elements, where $\kappa_n = d_n - d_{n-1}$ and $n = 1, 2, \ldots, N-1$. Since the total array size remains fixed at $2A$, the kerf width κ_n is constrained within the range $[0, 2A - 2Na]$. Consequently, optimizing the vector $\mathbf{d} = [d_0, d_1, \ldots, d_{N-1}]^{\mathrm{T}}$ is equivalent to optimizing the element position vector $\mathbf{y}$.

The implementation of SA procedure is illustrated in Fig. 6.22. The process is initiated by selecting an initial temperature, denoted as T_0, which is set high enough to increase the likelihood of accepting the first perturbation. The initial position and weight coefficient vectors are arbitrarily assigned, and then, with each iteration, the temperature progressively decreases. An exponential schedule is adopted here [Bai et al., 2009]

$$T_j = \gamma T_{j-1}, \quad j = 1, 2, \ldots, j_{\max}, \tag{6.25}$$

where the cooling factor $\gamma \in (0,1)$, j is the iteration index, and $j_{\max}$ is the maximal iteration number.

Algorithm: Simulated Annealing

Input : $T_0, \mathbf{d}, \mathbf{y}, \mathbf{w}, w_{\min}, w_{\max}, \gamma, \sigma_d, \sigma_w$

Output: $E_{\text{best}}, \mathbf{y}_{\text{best}}, \mathbf{w}_{\text{best}}$

```
d†_0 = 0; d†_n = κ_tot; y†_1 = a − A ;            /* Initialization */
E_best = E = J(y, w);
for j = 1 to j_max do                             /* Iterations */
    T_j = γ T_{j−1} ;                             /* Cooling */
    /* Perturb the kerf widths                                      */
    for n = 1 to N − 2 do
        d†_n = mod (N(d_n, σ_d^2), κ_tot);
    /* Obtain y-coordinates of the elements                         */
    for n = 1 to N do
        y†_n = y†_{n−1} + 2a + d†_n − d†_{n−1}
    /* Perturb the weight coefficients                              */
    for n = 1 to N do
        w†_n = mod (N(w_n − w_min, σ_w^2), w_max − w_min) + w_min;
    E† = J(y†, w†) ;                              /* Obtain new energy */
    ΔE = E† − E ;                                 /* Energy difference */
    if (ΔE < 0) or (U(0,1) < e^(−ΔE/T_j)) then
        d = d†; w = w†; E = E† ;                  /* Accept result */
    if E† < E_best then                           /* Update best result */
        y_best = y†; w_best = w†; E_best = E†
```

Figure 6.22 Implementation of the SA algorithm for optimizing the beam pattern generated by a steerable PAL. Extracted from [Zhong et al., 2023b, Fig. 2].

At each iteration, the values $d_1, \ldots, d_{N-2}$ are perturbed as $d_1^\dagger, \ldots, d_{N-2}^\dagger$ subject to a normal distribution $\mathcal{N}(d_n, \sigma_d^2)$ with a mean of d_n and a standard deviation of σ_d. The modulo operation $\bmod\left(\mathcal{N}(d_n, \sigma_d^2), \kappa_{\text{tot}}\right)$ is used to ensure the result falls within

κ_{tot}, where mod (ξ_1, ξ_2) is defined by the floor function $\lfloor\cdot\rfloor$ as $\xi_1 - \xi_2\lfloor\xi_1/\xi_2\rfloor$. The perturbed results $d_1^\dagger, \ldots, d_{N-2}^\dagger$ are then sorted to maintain a non-decreasing order. Subsequently, the new position vector $\mathbf{y}^\dagger$ is derived based on the perturbed vector $\mathbf{d}^\dagger$. Meanwhile, the weight coefficients $w_1, \ldots, w_N$ are perturbed to $w_1^\dagger, \ldots, w_N^\dagger$, following a normal distribution with a standard deviation of σ_w. The objective function with the perturbed results is evaluated as $E^\dagger = J(\mathbf{y}^\dagger, \mathbf{w}^\dagger)$. If the energy decreases ($\Delta E < 0$), the newly perturbed parameters are accepted. Conversely, if the energy increases ($\Delta E > 0$), the parameters are accepted with a probability that depends on the temperature at the current iteration, T_j. In this section, parameters are accepted when a random number generated from a standard uniform distribution $\mathcal{U}(0,1)$ is less than $\mathrm{e}^{-\Delta E/T_j}$. Lower temperatures correspond to a decreased likelihood of accepting a worse configuration.

6.4.4 PERFORMANCE OF THE OPTIMAL SPARSE RANDOM ARRAY

For the sake of simplicity, some parameters are fixed and are detailed in Table 6.1. Without loss of generality, the directivities discussed in the following sections have been normalized to a unity value in the steering direction φ_0. The merits of employing a sparse random array are exemplified in Fig. 6.23, where directivities are calculated according to an optimal array configuration after optimizing both the element position and weight coefficients using the SA algorithm. In this specific setup, the average interelement spacing equals two wavelengths of the ultrasound, i.e., $s = 17.15$ mm, while all other parameters remain consistent with those utilized in Fig. 6.21. The optimal parameters are presented in Table 6.2.

In Figs. 6.23(a–b), it becomes apparent that only the grating lobes around 30° have been retained, and their magnitudes have decreased. Furthermore, in Fig. 6.23(c),

Table 6.1
The parameters used in SA. Extracted from [Zhong et al., 2023b, Table I].

Symbol	Definition	Value
a	Half width of the element	5 mm
$j_{\max}$	Maximal iteration numbers	10,000
s	Average interelement spacing	17.15 mm
T_0	Initial temperature in SA	10
$w_{\min}$	Minimum of the weight coefficient	0.2
$w_{\max}$	Maximum of the weight coefficient	2
γ	Decay rate in the cooling schedule	0.95
σ_d	Standard deviation for position	$\kappa_{\text{tot}}/6$
σ_w	Standard deviation for weight coefficients	0.3
φ_0	Steering angle	70°
$\Delta\varphi$	Beam width of the main lobe	25°

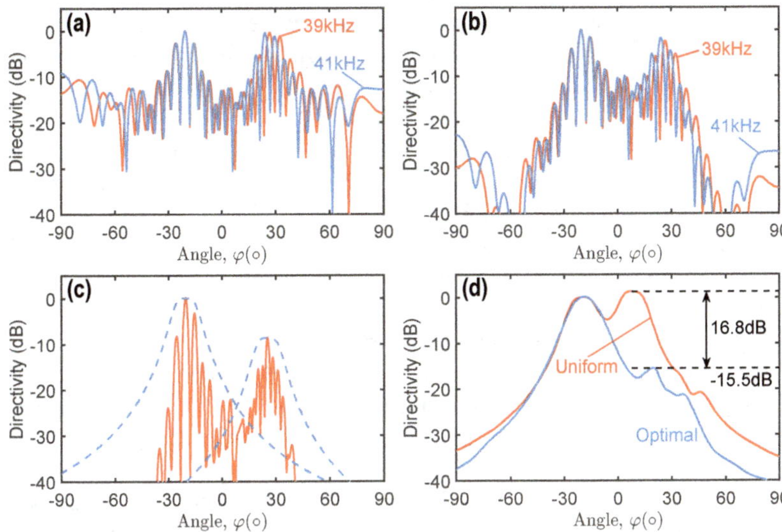

Figure 6.23 The directivities generated by an optimal array, where the element number $N = 8$, the audio frequency $f_a = 2$ kHz, and the average ultrasound frequency $f_u = 40$ kHz. (a) the discrete point source directivity and (b) the total array directivity for ultrasound; ——, $f_1 = 39$ kHz; ——, $f_2 = 41$ kHz. (c) ——, the product directivity of ultrasound; - - -, Westervelt's directivity. (d) The audio sound directivity. Extracted from [Zhong et al., 2023b, Fig. 4].

Table 6.2

The position and weight coefficients for the uniform and optimal array configurations presented in Figs. 6.21 and 6.23. Extracted from [Zhong et al., 2023b, Table II].

Element index, n	Position, x_n (mm)		Weight coefficients, w_n	
	Uniform array	Optimal array	Uniform array	Optimal array
1	−60	−60	1	1.99
2	−43	−49	1	1.17
3	−26	−39	1	1.15
4	−9	−28	1	0.92
5	9	−17	1	0.62
6	26	40	1	1.19
7	43	50	1	1.73
8	60	60	1	1.51

it is evident that the peaks in the two ultrasonic waves around 30° do not align, leading to further suppression of their product. In this specific configuration, the PSL is reduced to −15.5 dB, as illustrated in Fig. 6.23(d). When compared to the directivity achieved using the uniform array, this represents a substantial suppression of the grating lobe by 16.8 dB. These results affirm the feasibility of using the SA algorithm to effectively mitigate the grating lobes generated in a uniform array.

Figure 6.24 provides the optimized PSL as a function of the number of SA iterations, first after optimizing only the position and then after optimizing both the position and weight coefficients, while maintaining parameters consistent with those utilized in Fig. 6.23. These results have been derived from 100 trials, and from this point forward, the light blue and dark blue regions respectively represent the 5% to 95% and 25% to 75% empirical Gaussian quantiles of the PSL. The blue solid line signifies the median value. For computational reference, on a personal computer featuring an AMD Ryzen™ Threadripper™ 3960X CPU with 256 GB of RAM, the calculation time spans approximately 160 s for 10,000 iterations. It is apparent that the PSL converges toward smaller values in both cases. When optimizing solely the position, as shown in Fig. 6.24(a), the median and best results reached −12.9 dB and −14.4 dB, respectively. By optimizing both the position and weight coefficients, these values can be further reduced to −13.4 dB and −15.5 dB, respectively. While weight coefficients may not eliminate grating lobes, they do provide additional degrees of freedom to enhance suppression performance by reducing sidelobe levels [Murino et al., 1996]. Henceforth, all subsequent results aim to optimize both the position and weight coefficients for improved performance.

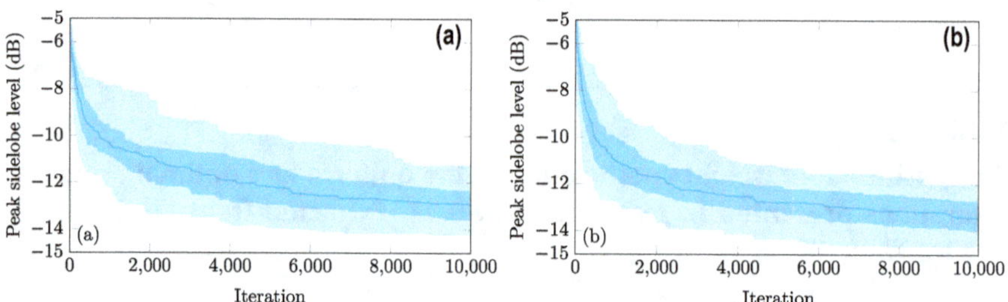

Figure 6.24 The optimized PSL as a function of the number of SA iterations based on 100 trials where the element number $N = 8$, the audio frequency $f_a = 2\,\text{kHz}$, and the average ultrasound frequency $f_u = 40\,\text{kHz}$. (a) optimizing only the position; (b) optimizing both the position and weight coefficients. Here and in the sequel, the light and dark blue regions indicate the 5% to 95% and 25% to 75% empirical Gaussian quantiles of the PSL, respectively. The blue solid line denotes the median value. Extracted from [Zhong et al., 2023b, Fig. 5].

6.4.4.1 Effects of the Audio Frequency

Figure 6.25 presents the statistical results of the PSL obtained after 10,000 iterations, derived from 100 independent trials at various audio frequencies ranging from 250 Hz to 8 kHz. The black solid line represents the PSL obtained using the uniform array. It is evident that the PSL generated by a uniform array generally decreases with increasing audio frequency, reaching its minimum at approximately 5.6 kHz. In contrast, the PSL generated by the optimal array decreases with increasing frequency and eventually stabilizes at around −20 dB.

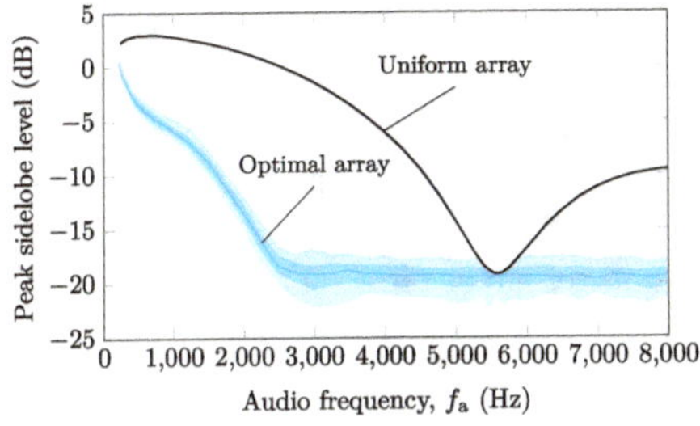

Figure 6.25 The statistical results of the PSL of the optimal array based on 100 trials after 10,000 iterations, where the element number $N = 8$, and the average ultrasound frequency $f_u = 40$ kHz. The black solid line represents the PSL obtained using the uniform array. Extracted from [Zhong et al., 2023b, Fig. 6].

Figure 6.26(a) displays the audio sound directivity at 1 kHz. In comparison to the result at 2 kHz in Fig. 6.23(d), it is noticeble that the PSL is only suppressed to −6.7 dB, indicating that it is more challenging to suppress grating lobes at lower frequencies. The underlying reason can be observed in Fig. 6.26(b), where the Westervelt's directivities are presented at various audio frequencies, all maintaining the same average ultrasonic frequency of 40 kHz. The half power beam width (HPBW) of the Westervelt's directivity expands from 7.3° to 10.3°, then to 14.6°, and finally to 20.6° as the frequency decreases from 8 kHz to 4 kHz, 2 kHz, and 1 kHz, respectively. This increase in HPBW at lower frequencies, as described by Eq. (2.53), leads to more pronounced aliasing effects, subsequently deteriorating the performance of grating lobe suppression. In contrast, within the high-frequency range, the Westervelt's directivity exhibits a smaller HPBW, allowing for more effective optimization of audio sound directivity.

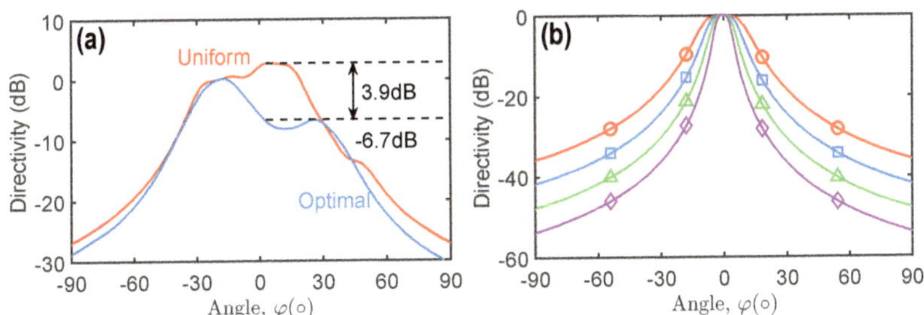

Figure 6.26 (a) The comparison of the audio sound directivity generated by a uniform array and an optimal array at the audio frequency $f_a = 1$ kHz, where the element number $N = 8$. (b) The Westervelt's directivity at different audio frequencies. The average ultrasound frequency $f_u = 40$ kHz. —o—, 1 kHz; —□—, 2 kHz; —△—, 4 kHz; —◇—, 8 kHz. Extracted from [Zhong et al., 2023b, Fig. 7].

An intriguing observation emerges when considering that the PSL generated by a uniform array reaches its minimum at approximately 5.6 kHz. With increasing audio frequency, the angular separation between the grating lobes generated by the two ultrasonic waves widens. Consequently, the product directivity of these two ultrasonic waves decreases, resulting in a lower level of audio sound directivity. This phenomenon is recognized as the grating lobe elimination phenomenon in a steerable PAL and is discussed comprehensively in Shi and Gan, 2011. However, the PSL generated by the optimal array, as depicted in Fig. 6.25, consistently remains lower than that achieved without optimization (uniform array) across the frequency spectrum ranging from 250 Hz to 8 kHz. This indicates that, for the parameters employed in Fig. 6.25, the uniform array never represents the optimal configuration. Hence, by harnessing the SA algorithm for array optimization, the suppression of grating lobes can consistently be achieved.

6.4.4.2 Effects of the Ultrasound Frequency

Figure 6.27 presents statistical results of the PSL for an optimal array after 10,000 iterations based on 100 independent trials conducted at various average ultrasound frequencies ranging from 30 kHz to 80 kHz. Notably, the PSL generally demonstrates an upward trend with increasing ultrasound frequency, regardless of whether it is a uniform or optimal array configuration. In Fig. 6.28(a), a comparison is drawn between the directivity curves achieved using uniform and optimal arrays when the ultrasound frequency is set at 50 kHz. Compared to Fig. 6.23(d), where the ultrasound frequency is 40 kHz, it becomes evident that the suppression of the grating lobe has diminished from 16.8 dB to 11.7 dB. This reduction can be explained by examining the Westervelt's directivity at various ultrasound frequencies, as depicted in Fig. 6.28(b). As the ultrasound frequency increases, the Westervelt's directivity becomes broader, as indicated by Eq. (2.53). The HPBW expands from 11.7° to 14.6°, then to 18.9°, and finally to 22.3° as the frequency increases from 30 kHz to 40 kHz, 60 kHz, and 80 kHz, respectively. This behavior is attributed to the generation of a narrow audio beam by the effective virtual source, which is influenced by the absorption length, $1/\alpha_{\mathrm{u}}$. At smaller absorption lengths, the effective length of

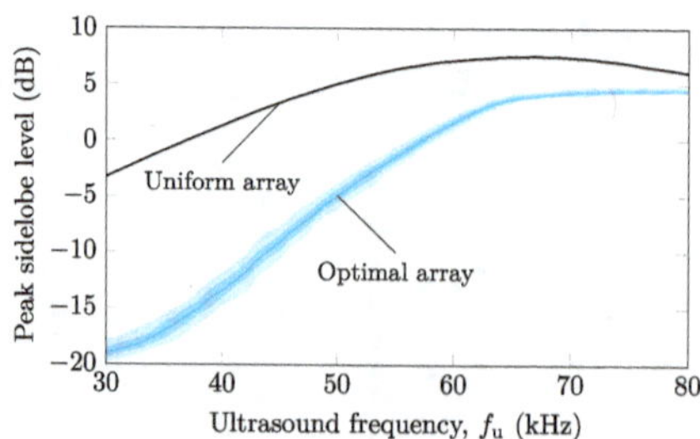

Figure 6.27 The statistical results of the PSL of the optimal array based on 100 trials after 10,000 iterations at different average ultrasonic frequencies, where the element number $N = 8$, and the audio frequency $f_{\mathrm{a}} = 2$ kHz. Extracted from [Zhong et al., 2023b, Fig. 8].

the virtual source remains limited, making it incapable of generating a narrow beam. Consequently, suppressing the grating lobe becomes more challenging at higher ultrasound frequencies.

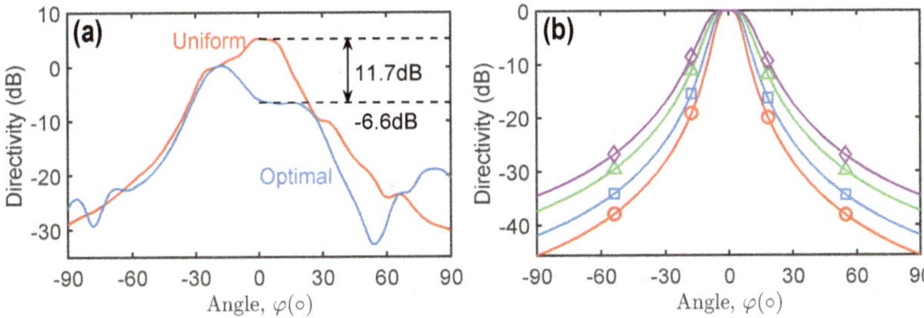

Figure 6.28 (a) The comparison of the audio sound directivity generated by a uniform array and an optimal array when the average ultrasound frequency $f_u = 50\,\text{kHz}$, where the element number $N = 8$. (b) The Westervelt's directivity at different audio frequencies. The audio frequency $f_a = 2\,\text{kHz}$. —o—, 30 kHz; —□—, 40 kHz; —△—, 60 kHz; —◇—, 80 kHz. Extracted from [Zhong et al., 2023b, Fig. 9].

6.4.4.3 Effects of the Number of Elements

Figure 6.29 showcases the statistical results of the PSL for both uniform and optimal arrays with varying numbers of elements, with an audio frequency of $f_a = 2$ kHz and an average ultrasound frequency of $f_u = 40\,\text{kHz}$. An intriguing observation arises as the PSL for the uniform array exhibits fluctuations with changing element numbers. It initially decreases as the number of elements increases, reaching a local minimum at $N = 21$, and then rises again until $N = 30$. This behavior differs from that of a uniform array composed of conventional loudspeakers, where the PSL is known to be independent of the number of elements. The explanation for this discrepancy becomes apparent when comparing Figs. 6.21 and 6.30. In Fig. 6.21, the directivities of both ultrasound and audio sound are presented for a uniform array with $N = 30$. Although the angular separation of the grating lobes around 40° for the two ultrasound waves remains consistent at 3.8°, these grating lobes become more sharply as

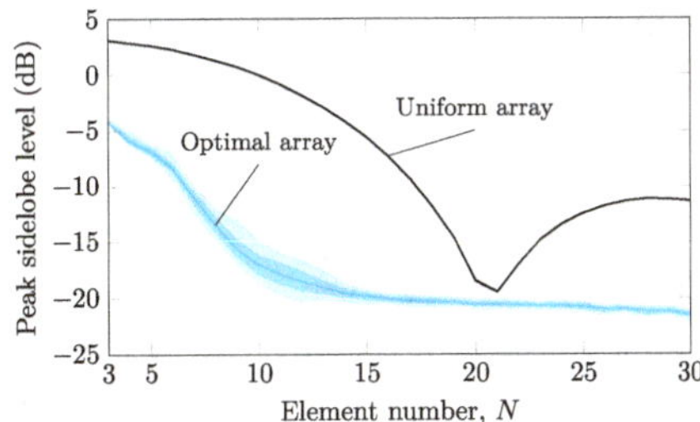

Figure 6.29 The statistical results of the PSL of the optimal array based on 100 trials after 10,000 iterations for the array consisting of N elements, where the audio frequency $f_a = 2\,\text{kHz}$ and the average ultrasound frequency $f_u = 40\,\text{kHz}$. Extracted from [Zhong et al., 2023b, Fig. 10].

the number of elements increases from 8 to 30. This sharpening effect can be derived from Eq. (6.20). Consequently, the product directivity diminishes when compared to Figs. 6.21(c) and 6.30(c). This phenomenon aligns with the concept of grating lobe elimination, even though it was not explicitly discussed in Shi and Gan, 2011. It notably contributes to the improved suppression performance observed in the optimal array, as evidenced in Fig. 6.29, where the PSL of the optimal array consistently decreases as the number of elements increases. Meanwhile, it is evident that the uniform array can consistently be optimized to achieve lower PSLs. For instance, as illustrated in Fig. 6.30(d), the PSL can be further reduced by 11.0 dB when the number of elements is $N = 30$.

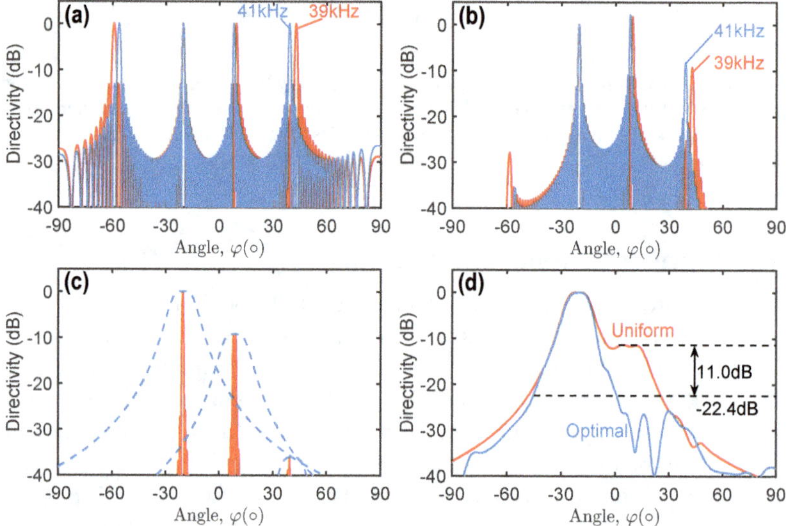

Figure 6.30 The directivities generated by a uniform array where the element number $N = 30$, the audio frequency $f_a = 2$ kHz, and the average ultrasound frequency $f_u = 40$ kHz. (a) the discrete point source directivity and (b) the total array directivity for ultrasound; ——, $f_1 = 39$ kHz; ——, $f_2 = 41$ kHz. (c) ——, the product directivity of ultrasound; - - -, Westervelt's directivity. (d) The audio sound directivity. Extracted from [Zhong et al., 2023b, Fig. 11].

The above results indicate the potential for enhancing suppression performance at low audio frequencies and/or high ultrasound frequencies, challenges that were previously demonstrated in Secs. 6.4.4.1 and 6.4.4.2. Figure 6.31 illustrates the audio sound directivity produced by an optimal array for two scenarios: one with 8 elements and the other with 30 elements. Upon comparing Figs. 6.31(a) and 6.26(a), it becomes evident that the PSL can be further reduced from −6.7 dB to −15.7 dB, marking a substantial improvement of 9.0 dB. This improvement is observed when the audio frequency is set at 1 kHz, and the average ultrasound frequency remains at 40 kHz. Similarly, a comparison between Figs. 6.31(b) and 6.28(a) reveals an even more significant reduction in PSL from −6.6 dB to −20.4 dB, signifying an impressive improvement of 13.8 dB. This improvement is observed when the audio frequency is set at 2 kHz, and the average ultrasound frequency is 50 kHz. In conclusion, augmenting the number of array elements has the potential to enhance the suppression of grating lobes at both low audio and high ultrasound frequencies.

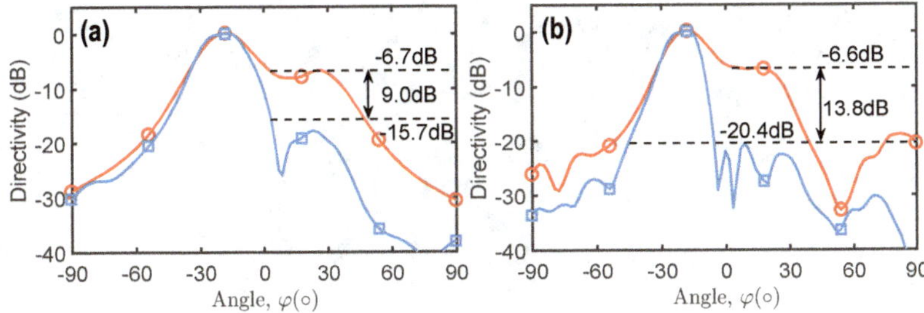

Figure 6.31 The comparison of the audio sound directivity generated by the optimal array when the element number $N = 8$ or 30. (a) the audio frequency $f_a = 1$ kHz and the average ultrasound frequency $f_u = 40$ kHz; and (b) the audio frequency $f_a = 2$ kHz and the average ultrasound frequency $f_u = 50$ kHz. —o—, $N = 8$; —□—, $N = 30$. Extracted from [Zhong et al., 2023b, Fig. 12].

6.4.5 EXPERIMENTAL VALIDATION

To empirically validate the theories presented in previous sections, eight steerable PALs were fabricated, employing both uniformly and optimally distributed elements to produce audio sound at either 1 kHz or 2 kHz. Two of these are illustrated in Figs. 6.32(a) and (b). Each PAL consists of either 8 or 16 channels, with each channel incorporating 8 circular ultrasonic emitters (Murata MA40S4S). These emitters possess a resonance frequency of 40 kHz and a diameter of 10 mm. Across all PALs, the average interelement spacing is maintained at two wavelengths of ultrasound, equivalent to $s = 17.15$ mm. For the PAL configurations with 8-channel at 2 kHz, the position and weight coefficients have been adopted from the simulation results presented in Table 6.2. Meanwhile, for the 8-channel array at 1 kHz, the 16-channel array at 1 kHz, and the 16-channel array at 2 kHz, these specific parameters are detailed in Tables 6.3, 6.4, and 6.5, respectively.

The implementation of the steerable PAL involves the utilization of a field-programmable gate array (FPGA, Xilinx XC7A100T, San Jose, CA), in conjunction with metal-oxide semiconductor field-effect transistor (MOSFET) drivers (Microchip MIC4127, Chandler, AZ). The FPGA is programmed to generate a set of multi-channel independent rectangular pulse signals, which are subsequently channeled into the corresponding MOSFET drivers [Zhong et al., 2022d]. The output from these drivers is connected to the positive terminals of the ultrasonic emitters. Each channel's signal comprises two pulse signals oscillating at two ultrasonic frequencies: 39 kHz and 40.98 kHz (or 39.06 kHz). This design choice ensures the generation of an audio sound wave at a frequency of 1.98 kHz (or 1 kHz). The FPGA boasts a clock rate of 50 MHz, enabling precise control over phase differences between channels through the application of high-resolution time shifts, with a resolution of 0.02 μs. Additionally, the weight coefficients for each channel can be effectively regulated via input voltage adjustments for the MOSFET drivers.

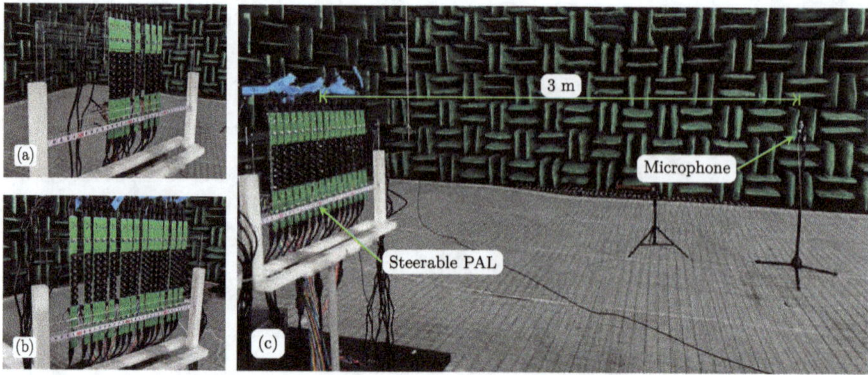

Figure 6.32 The prototypes of the (a) 8-channel and (b) 16-channel steerable PAL with optimally distributed elements generating audio sound at (a) 1 kHz and (b) 2 kHz, respectively, where each channel consists of 8 ultrasonic elements. (c) Photo of the experimental setup with a uniform 16-channel array. Extracted from [Zhong et al., 2023b, Fig. 13].

Table 6.3

The position and weight coefficients for the uniform and optimal 8-channel array configurations generating audio sound at 1 kHz.

Element index, n	Position, x_n (mm)		Weight coefficients, w_n	
	Uniform array	Optimal array	Uniform array	Optimal array
1	−60	−60	1	0.64
2	−43	−49	1	1.88
3	−26	−39	1	1.82
4	−9	0	1	0.53
5	9	30	1	0.47
6	26	40	1	0.93
7	43	50	1	1.54
8	60	60	1	1.34

The experiments were conducted in a full anechoic room with dimensions of 11.4 m × 7.8 m × 6.7 m (height). The relative humidity and temperature were 52% and 26°C, respectively. A photo of the experiment setup is presented in Fig. 6.32(c). It is important to note that the audio sound directivity for a PAL is typically defined within its inverse-law far field, a region typically situated more than 10 m away from the source (Sec. 3.2). However, conducting measurements at such a considerable distance can present challenges in experiments, including alignment errors and constraints imposed by the limited size of the anechoic room. As a practical compromise, a condenser microphone (Brüel & Kjær Type 4135) was used to measure the sound pressure at 3 m away from the PAL [Zhong et al., 2022d]. The signal is

Table 6.4
The position and weight coefficients for the uniform and optimal 16-channel array configurations generating audio sound at 1 kHz.

Element index, n	Position, x_n (mm)		Weight coefficients, w_n	
	Uniform array	Optimal array	Uniform array	Optimal array
1	−60	−129	1	1.85
2	−43	−118	1	1.82
3	−26	−106	1	1.65
4	−9	−86	1	1.00
5	9	−71	1	0.59
6	26	−56	1	0.54
7	43	−45	1	1.20
8	60	−32	1	0.45
9	60	23	1	0.29
10	60	58	1	0.35
11	60	75	1	0.58
12	60	86	1	1.75
13	60	97	1	1.91
14	60	107	1	1.77
15	60	118	1	1.91
16	60	129	1	1.57

conditioned (Brüel & Kjær Type 2690) and analyzed by a PULSE analyzer (Brüel & Kjær Type 3160). To avoid spurious sound induced by the intensive ultrasounds, the microphone was covered by a piece of thin plastic film.

Figure 6.33 presents the directivity patterns generated by both uniform and optimal array configurations with 8 channels, with measurements taken in the angular range from −90° to 90° and a resolution of 1°. In Fig. 6.34, corresponding results for the 16-channel PALs are displayed. To facilitate comparison, the measured results have been normalized to the SPL at the main lobe. For a uniform 8-channel array generating audio sound at 2 kHz, as depicted in Fig. 6.33(b), it is noteworthy that the directivity exhibits three distinct local peaks at −20°, 9°, and 40°, with magnitudes of 0 dB, −0.9 dB, and −15.6 dB, respectively. In contrast, the simulation results shown in Fig. 6.21(d) predict three local peaks at −20°, 9.1°, and 45.7°, with magnitudes of 0 dB, 1.3 dB, and −20.8 dB, respectively. While the magnitudes of the two grating lobes do not precisely match those from the simulation results, it is evident that the predicted locations of these three local peaks align quite well with the experimental measurements.

The directivity measured from the optimal 8-channel array configuration Fig. 6.33(b) reveals an 8 dB reduction in the grating lobe at 9°. This enhancement can be elucidated by examining the measured ultrasound directivities for both the

Table 6.5
The position and weight coefficients for the uniform and optimal 16-channel array configurations generating audio sound at 2 kHz.

Element index, n	Position, x_n (mm)		Weight coefficients, w_n	
	Uniform array	Optimal array	Uniform array	Optimal array
1	−60	−129	1	1.25
2	−43	−111	1	1.98
3	−26	−95	1	1.05
4	−9	−71	1	0.62
5	9	−56	1	0.30
6	26	−44	1	1.03
7	43	−33	1	1.22
8	60	−18	1	1.81
9	60	2	1	0.24
10	60	19	1	1.03
11	60	31	1	1.35
12	60	56	1	0.30
13	60	74	1	1.42
14	60	85	1	1.13
15	60	102	1	1.81
16	60	129	1	1.54

uniform and optimal array configurations, as displayed in Figs. 6.33(d) and (f), respectively. In the case of the uniform array, it is evident that the peaks around 9° in the two ultrasound directivities closely coincide with each other, resulting in a significant product directivity that contributes to the presence of a grating lobe in this direction. In contrast, for the optimal array shown in Fig. 6.33(f), the sparse array design leads to a null in one of the ultrasound directivities around 9°, precisely where the grating lobe from the other ultrasound directivity emerges. This observed behavior aligns well with the simulation results presented in Figs. 6.21(b) and 6.23(b).

At the lower audio frequency of 1 kHz, the performance of sidelobe reduction exhibits a decline, as evident in Fig. 6.33(a). In comparison to the performance at 2 kHz, the PSL increases from −8.8 dB to −3.1 dB. This behavior aligns with the simulation results presented in Figs. 6.25 and 6.26, confirming the deterioration in performance as the audio frequency decreases. This degradation in performance is similarly observed in the case of the 16-channel array, as seen by comparing Figs. 6.34(a) and (b), where the PSL of the audio directivity increases from −12.4 dB to −9.6 dB as the audio frequency decreases from 2 kHz to 1 kHz. The underlying reason for this trend is the broadening of the Westervelt's directivity at lower audio frequencies. Fortunately, this deterioration in performance can be mitigated to some extent by increasing the number of array elements. By comparing Figs. 6.33(a) and 6.34(a), it is

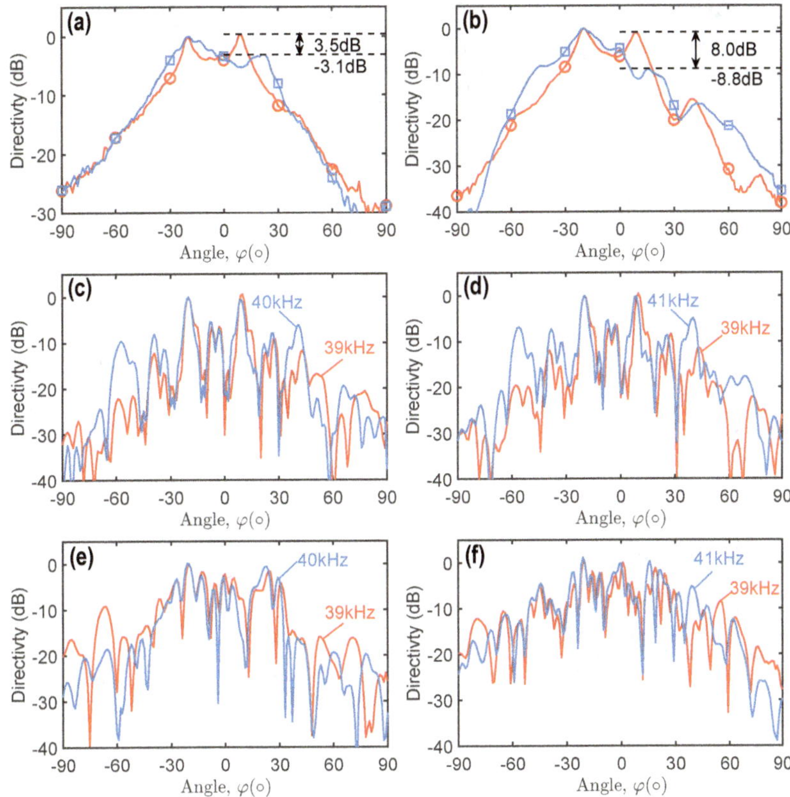

Figure 6.33 The directivities measured at 3 m generated by 8-channel steerable PALs. The ultrasound directivities generated by the (c) uniform and (e) optimal array for the audio sound directivity shown in (a) at 1 kHz. The ultrasound directivities generated by the (d) uniform and (f) optimal array for the audio sound directivity shown in (b) at 2 kHz. —○— , uniform array; —□— , optimal array. Extracted from [Zhong et al., 2023b, Fig. 14].

evident that doubling the number of elements results in a reduction of the PSL from −3.1 dB to −9.6 dB at 1 kHz.

The experimental results discussed above validate the suppression of sidelobes and grating lobes in the audio sound directivity of a PAL through the utilization of a sparse random array. However, it is essential to note that the achieved performance falls short of the predicted results. For example, the measured PSL of the optimal 8-channel array at 2 kHz, as shown in Fig. 6.33(b), is −8.8 dB, whereas the prediction in Fig. 6.23(d) suggests a PSL of −15.5 dB. Several factors could contribute to this disparity. Firstly, the direct CMD employed in this case for determining the far field directivity may be inaccurate at larger aperture sizes and audio frequencies. This inaccuracy arises due to the neglect of the aperture effect of audio sound [Zhong et al., 2023d]. Secondly, the validity of the convolution model is restricted to 2D physical models, assuming that the array dimension without phase modulation is significantly

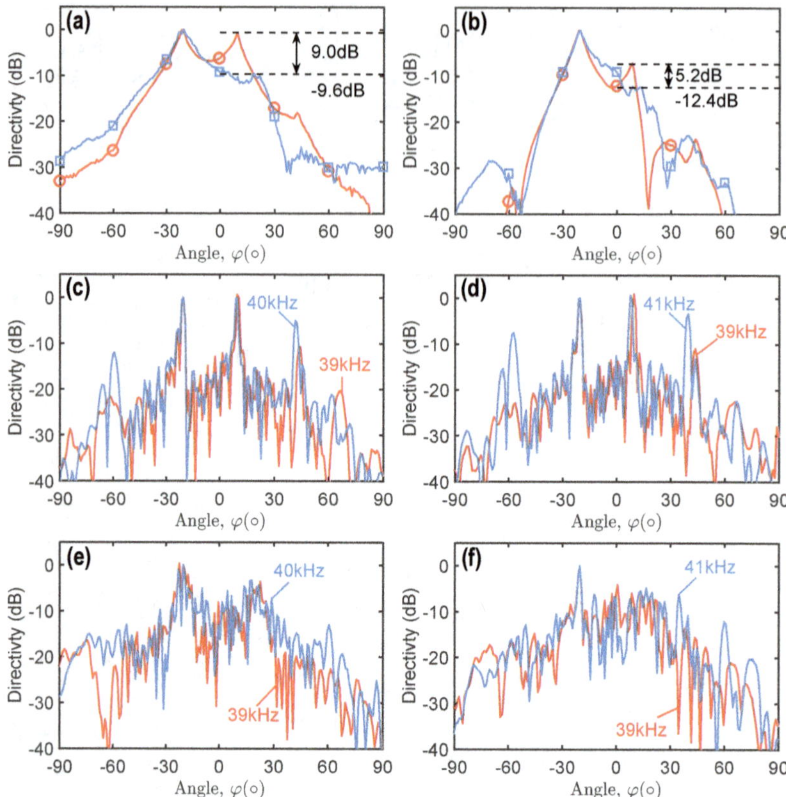

Figure 6.34 The directivities measured at 3 m generated by 16-channel steerable PALs. The ultrasound directivities generated by the (c) uniform and (e) optimal array for the audio sound directivity shown in (a) at 1 kHz. The ultrasound directivities generated by the (d) uniform and (f) optimal array for the audio sound directivity shown in (b) at 2 kHz. —○— , uniform array; —□— , optimal array. Extracted from [Zhong et al., 2023b, Fig. 15].

larger than both the ultrasound and audio sound wavelengths [Zhong et al., 2023d]. However, the dimension of the experimental prototypes (8 cm) is smaller than the latter (e.g., 34.3 cm at 1 kHz). Furthermore, audio sound directivity generated by a PAL is conventionally defined in the inverse-law far field, typically situated more than 10 m away from the source (Sec. 3.2). Measuring the directivity at such distances poses challenges due to alignment errors and the limited space within the anechoic room. Consequently, directivity measurements were conducted at a distance of only 3 m during the experiments. It is worth noting that the audio sound can exhibit a broader directivity in the near field, as demonstrated in Sec. 3.2. A promising avenue for improvement might involve optimizing the directivity in the near field, but this currently incurs a substantial computational cost. Lastly, discrepancies in the amplitude and phase responses among ultrasonic emitters in each channel can also contribute to errors. These issues require further in-depth research.

6.4.6 REMARKS

Suppression of grating lobes in a steerable PAL holds significant practical relevance, primarily because the interelement spacing in phased array PALs is often greater than half the wavelength of ultrasound. This section introduces optimal designs for mitigating sidelobes generated by a steerable PAL, employing a sparse random array technique akin to its previous applications in antenna [Kurup et al., 2003] and medical ultrasound [Goss et al., 1996] arrays. The optimization process involves fine-tuning the position and weight coefficients of the array elements using the SA algorithm. The ultimate goal is to minimize the peak sidelobe level in the directivity pattern. While the SA method is utilized as the optimization algorithm in this section, it is important to highlight that alternative algorithms can also be applied, including genetic algorithms [Xue et al., 2022], differential evolution algorithms [Murino et al., 1996], and particle swarm optimization [Khodier and Christodoulou, 2005].

This investigation demonstrates that sidelobes, including grating lobes, generated by a steerable PAL can be effectively suppressed, even if the interelement separation does not meet the Nyquist criterion. However, it is worth noting that the suppression performance degrades at lower audio and higher ultrasound frequencies due to the influence of the Westervelt's directivity. Nevertheless, this performance can be enhanced by increasing the number of array elements. Sidelobe suppression is achieved by optimally spacing the array elements and finely weighting the emitted waves. This strategic arrangement ensures that the emitted sound waves arrive randomly at the grating lobes while maintaining in-phase arrivals at the main lobe.

While the effects of ultrasound frequency have been explored through simulations (as illustrated in Fig. 6.28), experimental verification remains pending due to limitations in the response of the ultrasonic emitters. Future research endeavors might aim to explore this aspect further, especially considering that various ultrasonic frequencies, such as 60 kHz and 80 kHz. It is important to note that the positions and weight coefficients of array elements are optimized specifically for a particular audio frequency in this section. To address sidelobe levels in the context of a wideband audio signal, the wideband directivity can be employed as the objective function in Eq. (6.23). While the sparse random array technique remains applicable to wideband signals, it is crucial to acknowledge that the suppression performance may be affected by additional optimization constraints.

6.5 BEAM FOCUSING

6.5.1 PROBLEM DESCRIPTION

The key of the focusing PAL is to focus all ultrasonic waves onto a target focal point so that the audio sound around this point is amplified. There are various kinds of ways to realize a focusing PAL, such as convex and concave lens [Lucas et al., 1983], phased arrays [Ochiai et al., 2017], and gradient index phononic crystals [Červenka and Bednařík, 2021]. All of them aim to manipulate the wave propagation so that all wave fronts of ultrasound arrive at the focal point at the same time, where the largest constructive interference between waves happen.

Suppose the sound is generated by a baffled planar source with the velocity profile given by Eq. (2.28). For an arbitrary source point $\mathbf{r}_\mathrm{s}$ on the source plane, the elapsed time when the wavefront emitted from this point arrives at the focal point is $|\mathbf{r}_\mathrm{f}-\mathbf{r}_\mathrm{s}|/c_0$, where $\mathbf{r}_\mathrm{f}$ denotes the location of the focal point. The velocity profile needs to compensate this time delay to realize the beam focusing so that

$$v_{i,z}(\mathbf{r}_\mathrm{s}) = v_0 \exp\left(-\mathrm{i}\Re k_i |\mathbf{r}_\mathrm{f}-\mathbf{r}_\mathrm{s}|\right). \tag{6.26}$$

In this section, a circular PAL with a radius of a is used to investigate the physical properties of a focusing PAL and the sketch is shown in Fig. 6.35. For simplicity, the focal point is set on the radiator axis to achieve the axisymmetric focusing effect. A Cartesian coordinate system (x,y,z) is established with its origin, O, at the center of the PAL, and the positive z-axis pointing to the focal point. The spherical coordinate system (r,θ,φ) is established with respect to the Cartesian coordinates for further calculations, where r, θ, and φ are the radial distance, zenithal angle, and azimuthal angle, respectively. In such a case, the velocity profile given by Eq. (6.26) is simplified as

$$v_{i,z}(\mathbf{r}_\mathrm{s}) = v_0 \exp\left(-\mathrm{i}\Re(k_i)\sqrt{r_\mathrm{s}^2+F^2}\right). \tag{6.27}$$

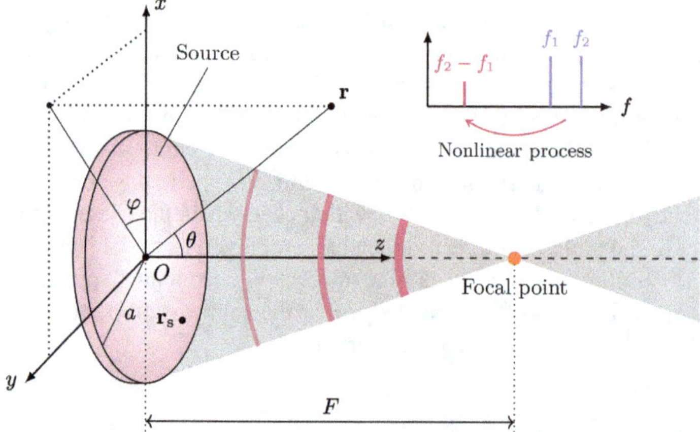

Figure 6.35 Sketch of a circular PAL focusing on a focal point located on the radiator axis. Extracted from [Zhong et al., 2022d, Fig. 1].

6.5.2 INFLUENCE OF LOCAL EFFECTS

For a conventional PAL, the generated ultrasonic waves are collimated because $k_i a \gg 1$. They can be approximated by plane waves, so the local nonlinear effects are negligible unless at the locations close to the PAL as analyzed in Chap. 3. However, this approximation is violated and the influence of local effects need to be investigated at first.

Figure 6.36 compares the SPL for audio sound generated by a PAL using the Westervelt and Kuznetsov equations at 250 Hz, 500 Hz, and 1 kHz, with a focal length of 0.2 m, 2 m, and without focusing. It is noted that the local effects are included in the modeling based on the Kuznetsov equation, while neglected in the Westervelt equation as discussed in Sec. 1.3. The dashed lines in the figures denote the transition distance (1.16 m) from the near field to the Westervelt far field. When the wave propagation is modeled by the Westervelt equation, the audio sound can be seen as the generation of an infinitely large virtual volume source with the source density proportional to the product of two ultrasound pressure $p_1^*(\mathbf{r})p_2(\mathbf{r})$ as evident by Eq. (1.106). The integration over such a virtual source represents the cumulative effects, which smooth the calculated sound pressure as shown in Fig. 6.36. It is observed that the difference obtained using two equations is large in the near field but negligible in the Westervelt far field for all cases.

This can be explained by analyzing the axial sound pressure and the Lagrangian density of the ultrasound at 40 kHz, as shown in Fig. 6.37. When the Kuznetsov equation is used, the Lagrangian density of ultrasound, representing the local (non-cumulative) effects, is considered, which can be considered as an algebraic correction to the results obtained using the Westervelt equation. Because the ultrasound significantly fluctuates in the near field as shown in Fig. 6.37, highly oscillatory audio sound pressure can be observed in the near field as shown in Fig. 6.36. The transition distance is the location of the first maxima of the ultrasound SPL, where the Lagrangian density also reaches around its maximum. Therefore, the difference obtained using the two equations is negligible beyond the transition distance.

In the near field, the ultrasound pressure is large and fluctuates significantly, so the local effects are strong which cannot be captured by the Westervelt equation. In the Westervelt far field, the Westervelt equation is accurate enough even for the focusing PAL. For example, the SPL difference obtained using the two equations is less than 0.1 dB and 0.4 dB at 250 Hz when the focal length is 0.2 m and 2 m, respectively. For a focusing PAL, the source density of audio sound is increased around the focal point as it is proportional to the product of the local ultrasound pressure. The cumulative effects captured by the Westervelt equation are therefore increased. For example, the SPL obtained using the Westervelt equation at 500 Hz increases from 51.8 dB to 60.8 dB after the focusing on the focal point of 0.2 m. However, it is interesting to note that the focusing behavior influences more on the local effects, resulting in a large increment in the SPL difference obtained using two equations near the focal point. For example, the SPL difference at 0.2 m is increased from 0.2 dB to 17.3 dB at 500 Hz after the focusing. At lower audio frequencies, this increment is more prominent. As the frequency decreases from 1 kHz to 500 Hz and 250 Hz, the SPL difference increases from 3.4 dB to 17.3 dB and 29.3 dB at 0.2 m, respectively.

When the focal length is 0.2 m, it can be found in Figs. 6.36(b), (e), and (h) that there is a peak at 0.2 m in SPL for both equations which corresponds to the focal

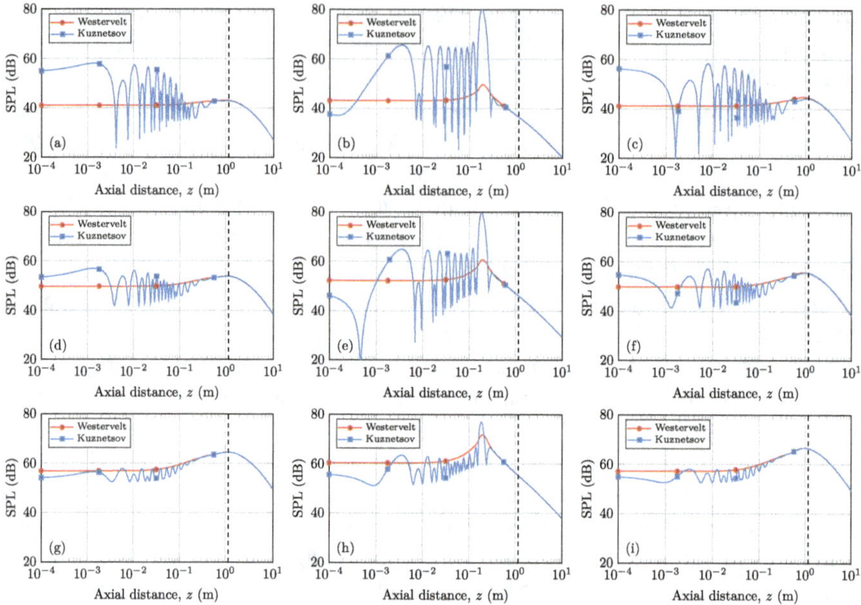

Figure 6.36 Axial audio SPL obtained using the Westervelt and Kuznetsov equations without focusing (left column) and with a focal length of 0.2 m (middle column) and 2 m (right column) at 250 Hz (top row), 500 Hz (middle row), and 1 kHz (button row). —●—, Westervelt equation; —■—, Kuznetsov equation; - - - , the transition distance. Extracted from [Zhong et al., 2022d, Fig. 2].

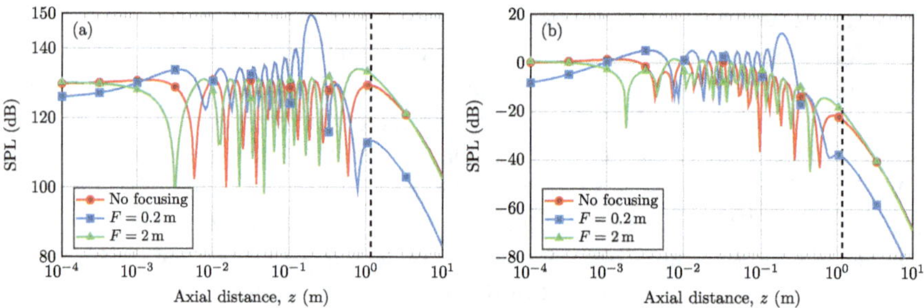

Figure 6.37 Axial (a) SPL and (b) the level of normalized Lagrangian density at 40 kHz, with a focal length of 0.2 m and 2 m, and without focusing. The dashed line denotes the transition distance. Extracted from [Zhong et al., 2022d, Fig. 3].

point. However, when the PAL is focusing on 2 m which is beyond the transition distance, Figs. 6.36(c), (f), and (i) show that the SPL trend is like that of without focusing. It is observed in Fig. 6.37 that the ultrasound pressure reaches the last maxima around 0.8 m instead of 2 m. This is known as the focal shift phenomenon due to the diffraction limit, and the location of the true focal point can be estimated by [Makov et al., 2008, Eq. (15)]

$$z_{\text{true}} = \frac{3\pi\mathscr{D}_{\text{c}}}{2\pi N_{\text{F}} + \sqrt{\pi^2 N_{\text{F}}^2 + 72}}, \tag{6.28}$$

where $\mathscr{D}_{\text{c}} = a^2/\lambda_i$ is the critical distance and the *Fresnel number* is defined as

$$N_{\text{F}} = \frac{\mathscr{D}_{\text{c}}}{F}. \tag{6.29}$$

The focusing PAL is unable to focusing the ultrasound beyond the transition distance. In the limiting case when the focal length is infinity, the ultrasound pressure reaches the last maxima at the transition distance. Therefore, the audio SPL difference obtained using two equations would be insignificant beyond the transition distance even if the focal point is located there. For example, as shown in Fig. 6.36 when the focal length is 2 m, the SPL difference beyond the transition distance is below 0.4 dB, 0.2 dB, and 0.1 dB at 250 Hz, 500 Hz, and 1 kHz, respectively. It means that the Kuznetsov equation must be used when the field point is located before the transition distance, and the results presented in Lucas et al., 1983 for underwater PAAs cannot be used for investigating a focusing PAL where the paraxial approximation of the Westervelt equation was adopted. Therefore, all the subsequent results are obtained using the Kuznetsov equation.

6.5.3 FOCUSING AUDIO SOUND FIELD

The audio sound generated by a conventional PAL without focusing and a focusing PAL with a focal length of 0.2 m and 2 m at 250 Hz, 500 Hz, and 1 kHz are presented in Fig. 6.38. It is clear that the audio beam generated by a conventional PAL is narrow and decays slowly along the radiator axis at all frequencies. For a focusing PAL with a focal length of 0.2 m, the energy is focused on the focal point. Beyond this point, the audio beam decays more rapidly than that without focusing. It shows an energy transfer from the far field to the near field. This is a merit for some applications of PALs. For a conventional PAL, the reflection of the audio sound is strong due to slow decay rate and its interference with the incident sound is annoying when it is used in a room, as discussed in Sec. 4.2. For a focusing PAL with a focal length of 2 m, the audio sound is slightly amplified when compared to that generated by a conventional PAL. The reason is that the ultrasound wave is ineffectively focused beyond the transition distance, which is 1.16 m calculated by Eq. (3.9). Instead, the sound pressure at the true focal point (0.8 m) is increased.

Figure 6.39 shows the audio SPL on the radiator axis ($\theta = 0$) generated by a PAL with and without focusing at 250 Hz, 500 Hz, 1 kHz, and 2 kHz. In the near field, the audio sound field fluctuates with the axial distance in all cases. The audio sound around the focal point is generally amplified when compared to that without focusing. For example, when the PAL is focused on 0.3 m, the SPL is increased from 42.1 dB to 64.8 dB at the focal point when the audio frequency is 250 Hz. It is observed that as the focal point becomes closer to the radiation surface, the SPL

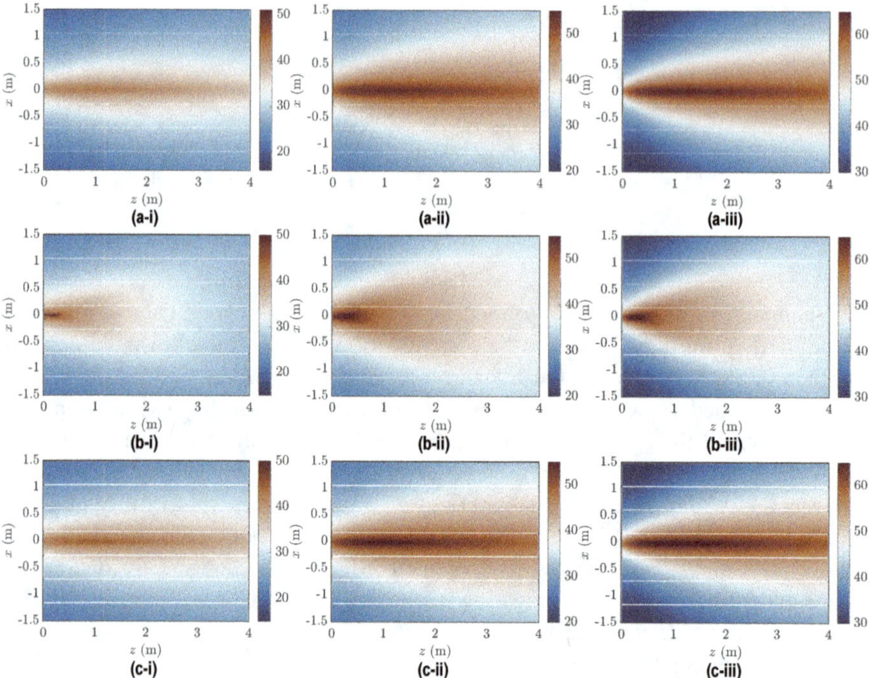

Figure 6.38 Audio SPL distributions generated by a conventional PAL (a) without focusing and a focusing PAL with a focal length of (b) 0.2 m and (c) 2 m on its radiator axis, at (i) 250 Hz, (ii) 500 Hz, and (iii) 1 kHz. Extracted from [Zhong et al., 2022d, Fig. 4].

increment around the focal point becomes larger. For example, the SPL increment increases from 1.9 dB to 41.0 dB when the focal point changes from 1 m to 0.2 m at 250 Hz. This is because the *focusing gain*, defined by the ratio of the Rayleigh distance to the focal length, expressed as [Jing et al., 2011]

$$G = \frac{\mathscr{D}_{\mathrm{R}}}{F}, \tag{6.30}$$

becomes larger resulting in a larger ultrasound field at the focal point [Makov et al., 2008]. Therefore, both the cumulative and local effects given are more significant around the focal point. It is also noted that the amplification is more prominent at lower audio frequencies. The reason is that the audio sound pressure determined by the cumulative effects drops about 12 dB as the audio frequency is halved, while that determined by the local effects changes a little. Although the contribution from the cumulative effects is neglected, that from the local effects is retained.

In the Westervelt far field, there is no fluctuations for the audio SPL, and it decreases as the axial distance increases. It can be found that the audio sound generated by a focusing PAL is smaller than that by a conventional PAL without focusing. As the focal point becomes closer to the radiation surface, the audio SPL is smaller in

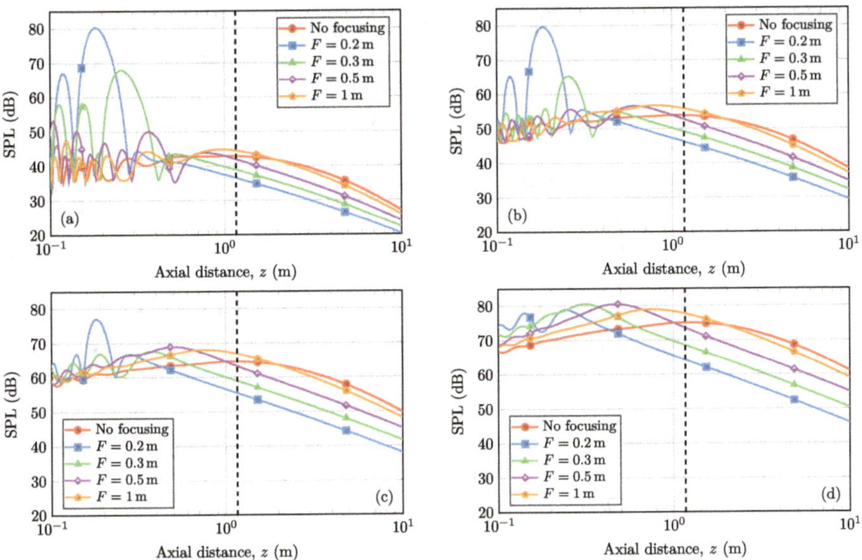

Figure 6.39 Axial audio SPL generated by a PAL with and without focusing at (a) 250 Hz, (b) 500 Hz, (c) 1 kHz, and (d) 2 kHz. - - - , transition distances. Extracted from [Zhong et al., 2022d, Fig. 5].

the Westervelt far field. For example, the audio SPL generated by a conventional PAL without focusing is 61.1 dB at 10 m and 2 kHz, while it respectively decreases to 46.0 dB, 50.4 dB, 55.0 dB, and 59.0 dB, when the focal length is 0.2 m, 0.3 m, 0.5 m, and 1 m. The ultrasound pressure on the radiator axis ($\rho = 0$) and the focal plane ($z = F$) can be respectively approximated by [Lucas and Muir, 1982, Eqs. (19) and (24)]

$$p_i(\rho = 0, \varphi, z) = \frac{p_0}{1 - z/F}\left[1 - \mathrm{e}^{\mathrm{i}\mathscr{D}_{\mathrm{R}}(1/z - 1/F)}\right]\mathrm{e}^{\mathrm{i}k_i z}, \tag{6.31}$$

$$\text{and } p_i(\rho, \varphi, z = F) = -\mathrm{i}p_0 G \operatorname{jinc}(k_i a \rho / F)\mathrm{e}^{\mathrm{i}k_i F}, \tag{6.32}$$

where the jinc function is defined by Eq. (2.84). As the focal point becomes closer to the radiation surface, the focal length decreases and the focusing gain becomes larger. Therefore, the ultrasound pressure in the far field decreases according to Eq. (6.31), while it increases in the focal plane according to Eq. (6.32). It demonstrates that more energy is transferred from the far field to the near field as the focusing gain becomes larger.

Figure 6.40 shows the SPL increment, which is defined as the SPL with focusing minus the SPL without focusing, at the focal point after the focusing as a function of the focal length at different audio frequencies. The SPL increment generally increases with some fluctuations as the focal point moves close to the PAL. The increment is smaller at higher audio frequencies. The reason is that the audio sound pressure caused by the cumulative effects increases by approximately 12 dB as the audio frequency is doubled, while the local effects change a little at different audio frequencies. Consequently, the influence of the local effects is weaker at higher audio frequencies. As the focal length increases, the SPL increment decreases rapidly at all frequencies resulting from smaller focusing gains. When the focal point is beyond the transition distance, the ultrasound is inefficiently focused on the focal point, and the SPL increment decreases and approaches to 0 dB. In the limiting case when the focal length is infinity, the radiation can be seen as a conventional PAL without focusing. The SPL increment can even be negative especially at low audio frequencies indicating that the SPL is decreased at the focal point after the focusing. For example, the SPL increment is −6 dB when the focal length is 0.5 m at 250 Hz, and the SPL as a function of the axial distance can be found in Fig. 6.39(a) for this case. This phenomenon is caused by the destructive interference between the cumulative and local effects. At small focal lengths, the local effects are dominant, while at large focal lengths, the cumulative are dominant. When the cumulative and local effects are comparable for a moderate focal length, the destructive interference between them can happen resulting in an audio SPL decrement.

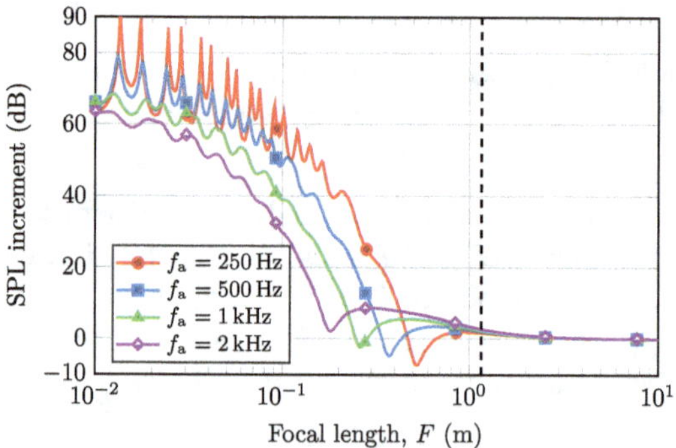

Figure 6.40 The SPL increment at the focal point after the focusing as a function of the focal length at 250 Hz, 500 Hz, 1 kHz, and 2 kHz. - - - , transition distance. Extracted from [Zhong et al., 2022d, Fig. 6].

The conventional PAL suffers from the poor low-frequency response because the SPL decreases about 12 dB as the frequency is halved. The above analysis shows the low-frequency audio sound can be amplified near the listener by using the focusing PAL with the focal point being around the listener. The mechanism is that focusing the ultrasound on the listener improves both the cumulative and local effects, so the audio sound field is amplified. However, it should be noted that there exists a focal shift phenomenon which means the true focal location with the maximal sound pressure shifts relative to the geometric focal point. In addition, the audio SPL at the focal point might be decreased at the focal point after the focusing. Therefore, the focusing PAL should be designed according to a specific audio application.

6.5.4 EXPERIMENTAL OBSERVATION OF THE BEAM FOCUSING

6.5.4.1 Prototype Fabrication

To validate the above findings by experiments, a prototype of a circular focusing PAL with a radius of 0.1 m was fabricated as shown in Fig. 6.41. The prototype consists of 367 circular ultrasonic emitters (Murata MA40S4S [Murata Manufacturing, 2017]), which has a resonant frequency of 40 kHz and a radius of 5 mm. The ultrasonic emitters are arranged in a compact hexagonal array as shown in Fig. 6.41(b) to minimize the interelement separation. To approximate the axisymmetric phase distribution as assumed in the theory, the emitters with the same distance to the centroid of the PAL are grouped together as a channel and assigned the same phase. There are 37 channels in total and the channel number is marked inside each emitter in Fig. 6.41(b). The distance between the centroid of the PAL and all emitters allocated to a same channel, as well as the number of emitters for each channel, are listed in Table 6.6.

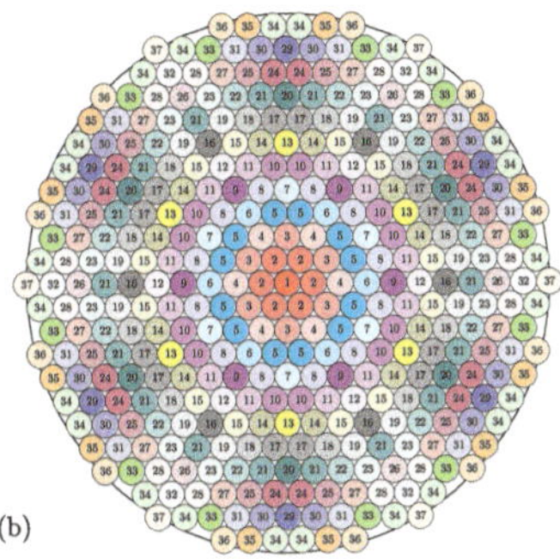

Figure 6.41 The prototype of a 37-channel circular focusing PAL with a radius of 0.1 m consisting of 367 circular ultrasonic emitters with a radius of 5 mm: (a) photo; (b) sketch, where the number inside the small circles denotes the channel number. Extracted from [Zhong et al., 2022d, Fig. 7].

The focusing PAL is constructed with an FPGA (Xilinx XC7A100T, San Jose, CA) and MOSFET drivers (Microchip MIC4127, Chandler, AZ). The logical diagram of the circuits is the same as Fig. 8 in Marzo et al., 2018. The FPGA is programmed to generate 37-channel independent rectangular pulse signals which are fed into the MOSFET drivers. The output of them is connected to the positive pins of ultrasonic emitters. The generation of the pulse signal for one channel is illustrated in Fig. 6.42, where two ultrasonic frequencies at 40.064 kHz and 39.936 kHz are used to generate an audio sound wave at 128 Hz. The FPGA has a clock rate of 50 MHz, which means the clock cycle (the duration of one pulse) is 0.02 μs.

Based on the clock signal, two signals are generated first. The first signal aims to radiate the ultrasound at 40.064 kHz. A sequence of rectangle signals is generated with the fundamental frequency of 25 MHz, and the total duration is 12.48 μs. Followed by them is a sequence of logic low lasting for 12.48 μs. This signal has the spectral components at 40.064 kHz (= 50 MHz/1248) and 25 MHz with higher harmonic components. Because the ultrasonic emitters have narrow bandwidths centered at the resonant frequency, only the frequency component at 40.064 kHz is

Table 6.6

The number of emitters in each channel and the distance to the centroid of the PAL. Extracted from [Zhong et al., 2022d, Table I].

Channel number	Number of emitters	Distance to the centroid ($\times$ 5 mm)
1	1	0
2	6	2
3	6	$2\sqrt{3} \approx 3.464$
4	6	4
5	12	$2\sqrt{7} \approx 5.292$
6	6	6
7	6	$4\sqrt{3} \approx 6.928$
8	12	$2\sqrt{13} \approx 7.211$
9	6	8
10	12	$2\sqrt{19} \approx 8.718$
11	12	$2\sqrt{21} \approx 9.165$
12	6	10
13	6	$6\sqrt{3} \approx 10.392$
14	12	$4\sqrt{7} \approx 10.583$
15	12	$2\sqrt{31} \approx 11.136$
16	6	12
17	12	$2\sqrt{37} \approx 12.166$
18	12	$2\sqrt{39} \approx 12.490$
19	12	$2\sqrt{43} \approx 13.115$
20	6	$8\sqrt{3} \approx 13.856$
21	18	14
22	12	$4\sqrt{13} \approx 14.422$
23	12	$2\sqrt{57} \approx 15.100$
24	12	$2\sqrt{61} \approx 15.620$
25	12	$6\sqrt{7} \approx 15.875$
26	6	16
27	12	$2\sqrt{67} \approx 16.371$
28	12	$2\sqrt{73} \approx 17.088$
29	6	$10\sqrt{3} \approx 17.321$
30	12	$4\sqrt{19} \approx 17.436$
31	12	$2\sqrt{79} \approx 17.776$
32	6	18
33	12	$4\sqrt{21} \approx 18.330$
34	24	$2\sqrt{91} \approx 19.079$
35	12	$2\sqrt{93} \approx 19.287$
36	12	$2\sqrt{97} \approx 19.698$
37	6	20

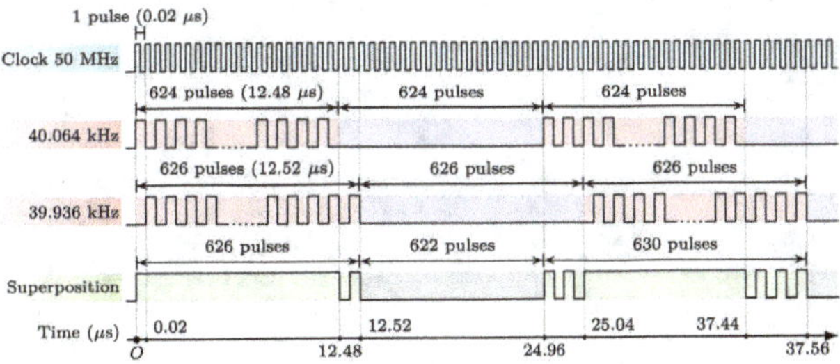

Figure 6.42 Timing diagram of signals for generating the audio sound at 128 Hz. Extracted from [Zhong et al., 2022d, Fig. 8].

radiated by the emitter [Marzo et al., 2018]. The same process is applied for the second signal aiming to radiate the ultrasound at 39.936 kHz. The superposition of the above two signals is fed into the ultrasonic emitters, and the ultrasound waves at two frequencies of 40.064 kHz and 39.936 kHz are therefore generated. The amplitude of the generated sound can be controlled by the input voltage for MOSFET drivers. The phase difference between channels can be realized by the time shifting of signals with a high resolution of only 0.02 μs (50 MHz). The advantage of this technique is that no digital-analog converters and the implementation of signal processing algorithms are required. The FPGA can also be replaced by inexpensive microcontrollers [Hahn et al., 2021; Marzo et al., 2018].

6.5.4.2 Experimental Setup

The experiments were conducted in a full anechoic room with dimensions of 11.4 m × 7.8 m × 6.7 m (height). The relative humidity and temperature were 68% and 13°C, respectively. The photo of the experiment setup is presented in Fig. 6.43. A condenser microphone Brüel & Kjær Type 4135 was used to measure the sound pressure. The signal is conditioned by a Brüel & Kjær Type 2690 conditioner and analyzed by a PULSE analyzer (Brüel & Kjær Type 3160). To avoid spurious sound induced by the intensive ultrasound, the microphone was covered by a piece of small and thin plastic film. As shown in Fig. 6.43, the sound field in a rectangular area was measured with dimensions of 1 m × 2 m. The PAL surface is perpendicular to the measurement plane. The microphone was mounted at the same height of the PAL center, and it is scanned by three stepper motors which are controlled by personal computers.

6.5.4.3 Measured Results

The measurement results in the rectangular area and the corresponding simulation results at 512 Hz without focusing and with a focal length of 0.2 m and 1 m are presented in Fig. 6.44. Since the sound field is more complicated in the near field, the spacing between the measurement points was set to 1 cm in the region of

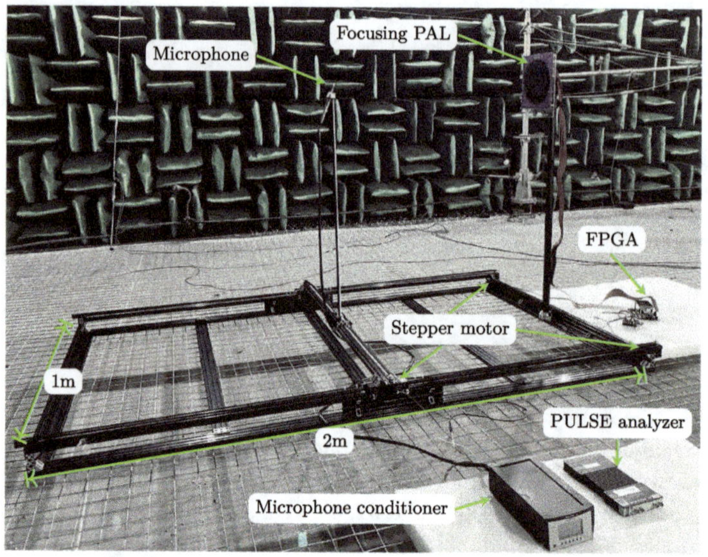

Figure 6.43 Photo of the experiment setup with a focusing PAL in a full anechoic room. Extracted from [Zhong et al., 2022d, Fig. 9].

$-0.15\,\text{m} \leq x \leq 0.15\,\text{m}$ and $0 \leq z \leq 0.5\,\text{m}$, while the spacing of 4 cm was chosen in other regions. The predictions are in good agreement with the measured sound field.

The audio sound is highly directional for a conventional PAL without focusing, and it is focused in the near field after the focusing. It is noted the location of the true focal location which has the maximal SPL when the focal length is 1 m is shifted to about 0.66 m. This is due to the focal shift phenomenon and can be estimated from Eq. (6.28). However, some discrepancies between measurements and simulation results are observed at the location close to the radiation surface and the true focal location. This might be caused by the spatial aliasing effect as the size of the ultrasonic emitters (10 mm) exceeds one half of the ultrasonic wavelength (4.3 mm). In addition, misalignments and individual deviations of magnitude and phase responses of ultrasonic emitters may also affect the generated sound field [Marzo et al., 2018].

Figure 6.45 compares the measurements and simulation results of the audio SPL along the radiator axis ($\theta = 0$) at 512 Hz and 1024 Hz with a focal length of 0.2 m, and at 512 Hz with a focal length of 1 m. In Figs. 6.45(a) and (b), although the focusing on the focal point is well predicted by the simulation, the measurements are lower and smoother than predictions. This can be attributed to the fact that the sound waves are insufficiently focused due to the discretized shaded phased array rather than a continuous profile given by Eq. (6.26). The acoustic filter used for removing the spurious sound also reduces audio component resulted from the local effects to some extent due to the reduction of the ultrasound pressure at the measurement point, but this does not affect the main discovery. Moreover, the highly oscillating

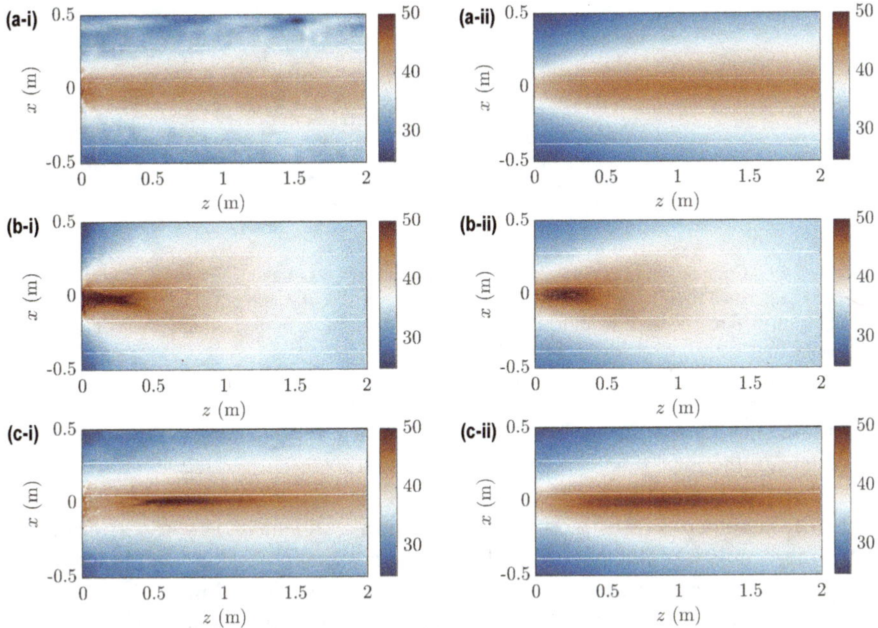

Figure 6.44 (i) Measurements and (ii) simulation results of the audio sound field at 512 Hz, (a) without focusing, and with a focal length of (b) 0.2 m and (c) 1 m. Extracted from [Zhong et al., 2022d, Fig. 10].

behavior of the sound field close to the radiation surface were not fully captured in the measured data which might because the size of the microphone (0.64 cm) is larger than the spacing of adjacent peaks of the predicted curve. The generated sound field changes rapidly even with a small deviation from the radiator axis. Figure 6.45(c) shows that the predictions at $\theta = 1.5^\circ$ agree better with measurements than the predictions at $\theta = 0$ representing the radiator axis. It means the imperfect positioning of the PAL and the microphone could result in large measurement errors. If the measurement plane is imperfectly perpendicular to the PAL surface, the true focal point deviates more from the measurement plane as the focal point moves farther away, so the measurement error is larger with a larger focal length. Nevertheless, a focusing PAL has been made in the laboratory, and the audio sound field can be well predicted by the methods presented in this book.

For comparison, the measured audio SPLs generated by a conventional PAL without focusing are also presented in Fig. 6.45. It can be observed that the audio SPL around the focal point is increased in all cases. For example, Fig. 6.45(b) shows that the measured increment of the audio SPL at the focal point (0.2 m) after the focusing is 7.1 dB at 1024 Hz. When the audio frequency decreases to 512 Hz, as shown in Fig. 6.45(a), the measured SPL increment at the focal point increases to 15.7 dB. The increment due to focusing is larger at lower audio frequencies. As discussed in Sec. 6.5.3, this is because the audio sound pressure determined by cumulative effects drops about 12 dB as the audio frequency is halved, while that determined by

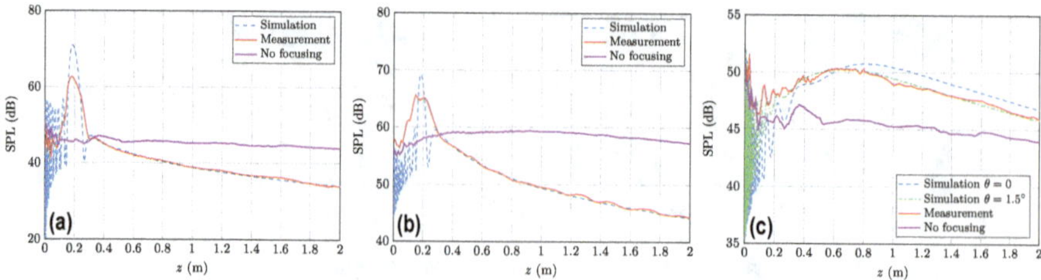

Figure 6.45 Comparison of measurements and simulation results of audio SPL along the radiator axis (a) at 512 Hz with a focal length of 0.2 m; (b) at 1024 Hz with a focal length of 0.2 m; (c) at 512 Hz with a focal length of 1 m. The thick solid lines represent the measurements of the audio SPL without focusing. Extracted from [Zhong et al., 2022d, Fig. 11].

local effects changes only a little at different audio frequencies. Consequently, the influence of the local effects is more significant at lower audio frequencies.

When the focal point moves to 1 m, as shown in Fig. 6.45(c), measurement shows the SPL increment is smaller and the maximal increment is only 4.7 dB at 0.6 m. The improvement of the audio sound response is larger when the focal point moves closer to the radiation surface. However, the sound pressure at almost all points ranging from 0 to 2 m is amplified, which agrees with the simulation results presented in Fig. 6.39. It is also observed that the audio SPL decays more rapidly at large distances after focusing. For example, the audio SPL at 2 m decreases from 44.0 dB and 57.3 dB, to 33.9 dB and 44.2 dB, at 512 Hz and 1024 Hz, respectively, after focusing with a focal length of 0.2 m. This can be an advantage in applications because the reflections of the audio beam would be smaller if there exists a reflecting surface behind the listener.

6.5.5 REMARKS

The concept and the realization of a focusing PAL are introduced in this section. The results show that the focusing PAL has the advantage of improving the low-frequency response. However, it is noted that there are safety concerns on intensive ultrasonic exposures caused by the PAL. The exposure would be more intensive around the focal point. Existing guidelines for the safe usage of airborne ultrasound cannot be readily applied for a PAL because the ultrasound waves emitted by the PAL has a relatively narrow spectral band [Gan et al., 2012]. Some studies pointed out that the temporary hearing loss should not occur for levels below 140 dB when exposed to the sufficiently high-frequency airborne ultrasound [Pompei, 2002], but the specifically allowed ultrasound exposure level for the PAL is still unclear at present (see Sec. 8.5 for details). The ultrasound level is measured around 127 dB at the focal point for the cases in Figs. 6.45(a) and (b), where the focal length is 0.2 m. In real applications, one can reduce the focusing gain and/or the ultrasound pressure level to reduce the ultrasound exposure.

6.6 SUMMARY

The steerable and focusing PALs bring more possibilities and flexibilities in audio applications. The steerable PAL has the advantage of steering the highly directional audio beam to a desired direction without the need to mechanically rotate the source, while the focusing PAL benefits improving the low-frequency response around the focal point.

In Sec. 6.1, a brief introduction on both steerable and focusing PALs is presented. Then the 2D steerable PAL is analyzed and discussed in Sec. 6.2, which is usually realized using a linear phased array PAL and enables the beam steering in one dimension. In Sec. 6.2.3, the far field directivities generated by a 2D steerable PAL are modeled using the the modified CDM. The modified CDM is unable to predict the audio sound in the near field, which is then resolved in Sec. 6.2.4 using the CWE. Section 6.3 aims to present the 3D counterpart of the steerable PAL, which can steer the directional audio beam to both azimuthal and zenithal directions. The far field directivities generated by a 3D steerable PAL are presented in Sec. 6.3.2. The 3D circular steerable PAL is modeled using the SWE and discussed in Sec. 6.3.3. All simulation results demonstrate the capability of steerable PALs to electronically steer the highly directional audio beam to a specified direction. However, the implementation of the steerable PAL using arrays of ultrasonic emitters gives rise to grating lobes due to the fact that typical emitter sizes are larger than the ultrasound wavelength. These grating lobe issues and their associated suppression techniques are addressed in Sec. 6.4.

In Sec. 6.5, the audio sound field generated by a focusing PAL is investigated both numerically and experimentally. The computationally efficient method, i.e., the SWE, described in Sec. 2.6 is used to calculate the quasilinear solution of both the Westervelt and Kuznetsov equations. The accuracy of the generated sound field using the Westervelt equation is initially examined in Sec. 6.5.2, and the results are then compared with those obtained using the Kuznetsov equation. It is found that the focusing on the true focal location deteriorates the prediction accuracy using the Westervelt equation due to the strong and complicated local effects. To validate the numerical results, a 37-channel focusing PAL prototype was constructed and the performance is presented in Sec. 6.5.4. Both the simulation and experimental results show that the audio sound pressure around the true focal location is increased, and the increment becomes larger at low audio frequencies and when the focal point moves close to the radiation surface, which is attributed to the strong local effects in the near field. It has been shown that a focusing PAL can improve the poor low-frequency response for a conventional PAL without focusing, and the generated audio sound decays rapidly with the distance in the far field when compared to that without focusing.

7 Active Noise Control with PALs

7.1 INTRODUCTION

7.1.1 ANC SYSTEMS AND SPILLOVER EFFECTS

Noise pollution is becoming more serious nowadays [Hogan and Latshaw, 1973], and exposures to high levels of noise would cause health issues to human beings, such as the hearing loss, cardiovascular diseases, cognitive impairment, and tinnitus [Organization, 2011; Münzel et al., 2014]. Noise can be mitigated at the noise source, along the wave propagation path, and at the human ears (receivers), where different noise control techniques can be applied.

Passive noise control approaches include installation of enclosures covering noise sources [ISO 15667, 2000; ISO 11546-1, 2009], building sound barriers and insulation walls in highways, around airports, and construction sites [Kurze, 1974; Tong et al., 2015], and wearing earplugs and earmuffs around the human ears [Gerges, 2012]. The size of the devices and/or materials used in passive control techniques (e.g., porous absorbers [Allard and Atalla, 2009; Padhye and Nayak, 2016] and Helmholtz resonators [Cai and Mak, 2016]) is related to the wavelength of noise to be controlled and is usually large at low frequencies, so their applications are limited in the low-frequency range due to the weight and volume constraints. *Active noise control (ANC)* is a method to mitigate noise at target regions by introducing additional sound sources (called *secondary sources*) [Nelson and Elliott, 1992; Qiu, 2019]. It provides an alternative solution to control the low-frequency noise. The theory and physical mechanism of ANC methods have been well established and investigated, and successful applications include the ANC systems in headphones [Ang et al., 2017], headrests in cars [Jung et al., 2018; Elliott et al., 2018], domestic windows enabling natural ventilation [Lam, 2019; Lam et al., 2020; Lam et al., 2021].

Figure 7.1 shows a typical structure of an ANC system. In practice, the unwanted noise is often time varying, so an adaptive controller is required to process the real time audio signals. The reference sensors (e.g., microphones, tachometers, or accelerometers) are placed near the noise source (called the *primary source*) to capture the noise signal, which is called the *reference signal*. The error sensors (e.g., microphones) are used to obtain the residual noise level at the error locations. The reference signals are filtered by control filters in the controller to generate the real-time anti-noise signal for secondary sources, resulting in a noise reduction at error locations. There are various kinds of algorithms to obtain the control filters. For example, the filtered-x least mean square (FxLMS) algorithm is designed to minimize the sum of the square of sound pressure at all error locations [Elliott, 2000]. It has

DOI: 10.1201/9781003354994-7

low computational cost, stable performance, and commonly employed in commercial ANC controllers.

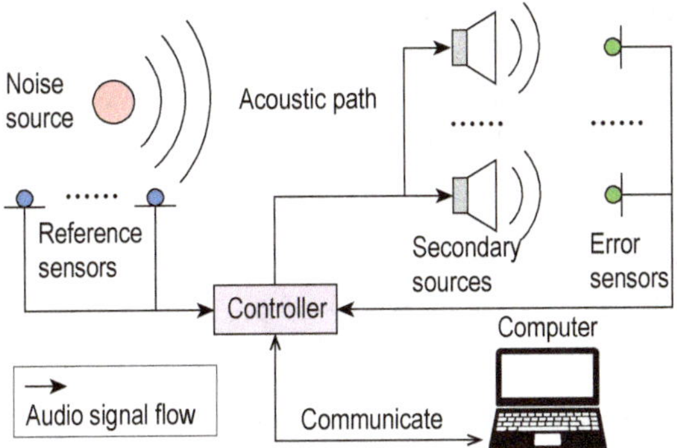

Figure 7.1 Structure of a multi-channel ANC system.

There are two strategies of ANC systems, namely global and local control [Nelson and Elliott, 1992]. The global control aims to reduce the total sound power radiated from the noise and secondary sources [Boodoo et al., 2015]. Although the radiation of the system can be globally mitigated when the distance between secondary and noise sources is small, the noise reduction performance deteriorates significantly when the distance is larger than half wavelength corresponding to the frequency at which the noise to be controlled [Zhong et al., 2019a; Zhong et al., 2019b; Zhong et al., 2020a]. When global control cannot be achieved, local control can be used to mitigate the noise in a particular region and form a so-called *quiet zone*, which is defined as the region where the noise reduction is larger than 10 dB [Guo and Pan, 1998; Elliott and Cheer, 2015; Shi et al., 2020]. When conventional loudspeakers are used as secondary sources in ANC systems, the noise in some other areas outside the quiet zone can be amplified because conventional loudspeakers are approximately omnidirectional at low frequencies. This phenomenon is referred to as the *spillover effect* in the literature [Guo et al., 1997; Kidner et al., 2006; Tanaka and Tanaka, 2010]. The spillover effect is annoying in practical applications. For example, when a quiet zone is created in the driver's seat inside a car cabin, the occupants sitting in other seats experience noise amplification [Jung, 2018; Cheer, 2012]. It is therefore desirable to manipulate the secondary sound waves so that they propagate only in the direction of the quiet zone.

The spillover effect makes the ANC system complicated. For example, for the binaural ANC system presented in Tanaka et al., 2017, two secondary sources are introduced to mitigate the noise at two ears. The secondary field at each ear is the superposition of the sound waves radiated by two secondary loudspeakers, so there are two crosstalk secondary paths between loudspeakers and the contralateral ears. If the secondary source is omnidirectional, the crosstalk secondary paths need to be taken into account in control algorithms, which increases the computational cost in ANC controllers.

7.1.2 ANC USING DIRECTIONAL LOUDSPEAKERS

The sound waves radiated by directional sources can focus on one direction, so have small effects in other directions. Owing to this feature, using directional sources in ANC systems can mitigate the spillover effect [Mangiante, 1977; Chen et al., 2011; Hu and Tang, 2019] as well as reduce the crosstalk secondary paths [Tanaka et al., 2014; Tanaka et al., 2017]. PALs are a special kind of loudspeakers, which has sharp radiation directivity when compared to conventional loudspeakers.

The type of secondary source in an ANC system is crucial for delivering high levels of noise reduction [Bolton et al., 1995; Qiu and Hansen, 2000]. It is common to use a single monopole, or an array of monopoles, as secondary sources in ANC systems. However, directional sources can be used as secondary sources in multiple channel ANC systems to improve the control performance. For example, tripole secondary sources with a cardioid radiation pattern have been used to reduce noise source radiation [Mangiante, 1977]. In 2011, Chen et al. designed a unidirectional source consisting of two closely located loudspeakers with pre-adjusted phase difference [Chen et al., 2011]. For an active noise barrier system in their work, both numerical and experimental results from 160 Hz to 300 Hz show that the noise reduction performance can be improved significantly by replacing monopole sources with unidirectional sources.

In 2019, Hu and Tang proposed a directional source consisting of a central circular core enclosed within an annulus. It has been demonstrated that the proposed source is much more directional than a conventional piston source, even though its size is much smaller than the latter if the magnitude and the phase of the two parts are optimally designed [Hu and Tang, 2019]. Hu and Tang used these directional sources to cancel the noise generated by a finite length coherent line source. The numerical results show that their proposed directional sources can make the ANC system more compact and improve noise reduction performance, when compared to conventional piston sources [Hu and Tang, 2019]. These directional secondary sources have been chosen because they radiate only in the direction of the target region and have less effect on the other areas, so can reduce the effects of noise amplification outside of the quiet zone.

In 2005, Brooks et al. investigated the feasibility of using PALs in ANC systems as secondary sources [Brooks et al., 2005]. They pointed out two concerns in practical applications which might adversely affect the performance of such ANC systems with PALs. The first is that the frequency response of a PAL is poor at low frequencies, so the noise reduction performance deteriorates significantly as the frequency decreases. The second is that the demodulated audio sound in air varies quickly as a function of time, so the performance controlling wideband noise is degraded. It is noted that the low-frequency response of PALs may be improved by focusing the sound beams [Lucas et al., 1983; Saito and Muir, 1986; Saito and Tsubono, 1998; Červenka and Bednařík, 2021].

The quiet zone is usually generated around a fixed error microphone for an ANC system. To overcome this limitation, Kidner et al. proposed an ANC system using PALs to create a moving local quiet zone with the aid of virtual sensing techniques

[Kidner et al., 2006]. Their results show that the quiet zone created by the PAL can be extended further in the radial direction when compared to a conventional piston source. The reason is that the sound wave generated by a PAL decays slowly, resulting in a better matching between the primary (noise) and secondary sound fields. Recently, an alternative was proposed to sense the error signal remotely, where a custom-made low-mass membrane pick-up from a retroreflective film and a laser Doppler vibrometer were used to form a remote sensing apparatus to determine the acoustic information with minimum obstructions to the person [Zhong et al., 2020f]. Their results also show that such an ANC system using one PAL can achieve similar overall noise reductions up to 6 kHz at the ear location.

Despite the advantages of using PALs in ANC systems, the target area is limited along the radiation direction of the PAL due to its sharp directivity. To cope with this problem, a steerable PAL employing the phased array technique was proposed by Tanaka and Tanaka in 2010 [Tanaka and Tanaka, 2010]. Both numerical and experimental results show similar noise reduction performance can be achieved around the target position while the noise amplification (spillover effect) in the surrounding areas is negligible for the case with the PAL. This successful realization of the steerable PAL is an attractive technique for tracking control of a moving target without rotating the PAL mechanically.

Tanaka and Tanaka continued their work and proposed a focused PAL to achieve a mathematically trivial global control of the sound radiated by a point source [Tanaka and Tanaka, 2011]. It is known that the global control is limited when the separation of the primary and secondary sources is larger than the half wavelength of the sound (e.g., 0.17 m at 1 kHz) [Nelson and Elliott, 1992]. However, they obtained global control with a focused PAL even when the distance between primary and secondary sources is larger than the half wavelength. The global sound power control using PALs in an ANC system has also been investigated numerically by Ye et al. [Ye et al., 2012]. They developed a framework to analyze the sound power output by a PAL in ANC systems. A useful conclusion is that the minimization of the total power output of the system is the same as the minimization of the power output only by the PAL due to its sharp directivity. But the power output of the PAL was calculated based on a far field solution in [Ye et al., 2012], which is not accurate.

PALs have been used in multi-channel ANC systems. For example, a 2-channel ANC system using PALs was proposed to reduce the binaural factory noise at human ears in 2017 [Tanaka et al., 2017]. Two advantages of using PALs were demonstrated in multi-channel ANC systems. First, the noise at target points can be reduced while the sound pressure in other areas is not affected. Second, the crosstalk between secondary paths is negligible so the crosstalk cancellation technique is not required. The quiet zone created by multiple PALs has been investigated recently [Zhong et al., 2022e], where simple empirical formulas are presented to estimate the size of the quiet zone. The results show that the size of the quiet zone generated by PALs is similar to that observed with conventional omnidirectional loudspeakers while the effects of using PALs on the sound field outside the target zone is much smaller due to their sharp radiation directivity and slow decay rate along the propagation distance.

The length-limited PAL has been used for ANC. Shi and Gan proposed a theoretical framework based on the KZK equation for a length-limited PAL that has a concentrically-nested structure to cancel the noise in an ANC system [Shi and Gan, 2013]. The speaker is made up of an inner hexagonal array of 36 emitters with a center frequency of 40 kHz and an outer annulus of 36 emitters with a center frequency of 25 kHz. Based on this framework, Lam et al. investigated the feasibility of using a length-limited PAL in an ANC system [Lam et al., 2014]. Their experimental results show that a noise source at 1.5 kHz in a pipe is reduced by 18.4 dB at 8 m. A preliminary numerical study on the formation of a quiet zone using length-limited PAL was also presented in Wang et al., 2019.

7.2 ANC USING ONE PAL

7.2.1 PROBLEM DESCRIPTION AND THEORETICAL FRAMEWORK

The noise pressure distribution in space is often termed as the *primary field* in the realm of ANC, and it is denoted as $p_\mathrm{p}(\mathbf{r},t)$ in this book. Here, $\mathbf{r}=(x,y,z)$ are coordinates of the field (observation) point and t is the time variable. In this book, the time harmonic field is considered, i.e., $p_\mathrm{p}(\mathbf{r},t)=p_\mathrm{p}(\mathbf{r})\mathrm{e}^{-\mathrm{i}\omega t}$, and the time harmonic term, $\mathrm{e}^{-\mathrm{i}\omega t}$ is omitted for simplicity. The primary field can be generated by arbitrary sound sources. The simplest case is that generated by a point monopole located at $\mathbf{r}_\mathrm{p}=(x_\mathrm{p},y_\mathrm{p},z_\mathrm{p})$ as

$$p_\mathrm{p}(\mathbf{r})=Q_\mathrm{p}Z(\mathbf{r},\mathbf{r}_\mathrm{p}). \tag{7.1}$$

Here, Q_p is the *source strength* of the primary source, which has dimensions of volume velocity. $Z(\mathbf{r},\mathbf{r}_\mathrm{p})$ is the transfer function between $\mathbf{r}$ and $\mathbf{r}_\mathrm{p}$, which is determined by the acoustic environment. In free space, the transfer function is

$$H(\mathbf{r},\mathbf{r}_\mathrm{p})=\frac{-\mathrm{i}\rho_0\omega}{4\pi|\mathbf{r}-\mathbf{r}_\mathrm{p}|}\mathrm{e}^{\mathrm{i}k|\mathbf{r}-\mathbf{r}_\mathrm{p}|}. \tag{7.2}$$

Here, ρ_0 is the ambient static density, $\omega=2\pi f$ is the angular frequency, f is the frequency, and k is the wavenumber.

The primary objective of the single-channel ANC system is to introduce another loudspeaker, known as the *secondary source* [Nelson and Elliott, 1992; Elliott, 2000] or *control source* [Hansen et al., 2012], to mitigate the noise at a target point (*error point*). The secondary pressure field generated by the secondary source is expressed as

$$p_\mathrm{s}(\mathbf{r})=Q_\mathrm{s}Z(\mathbf{r},\mathbf{r}_\mathrm{s}). \tag{7.3}$$

Here, Q_s is the source strength of the secondary source. $Z(\mathbf{r},\mathbf{r}_\mathrm{s})$ is the transfer function between $\mathbf{r}$ and and the coordinates of the secondary source, i.e., $\mathbf{r}_\mathrm{s}=(x_\mathrm{s},y_\mathrm{s},z_\mathrm{s})$.

When conventional loudspeakers are used as secondary sources, they are typically modeled as point monopoles [Nelson and Elliott, 1992; Elliott, 2000]. Consequently, the source strength Q_s has dimension of volume velocity and the transfer function has the same form of Eq. (7.2). When PALs are used as secondary sources, the equivalent source strength is determined by the on-surface ultrasound pressure, so it typically takes the form of [Tanaka and Tanaka, 2010, Eq. (16)]

$$Q_s = p_{10}^* p_{20}. \tag{7.4}$$

Here, p_{i0} is the on-surface ultrasound pressure at frequency f_i, $i = 1, 2$. Other forms are also used to include the on-surface vibration velocity amplitude, $v_{10}^* v_{20}$ [Zhong et al., 2022e]. The relations are $p_{i0} = \rho_0 c_0 v_{i0}$. The transfer function is then

$$Z(\mathbf{r}, \mathbf{r}_s) = p_a(\mathbf{r}) / (p_{10}^* p_{20}). \tag{7.5}$$

Here, $p_a(\mathbf{r})$ is the audio sound pressure generated by the PAL and can be calculated using various kinds of algorithms as discussed in Chap. 2.

At an arbitrary observation point $\mathbf{r}$, the total audio sound pressure contributed by both primary and secondary sources is

$$p_t(\mathbf{r}) = p_p(\mathbf{r}) + p_s(\mathbf{r}) = p_p(\mathbf{r}) + Q_s Z(\mathbf{r}, \mathbf{r}_s). \tag{7.6}$$

To minimize the sound pressure at the error point, it is equivalent to minimize the squared pressure amplitude [Hansen et al., 2012, Eq. (8.2.6)]

$$\mathcal{J} = |p_t(\mathbf{r}_e)|^2 + \beta_0 |Q_s|^2. \tag{7.7}$$

Here, $\mathbf{r}_e = (x_e, y_e, z_e)$ are coordinates of the error point. A positive real number β_0 is commonly introduced to constrain the output of the secondary source [Kirkeby et al., 1996]. Substituting Eq. (7.6) into Eq. (7.7) yields

$$\mathcal{J} = Q_s^* A Q_s + Q_s^* B + B^* Q_s + C, \tag{7.8}$$

where coefficients are

$$A = Z^*(\mathbf{r}_e, \mathbf{r}_s) Z(\mathbf{r}_e, \mathbf{r}_s) + \beta_0, \quad B = Z^*(\mathbf{r}_e, \mathbf{r}_s) p_p(\mathbf{r}_e), \quad C = |p_p(\mathbf{r}_e)|^2. \tag{7.9}$$

Equation (7.8) shows that the cost function is a real quadratic function of the secondary source with a complex source strength of Q_s. Furthermore, it is clear that $A > 0$. The cost function then achieves the minimal at the optimal source strength of [Hansen et al., 2012, Eq. (8.2.12)]

$$Q_{s,opt} = -A^{-1} B = -\frac{Z^*(\mathbf{r}_e, \mathbf{r}_s) p_p(\mathbf{r}_e)}{Z^*(\mathbf{r}_e \mathbf{r}_s) Z(\mathbf{r}_e \mathbf{r}_s) + \beta_0}. \tag{7.10}$$

The minimal cost function (squared acoustic pressure amplitude) at the error point is found to be [Hansen et al., 2012, Eq. (8.2.14)]

$$\mathcal{J}_{opt} = C + B^* Q_{s,opt} = C - B^* A^{-1} B. \tag{7.11}$$

The total audio sound field under the control is then

$$p_t(\mathbf{r}) = p_p(\mathbf{r}) + Z(\mathbf{r}, \mathbf{r}_s) Q_{s,opt}. \tag{7.12}$$

7.2.2 CONTROL PERFORMANCE

7.2.2.1 ANC Using a Conventional PAL

The above theory demonstrates that the primary sound pressure at the error point can be mitigated by introducing another source. However, the sound pressure in other regions is not guaranteed to be mitigated. In practical applications, the noise in some other regions can unfortunately be amplified, which is referred to the spillover effect in the literature [Guo et al., 1997; Kidner et al., 2006; Tanaka and Tanaka, 2010].

To illustrate the spillover effect, Fig. 7.2 is presented showing the physical configuration of a single-channel ANC system where a conventional omnidirectional source is used as the secondary source. When a secondary source (denoted by a cross mark) is used to cancel the noise radiated by a point source (denoted by a circle mark) at an error location (denoted by a square mark), the quiet zone is formed around the error location. The spillover effect can be observed, showing that the sound pressure in some other areas is amplified due to the omnidirectional sound waves generated by the conventional secondary loudspeakers. This is undesirable in real applications. Moreover, this might increase the instability of the ANC system when reference signal is used, which might be contaminated by the secondary source signal received by the reference sensors [Tanaka and Tanaka, 2010]. As shown in Fig. 7.3, placing the secondary source close to the target point can mitigate the increase of total energy in the other areas [Joseph et al., 1994; David and Elliott, 1994], but it reduces the quiet zone size [Guo and Pan, 1998].

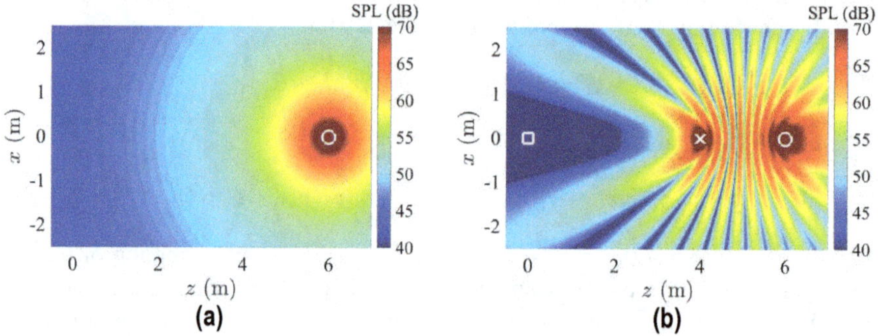

Figure 7.2 Sound fields generated by a point source located at $(x,z) = (0, 6\,\text{m})$ at 1 kHz: (a) primary (noise) field; (b) the noise at the $(x,z) = (0,0)$ is controlled by introducing a secondary omnidirectional point source at $(x,z) = (0, 4\,\text{m})$. Circle, primary source; cross, secondary source; square, error location.

After replacing the omnidirectional secondary source in Fig. 7.3 by a PAL, the noise reduction performance is shown in Fig. 7.4. By comparing it to Figs. 7.2 and 7.3, it is clear that the spillover effect is reduced.

The spillover effect with a single-channel ANC system has been observed in experiments as shown in Fig. 7.5 [Tanaka and Tanaka, 2010]. In the experiments, as shown in Figs. 7.5(b) and (c), the primary sound pressure at the error point is suppressed by 14.4 dB and 13.7 dB by using a point monopole and a PAL, respectively.

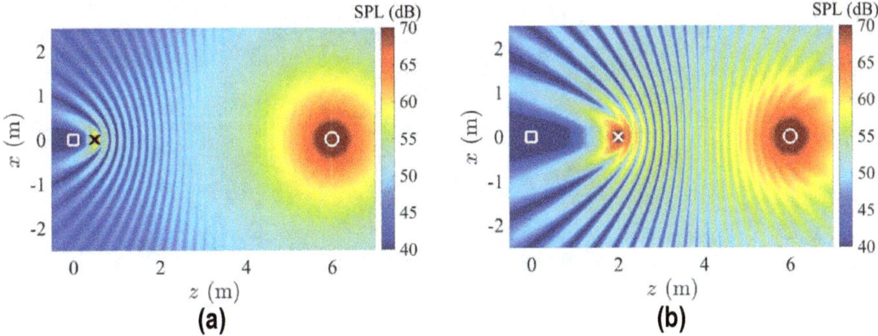

Figure 7.3 Sound fields generated by a point source located at $(x,z) = (0,6\,\text{m})$ at 1 kHz and the sound at the $(x,z) = (0,0)$ is controlled by a secondary omnidirectional point source at: (a) $(x,z) = (0,0.5\,\text{m})$; and (b) $(x,z) = (0,2\,\text{m})$. Circle, primary source; cross, secondary source; square, error location.

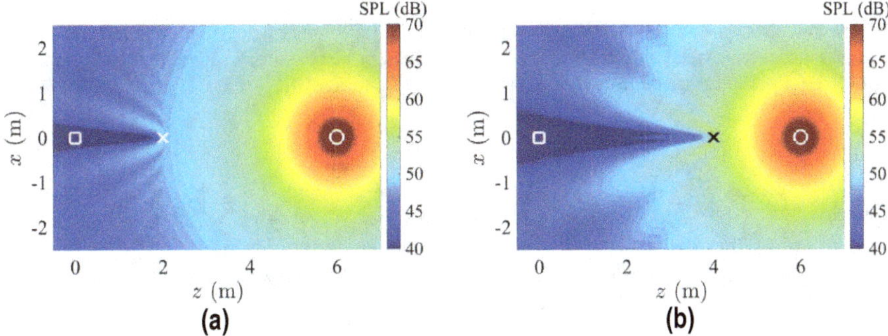

Figure 7.4 Sound fields generated by a point source located at $(x,z) = (0,6\,\text{m})$ at 1 kHz and the sound at the $(x,z) = (0,0)$ is controlled by a PAL at: (a) $(x,z) = (0,0.5\,\text{m})$; and (b) $(x,z) = (0,2\,\text{m})$. Circle, primary source; cross, secondary source; square, error location.

In addition, the spillover effect was quantitative by a performance index (PI) which is defined as the ratio of the averaged sound pressure in the sound field of interest (hemi-circle in the figures) after control to that before control. The PI in Fig. 7.5(b) due to a point monopole indicates 50.6% after control, while PI is only 4.13% (12.3 times smaller) for Fig. 7.5(c) due to the PAL, demonstrating the advantage of using PALs to mitigate the spillover effects.

7.2.2.2 ANC Using a Steerable PAL

The steerable PAL (see Chap. 6 for details) implemented by phased array has also been utilized to control the noise at the error point [Tanaka and Tanaka, 2010]. Incorporating this technique in ANC systems enables the generation of a moving quiet zone without mechanically rotating the PAL. This is particularly useful if the error point (such as a human ear) is moving in practical applications.

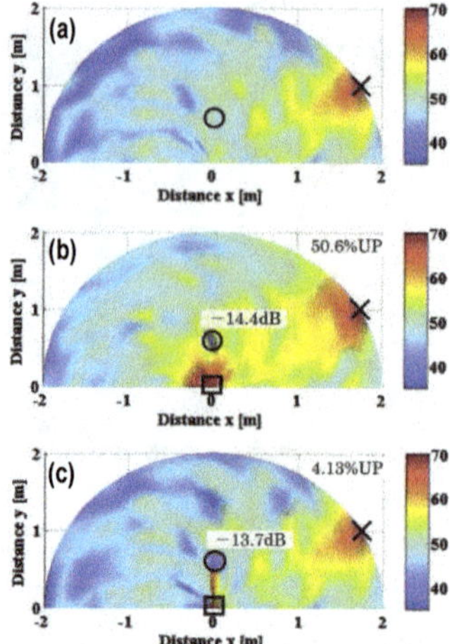

Figure 7.5 Measured audio SPL distributions at 1.5 kHz. (a) Primary field; (b) total field controlled by a point monopole; (c) total field controlled by a PAL. Cross, primary sources; squares, secondary sources; circles, error point. Extracted from [Tanaka and Tanaka, 2010, Fig. 10].

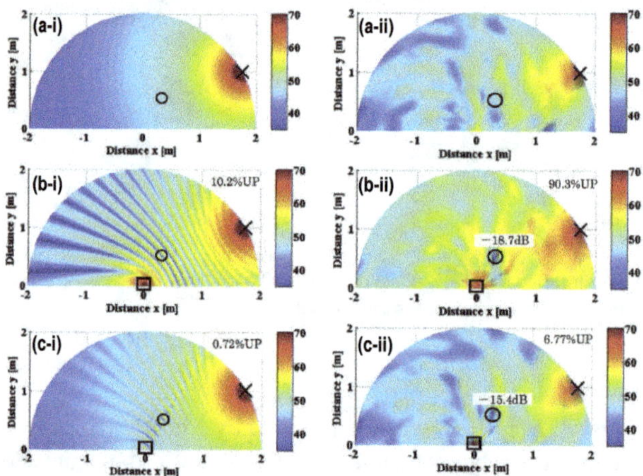

Figure 7.6 Audio SPL distributions at 1.5 kHz. (i) Simulations; (ii) experiments. (a) Primary field; (b) total field controlled by a point monopole; (c) total field controlled by a steerable PAL. Cross, primary sources; squares, secondary sources; circles, error point. Extracted from [Tanaka and Tanaka, 2010, Figs. 11 and 12].

Figure 7.6 presents both simulation and experimental results for the audio sound field distributions controlled by a conventional point monopole and a steerable PAL. The error point is located at a distance of 0.6 m with an azimuthal angle of 60°. In simulation results presented in Fig. 7.6, the fluctuated sound pressure observed as a striping is seen everywhere in the whole sound field in Fig. 7.6(b-i) where a point monopole is employed. In contrast, the strong stripe pattern is seen only along the radiation direction by the steerable PAL as evident in Fig. 7.6(c-i). As with the spillover effect, PI is up to 10.2% for the point monopole in Fig. 7.6(b-i), while it is only 0.72% for the PAL in Fig. 7.6(c-i). In the experiments, the primary sound field at the error point is suppressed by 18.7 dB and 15.4 dB with a point monopole and a PAL, respectively, demonstrating a similar control performance. However, the PI with the point monopole is 90.3% which is 13.3 times larger than that with a steerable PAL (6.77%).

7.2.2.3 ANC Using a Focusing PAL

The focusing PAL is typically used to improve the low audio frequency response around the focal point (see Chap. 6 for details). Nevertheless, it also shows the application potential in ANC systems. A 2D focusing PAL was implemented to achieve the *global* ANC [Tanaka and Tanaka, 2011]. Unlike the *local* ANC discussed in previous sections, the global ANC aims to mitigate the radiation of primary sources in all directions. In real implementations, the total sound power [Nelson et al., 1987; Pan et al., 1992; Snyder and Tanaka, 1995] in open space, and the total acoustic potential energy [Bullmore et al., 1987; Li and Hodgson, 2005] in closed space are commonly used as the cost function to minimize. In contrast, the square sound pressure amplitude [Eq. (7.7)] is commonly used in the local ANC.

The advantage of the global ANC is clear. However, it is hard to achieve a high performance global ANC using conventional omni-directional sources. For example, when two point sources are placed in free space, one is used for a primary source and the other for a secondary source; the global control performance for minimizing the total sound power output is dependent on the distance between two sources. Specifically, the minimal total sound power output under the optimal control is [Nelson and Elliott, 1992, Eq. (8.4.3)]

$$W_{\mathrm{opt}} = W_{\mathrm{p}}\left[1 - \mathrm{sinc}^2(kd)\right]. \tag{7.13}$$

Here, W_{p} is the sound power output of the primary source, k is the wavenumber, and d is the source separation. The optimal secondary source strength is [Nelson and Elliott, 1992, Eq. (8.4.2)]

$$Q_{\mathrm{s,opt}} = -Q_{\mathrm{p}}\,\mathrm{sinc}(kd). \tag{7.14}$$

Figure 7.7 presents the performance of minimizing the total sound power generated by a point primary source using a point secondary source as a function of the normalized source separation (d/λ), where λ is the wavelength. As illustrated in this figure, when the separation distance between the two sources is increased relative to the acoustic wavelength, the minimal total sound power under the optimal control decreases and achieves to 0 dB at $d = \lambda/2$. To achieve a noise reduction in the total sound power more than 9 dB, the source separation is required to be less than 1/10 wavelength. For source separations larger than half wavelength ($d > \lambda/2$), there is very little that can be done with the secondary source to reduce the total sound power with a maximum reduction of only 0.2 dB. Consequently, to achieve better global control performance, the secondary source is required to placed as close

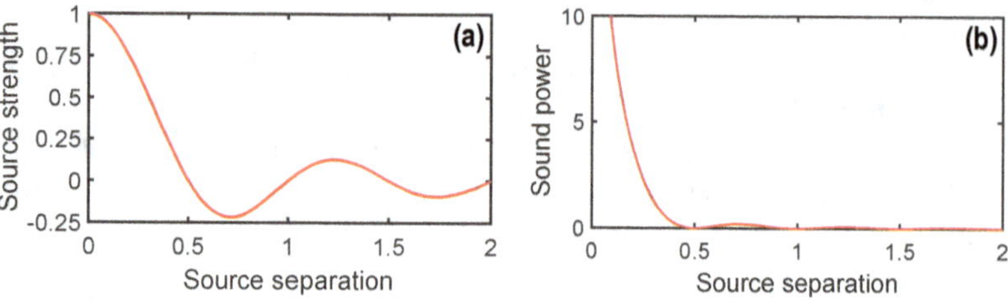

Figure 7.7 Performance of minimizing the total sound power generated by a point primary source using a point secondary source as a function of the normalized source separation (d/λ). (a) The normalized optimal secondary source strength $-Q_{s,opt}/Q_p$. (b) The maximum achievable noise reduction (dB) in total sound power radiated $10\lg(W_{opt}/W_p)$.

to the primary source. However, there are many practical cases where a control loudspeaker cannot be placed close enough to a primary noise source. When the operating frequency becomes higher, it becomes more difficult to achieve satisfactory global control performance due to the shorter wavelength. For example, the control loudspeaker needs to be placed as close as 34 mm (1/10 wavelength) away from the primary source to obtain 9 dB reduction in total sound power at 1 kHz.

Instead of placing an actual secondary source close to the primary source, one may create a virtual sound source just on the primary source. This can be realized by using a focusing PAL [Tanaka and Tanaka, 2011]. Consider a baffled primary source, by setting the focal point coincide with the primary source, the reflected wave generated by the focusing PAL behaves as if it was radiated from the virtual sound source created just on the primary source. Consequently, the total sound power output can be largely reduced even the distance between the primary source and the focusing PAL is more than half wavelength. Figure 7.8 presents both simulation

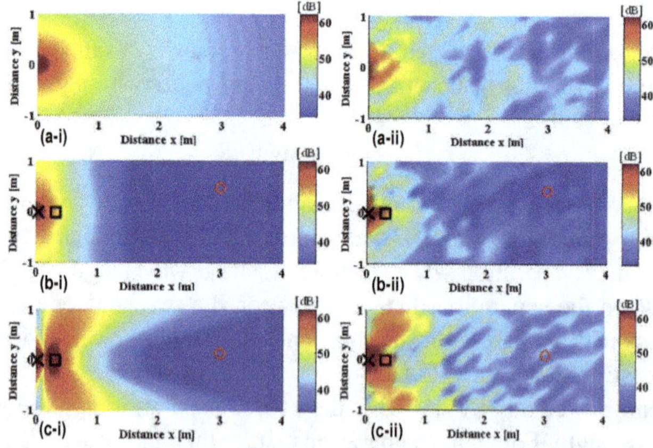

Figure 7.8 Audio SPL distributions at 1 kHz. The source separation is 0.3 m (0.87 wavelength). (i) Simulations; (ii) experiments. (a) Primary field; (b) total field controlled by a focusing PAL; (c) total field controlled by a point source. Cross, primary sources; squares, secondary sources; circles, error point. Extracted from [Tanaka and Tanaka, 2011, Figs. 3 and 8].

and experimental results at 1 kHz, where conventional point sources and focusing PALs are used. It is clear that a much more improved global control performance is achieved with the focusing PAL.

7.2.3 CONTROL OF BROADBAND NOISE COMBINING WITH A REMOTE ERROR SENSING STRATEGY

Previous sections discuss the ANC performance with regard to the pure-tone noise source. In practical applications, the noise is typically a transient broadband signal. This section introduces an ANC system using a PAL to cancel a broadband noise at a human's ear. In addition, a custom-made low-mass membrane pick-up from a retroreflective film and a laser Doppler vibrometer (LDV) can be used to form a remote sensing apparatus to determine the acoustic information with minimum obstructions to the person. An experiment was designed to test the noise reduction performance of such a system [Zhong et al., 2020f].

The experiments were conducted in a hemi-anechoic room with dimensions of 7.20 m × 5.19 m × 6.77 m (height). The schematic diagram and the photos of the experiment setup are shown in Figs. 7.9 and 7.10. A broadband primary noise (1 kHz to 6 kHz) was generated by a conventional loudspeaker (Genelec 8010A) at 6 m away from a head and torso simulator (HATS, Brüel & Kjær Type 4128). The custom-made membrane was placed in the left synthetic ear of the HATS, and the radius and thickness of the membrane were 10.5 mm and 0.1 mm, respectively. The LDV (Polytec NLV-2500-5) was placed at a stand-off distance of 0.7 m away from the membrane in the left ear. All the equipment was at the same height during the experiments. The error signal was obtained by measuring the vibration of the membrane and was then fed into a commercial ANC controller (Antysound TigerANC WIFI-Q), where the FxLMS algorithm is used to obtain the control filter.

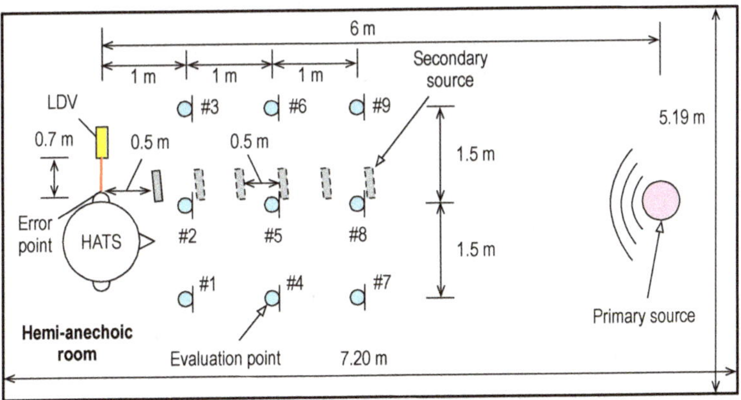

Figure 7.9 Schematic diagram of the experiment setup. Extracted from [Zhong et al., 2020f, Fig. 1(a)].

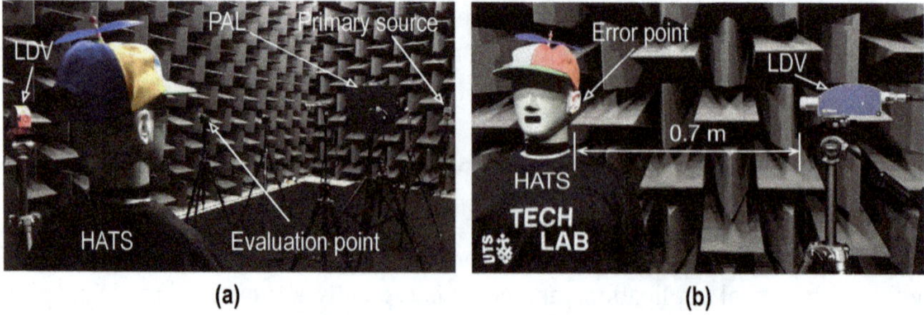

Figure 7.10 (a) A photo of the experiment setup in the hemi-anechoic room. (b) A photo of the LDV error sensing system. Extracted from [Zhong et al., 2020f, Figs. 1(b–c)].

A PAL (Holosonics Audio Spotlight AS-16i with the surface size of 40 cm × 40 cm) was used as the secondary source, and the performance of the ANC system using it was compared with that using a conventional omnidirectional loudspeaker (Genlec 8010A). Six groups of experiments were carried out, where the secondary source was placed in front of the error location at the distance from 0.5 m to 3 m with an interval of 0.5 m. To investigate the effects of the secondary source on the sound fields in the other areas, 9 evaluation microphones (Antysound Anty M1212) were placed in front of the HATS as shown in Figs. 7.9 and 7.10. The acoustic signals at all the microphones and the HATS were recorded with a Brüel & Kjær PULSE system with a sampling rate of 12.8 kHz. The fast Fourier transform (FFT) analyzer in PULSE LabShop was used to obtain the FFT spectrum. The frequency span was set to 6.4 kHz with 6400 lines and the averaging type is linear with 66.67% overlap and 30 s duration. The noise signal is directly fed into the controller as the reference signal. All the signals fed into the primary loudspeakers are correlated.

Figure 7.11 shows the SPLs measured by the left ear simulator of the HATS with and without ANC when the distance between the secondary source and the error location was 1 m. The overall noise reductions from 1 kHz to 6 kHz measured by the HATS were 18.7 dB and 17.8 dB when the conventional loudspeaker and the PAL were used as the secondary source, respectively. Both types of loudspeakers can reduce the noise at the ear effectively. There are two troughs near 1 kHz and 1.6 kHz on the curve of the primary noise without ANC in Fig. 7.11(b), which might be caused by the scattering effects of the square PAL used in the experiments.

To investigate the effects of the secondary source on the sound fields in the other areas, the SPLs at two typical evaluation locations #2 (closest to the secondary source) and #7 (farthest away from the secondary source) with and without ANC are presented in Fig. 7.12, where the distance between the secondary source and the error location was again 1 m. Both types of loudspeakers had little effect on the SPLs at point #7 because it was away from the secondary source as shown in Fig. 7.9. However, at point #2 which was close to the conventional loudspeaker, the SPLs changed significantly with ANC on due to the omnidirectional secondary source, and the overall noise reduction from 1 kHz to 6 kHz of the ANC system was −4.9 dB indicating that the overall sound energy at this point increased with ANC. The overall noise

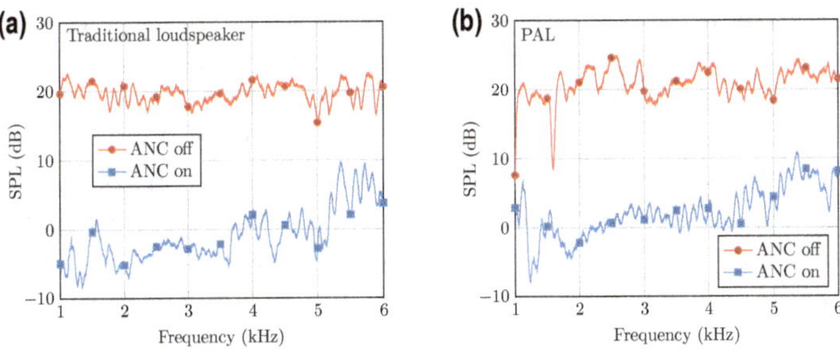

Figure 7.11 Sound fields measured by the left ear simulator of the HATS with and without ANC, where the secondary source was (a) a conventional loudspeaker and (b) a PAL, at a distance of 1 m from the error sensing point. Extracted from [Zhong et al., 2020f, Fig. 2].

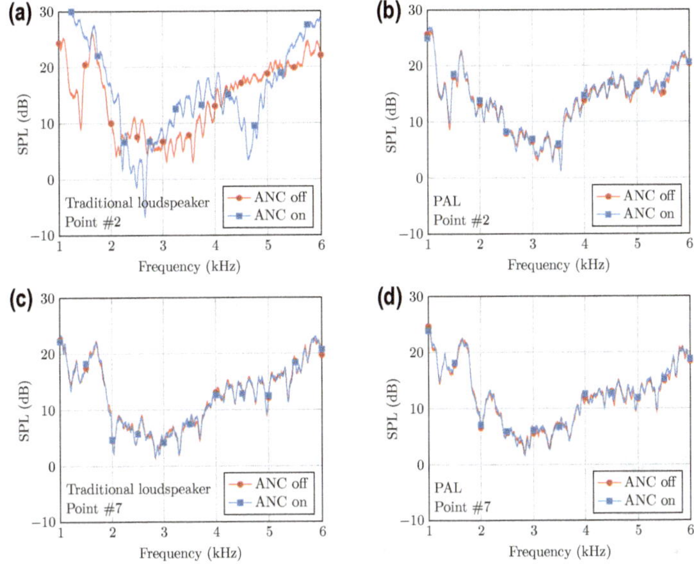

Figure 7.12 Sound fields at point #2 when the secondary source was (a) a conventional loudspeaker and (b) a PAL; and at point #7 when the secondary source was (c) a conventional loudspeaker and (d) a PAL. The distance between the secondary source and the error location was 1 m. Extracted from [Zhong et al., 2020f, Fig. 3].

reduction of the ANC system using the PAL at point #2 was only −0.2 dB due to its sharp directivity. Therefore, using a PAL has little effect on the sound fields in the other areas for an ANC system.

Figure 7.13 shows the overall noise reductions from 1 kHz to 6 kHz measured by the left ear simulator of the HATS and at the 9 evaluation locations with ANC on when the distance between the secondary source and the error location, which is denoted by d_{se}, was varied from 0.5 m to 3 m. Figure 7.13(a) shows that the noise reductions using the PAL were similar to those when using the conventional loudspeaker, and are generally between 18 dB and 20 dB. Figure 7.13(b) demonstrates that the noise reduction levels at all evaluation locations were generally increased as the distance between the conventional loudspeaker and the error location increased. Figure 7.13(c) indicates that the sound pressures at evaluation locations were almost unchanged with the PAL except at points #2 and #5. The noise reduction levels at these two points were negative, indicating that the sound pressure increased with ANC on. The reason is that the two points were close to the radiator axis of the PAL as shown in Fig. 7.9, and the SPL variation around them was large with and without ANC. The audio sound waves generated by the PAL decay slowly by distance, so the amplitude of the generated secondary sounds changes a little when the distance between the PAL and the error location increases. The sound pressure in the other areas was less affected by the ANC system when the PAL was used at far away from the ear.

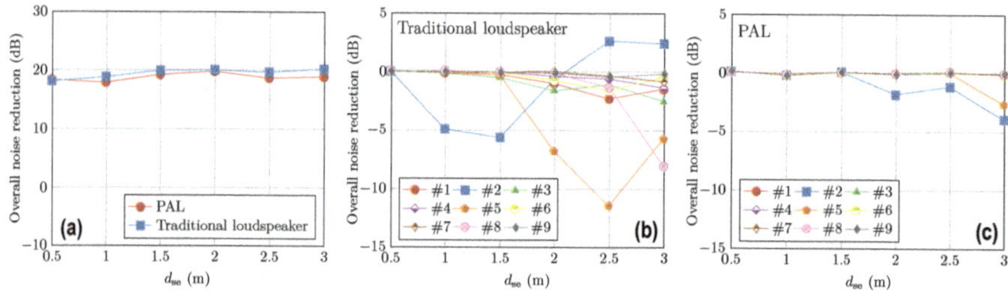

Figure 7.13 Overall noise reductions from 1 kHz to 6 kHz (a) at the left ear of the HATS, and at the evaluation locations, where the secondary source was (b) a conventional loudspeaker and (c) a PAL. Extracted from [Zhong et al., 2020f, Fig. 4].

7.2.4 REMARKS

The reader is introduced to a single-channel ANC system using only one secondary source to create a quiet zone in this section. The quiet zone generated by either a conventional omnidirectional loudspeaker or a PAL is demonstrated in Sec. 7.2.2. Then a remote error sensing strategy using an LDV is integrated into the ANC system in Sec. 7.2.3. An advantage of this system is that the PAL and the LDV can be compactly placed away from the person without deteriorating the broadband noise reduction performance. The performance of the ANC system was compared with a similar ANC system albeit using a conventional omnidirectional loudspeaker. The results demonstrate that the overall noise reductions from 1 kHz to 6 kHz at the person's ear were similar with both types of loudspeakers. The sound pressure levels in the other areas were almost unchanged when the PAL was placed away from the ear in the ANC system, while the overall sound pressure levels became higher with the conventional loudspeaker being used at a large distance from the ear.

7.3 BINAURAL ANC USING TWO PALS

The most important feature of PALs is the ability of generating highly directional audio beams. This feature outperforms the conventional omni-directional loudspeakers to mitigate the spillover effect as discussed in Sec. 7.2. Another advantage of the sharp directivity is that the crosstalk between PALs is small when the audio beams generated by them are not intersected by each other. This benefit removes the requirement of the crosstalk cancelation in practical implementations of ANC systems, consequently saves the computational cost of the real-time signal processor.

A typical application to demonstrate this property is the binaural ANC system. As shown in Fig. 7.14, in this dual-channel ANC system, two secondary sources are used and each one is principally in charge of the control of one error point [Tanaka et al., 2017]. The typical factory noise was used as the primary source in the experiments, where impulsive components are generated by the packaging machine. The head and torso simulator (HATS) mimics the worker in a stationary position. The left and right ears of the HATS were used as the error microphones. The distances from the noise source and PAL to the error microphones are roughly 200 cm and 150 cm, respectively. When the secondary sources are omni-directional, crosstalk becomes an unavoidable problem and brings extra computational cost in the implementation, except for the spillover problem.

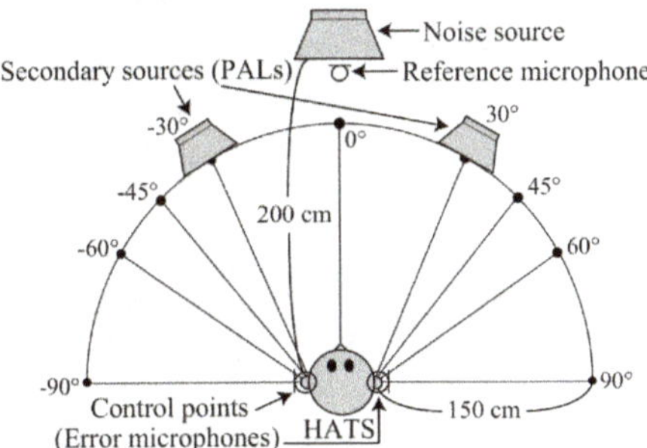

Figure 7.14 Sketch of the binaural ANC system using two PALs as secondary sources to control noise at two ears of a head and torso simulator (HATS). One noise source is located at 0°. Extracted from [Tanaka et al., 2017, Fig. 6].

Figure 7.15 shows the noise reduction performance at two ears of the HATS. Here, curves labeled by "ANC off" denote the primary noise signal sensed by two microphones of the HATS. Curves labeled by "ANC on" indicate the measured results under the optimal control of two PALs and crosstalk secondary path models are included. In this case, 10.4 dB and 11.6 dB noise reductions are achieved at left and right ears, respectively. Interestingly, when the crosstalk secondary path models are moved, the noise reduction performance in both ears is not compromised. As the crosstalk secondary path models are not required, the computational cost is reduced, which demonstrate the advantage of using PALs as secondary sources.

In practical applications, the noise can come from multiple directions. Figure 7.16 presents the control performance of the PAL-based binaural ANC system when two

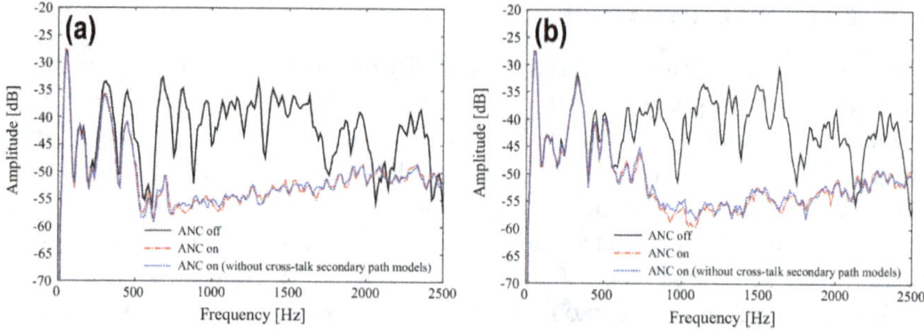

Figure 7.15 Experimental results of the dual-channel binaural ANC system when one noise source is placed at 0°. (a) Left ear; (b) right ear. Extracted from [Tanaka et al., 2017, Fig. 8].

noise sources are placed in ±30° directions. It is observed that the noise reduction performance is degraded. Specifically, the noise reductions at the left and right ears are 6.8 dB and 7.3 dB, respectively. However, it is interesting to note that the noise reduction performance is not comprised even the crosstalk secondary path models are removed. Consequently, above analysis demonstrates that the crosstalk path models can be removed to save the computational cost when using two PALs to create two quiet zones in the vicinity of two ears of a worker. This is mainly due to the narrow directivity of the audio beam generated by PALs.

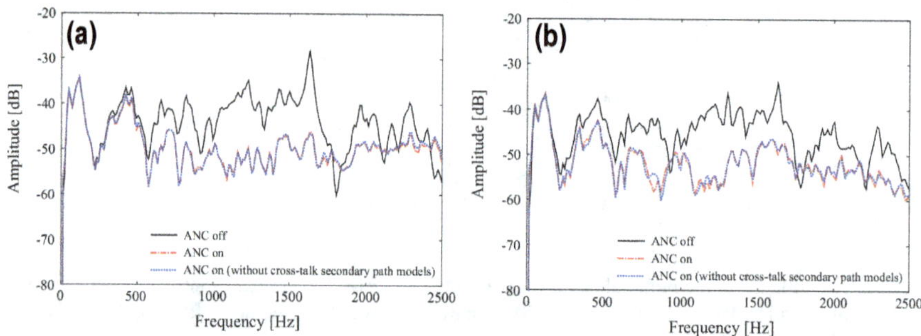

Figure 7.16 Experimental results of the dual-channel binaural ANC system when two noise sources are placed at ±30°. (a) Left ear; (b) right ear. Extracted from [Tanaka et al., 2017, Fig. 12].

7.4 ANC USING MULTIPLE PALS

7.4.1 PROBLEM DESCRIPTION

The challenge to be addressed with ANC involves maximizing the size of the quiet zone. Section 7.2 has demonstrated the advantage of using PALs in a single-channel ANC system. However, the created quiet zone size is limited. For one secondary source, the size of the quiet zone is only about 1/10 of the wavelength in a diffuse sound field [Guo and Pan, 1998; Elliott et al., 1988]. For example, when the frequency is 4 kHz, the size of the quiet zone would be only 8.6 mm in a diffuse field, which is too small in practical applications. If one intends to create a quiet zone with a size of 10 cm, the cutoff frequency would be as low as 343 Hz when one secondary source is used in a diffuse field.

A larger quiet zone can be created by using multiple secondary sources surrounding the target region [Zhang et al., 2019]. In a free field, the quiet zone size is proportional to the number of secondary sources [Guo et al., 1997]. The experimental measurements show that more than 20 dB of noise reduction can be obtained inside a sphere with a radius of 0.3 m at frequencies from 100 Hz to 500 Hz when using 30 secondary sources on a spherical surface [Epain and Friot, 2007]. In an ordinary room, the experimental results with 16 secondary sources distributed over a cylindrical surface demonstrate that a cylindrical quiet zone with a height of 0.2 m and a radius of 0.2 m is possible below 550 Hz [Zou et al., 2007]. In these ANC systems, omnidirectional loudspeakers were used as secondary sources. Although the noise inside the target region is reduced, the sound pressure outside the region often increases, i.e., the spillover effect happens [Guo et al., 1997; Tanaka and Tanaka, 2010]. The noise amplification outside the target areas can be mitigated by optimizing some parameters, such as the separation between the secondary sources and the distance between secondary sources and error sensors [David and Elliott, 1994; Joseph et al., 1994; Guo et al., 1997].

Figures 7.17 and 7.18 show the schematic diagram of two typical configurations to be investigated here. The primary sources consist of N_p point monopoles randomly located on the xOy plane, or in 3D space. They are assumed to be harmonic with a frequency of f_a. The number of secondary sources is N_s, and they are uniformly distributed on a circle on the xOy plane, or a sphere with a radius of R_s, as shown in Figs. 7.17 and 7.18, respectively. For all cases, the target zone to be controlled is a two-dimensional interior region of a circle on the xOy plane with a radius of R_0 centered at the origin of the rectangular coordinate system $Oxyz$.

Both 2D and 3D configurations are investigated, where the primary sources are assumed to be point monopoles located randomly on a two-dimensional plane, or in three-dimensional space. The 2D configuration shown in Fig. 7.17 is denoted by *2D secondary sources* where N_s secondary sources ($N_\text{s} = 8$ in the figure) are evenly placed on a circle and the azimuthal angle at the i-th source is $\varphi_{\text{s},i} = 2\pi(i-1)/N_\text{s}$, $i = 1, 2, \ldots, N_\text{s}$. The three-dimensional configuration shown in Fig. 7.18 is denoted by *3D secondary sources*, where N_s secondary sources ($N_\text{s} = 6$ in the figure) are evenly placed on a spherical surface and the minimal distance between arbitrary two secondary sources is denoted by d_min. The secondary source locations are obtained by maximizing d_min, and the source coordinates for different N_s can be found in [Sloane et al., 2021]. The radius of the secondary sources is set to be $R_\text{s} = 1.5$ m in all the simulations that follow.

The quiet zone size, L, is obtained by using an iterative procedure. The initial lower (R_1) and upper (R_2) values are set to 0 and 1.5 m, respectively. The NR inside the circular target zone is calculated when the radius of the target zone R_0 is set

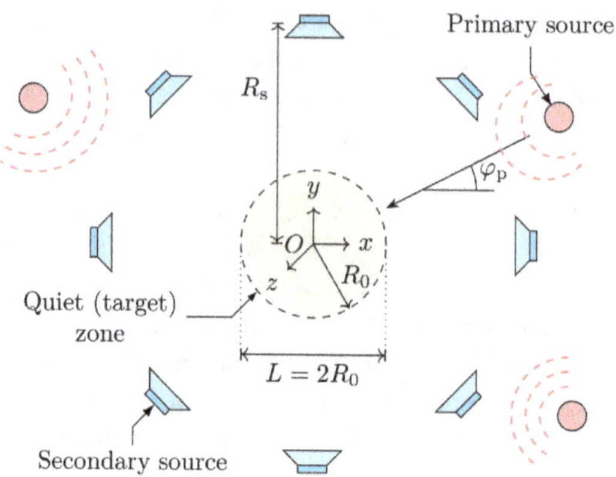

Figure 7.17 Sketch of an ANC system where the primary and secondary sources are located on the *xOy* plane. Extracted from [Zhong et al., 2022e, Fig. 1(a)].

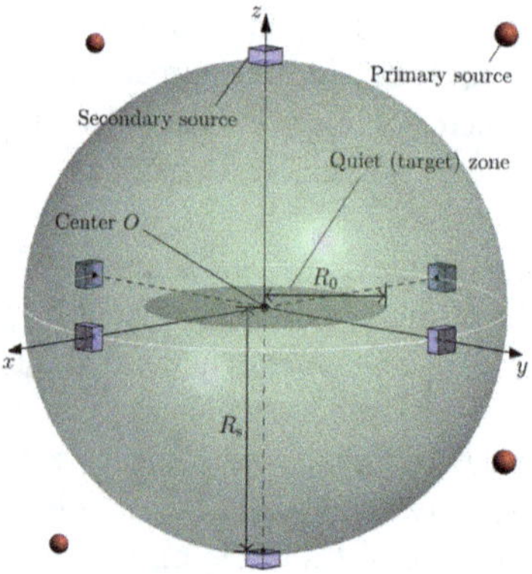

Figure 7.18 Sketch of an ANC system where the primary and secondary sources are located in three-dimensional space. Extracted from [Zhong et al., 2022e, Fig. 1(b)].

as $(R_1 + R_2)/2$. If NR < 10 dB, the upper value R_2 is updated as $(R_1 + R_2)/2$. If NR > 10 dB, the lower value R_1 is updated as $(R_1 + R_2)/2$. The iteration stops when $(R_2 - R_1)$ is less than 1/20 of the wavelength and $(R_1 + R_2)$ is chosen as the quiet zone size L. All simulation results presented here have followed this procedure. Each PAL is assumed to be circular with a radius of 0.1 m and a uniform surface velocity profile for the ultrasound. The lower ultrasound frequency is set as 64 kHz, which is a typical carrier frequency for commercial PALs [Holosonics, 2019]. The sound attenuation coefficients at both ultrasonic and audio sound frequencies are calculated according to ISO 9613-1 with a relative humidity of 60% and a temperature of 25°C.

7.4.2 THEORETICAL FRAMEWORK

Suppose the noise is generated by N_p primary sources in 3D space. For a multi-channel ANC system, the sound pressure at N_e error locations is controlled by N_s secondary sources. The summation of the square of the sound pressure at each error location is chosen here, which yields [Kirkeby et al., 1996; Qiu, 2019]

$$\mathcal{J} = \mathbf{p}_{et}^{H}\mathbf{p}_{et} + \beta_0 \mathbf{Q}_{s}^{H}\mathbf{Q}_{s}, \tag{7.15}$$

where $\mathbf{p}_{et} = \left[p_t(\mathbf{r}_{e,1}), p_t(\mathbf{r}_{e,2}), \dots, p_t(\mathbf{r}_{e,N_e})\right]^T$ is the total sound pressure vector, $p_t(\mathbf{r}_{e,i})$ is the total sound pressure at the i-th error location located at $\mathbf{r}_{e,i}$, and $i = 1,2,\dots,N_e$. The total sound pressure $\mathbf{p}_{et}$ is the superposition of the sound pressure generated by N_p primary sources and N_s secondary ones, i.e.,

$$\mathbf{p}_{et} = \mathbf{p}_{ep} + \mathbf{p}_{es}, \tag{7.16}$$

where $\mathbf{p}_{ep} = [p_p(\mathbf{r}_{e,1}), p_p(\mathbf{r}_{e,2}), \dots, p_p(\mathbf{r}_{e,N_e})]^T$ and $\mathbf{p}_{es} = [p_s(\mathbf{r}_{e,1}), p_s(\mathbf{r}_{e,2}), \dots, p_s(\mathbf{r}_{e,N_e})]^T$ are the sound pressure vectors at error locations radiated by primary and secondary sources, respectively. $p_p(\mathbf{r}_{e,i})$ and $p_s(\mathbf{r}_{e,i})$ are sound pressure at the i-th error location generated by N_p primary and N_s secondary sources, respectively. In addition, β_0 in Eq. (7.15) is a real number to constrain the outputs of secondary sources [Kirkeby et al., 1996], $\mathbf{Q}_s = \left[Q_{s,1}, Q_{s,2}, \dots, Q_{s,N_s}\right]^T$ is the source strength vector of the secondary sources, and $Q_{s,i}$ is the source strength of the i-th secondary source, $i = 1,2,\dots,N_s$.

The sound pressure vector generated by the secondary sources can be represented as

$$\mathbf{p}_{es} = \mathbf{Z}_{es}\mathbf{Q}_s, \tag{7.17}$$

where $\mathbf{Z}_{es}$ is a matrix of the acoustic transfer functions from N_s secondary sources to N_e error locations reading

$$\mathbf{Z}_{es} = \begin{bmatrix} Z(\mathbf{r}_{e,1}, \mathbf{r}_{s,1}) & Z(\mathbf{r}_{e,1}, \mathbf{r}_{s,2}) & \cdots & Z(\mathbf{r}_{e,1}, \mathbf{r}_{s,N_s}) \\ Z(\mathbf{r}_{e,2}, \mathbf{r}_{s,1}) & Z(\mathbf{r}_{e,2}, \mathbf{r}_{s,2}) & \cdots & Z(\mathbf{r}_{e,2}, \mathbf{r}_{s,N_s}) \\ \vdots & \vdots & \ddots & \vdots \\ Z(\mathbf{r}_{e,N_e}, \mathbf{r}_{s,1}) & Z(\mathbf{r}_{e,N_e}, \mathbf{r}_{s,2}) & \cdots & Z(\mathbf{r}_{e,N_e}, \mathbf{r}_{s,N_s}) \end{bmatrix}, \tag{7.18}$$

$Z(\mathbf{r}_{e,i}, \mathbf{r}_{s,j})$ is the acoustic transfer function from the i-th error location to the j-th secondary source, $i = 1,2,\dots,N_e$, and $j = 1,2,\dots,N_s$,

To minimize the cost function given by Eq. (7.15), the optimized source strengths of the secondary sources are obtained as [Nelson and Elliott, 1992; Qiu, 2019]

$$\mathbf{Q}_{s,opt} = -(\mathbf{Z}_{es}^{H}\mathbf{Z}_{es} + \beta_0\mathbf{I})^{-1}\mathbf{Z}_{es}^{H}\mathbf{p}_{ep}, \tag{7.19}$$

where $\mathbf{I}$ is an identity matrix of size N_s. After obtaining the optimal secondary source strengths, the total sound pressure with control can then be calculated as

$$\mathbf{p}_{et,opt} = \mathbf{p}_{ep} + \mathbf{Z}_{es}\mathbf{Q}_{s,opt}. \tag{7.20}$$

For comparison, either conventional omnidirectional loudspeakers or PALs are adopted as secondary sources. The conventional loudspeakers are usually modeled as point monopoles, where the sound pressure is given as

$$p(\mathbf{r}) = -\mathrm{i}\rho_{\mathrm{air}}\omega_{\mathrm{a}}Q_0 \frac{\mathrm{e}^{\mathrm{i}k_{\mathrm{a}}|\mathbf{r}-\mathbf{r}_{\mathrm{s}}|}}{4\pi|\mathbf{r}-\mathbf{r}_{\mathrm{s}}|}, \tag{7.21}$$

ρ_{air} is the air density, $\omega_{\mathrm{a}} = 2\pi f_{\mathrm{a}}$ is the angular frequency, $k_{\mathrm{a}} = \omega_{\mathrm{a}}/c_{\mathrm{air}}$ is the wavenumber, Q_0 is the source strength, and $|\mathbf{r}-\mathbf{r}_{\mathrm{s}}|$ is the distance between the field point $\mathbf{r}$ and the source point $\mathbf{r}_{\mathrm{s}}$. Equation (7.21) shows that the acoustic transfer path between $\mathbf{r}$ and $\mathbf{r}_{\mathrm{s}}$ for this source is

$$Z(\mathbf{r},\mathbf{r}_{\mathrm{s}}) = -\mathrm{i}\rho_{\mathrm{air}}\omega_{\mathrm{a}} \frac{\mathrm{e}^{\mathrm{i}k_{\mathrm{a}}|\mathbf{r}-\mathbf{r}_{\mathrm{s}}|}}{4\pi|\mathbf{r}-\mathbf{r}_{\mathrm{s}}|}. \tag{7.22}$$

The audio sound pressure radiated by a PAL can be calculated using the methods presented in Chap. 2. The created quiet zone is usually located in the Westervelt far field of the PAL, so the local effects can be neglected to simplify the calculation. The audio sound field generated by the PAL can then be considered as a superposition of the pressure radiated by virtual sources in air with the source density function proportional to the sound pressure of ultrasound. This can be calculated as

$$p(\mathbf{r}) = -\mathrm{i}\rho_0\omega_{\mathrm{a}}u_1^*u_2 \iiint_V q(\mathbf{r}_{\mathrm{v}}) \frac{\mathrm{e}^{\mathrm{i}k_{\mathrm{a}}|\mathbf{r}-\mathbf{r}_{\mathrm{v}}|}}{4\pi|\mathbf{r}-\mathbf{r}_{\mathrm{v}}|} \mathrm{d}^3\mathbf{r}_{\mathrm{v}}, \tag{7.23}$$

where u_1 and u_2 are the amplitude of the vibration velocity on the transducer surface for the ultrasound f_1 and f_2 ($f_2 > f_1$), respectively, the integration range V represents the whole three-dimensional space, and $q(\mathbf{r}_{\mathrm{v}})$ is the source density function at the virtual source point $\mathbf{r}_{\mathrm{v}}$ determined by the ultrasound pressure. The radiated audio sound pressure can be changed by changing the values of u_1 and u_2 of the ultrasound, so that $u_1^*u_2$ is defined as the source strength of a PAL, which is equivalent to Q_0 in Eq. (7.21). For a circular PAL, Eq. (7.23) can be efficiently calculated using the spherical wave expansion method presented in Sec. 2.6. Equation (7.23) shows that the acoustic transfer path between $\mathbf{r}$ and $\mathbf{r}_{\mathrm{s}}$ for this source is

$$Z(\mathbf{r},\mathbf{r}_{\mathrm{s}}) = -\mathrm{i}\rho_0\omega_{\mathrm{a}} \iiint_V q(\mathbf{r}_{\mathrm{v}}) \frac{\mathrm{e}^{\mathrm{i}k_{\mathrm{a}}|\mathbf{r}-\mathbf{r}_{\mathrm{v}}|}}{4\pi|\mathbf{r}-\mathbf{r}_{\mathrm{v}}|} \mathrm{d}^3\mathbf{r}_{\mathrm{v}}, \tag{7.24}$$

where $\mathbf{r}_{\mathrm{s}}$ is assumed to be the location at the centroid of the PAL.

7.4.3 QUIET ZONE AND RELATED METRICS

To generate a desired quiet zone denoted as Γ, the approach is to mesh this region and all grid points are chosen as the error locations. The mesh size of the region Γ is required to be much less than the wavelength, e.g., 1/20 of wavelength. The noise reduction (NR) inside the target zone Γ is defined as

$$\mathrm{NR} = 10\log_{10}\left(\frac{\mathbf{p}_{\mathrm{ep}}^{\mathrm{H}}\mathbf{p}_{\mathrm{ep}}}{\mathbf{p}_{\mathrm{et}}^{\mathrm{H}}\mathbf{p}_{\mathrm{et}}}\right) \quad (\mathrm{dB}). \tag{7.25}$$

Usually, the NR decreases as the size of the target zone increases, so that the size of the quiet zone L is defined as the diameter of the maximal size target zone satisfying that the NRs inside the zone are all greater than 10 dB.

To evaluate the spillover effect of secondary sources on the surround areas quantitatively, a measure called the *potential energy gain* is defined as

$$G = 10\log_{10}\left(\frac{\mathbf{p}_{\mathrm{vt}}^{\mathrm{H}}\mathbf{p}_{\mathrm{vt}}}{\mathbf{p}_{\mathrm{vp}}^{\mathrm{H}}\mathbf{p}_{\mathrm{vp}}}\right) \quad (\mathrm{dB}) \tag{7.26}$$

where $\mathbf{p}_{\mathrm{vt}} = \left[p_{\mathrm{t}}(\mathbf{r}_{\mathrm{v},1}), p_{\mathrm{t}}(\mathbf{r}_{\mathrm{v},2}), \ldots, p_{\mathrm{t}}(\mathbf{r}_{\mathrm{v},N_{\mathrm{v}}})\right]^{\mathrm{T}}$ is the total sound pressure vector at N_{v} evaluation locations, $p_{\mathrm{t}}(\mathbf{r}_{\mathrm{v},i})$ is the total sound pressure at the i-th evaluation locations which is located at $\mathbf{r}_{\mathrm{v},i}$, and $i = 1, 2, \ldots, N_{\mathrm{v}}$.

The total sound pressure $\mathbf{p}_{\mathrm{vt}}$ is the superposition of the sound pressure generated by N_{p} primary sources and N_{s} secondary sources, i.e.,

$$\mathbf{p}_{\mathrm{vt}} = \mathbf{p}_{\mathrm{vp}} + \mathbf{p}_{\mathrm{vs}}, \tag{7.27}$$

where $\mathbf{p}_{\mathrm{vp}} = [p_{\mathrm{p}}(\mathbf{r}_{\mathrm{v},1}), p_{\mathrm{p}}(\mathbf{r}_{\mathrm{v},2}), \ldots, p_{\mathrm{p}}(\mathbf{r}_{\mathrm{v},N_{\mathrm{e}}})]^{\mathrm{T}}$ and $\mathbf{p}_{\mathrm{vs}} = [p_{\mathrm{s}}(\mathbf{r}_{\mathrm{v},1}), p_{\mathrm{s}}(\mathbf{r}_{\mathrm{v},2}), \ldots, p_{\mathrm{s}}(\mathbf{r}_{\mathrm{v},N_{\mathrm{v}}})]^{\mathrm{T}}$ are the sound pressure vectors at evaluation locations radiated by the primary and secondary sources, respectively. $p_{\mathrm{p}}(\mathbf{r}_{\mathrm{v},i})$ and $p_{\mathrm{s}}(\mathbf{r}_{\mathrm{v},i})$ are sound pressure at the i-th evaluation locations generated by N_{p} primary and N_{s} secondary sources, respectively. Equation (7.26) is similar to the definition of NR given by Eq. (7.25) but defines the change in the acoustic potential energy in the evaluation region with control. The result $G > 0$ represents a gain, while $G < 0$ indicts a reduction, in the acoustic potential energy.

7.4.4 TWO-DIMENSIONAL CONFIGURATION

7.4.4.1 Simulation Results

Figure 7.19 shows the primary and total sound fields at 1 kHz under the optimal control by 8 PALs, or point monopoles, shown as black rectangles or circles, respectively. The primary source (not shown in the figure) is a single monopole located at an azimuthal angle of $\varphi_{\mathrm{p}} = 22.5^\circ$ with a distance to the origin of 4 m. The angle of 22.5° is selected because it is the angle of the bisector of the first and secondary sources, which is the worst case for the NR performance [Qiu, 2019]. The quiet zone size in Figs. 7.19(b) and (c) is 0.45 m and 0.46 m, respectively. Although the quiet zone sizes are similar, the energy gain associated with the point monopole sources is 6.9 dB, which is much larger than the −0.1 dB obtained with the PALs. The decrement of energy is mainly contributed from the noise reduction in the quiet zone, and it is roughly same for both kinds of loudspeakers. The sound pressure around the conventional loudspeakers located on the direction of the primary source is large, so that the noise amplification can be clearly observed. However, it is not observed when using PALs because they generate unidirectional secondary waves and these waves decay slowly along the propagation path. This is the reason the energy gain when using PALs is much smaller than that using conventional loudspeakers.

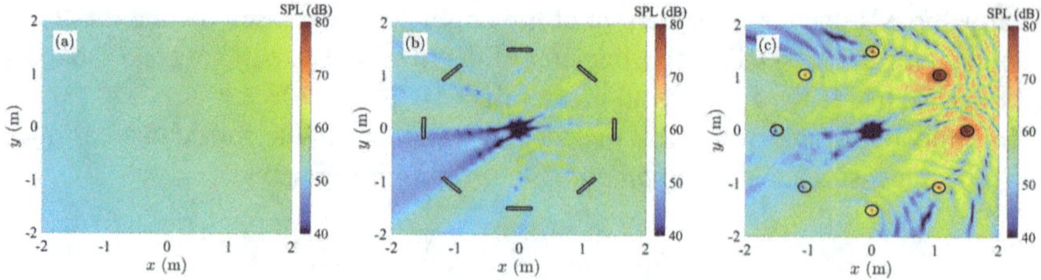

Figure 7.19 Audio sound fields at 1 kHz (a) for the primary noise comes from the direction $\varphi_p = 22.5°$, (b) under the optimal control with 8 PALs, and (c) under the optimal control with 8 point monopoles. Extracted from [Zhong et al., 2022e, Fig. 2].

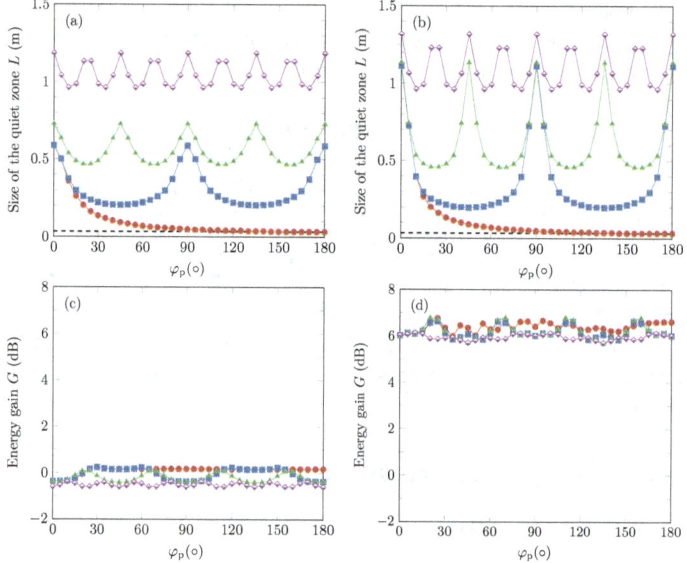

Figure 7.20 The quiet zone size and energy gain as a function of the primary source azimuthal angles at 1 kHz for different numbers of secondary sources: (a) and (b) the quiet zone size created by PALs and point monopoles, respectively; (c) and (d) the energy gain caused by PALs and point monopoles, respectively. —●—, $N_s = 1$; —■—, $N_s = 4$; —▲—, $N_s = 8$; —◇—, $N_s = 16$; - - -, $\lambda/10$. Extracted from [Zhong et al., 2022e, Fig. 3].

Figure 7.20 shows the quiet zone size and energy gain as a function of the azimuthal angle of a primary source 4 m away from the origin at 1 kHz for various numbers of secondary sources. The peaks and valleys in Fig. 7.20 (a) and (b) are associated with the location of the secondary sources. It is clear that the quiet zone size increases when the primary wave angle approaches that of the secondary source due to better matching of the wave fronts from the primary and secondary sound waves [Guo et al., 1997]. In most cases when there is one secondary source, the difference between the size of the quiet zone for the PAL and the point monopole secondary sources is less than 10% when the primary source angle is greater than 20°. The minimal quiet zone size is also 0.035 m when the primary source angle is at 180°, and the size is about a tenth of the wavelength.

It is shown in Fig. 7.20 that increasing secondary source number enlarges the size of the quiet zone. For example, the minimal quiet zone size is 0.035 m, 0.2 m, 0.46 m, and 0.97 m when the secondary source number is 1, 4, 8, and 16, respectively. The primary source azimuthal angles with minimal quiet zone sizes are the ones with their bisector between two adjacent secondary sources. Although the quiet zone size created by both PAL and point monopole secondary sources are approximately the same in most cases, the energy gain caused by the point monopoles are generally above 6 dB which is larger than the one caused by PALs which is around 0 dB.

To understand the performance of the ANC systems under complex acoustic environments, 8 primary sources are placed on the same plane as the secondary sources in the next examples, where the distance between each primary source and the origin is randomly and uniformly set between 3.5 m and 4.5 m; the azimuthal angle is randomly and uniformly set between 0 and 360°; the source strength is randomly and uniformly set between 0.75×10^{-4} m^3/s and 1.25×10^{-4} m^3/s; and this configuration is denoted here as *2D primary sound field*. Figure 7.21 shows the results of one trial of the 2D primary sound field and the total sound field controlled by 8 secondary sources. The quiet zone size created by both sources is 0.53 m, while the energy gain is −0.2 dB with the PALs, and 6.5 dB with the point monopoles.

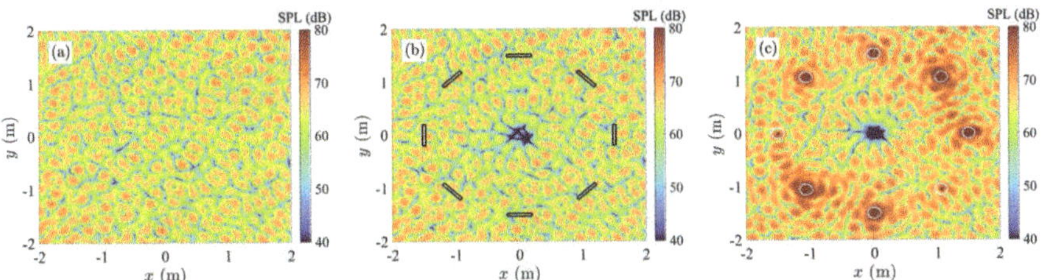

Figure 7.21 Audio sound fields at 1 kHz (a) generated by 8 point monopoles randomly located on the *xOy* plane, (b) under the optimal control with 8 PALs, and (c) under the optimal control with 8 point monopoles. Extracted from [Zhong et al., 2022e, Fig. 4].

Figure 7.22 shows the quiet zone size and energy gain based on 100 trials of random 2D primary sound fields generated by 8 point monopoles randomly located on the *xOy* plane. The quiet zone size decreases as the frequency increases in all cases and is about $0.75\lambda, 1.5\lambda$, and 3λ under optimal control conditions with 4, 8, and 16 point monopoles, respectively, where λ is the wavelength of the sound at the corresponding frequency. The quiet zone size generated by point monopoles is larger than the one generated by PALs especially at the low frequencies; however, the difference between them becomes negligible at the middle and high frequencies. For example, the quiet zone sizes are 1.33 m and 1.21 m respectively for 8 monopoles and PALs at 400 Hz, while both become about 0.55 m at 1 kHz. The energy gain caused by the point monopoles is always larger than the one caused by the PALs, which increases only slightly as the frequency increases.

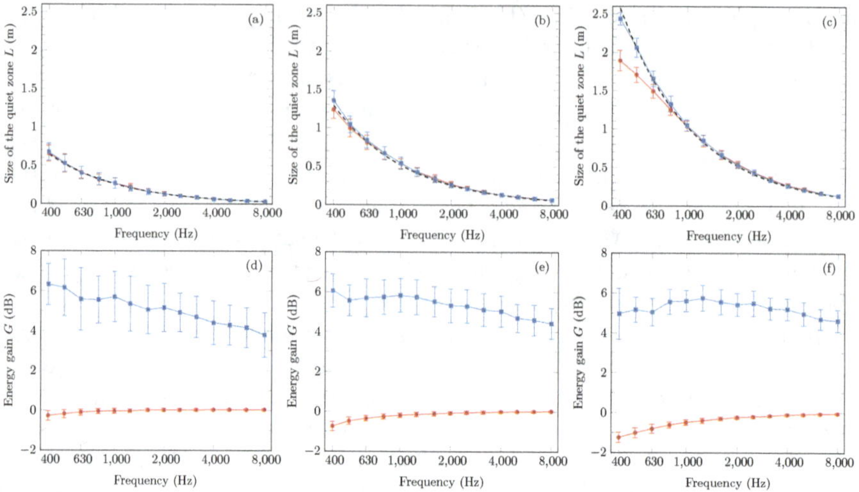

Figure 7.22 For random 2D primary sound fields under the optimal control with the 2D secondary sources, (a), (b), and (c) the quiet zone size when the secondary source number is 4, 8, and 16, respectively; (d), (e), and (f) the energy gain when the secondary source number is $N_s = 4, 8$, and 16, respectively, where the value and error bar are the mean value and standard deviation of 100 random trials, and λ is the wavelength. —●—, PAL; —■—, monopole; - - -, 0.75λ, 1.5λ, and 3λ for (a), (b), and (c), respectively. Extracted from [Zhong et al., 2022e, Fig. 5].

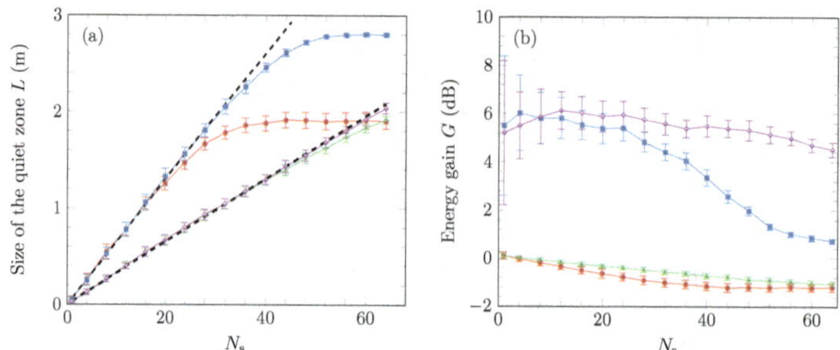

Figure 7.23 For random 2D primary sound fields under the optimal control with 2D secondary sources at 1 kHz and 2 kHz: (a) the quiet zone size as a function of secondary source number; (b) the energy gain as a function of secondary source number. —●—, PAL at 1 kHz; —■—, monopole at 1 kHz; —▲—, PAL at 2 kHz; —◇—, monopole at 2 kHz. Extracted from [Zhong et al., 2022e, Fig. 6].

Figure 7.23 shows the quiet zone size and energy gain at different numbers of secondary sources at 1 kHz and 2 kHz. It is clear the quiet zone size increases as the number of secondary sources increases. When the secondary source number becomes large, the quiet zone size increases slowly due to the limitations of the size of secondary sources (3 m) and the narrow beam width of sound radiated by PALs. As shown in Fig. 7.23(a), when the secondary source number is, respectively, less than 20 and 48 at 1 kHz and 2 kHz, the quiet zone size is approximately proportional to the secondary source number and can be estimated by the following formula

$$L = 0.19\lambda N_s. \tag{7.28}$$

For the ANC systems with N_s secondary sources, the arc length between adjacent secondary sources can be obtained by dividing the circumference of the generated quiet zone to give a length of $\pi L/N_s$. Using the L value in Eq. (7.28), the arc length is about 0.6λ which indicates that the separation between secondary sources is about one half of the wavelength. This agrees with the remarks in Qiu, 2019 and Nelson and Elliott, 1992. Figure 7.23(b) indicates that the energy gain can also be reduced by introducing more secondary sources and the reduction is more significant at lower frequencies.

In the following simulations, the primary noise comes from multiple directions in three-dimensional space and the configuration is denoted by *3D primary sound field*, even though the secondary sources are still located on the two-dimensional xOy plane. Figure 7.24 shows the quiet zone size and energy gain based on 100 random trials of 3D primary sound fields under the optimal control of 8 secondary sources in the xOy plane. Compared with Fig. 7.22(b), the mean value of the quiet zone size decreases from 1.5λ to 0.75λ, while the standard deviation shows no significant changes. Compared with Fig. 7.22(e), the energy gain caused by the PALs becomes positive at most frequencies, while it is still much less than the one caused by the point monopoles.

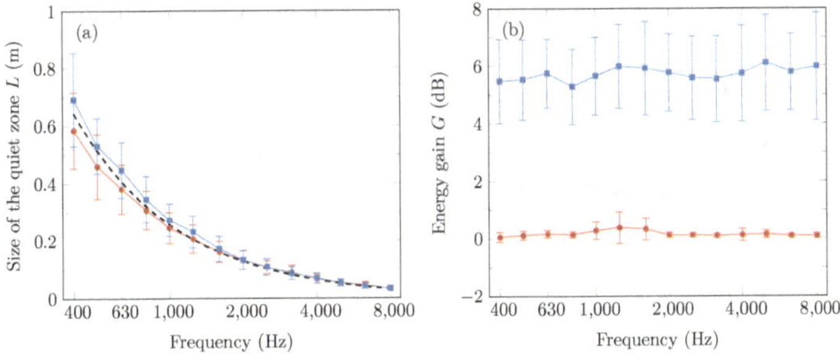

Figure 7.24 For random 3D primary sound fields under the optimal control with eight 2D secondary sources in the plane xOy, (a) the quiet zone size as a function of frequency; (b) the energy gain as a function of frequency. —●— , PAL; —■— , monopole; - - - , 0.75λ. Extracted from [Zhong et al., 2022e, Fig. 7].

Figure 7.25 shows the quiet zone size and energy gain based on 100 random trials of 3D primary sound fields using different numbers of 2D secondary sources. When the secondary source number N_s is small, the quiet zone size increases as N_s increases, but it changes slightly when $N_s > 10$. At 1 kHz, the quiet zone size generated by using the point monopoles and PALs approaches 0.29 m and 0.27 m, respectively, at large secondary source numbers. At 2 kHz, the quiet zone size approaches to 0.14 m. The maximal quiet zone size is about 0.75λ no matter how many secondary sources are used. When $N_s < 8$, the quiet zone size can be estimated as

$$L = 0.095\lambda N_s \tag{7.29}$$

which is a half of that in Eq. (7.28). The reason for having a small quiet zone size is that the primary noises coming out of the xOy plane cannot be effectively controlled because no secondary sources are placed in these directions. Furthermore, the energy gain caused by both types of secondary sources decreases with an increasing number of secondary sources, as shown in Fig. 7.25(b).

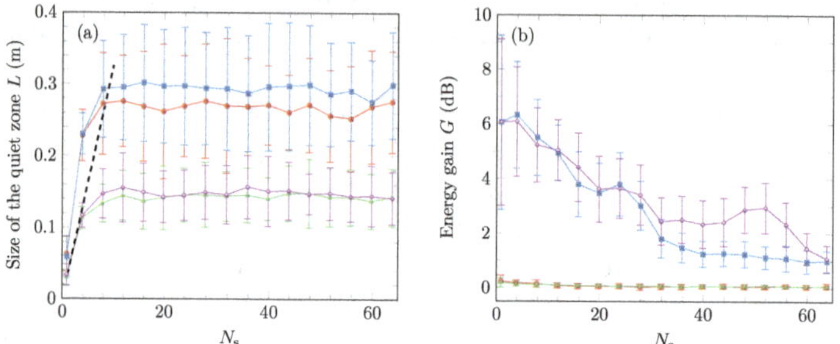

Figure 7.25 For random 3D primary sound fields controlled by 2D secondary sources at 1 kHz and 2 kHz: (a) the quiet zone size as a function of secondary source number; (b) the energy gain as a function of secondary source number. —●—, PAL at 1 kHz; —■—, monopole at 1 kHz; —▲—, PAL at 2 kHz; —◇—, monopole at 2 kHz; - - -, $0.095\lambda N_s$. Extracted from [Zhong et al., 2022e, Fig. 8].

7.4.4.2 Experimental Validation

The experimental results are presented in this subsection to validate the above simulations. The experiments were conducted in a full anechoic room with dimensions of 11.4 m × 7.8 m × 6.7 m (height). The sketch and photo of the experimental setup and equipment are shown in Figs. 7.26 and 7.27, respectively. All equipment was placed at the same height. Four commercial PALs (Holosonics Audio Spotlight AS-16i) were used and evenly located on a circle with a radius of 2.2 m, which is a 2D configuration with 4 secondary sources. The PAL has a surface size of 0.4 m × 0.4 m and a carrier frequency of 64 kHz. Four custom-made conventional loudspeakers with dimensions of 20.7 cm × 18.7 cm × 11.6 cm were used as primary sources and they were placed on a circle with a radius of 2.5 m.

An error microphone array was placed at the center with a size of 0.55 m × 0.55 m and a grid separation of 0.05 m. All microphones are electret microphones with a sensitivity of about 30 mV/Pa and are of the same type BAST M1212 (No. 23 Xixiaofu, Haidian, Beijing). The sound pressure at microphones was sampled with a Brüel & Kjær PULSE system (the analyzer 3053-B-120 equipped with the front panel UA-2107-120). The fast Fourier transform analyzer in PULSE LabShop was used to obtain the FFT spectrum. The frequency span was set to 6.4 kHz, with 6400 lines and the averaging type is linear with 66.67% overlap and 30 s duration. All microphones were covered by a piece of small and thin plastic film to avoid spurious sound.

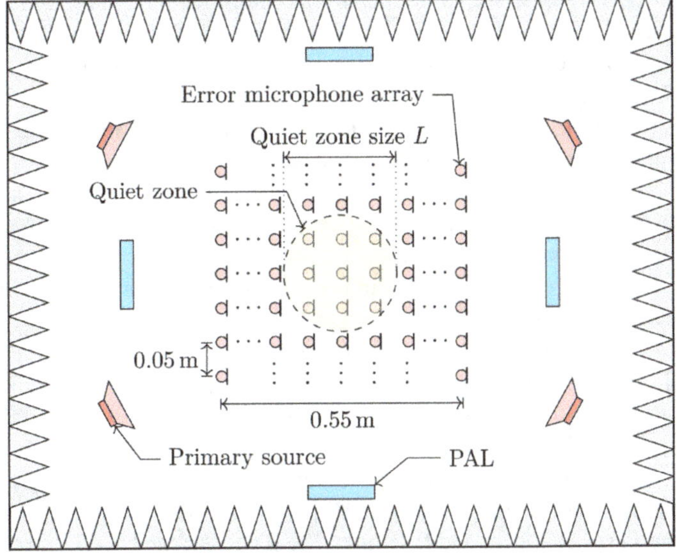

Figure 7.26 Sketch of the experiment setup in a full anechoic room. Extracted from [Zhong et al., 2022e, Fig. 11].

Figure 7.27 Photo of the experiment setup in a full anechoic room. Extracted from [Zhong et al., 2022e, Fig. 12].

A custom-made ANC controller was used in the experiment which has a multicore digital signal processor (type TMS320C6678F, Texas Instruments). All of the error microphones used in the experiments were connected into the ANC controller via a multi-channel pre-amplifier. The output signals for the secondary sources were calculated based on a multi-channel FxLMS algorithm and fed into the PALs directly. The primary sources were played with pure tone sound, and the signal was also fed to the controller as the reference signal. All the signals fed into the primary loudspeakers are correlated.

The quiet zone size in the experiments was determined as follows. First, simulations were used to estimate the quiet zone size in Fig. 7.26. Second, all microphones inside this circle were connected to the controller as the error sensors. When the number of microphones exceeds 24, only 24 of them were selected and then uniformly distributed with a separation of adjacent microphones no less than one sixth of the wavelength. This tradeoff is due to the limitations of the number of input channels available and the computational ability of the controller. Third, a tonal primary sound field was generated, and the ANC process started. Finally, if the measured noise reduction at these microphones was less (or larger) than 10 dB, the size was decreased (or increased) a small amount until a quiet zone size was found where the noise reduction was close to 10 dB.

Figure 7.28 compares the experimental measurements with predictions obtained using Eq. (7.28) for 2 and 4 PALs as secondary sources, for 1/3 octave center frequencies from 400 Hz to 4 kHz. The experimental results are generally in accordance with predictions from 500 Hz to 4 kHz, which demonstrates that PALs are able to create a quiet zone in real multi-channel ANC systems like conventional loudspeakers. The measured quiet zone size at low frequencies is lower than expected. For example, the measured size is 0.27 m and 0.48 m at 400 Hz with 2 and 4 PALs, respectively, which is lower than 0.32 m and 0.65 m in predictions. This might be caused by the poor low-frequency response of the PAL. The measured size above 630 Hz is usually slightly larger than predictions with 4 PALs. The reason might be that only 4 primary sources were used in the experiments which is not ideal to simulate a random primary sound field. The measured size is lower than prediction at higher frequencies when using 2 PALs. This can be attributed to that the grid separation (5 cm in the experiments) of the microphone array is not fine enough to identify the exact quiet zone size. Finally, it is noted that the experiment was done only for the 2D configuration due to the practical difficulties and the number of PALs available. However, over most of the frequency range the predictions agree well with the experimental measurements.

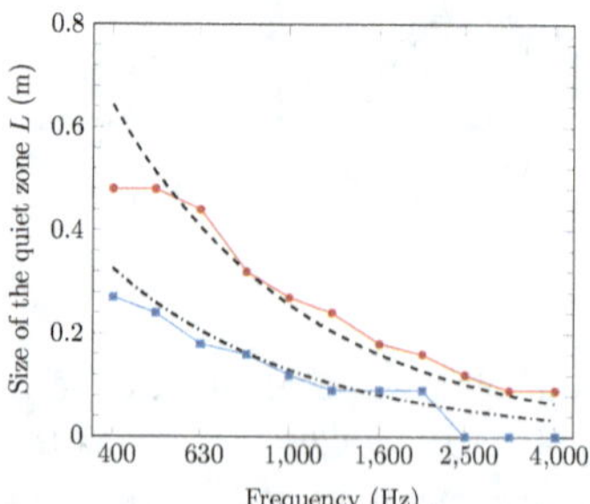

Figure 7.28 Experimental measurements and predictions of SPL for 2 and 4 PALs as secondary sources. —●—, experimental measurements for 4 PALs; —■—, experimental measurements for 2 PALs; - - - , predictions 0.75λ; −·−·−, predictions 0.38λ. Extracted from [Zhong et al., 2022e, Fig. 13].

7.4.5 THREE-DIMENSIONAL CONFIGURATION

Figure 7.29 shows the quiet zone size and energy gain when random 3D primary sound fields are optimally controlled by 8 and 20 secondary sources located in three-dimensional space at different frequencies. Compared with Fig. 7.24(a), the quiet zone size increases from 0.75λ to λ by using a 3D instead of 2D secondary sources, even with the same number of secondary sources ($N_s = 8$). Therefore, the quiet zone size is not limited by 0.75λ, as was the case for 2D secondary sources, but may be increased to 2.2λ when the secondary source number is 20. For the monopoles, although the quiet zone is enlarged, the energy gain is also increased by more than 2–4 dB as the number of secondary sources increased from 8 to 20 as shown in Fig. 7.29(b).

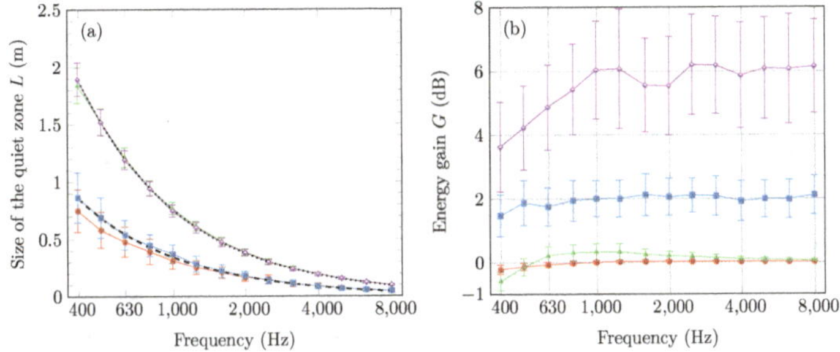

Figure 7.29 For random 3D primary sound fields under the optimal control with the 3D secondary sources, (a) the quiet zone size as a function of frequency; (b) the energy gain as a function of frequency. —●—, PAL when $N_s = 8$; —■—, monopole when $N_s = 8$; —▲—, PAL when $N_s = 20$; —◇—, monopole when $N_s = 20$; - - -, λ; ······, 2.2λ. Extracted from [Zhong et al., 2022e, Fig. 9].

Figure 7.30 shows the quiet zone size and energy gain when random 3D primary sound fields are optimally controlled by the 3D secondary sources at 1 kHz and 2 kHz. The quiet zone size is observed to be approximately proportional to the square root of the secondary source number, and can be estimated by

$$L = 0.55\lambda\sqrt{N_s} \tag{7.30}$$

when the secondary source number N_s is less than 120. To control the 3D primary sound field, the secondary sources are distributed evenly on a spherical surface. Suppose there is a smaller sphere with a diameter of L centered at the origin and enclosed by the secondary sources. The area of this sphere divided by N_s is $\pi L^2/N_s$, and by using the value for L from Eq. (7.30), the area that can be controlled by each secondary source is then $0.95\lambda^2$. In other words, the size of the area controlled by each secondary source is about one wavelength. As for the energy gain, the gain from using the monopoles varies significantly at different secondary source numbers, whereas the gain from the PALs is much smaller for both the mean value and standard deviation.

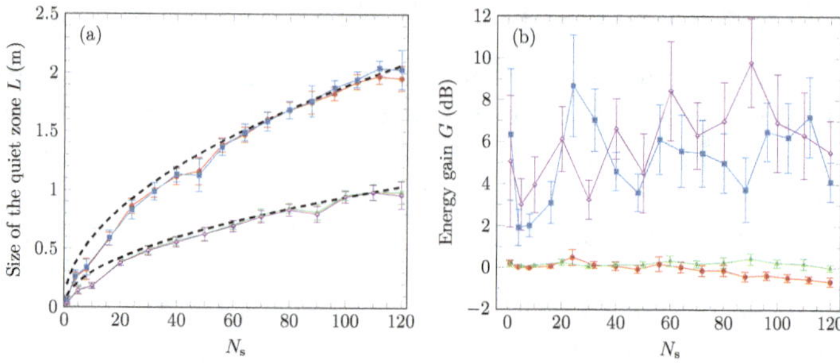

Figure 7.30 For random 3D primary sound fields under the optimal control with the 3D secondary sources, (a) the quiet zone size as a function of secondary source number; (b) the energy gain as a function of secondary source number. —●—, PAL at 1 kHz; —■—, monopole at 1 kHz; —▲—, PAL at 2 kHz; —◇—, monopole at 2 kHz; - - -, $0.55\lambda\sqrt{N_S}$. Extracted from [Zhong et al., 2022e, Fig. 10].

7.4.6 REMARKS

The reader is introduced to ANC systems employing multiple PALs in this section. The quiet zone size generated by PALs is found to be similar to that observed with conventional omnidirectional loudspeakers. However, the spillover effects of using PALs as secondary sources are much smaller than that with conventional omnidirectional loudspeakers, indicating that they can create a quiet zone around the target point with little side effect in other areas. This is because PALs are highly directional loudspeakers, whereas conventional loudspeakers are approximately omnidirectional in the low-frequency range. Therefore, PALs can be promising secondary sources in multichannel ANC systems, such as a virtual sound barrier system [Qiu, 2019]. However, it should be noted that the poor low-frequency response of PALs may limit their use in real applications at low frequencies.

In the study, all of the points inside the target zone to be controlled are chosen as error locations, which requires many error sensors and a high-performance digital signal processor in implementation. To reduce the number of error sensors, it is desirable to conduct further studies on the optimal error sensing strategy when using PALs.

7.5 SUMMARY

In ANC applications, the spillover effect can happen when the noise in some other areas is unwantedly amplified, albeit it is reduced in the targeted quiet zone. This can be particularly frustrating in practice, especially when conventional omnidirectional loudspeakers are employed as secondary sources. PALs, with their capability to generate highly directional audio beams, offer a promising alternative to mitigate such issues.

A brief introduction on ANC systems and the spillover effects are presented in Sec. 7.1. Section 7.2 delves into the performance of using a single conventional, steerable, or focusing PAL in a single-channel ANC system to create a quiet zone. A clear improvement in spillover effects is demonstrated when comparing PALs to conventional omnidirectional loudspeakers. Additionally, a remote error sensing strategy is presented, showcasing its compatibility with PALs to create a compact system for remote noise reduction near a human's ears. In Sec. 7.3, the binaural ANC system employing two PALs is introduced, highlighting PALs' ability to introduce less crosstalk between binaural channels compared to conventional loudspeakers.

Section 7.4 investigates the quiet zone generated by an ANC system using multiple PALs. The theoretical framework of multi-channel active control of sound fields and relevant metrics for characterizing noise reduction performance are provided. Two typical configurations are considered: 2D control, where secondary sources are uniformly distributed around the circumference of a circle on the same plane as primary sources, and 3D control, where secondary sources are uniformly distributed over a spherical surface. The size of the quiet zone generated by multiple PALs and the potential energy gain are compared to those using omnidirectional loudspeakers, with both simulation and experimental results demonstrating the advantages of multiple PALs in multi-channel ANC systems.

In this chapter, only the active control of primary field in free space is analyzed and discussed. While canceling noise in more complex environments like rooms holds practical significance, the utilization of PALs in such scenarios remains largely unexplored.

8 Implementation of PALs

8.1 INTRODUCTION

The implementation of PALs presents unique challenges in various aspects, including hardware, signal processing, measurement, and safety issues, because the fundamental physical mechanism of PALs differs from that of conventional loudspeakers. In Sec. 8.2, major components of PAL hardware are introduced, which include an ultrasonic emitter, a power amplifier, and a signal processor. Figure 8.1 shows a block diagram for implementing a PAL, where an audio signal undergoes preprocessing in a signal processor, and then a pure-tone ultrasonic signal is amplitude modulated with the preprocessed audio signal with different modulation algorithms. The modulated signal then passes through a power amplifier and is emitted by an ultrasonic emitter. Due to the nonlinear parametric effect, the audio sound is generated and demodulated in the air. Piezoelectric transducers were commonly used as ultrasonic emitters in early studies; however, these transducers have certain limitations, such as narrow bandwidth, low electroacoustic power conversion efficiency, and difficulties in miniaturization. To overcome these shortcomings, micromachined ultrasound transducers (MUTs) were developed. Details on the ultrasonic emitters used for PALs are presented in Sec. 8.2.1. Different power amplifiers can be employed in PAL implementations, which are discussed in Sec. 8.2.2. The implementation of PALs can be achieved through both analog and digital circuits, and the relevant signal processors are discussed in Sec. 8.2.3.

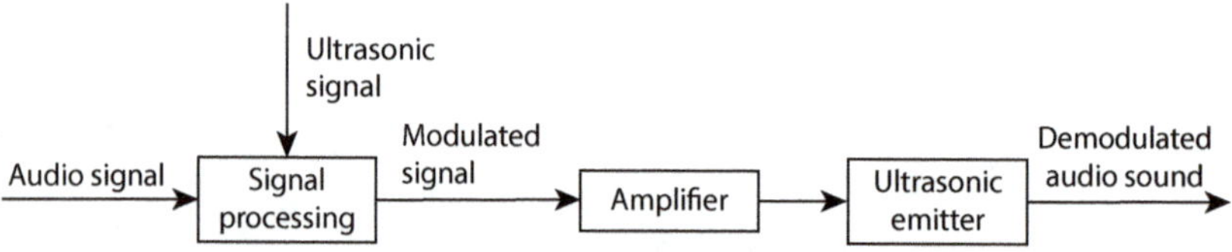

Figure 8.1 Block diagram of the implementation of a PAL. Extracted from [Gan et al., 2012, Fig. 11].

To achieve high-fidelity audio reproduction, signal processing techniques and modulation algorithms play a crucial role, which are discussed in Sec. 8.3. In practical PAL implementations, the demodulation of wideband audio signals in air is usually employed instead of pure-tone signals. Therefore, analyzing PAL reproduction performance requires applications of the transient sound theory. The quasilinear solution in the time domain is introduced in Sec. 8.3.1, which is served as the foundation for further derivations. Although the KZK equation in the time domain can be solved [Ji et al., 2011], it has been shown in Chap. 2 to be inaccurate even in the paraxial region. Hence, Sec. 8.3.1.1 begins with the quasilinear solution of the Westervelt equation in the time domain. Then, the widely used Berktay's far-field solution

DOI: 10.1201/9781003354994-8

for the signal preprocessing is derived based on the collimated beam assumption in Sec. 8.3.1.2. Followed these, the preprocessing methods as discussed in Sec. 8.3.2.

According to the Berktay's far-field solution, the sound pressure of the demodulated audio signal along the propagation axis is proportional to the second derivative of the square of the envelope of the amplitude-modulated ultrasonic carrier wave. Consequently, directly feeding the audio signal into a PAL introduces two major issues. Firstly, the amplitude of the on-axis audio sound pressure is proportional to the square of the audio sound frequency due to the presence of the second derivative with respect to time. This leads to a non-flat frequency response. The equalization of the frequency response is addressed in Sec. 8.3.2.1. Secondly, the demodulated audio sound experiences significant distortion due to the presence of the square of the envelope signal. The distortion performance, as well as several typical modulation algorithms, are discussed in Sec. 8.3.2.2. The phased array technique enhances the flexibility to control the directional audio beam produced by a PAL. The structure of the phased array PAL is discussed in detail in Sec. 8.3.3. Further analysis pertaining to the generation of pure-tone and wideband audio signals using the phased array technique is provided in Secs. 8.3.3.1 and 8.3.3.2, respectively.

The audible sound reproduced by a PAL results from the nonlinear interactions of intense ultrasound in the air, which can reach levels of up to 130 dB, particularly in the near field. Such high-intensity ultrasound exposure presents two significant challenges. Firstly, when intense ultrasound at two frequencies f_1 and f_2 is measured by microphones, the nonlinear intermodulation distortion of the microphone itself produces an additional unwanted signal at the difference frequency $f_a = f_2 - f_1$. This additional signal, known as *spurious sound*, is not actually the audio sound in the air and cannot be heard by human ears. Nonetheless, it hinders the accurate measurement of audio sound pressure. The concept of spurious sound in measurements, along with potential solutions, is introduced in Sec. 8.4. The second challenge relates to safety considerations due to exposure to such high-intensity ultrasonic waves, as discussed in Sec. 8.5. Therefore, an essential aspect of designing and developing a PAL is to explore methods of reducing the intensity of the carrier ultrasound without compromising the performance of the demodulated audio sound.

8.2 HARDWARE

8.2.1 ULTRASONIC EMITTERS

Piezoelectric transducers are commonly used as ultrasonic emitters in PALs due to their low cost and availability in the market. Figure 8.2(a) illustrates a typical structure of a conventional piezoelectric ultrasonic transducer, which consists of a piezoelectric material layer sandwiched between two thin, highly conductive electrode layers connected with electrical wires [Qiu et al., 2015]. The transducer utilizes the longitudinal vibration mode (d_{33}-mode) of the piezoelectric material, where the anti-resonant frequency is determined by the thickness of the piezoelectric layer and the longitudinal velocity of sound [ANSI/IEEE Std 176, 1987; Rupitsch, 2019]. Table 8.1 and Fig. 8.3 provide information on some commercial ultrasonic emitters available worldwide, including their dimensions and cited literature.

PALs are typically built using arrays of small radiators because it is difficult to design a single large radiator that can achieve a uniform vibration mode across the whole area [Kim et al., 2022a]. It is then important to have ultrasonic emitters with a uniform phase response, enabling the radiators to produce in-phase waves. Table 8.2 presents the phase response standard deviation of selected commercially available ultrasonic emitters [Marzo et al., 2017]. In the measurements, the emitters were excited with a 10 Vpp square-wave generated by an Agilent 33210A, and a wideband microphone (1/8 inch Brüel & Kjær calibrated microphone Type 4138-A-015) positioned 20 cm away, was used to measure the peak-to-peak amplitude and phase deviation with respect to a reference signal. The measurements were repeated for 10 emitters. The results show that the MA40S4S emitter from Murata exhibits the lowest phase response standard deviation of 3.8°. Consequently, a PAL comprising multiple MA40S4S emitters yields a more uniform radiation profile compared to other emitters. This is one of the major reasons why the MA40S4S emitter is widely used in the literature, as shown in Table 8.1.

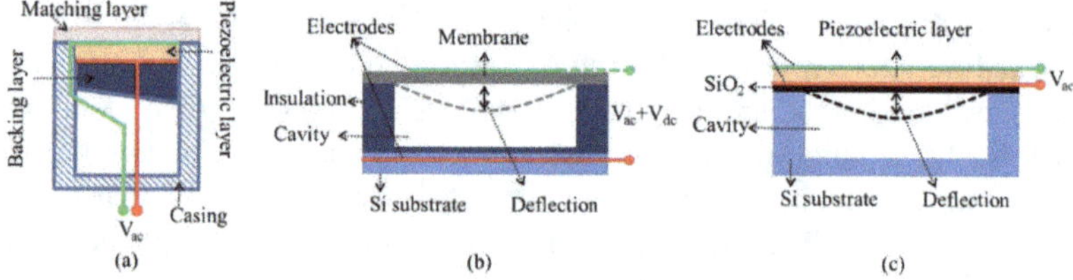

Figure 8.2 Typical cross-sectional structures of (a) conventional piezoelectric ultrasonic transducers, (b) CMUTs, and (c) PMUTs. Extracted from [Qiu et al., 2015, Fig. 1].

Conventional piezoelectric ultrasonic emitters have several limitations in PAL implementations. First, the bandwidth of a piezoelectric ultrasonic emitter is typically narrow due to its physical mechanisms. Figure 8.4 illustrates the measured frequency response of the MA40S4S emitter [Murata Manufacturing, 2017]. A standard condenser microphone (Brüel Kjær 4135) is positioned 30 cm away and on the radiator axis of the emitter, with an input voltage of 10 Vrms. It can be observed that the −3 dB bandwidth is only 1.8 kHz, which is significantly less than that of a typical conventional loudspeaker. Second, conventional piezoelectric ultrasonic emitters usually have low electroacoustic power conversion efficiency. This is primarily due to the substantial acoustic impedance mismatch between air (415 Rayls) and the emitter (> 34 MRayls) [Gallego-Juarez et al., 1978; Qiu et al., 2015].

Moreover, the size of piezoelectric ultrasonic emitters is typically larger than the ultrasound wavelength due to fabrication constraints at miniature dimensions. For instance, the MA40S4S emitter has a diameter of 10 mm, which exceeds the ultrasound wavelength, 8.6 mm, at a typical frequency of 40 kHz. Consequently, when implementing a PAL with an array of these kinds of emitters, undesired grating lobes may arise. In phased array PALs used for beam steering or focusing, the separation between adjacent ultrasonic emitters must adhere to strict requirements. For example, a separation distance less than half the wavelength is required based on the Nyquist

Table 8.1
Commercial PZT-based ultrasonic emitter examples.

Model	Company	Center frequency (kHz)	Diameter (mm)	Used in
Airducer AT-50	AirMar	50	57	[Hedberg et al., 2010; Johnson, 2016]
Airducer AT-75	AirMar	75	25	[Hedberg et al., 2010]
FBULS1007P-T	Ningbo Best Group Factory	40	10	[Marzo et al., 2017]
MA40H1S-R	Murata	40	5.2	[Jager, 2019; Okano and Kajikawa, 2022]
MA40S4S	Murata	40	10	[Marzo et al., 2017; Hahn et al., 2021; Jager, 2019; Marzo et al., 2018; Sayin et al., 2013; Sayin and Guasch, 2013; Ji and Yang, 2019; Shi, 2013; Tseng et al., 2018; Hirayama et al., 2019]
MCUSD16A40S12RO	MultiComp	40	16	[Marzo et al., 2017; Arnela et al., 2018]
MCUST10P40B07RO	MultiComp	40	16	[Marzo et al., 2017; Arnela et al., 2018; Arnela et al., 2021]
MSO-A1625H12T	Manorshi	25	16	[Zhong et al., 2023a]
MSO-A1640H10T	Manorshi	40	16	[Marzo et al., 2017]
MSO-P1040H07T	Manorshi	40	10	[Marzo et al., 2017]
MSO-P1640H10TR	Manorshi	40	10	[Marzo et al., 2017]
ST-203L	SensorTec	40	18	[Ju and Kim, 2010]
T4010A1	Nippon Ceramic	40	10	[Hoshi et al., 2010]
T4010B4	Nippon Ceramic	40	10	[Ochiai et al., 2017]
ZT40-16	Shanghai Nicera Sensor	40	16	[Ji et al., 2016; Ji and Yang, 2019]

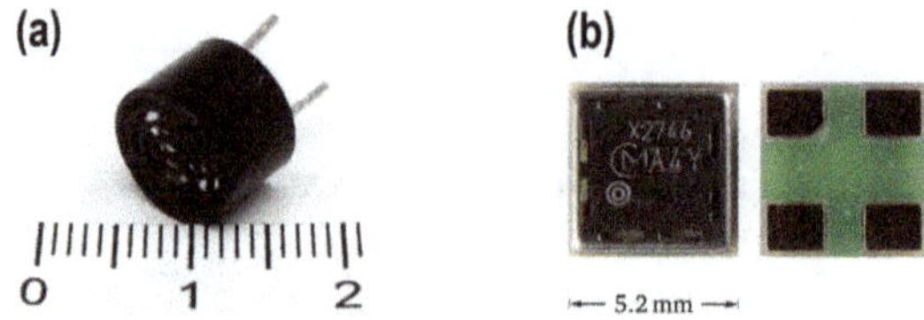

Figure 8.3 Photos of two selected commercial ultrasonic emitters. (a) Murata MA40S4S. (b) Murata MA40H1S-R. Extracted from [Jager, 2019, Fig. 2.17].

Table 8.2

Measured phase response standard deviation of selected commercially available ultrasonic emitters. The measurements were repeated for 10 items. Extracted from [Marzo et al., 2017, Table 1 in Supplementary Material].

Model	Phase response standard deviation (○)
FBULS1007P-T	13.9
MA40S4S	3.8
MCUSD16A40S12RO	18.3
MCUST10P40B07RO	33.1
MSO-A1640H10T	9.2
MSO-P1040H07T	13.9
MSO-P1640H10TR	8.7

criterion. The MA40H1S-R emitter, as shown in Fig. 8.3(b), has the smallest dimension among the commercial emitters, measuring 5.2 mm. Although it is smaller than the ultrasound wavelength, it still exceeds half the wavelength. One solution to address this issue is using waveguides, which shrink the aperture pitch below half the wavelength [Jager, 2019].

Recent developments have introduced micromachined ultrasound transducers (MUTs) as potential solutions to overcome many of the issues associated with conventional piezoelectric ultrasonic emitters. The MUT family includes capacitive micromachined ultrasonic transducers (CMUTs) [Wygant et al., 2009; Zhang et al., 2021] and piezoelectric micromachined ultrasonic transducers (PMUTs) [Qiu et al., 2015].

CMUTs, illustrated in Fig. 8.2(b), operate based on flexural vibrations resulting from a field-induced electrostatic attraction between a suspended membrane and the substrate (d_{31}- or d_{33}-mode excitation of a piezoelectric membrane). They function as miniaturized capacitors, consisting of a thin metallized suspended member positioned over a cavity with a rigid metallized substrate. When a DC voltage is applied across the electrodes, the membrane experiences deflection, attracted toward the substrate by electrostatic forces. The stiffness of the membrane provides a mechanical

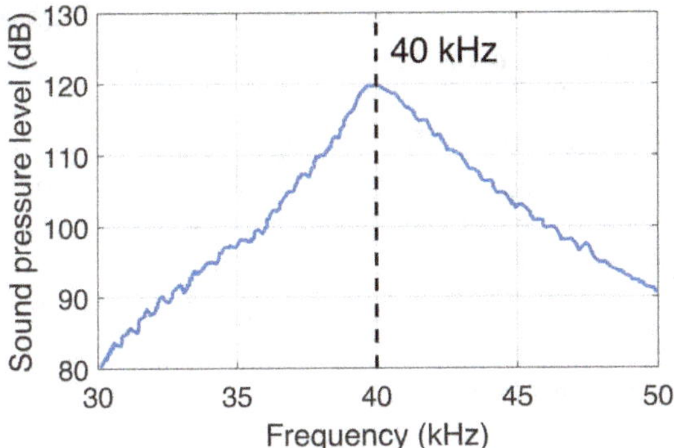

Figure 8.4 Measured frequency response of the MA40S4S emitter.

restoring force that opposes the attraction. Consequently, AC voltage input leads to membrane oscillations, generating ultrasound.

On the other hand, PMUTs, depicted in Fig. 8.2(c), rely on flexural vibrations caused by the lateral strain resulting from the piezoelectric effect of the membrane. PMUTs require at least one piezoelectric layer, along with a passive elastic layer, to enable the desired deflection of the membrane.

CMUTs have the advantage of low acoustic impedance and do not require an impedance match layer, allowing for wider bandwidths. For instance, CMUTs used in PALs have a −3 dB bandwidth of 5.3 kHz at a center frequency of 55 kHz [Wygant et al., 2009]. However, achieving high performance in practice requires applying a large DC bias near the collapse voltage (380 V in Wygant et al., 2009), which increases the risk of the device failure [Qiu et al., 2015]. In contrast, PMUTs do not require a large voltage bias and have fewer geometric and design constraints. Additionally, PMUTs can generate larger output signals, exhibit lower loss, and offer higher signal-to-noise ratios by implementing piezoelectric thin films [Trolier-McKinstry and Muralt, 2004; Lee et al., 2009]. PMUTs can efficiently convert electrical energy into acoustic waves, as demonstrated by Lee et al. with an electroacoustic efficiency of 21.9% [Lee et al., 2009]. Subsequently, they improved the efficiency to 58.4% [Je et al., 2013] and then to 71% [Je et al., 2015]. Furthermore, PMUTs have a wide bandwidth, as evidenced by a ±3 dB bandwidth of 17 kHz (from 88 to 105 kHz) [Je et al., 2015].

8.2.2 POWER AMPLIFIERS

After signal processors generate signals, power amplifiers are used to drive the emitters and provide sufficient power. In earlier PAL designs, the audio signal was typically modulated with the ultrasonic signal before being fed into a power amplifier. Class D amplifiers, known for their high efficiency in electroacoustic conversion, are frequently employed for this purpose [Stark, 2007]. In such cases, conventional audio amplifiers can be used to amplify the analog signal.

In an alternative PAL implementation, a pulse train or square wave signal with a fundamental frequency matching the resonant frequency of the ultrasonic emitter (typically ranging from 40 kHz to 80 kHz) is generated as the carrier signal. This can be achieved using cost-effective microcontrollers [Marzo et al., 2018; Hahn et al., 2021; Zhong et al., 2022d]. By driving the metal-oxide semiconductor field-effect transistor (MOSFET) drivers with a real-time audio signal, an amplitude modulation on the carrier signal is achieved, resulting in the generation of audio sound in the air. In this configuration, MOSFET drivers serve as power amplifiers.

8.2.3 SIGNAL PROCESSORS

In earlier studies, PALs were commonly implemented using analog circuits [Yoneyama et al., 1983]. As illustrated in Fig. 8.1, an analog ultrasonic signal is modulated by the analog audio signal and then passed to the power amplifier. In the analog signal system, a pure-tone ultrasonic signal can be generated using an arbitrary function generator [Ju and Kim, 2010]. However, due to the nonlinear nature of the PAL, complex signal processing methods are necessary to reduce the distortions in demodulated audio signals. Implementing such methods in an analog signal system can be challenging.

To address these challenges, digital signal systems were introduced for PAL implementation. In a typical digital signal system, the analog signal is first converted into digital format using an analog-to-digital converter (ADC). The obtained digital signal can then be processed using a digital signal processor (DSP) or a field-programmable gate array (FPGA), providing greater flexibility in signal processing. The processed digital signal is converted back to analog signal using a digital-to-analog converter (DAC) and then fed into the power amplifier. This digital signal system allows for sophisticated signal processing techniques to be applied with ease.

The implementation of a PAL using digital processing systems involves ultrasonic frequency oscillation and modulation. Traditional DSP chips typically process digital signals sequentially, leading to delays in the processed signals. FPGAs offer a programmable feature with logical cells that allow for parallel processing of signals, resulting in lower delays. Therefore, FPGA implementation is preferable for PAL applications. A detailed implementation using FPGAs can be found in Karnapi et al., 2004. Another FPGA implementation for a steerable PAL was proposed in Wu et al., 2012, which utilized fractional delay filters to enhance the precision of delay control.

Implementations of PALs based on digital signal systems require a high sampling frequency, which must be at least twice the ultrasonic frequency to adhere to the Nyquist criterion. For instance, a sampling frequency of 192 kHz is commonly adopted in literature, allowing for the processing of ultrasonic signals up to 96 kHz [Karnapi et al., 2004]. This sampling frequency is much higher compared to conventional loudspeakers, which only require processing of signals within the audible range (up to 20 kHz). Moreover, even higher sampling frequencies are necessary for phased array PALs, where precise control of time delays for each channel is crucial [Shi, 2013]. To overcome this limitation, the pulsed train signal technique has been recently proposed [Marzo et al., 2018]. For instance, for the steerable PAL as discussed in Chap. 6, the minimum steerable angle is limited by the sampling frequency. When the sampling frequency is 192 kHz, the sampling period is about 5.2 μs, and the minimum time delay is equal to the sampling period. When the emitter spacing is 5 mm, the minimum steering angle is found to be 20.96° [Okano and Kajikawa, 2022]. This approach involves generating a pulse train signal with the fundamental frequency matching the resonant frequency of the ultrasonic emitter as the carrier signal, which can be achieved using cost-effective microcontrollers. By modulating the carrier signal amplitude in real-time with an audio signal and powering the MOSFET drivers, the real-time audio sound is generated in the air. This technique enables high-resolution phase manipulation of different channels by simply temporal shifting of the pulse train, eliminating the need for a high sampling rate system. This implementation structure has been successfully used for the realization of steerable [Hahn et al., 2021] and focusing [Zhong et al., 2022d] PALs.

8.3 SIGNAL PROCESSING AND MODULATION ALGORITHMS

8.3.1 QUASILINEAR SOLUTIONS IN THE TIME DOMAIN

8.3.1.1 Quasilinear Solution of the Westervelt Equation

Consider the quasilinear solution of the Westervelt equation given by Eq. (1.95). If the Fourier transform pair is introduced as

$$\begin{cases} p(\mathbf{r},\omega) = \mathcal{F}[p(\mathbf{r},t)] = \displaystyle\int_{-\infty}^{\infty} p(\mathbf{r},t)\mathrm{e}^{\mathrm{i}\omega t}\,\mathrm{d}t, \\ p(\mathbf{r},t) = \mathcal{F}^{-1}[p(\mathbf{r},\omega)] = \dfrac{1}{2\pi}\displaystyle\int_{-\infty}^{\infty} p(\mathbf{r},\omega)\mathrm{e}^{-\mathrm{i}\omega t}\,\mathrm{d}\omega, \end{cases} \tag{8.1}$$

then Eq. (1.95) can be written as

$$\begin{cases} \left[\nabla^2 + \left(\dfrac{\omega}{c_0} + \mathrm{i}\alpha(\omega)\right)^2\right] p_{\mathrm{I}}(\mathbf{r},\omega) = \mathrm{i}\rho_0\omega q_{\mathrm{I}}(\mathbf{r},\omega), \\ \left[\nabla^2 + \left(\dfrac{\omega}{c_0} + \mathrm{i}\alpha(\omega)\right)^2\right] p_{\mathrm{II}}(\mathbf{r},\omega) = \mathrm{i}\rho_0\omega q_{\mathrm{II}}(\mathbf{r},\omega), \end{cases} \tag{8.2}$$

where $\alpha(\omega)$ is the absorption coefficient at the angular frequency of ω, the Fourier pairs are introduced as $p_{\mathrm{I}}(\mathbf{r},\omega) = \mathcal{F}[p_{\mathrm{I}}(\mathbf{r},t)]$, $p_{\mathrm{II}}(\mathbf{r},\omega) = \mathcal{F}[p_{\mathrm{II}}(\mathbf{r},t)]$, $q_{\mathrm{I}}(\mathbf{r},\omega) = \mathcal{F}[q_{\mathrm{I}}(\mathbf{r},t)]$, and $q_{\mathrm{II}}(\mathbf{r},\omega) = \mathcal{F}[q_{\mathrm{II}}(\mathbf{r},t)]$. It is worth noting that the first and second equations of Eq. (8.2), which govern the primary and secondary wave fields, respectively, exhibit the form of the inhomogeneous Helmholtz equation.

A typical PAL is commonly modeled by a baffled planar radiator positioned at the plane $z = 0$. The radiation profile of the PAL corresponds to the velocity in the z-direction, denoted as $v_z(\mathbf{r},t)$. Employing the image source method, the equivalent source density is expressed as $q_{\mathrm{I}}(\mathbf{r},t) = 2v_z(\mathbf{r},t)\delta(z)$, where $\delta(\cdot)$ is the Dirac delta function. Consequently, the primary wave field can then be solved through the Rayleigh integral as [Beranek and Mellow, 2019]

$$p_{\mathrm{I}}(\mathbf{r},\omega)=\frac{-\mathrm{i}\rho_0\omega}{2\pi}\iint_{-\infty}^{\infty}\frac{v_z(\mathbf{r}_\mathrm{s},\omega)}{|\mathbf{r}-\mathbf{r}_\mathrm{s}|}\mathrm{e}^{\mathrm{i}(\omega/c_0+\mathrm{i}\alpha(\omega))|\mathbf{r}-\mathbf{r}_\mathrm{s}|}\mathrm{d}x_\mathrm{s}\mathrm{d}y_\mathrm{s},\tag{8.3}$$

where $v_z(\mathbf{r},\omega)=\mathcal{F}[v_z(\mathbf{r},t)]$ is the spectrum of the vibration profile of the radiation source.

Applying the inverse Fourier transform to Eq. (8.3), the transient solution of the primary wave is obtained as [Cheng, 2019b, Eq. (2.5.62)]

$$p_{\mathrm{I}}(\mathbf{r},t)=\frac{\rho_0}{2\pi}\iint_{-\infty}^{\infty}\frac{1}{|\mathbf{r}-\mathbf{r}_\mathrm{s}|}\frac{\partial V(\mathbf{r},\mathbf{r}_\mathrm{s},t)}{\partial t}\mathrm{d}x_\mathrm{s}\mathrm{d}y_\mathrm{s},\tag{8.4}$$

where

$$V(\mathbf{r},\mathbf{r}_\mathrm{s},t)=\frac{1}{2\pi}\int_{-\infty}^{\infty}v_z(\mathbf{r},\omega)\exp\left[\mathrm{i}\omega\left(\frac{|\mathbf{r}-\mathbf{r}_\mathrm{s}|}{c_0}-t\right)-\alpha(\omega)|\mathbf{r}-\mathbf{r}_\mathrm{s}|\right]\mathrm{d}\omega.\tag{8.5}$$

Given that the PAL emits an ultrasonic wave with a narrow spectral band centered around the frequency f_u, it is reasonable to assume that the absorption coefficient, $\alpha(\omega)$, remains relatively constant within this bandwidth. Therefore, $\alpha(\omega)$ can be approximated by α_u, which represents the absorption coefficient at the center ultrasound frequency. With this approximation, Eq. (8.5) can be simplified to

$$V(\mathbf{r},\mathbf{r}_\mathrm{s},t)\approx\mathrm{e}^{-\alpha_\mathrm{u}|\mathbf{r}-\mathbf{r}_\mathrm{s}|}v_z(\mathbf{r},t-|\mathbf{r}-\mathbf{r}_\mathrm{s}|/c_0).\tag{8.6}$$

The secondary wave containing the audio sound can be solved as

$$p_{\mathrm{II}}(\mathbf{r},\omega)=\frac{-\mathrm{i}\rho_0\omega}{4\pi}\iiint_{-\infty}^{\infty}\frac{q_{\mathrm{II}}(\mathbf{r}_\mathrm{v},\omega)}{|\mathbf{r}-\mathbf{r}_\mathrm{v}|}\mathrm{e}^{\mathrm{i}(\omega/c_0+\mathrm{i}\alpha(\omega))|\mathbf{r}-\mathbf{r}_\mathrm{v}|}\mathrm{d}^3\mathbf{r}_\mathrm{v}.\tag{8.7}$$

The steps for calculating the secondary wave field is then as follows:

(1) calculate the vibration profile of the primary wave in the frequency domain by $v_z(\mathbf{r},\omega)=\mathcal{F}[v_z(\mathbf{r},t)]$,
(2) calculate $V(\mathbf{r},\mathbf{r}_\mathrm{s},t)$ by Eq. (8.5) or (8.6),
(3) calculate $p_\mathrm{I}(\mathbf{r},t)$ by Eq. (8.4) or by Eq. (8.3) with the inverse Fourier transform $p_\mathrm{I}(\mathbf{r},t)=\mathcal{F}^{-1}[p_\mathrm{I}(\mathbf{r},\omega)]$,
(4) calculate the source density for the secondary wave, $q_\mathrm{II}(\mathbf{r},t)$, by Eq. (1.96),
(5) calculate the source density in the frequency domain by $q_\mathrm{II}(\mathbf{r},\omega)=\mathcal{F}[q_\mathrm{II}(\mathbf{r},t)]$,
(6) calculate the secondary wave in the frequency domain by Eq. (8.7), and
(7) calculate the secondary wave in the time domain by $p_\mathrm{II}(\mathbf{r},t)=\mathcal{F}^{-1}[p_\mathrm{II}(\mathbf{r},\omega)]$.

It is worth mentioning that the secondary wave field obtained through the aforementioned process encompasses both the audio sound and the second harmonic of the primary ultrasound. Consequently, in order to extract the audio sound from the results, an appropriate low-pass filter is necessary. The source density for the secondary wave, as shown in Eq. (8.2), can be separated into the audio sound and the second harmonic of the ultrasound as

$$q_{\mathrm{II}}(\mathbf{r},t) = q_{\mathrm{a}}(\mathbf{r},t) + q'_{\mathrm{II}}(\mathbf{r},t), \tag{8.8}$$

where the latter one attenuates rapidly. In contrast, the audio sound waves typically undergo negligible attenuation resulted from the atmospheric absorption in air. Therefore, the audio sound is governed by the simplified equation as

$$\left(\nabla^2 - \frac{1}{c_0^2}\frac{\partial^2}{\partial t^2}\right) p_{\mathrm{a}}(\mathbf{r},t) = -\rho_0 \frac{\partial q_{\mathrm{a}}(\mathbf{r},t)}{\partial t}. \tag{8.9}$$

Applying the Fourier transform to Eq. (8.9), it has

$$\left[\nabla^2 + \left(\frac{\omega}{c_0}\right)^2\right] p_{\mathrm{a}}(\mathbf{r},\omega) = \mathrm{i}\rho_0 \omega q_{\mathrm{a}}(\mathbf{r},\omega). \tag{8.10}$$

Equation (8.9) is an inhomogeneous linear wave equation and can be solved using Green's functions. The solution is expressed as [Moffett and Mellen, 1977, Eq. (2); Yoneyama et al., 1983, Eq. (3)]

$$p_{\mathrm{a}}(\mathbf{r},t) = -\frac{\rho_0}{4\pi}\iiint_{-\infty}^{\infty} \frac{1}{|\mathbf{r}-\mathbf{r}_{\mathrm{v}}|}\frac{\partial}{\partial t}\left[q_{\mathrm{a}}\left(\mathbf{r}_{\mathrm{v}}, t - \frac{|\mathbf{r}-\mathbf{r}_{\mathrm{v}}|}{c_0}\right)\right] \mathrm{d}^3\mathbf{r}_{\mathrm{v}}. \tag{8.11}$$

where $t - |\mathbf{r}-\mathbf{r}_{\mathrm{v}}|/c_0$ is the retarded time. While Eq. (8.11) provides an accurate representation of the quasilinear solution for the Westervelt equation, obtaining numerical results based on this equation can be challenging. In the subsequent section, a simplified solution is derived by employing the collimated beam assumption as that adopted in Sec. 2.2.4.

8.3.1.2 Berktay's Far-Field Solution

The implementation of a PAL involves applying the amplitude modulation on a pure-tone ultrasonic signal, $\cos(\omega_{\mathrm{u}}t)$, with an envelope signal, $E(t)$. The envelope signal carries the audio signal that is intended to be reproduced. The modulated signal, which is then fed into the ultrasonic emitter, can be represented as

$$u(t) = E(t)\cos(\omega_{\mathrm{u}}t). \tag{8.12}$$

Utilizing the collimated beam assumption (see Sec. 2.2.4 for details), the primary sound pressure on the source surface can be expressed as

$$p_{\mathrm{I}}(x,y,0,t) = p_0 u(t)\mathcal{A}(\mathbf{r}) = p_0 E(t)\cos(\omega_{\mathrm{u}}t)\mathcal{A}(\mathbf{r}), \tag{8.13}$$

where p_0 is the on-surface pressure amplitude, and $\mathcal{A}(\boldsymbol{\rho})$ indicates the aperture area, specifying the effective region of the collimated beam. For instance, for a circular PAL with a radius of a, the aperture area $\mathcal{A}(\mathbf{r}) = \mathrm{H}(a-\rho)$, where $\mathrm{H}(\cdot)$ is the

Heaviside function, and $\rho = \sqrt{x^2+y^2}$. For a rectangular PAL, the aperture area is determined by Eq. (2.35).

Due to the high collimated nature of the radiated primary wave field, an approximation can be made for the sound pressure, given by

$$p_{\mathrm{I}}(\mathbf{r},t) = p_0 E(\tau)\mathrm{e}^{-\alpha_{\mathrm{u}} z}\cos(\omega_{\mathrm{u}}\tau)\mathcal{A}(\mathbf{r}), \tag{8.14}$$

which is the transient version of Eq. (2.57), where the retarded time $\tau \equiv t - z/c_0$ is used for simplicity. According to Eq. (1.96), the source density for the secondary wave field is

$$q_{\mathrm{II}}(\mathbf{r},t) = \frac{\beta p_0^2}{2\rho_0^2 c_0^4}\mathcal{A}(\mathbf{r})\mathrm{e}^{-2\alpha_{\mathrm{u}} z}\frac{\partial}{\partial t}\{E^2(\tau)[1+\cos(2\omega_{\mathrm{u}}\tau)]\}. \tag{8.15}$$

Since the component $\cos(2\omega_{\mathrm{u}}\tau)$ attenuates quickly, it can be disregarded, leading to the source density for the audio sound as

$$q_{\mathrm{a}}(\mathbf{r},t) = \frac{\beta p_0^2}{2\rho_0^2 c_0^4}\mathcal{A}(\mathbf{r})\mathrm{e}^{-2\alpha_{\mathrm{u}} z}\frac{\partial E^2(\tau)}{\partial t}. \tag{8.16}$$

By applying the Fourier transform to Eq. (8.16), it is obtained that

$$q_{\mathrm{a}}(\mathbf{r},\omega) = \frac{-\mathrm{i}\beta p_0^2\omega}{2\rho_0^2 c_0^4}\mathcal{A}(\mathbf{r})\mathrm{e}^{(-2\alpha_{\mathrm{u}}+\mathrm{i}\omega/c_0)z}\mathcal{F}\left[E^2(t)\right], \tag{8.17}$$

where the property $\mathcal{F}[E^2(t-z/c_0)] = \mathcal{F}[E^2(t)]\exp(\mathrm{i}\omega z/c_0)$ has been used.

Substituting Eq. (8.17) into Eq. (8.11) yields

$$p_{\mathrm{a}}(\mathbf{r},\omega) = -\frac{\beta p_0^2\omega^2}{8\pi\rho_0 c_0^4}\mathcal{F}\left[E^2(t)\right]\iiint_{-\infty}^{\infty}\frac{\mathrm{e}^{(-2\alpha_{\mathrm{u}}+\mathrm{i}\omega/c_0)z_{\mathrm{v}}+\mathrm{i}\omega/c_0|\mathbf{r}-\mathbf{r}_{\mathrm{v}}|}}{|\mathbf{r}-\mathbf{r}_{\mathrm{v}}|}\mathrm{d}^3\mathbf{r}_{\mathrm{v}}. \tag{8.18}$$

Similar to the technique used in Sec. 2.2.4, the audio sound pressure can be simplified as

$$p_{\mathrm{a}}(\mathbf{r},\omega) = -\frac{\beta p_0^2\omega^2 S_0}{8\pi\rho_0 c_0^4 r}\frac{\mathcal{F}\left[E^2(\tau)\right]}{2\alpha_{\mathrm{u}} - 2\mathrm{i}\omega/c_0\sin^2(\theta/2)}, \tag{8.19}$$

where θ is the zenithal angle in the spherical coordinate system. The on-axis audio sound pressure is obtained by setting $\theta = 0$ into Eq. (8.19) as

$$p_{\mathrm{a}}(0,0,z,t) = \frac{\beta p_0^2 S_0}{8\pi\rho_0 c_0^4\alpha_{\mathrm{t}} z}\frac{\partial^2}{\partial t^2}\left[E^2(t-z/c_0)\right], \tag{8.20}$$

where the total absorption coefficient $\alpha_{\mathrm{t}} \equiv 2\alpha_{\mathrm{u}}$. Equation (8.20) is widely known as the *Berktay's far-field solution*, initially derived and presented as Eq. (IV.9) in Berktay, 1965a. It establishes a simple closed-form expression, relating the envelope signal and the demodulated audio signal. This equation serves as the foundation for signal processing in PAL applications.

However, it is essential to note that several assumptions are made to derive Eq. (8.20), and its accuracy is limited in real-world applications. For instance, when the ultrasound pressure level is more than 130 dB, the assumption of weak nonlinearity becomes invalid (see Sec. 1.4.3). In such cases, the *Merklinger's far-field solution* is commonly used, which is expressed as [Merklinger, 1973; Merklinger, 1975]

$$p_a(0,0,z,t) = \frac{p_0 S}{4\pi c_0 \omega z} \frac{\partial^2}{\partial t^2} \left\{ E(t - z/c_0) \tan^{-1} \left[\frac{\mathcal{N}_{\mathrm{sf}}^{\mathrm{ab}} E(t - z/c_0)}{4} \right] \right\}, \tag{8.21}$$

where the *absorption-saturation number*, $\mathcal{N}_{\mathrm{sf}}^{\mathrm{ab}}$, characterizes the relative influence of absorption versus saturation, and is defined by Eq. (1.115).

When $\mathcal{N}_{\mathrm{sf}}^{\mathrm{ab}} E(\tau)/4$ is small, which indicates a weak nonlinearity, the approximation $\tan^{-1}\left[\mathcal{N}_{\mathrm{sf}}^{\mathrm{ab}} E(t)/4\right] \approx \mathcal{N}_{\mathrm{sf}}^{\mathrm{ab}} E(\tau)/4$ can be substituted into the Merklinger's solution, as shown by Eq. (8.21). This substitution results in the Berktay's solution, given by Eq. (8.20). For a detailed comparison of the signal processing techniques based on the Merklinger's and Berktay's solutions, the readers are referred to [Shi and Kajikawa, 2016a], where the differences between the two approaches are discussed.

8.3.2 PREPROCESSING METHODS

8.3.2.1 Equalization

Based on the Berktay's far field solution, the amplitude of the audio beam is proportional to the second derivative of the square of the envelope signal. This second-derivative factor introduces a frequency response slope of +12 dB per octave. To compensate for this slope and achieve a flat frequency response, an equalizer with a −12 dB per octave filter is necessary. The simplest approach is applying double integration on the audio signal prior to the amplitude modulation [Kite et al., 1998; Pompei, 1999]. The audio signal after the equalization is then

$$s^{\mathrm{eq}}(t) = \iint s(t)\mathrm{d}t. \tag{8.22}$$

A flat response within ±5 dB from 400 Hz to 16 kHz was observed in Pompei, 1999 using Eq. (8.22) for an ultrasonic frequency of 60 kHz.

It is important to note that the Berktay's solution serves as an approximation applicable in the inverse-law far field, typically located more than 10 m away from the PAL, as demonstrated in Chap. 3. In the near field, more accurate models are necessary to obtain precise equalization. For instance, simulations based on the KZK equation indicate that the gain slope of the frequency response of the demodulated audio sound is smaller than that predicted by the Berktay's solution, ranging from +6 dB to +12 dB per octave [Ji et al., 2011]. However, it should be acknowledged that the KZK equation also exhibits inaccuracies in the near field, and the utilization of the Westervelt and Kuznetsov equations is recommended. Furthermore, the frequency response of the audio sound generated by a PAL may vary at different locations and/or excitation levels, which distinguishes it from that of a conventional loudspeaker. Consequently, further research is necessary to develop more suitable equalizers for PAL applications, considering the specific characteristics and requirements of the system.

8.3.2.2 Modulation Algorithms

8.3.2.2.1 Distortion Performance Measures

When reproducing an audio signal, a loudspeaker system can experience nonlinear distortions, particularly when subjected to significantly large excitation. To assess the nonlinear distortion performance of the system, two primary measures are commonly employed, which are the *total harmonic distortion (THD)* and the *intermodulation distortion (IMD)* [Klippel, 2006]. Figure 8.5 illustrates the concept of THD and IMD using a two-tone excitation comprising fundamental components at frequencies f_a and f'_a ($f'_\mathrm{a} > f_\mathrm{a}$).

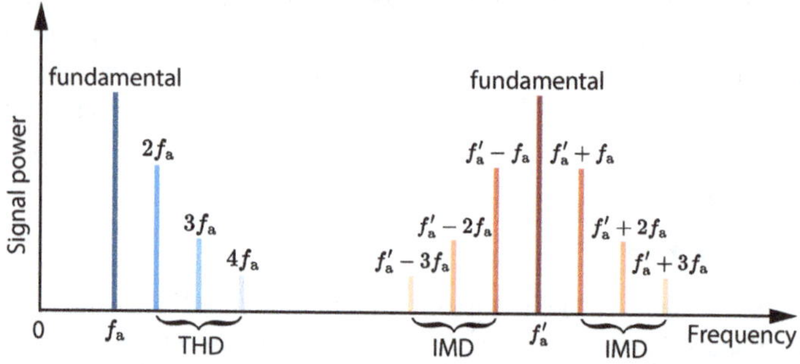

Figure 8.5 Signal power of a two-tone signal with its harmonic and intermodulation distortions.

THD quantifies the percentage of the total signal that is affected by distortion caused by harmonics. In Fig. 8.5, the components at frequencies such as $2f_\mathrm{a}, 3f_\mathrm{a}, 4f_\mathrm{a}$, etc., are harmonics of the fundamental component at frequency f_a. The International Electrotechnical Commission (IEC) and the Institute of Electrical and Electronics Engineers (IEEE) define two distinct approaches to calculate THD [Listen Inc., 2020]. At low THD values, the difference between the two definitions is minimal. However, for signals with high harmonic distortions, the two definitions yield different values [Shmilovitz, 2005; Listen Inc., 2020]. In evaluating the distortion performance of the demodulated audio sound produced by a PAL, the IEC definition is commonly used and therefore adopted in this book [Ji et al., 2011; Wang et al., 2023b].

According to the IEC definition, the n-th order THD is calculated by comparing the power sum of the distortion components to the total input power, which includes both the fundamental and distortion components. Specifically, it is given by [IEC 60268-5, 2007]

$$\mathrm{THD} = \sqrt{\frac{A^2(2f_\mathrm{a}) + A^2(3f_\mathrm{a}) + \cdots + A^2(nf_\mathrm{a})}{A^2(f_\mathrm{a}) + A^2(2f_\mathrm{a}) + \cdots + A^2(nf_\mathrm{a})}} \times 100\%, \tag{8.23}$$

where $A(f)$ is the magnitude of the signal at the frequency f.

The IMD arises from the linear combination of fundamental frequencies. When a loudspeaker is stimulated by a two-tone input at frequencies f_a and f'_a, intermodulation distortion can occur at the sum and difference frequencies of the two tones and their harmonics. These include components at frequencies $nf_a \pm n'f'_a$, where $n, n' = 1, 2, 3, \ldots$. Various definitions for IMD exist [Klippel, 2006; IEC 60268-5, 2007]. When evaluating the performance of the demodulated audio sound from a PAL, the commonly used definition is given by [Tan et al., 2010, Eq. (20)]

$$\mathrm{IMD} = \frac{\sqrt{\sum_{n,n'=1}^{\infty} A^2(nf_a \pm n'f'_a)}}{\sqrt{A^2(f_a) + A^2(f'_a) + \sum_{n,n'=1}^{\infty} A^2(nf_a \pm n'f'_a)}} \times 100\%. \tag{8.24}$$

8.3.2.2.2 *Double Sideband (DSB) Amplitude Modulation*

The *DSB amplitude modulation (AM)* is a fundamental modulation technique initially proposed in Yoneyama et al., 1983. The block diagram representing this modulation scheme (DSBAM) is depicted in Fig. 8.6, and the envelope is given by Yoneyama et al., 1983

$$E(t) = 1 + \mathcal{M}s(t) \tag{8.25}$$

where $s(t)$ is the audio signal, and a constant $\mathcal{M}$ is referred to as the *modulation index*. Substituting Eq. (8.25) into Eq. (8.13), the primary sound pressure on the source surface is derived as

$$p_I(x, y, 0, t) \propto u(t) = [1 + \mathcal{M}s(t)]\cos(\omega_u t). \tag{8.26}$$

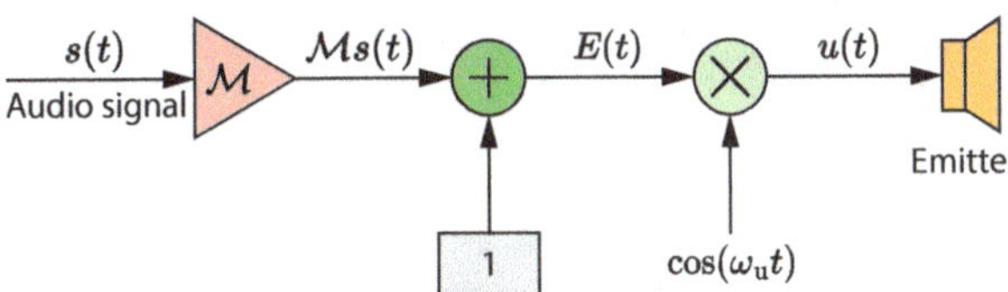

Figure 8.6 Block diagram of the DSBAM.

Equation (8.26) demonstrates the process of amplitude modulation where a carrier ultrasonic signal, $\cos(\omega_u t)$, is modulated by an envelope signal represented by Eq. (8.25). When a carrier signal undergoes amplitude modulation, it typically results in the creation of two sidebands, depicted as mirror images in Fig. 8.7. The signal components above the carrier frequency constitute the *upper sideband (USB)*, while those below the carrier frequency form the *lower sideband (LSB)*. To illustrate the generation of sidebands, a pure-tone audio signal, $s(t) = \cos(\omega_a t)$, is considered for simplicity. By using the trigonometric identity, the primary sound pressure on the source surface is proportional to

$$\underbrace{\cos(\omega_u t)}_{\text{carrier wave}} \underbrace{[1+\mathcal{M}s(t)]}_{\text{amplitude modulation}} = \underbrace{\frac{1}{2}\mathcal{M}\cos[(\omega_u-\omega_a)t]}_{\text{lower sideband}} + \underbrace{\cos(\omega_u t)}_{\text{carrier wave}} + \underbrace{\frac{1}{2}\mathcal{M}\cos[(\omega_u+\omega_a)t]}_{\text{upper sideband}}, \tag{8.27}$$

where the components at frequencies of $f_u - f_a$ and $f_u + f_a$ correspond to the LSB and USB, respectively. For a wideband audio signal with a maximum frequency of f_m, the LSB and USB consist of components within the frequency ranges $(f_u - f_m, f_u)$ and $(f_u, f_u + f_m)$, respectively, as illustrated in Fig. 8.7. Figure 8.8 specifically illustrate the spectral densities of several typical audio signals.

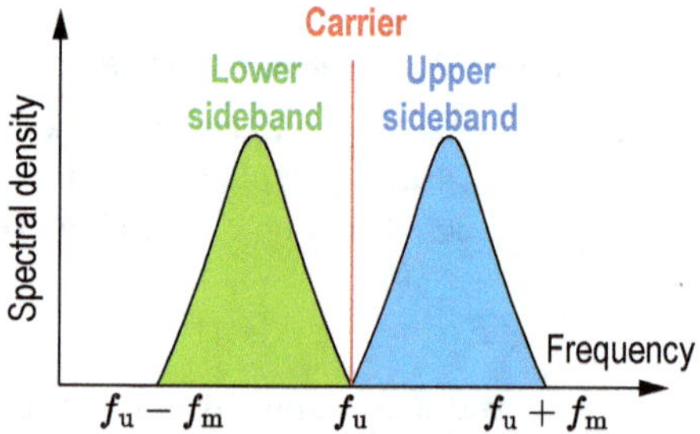

Figure 8.7 The spectral density of the amplitude modulation of a carrier ultrasonic signal by an audio signal with a maximum modulation frequency of f_m.

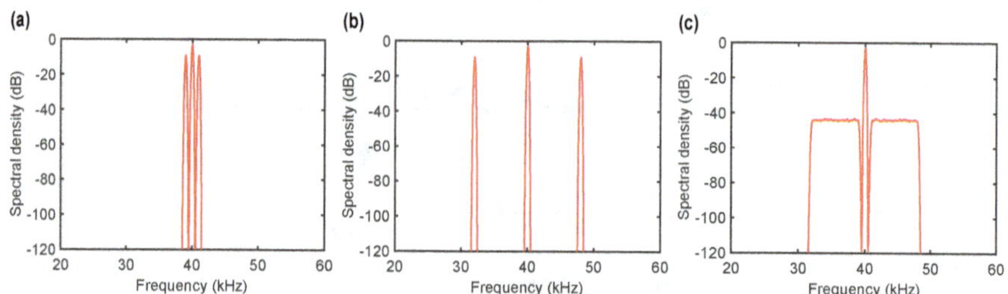

Figure 8.8 Spectral densities (dB) of different audio signals modulated by the DSBAM, where the modulation index is $\mathcal{M} = 1$. The audio signal is (a) a pure-tone signal at frequency $f_a = 1\,\text{kHz}$; (b) a pure-tone signal at frequency $f_a = 8\,\text{kHz}$; (c) a white noise signal filtered by a bandpass filter with upper and lower cutoff frequencies of 1 kHz and 8 kHz, respectively.

Substituting Eq. (8.25) into the Berktay's solution Eq. (8.20), the on-axis sound pressure is obtained as

$$p_a(0,0,z,t) \propto \mathcal{M}s''(\tau) + \mathcal{M}^2\left[s(\tau)s''(\tau) + s'^2(\tau)\right]. \tag{8.28}$$

In Eq. (8.28), the first term $\mathcal{M}s''(\tau)$ represents the fundamental component, while the distortion is generated by the terms $\mathcal{M}^2\left[s(\tau)s''(\tau) + s'^2(\tau)\right]$.

Considering a pure-tone audio signal of $s(t) = \cos(\omega_a t)$, the demodulated audio signal given by Eq. (8.28) becomes

$$p_a(0,0,z,t) \propto \mathcal{M}\cos(\omega_a t) + \mathcal{M}^2\cos(2\omega_a t). \tag{8.29}$$

Substituting Eq. (8.29) into Eq. (8.23), the THD is obtained as

$$\mathrm{THD_{DSB}} = \frac{\mathcal{M}}{\sqrt{1+\mathcal{M}^2}} \times 100\%. \tag{8.30}$$

Considering a two-tone audio signal of $s(t) = \cos(\omega_a t) + \cos(\omega_a' t)$, the IMD of the demodulated audio signal can be obtained as

$$\mathrm{IMD_{DSB}} = \frac{\mathcal{M}}{\sqrt{16+\mathcal{M}^2}} \times 100\%. \tag{8.31}$$

Equations (8.30) and (8.31) show that reducing the modulation coefficient $\mathcal{M}$ can reduce both THD and IMD, but it also reduces the amplitude of the audio sound pressure as shown by Eq. (8.28).

8.3.2.2.3 Single Sideband (SSB) Amplitude Modulation

The *SSB amplitude modulation* was initially introduced in 1985 [Kamakura et al., 1985]. The block diagram representing this modulation scheme (SSBAM) is depicted in Fig. 8.9. In SSBAM, the Hilbert transform of the audio signal is obtained and denoted as $\hat{s}(t)$. It can be expressed as [Wang et al., 2009]

$$\hat{s}(t) = s(t) * \frac{1}{\pi t}, \tag{8.32}$$

where the symbol $*$ represents the linear convolution operation. As a special case, for a cosine signal, $\cos(\omega_a t)$, the Hilbert transform yields a sine signal, $\sin(\omega_a t)$.

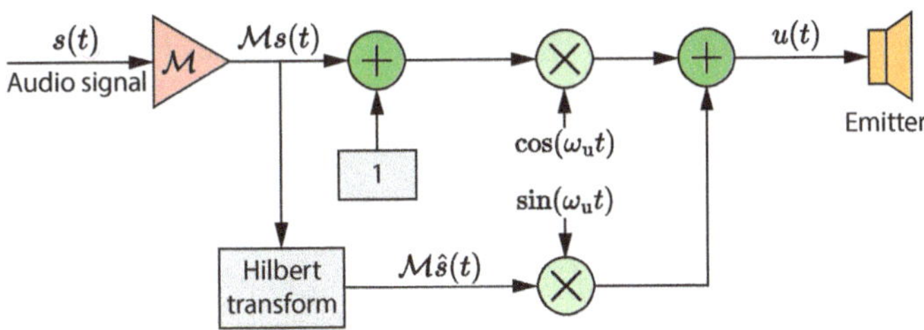

Figure 8.9 Block diagram of the SSBAM.

In the SSBAM, the primary sound pressure on the source surface is

$$p_I(x,y,0,t) \propto [1+\mathcal{M}s(t)]\cos(\omega_u t) \pm \mathcal{M}\hat{s}(t)\sin(\omega_u t). \tag{8.33}$$

Considering a pure-tone audio signal $\cos(\omega_a t)$, Eq. (8.33) can be simplified to

$$p_{\mathrm{I}}(x,y,0,t) \propto \cos(\omega_u t) + \frac{1\pm 1}{2}\mathcal{M}\cos\left[(\omega_u - \omega_a)t\right] + \frac{1\mp 1}{2}\mathcal{M}\cos\left[(\omega_u + \omega_a)t\right]. \tag{8.34}$$

It can be observed that when the positive sign is adopted in Eq. (8.33), the upper sideband component diminishes in Eq. (8.34). In contrast, when the negative sign is adopted in Eq. (8.33), the lower sideband component diminishes in Eq. (8.34). Therefore, the positive and negative signs in Eq. (8.33) represent the lower and upper SSBAMs, respectively.

The SSB amplitude modulation Eq. (8.33) can be rewritten as

$$\begin{aligned} p_{\mathrm{I}}(x,y,0,t) \propto & \sqrt{\mathcal{M}^2[s^2(t)+\hat{s}^2(t)]+2\mathcal{M}s(t)+1} \\ & \times \cos\left[\omega_u t \mp \mathrm{atan2}(\mathcal{M}\hat{s}(t), 1+\mathcal{M}s(t))\right], \end{aligned} \tag{8.35}$$

where atan2 is the 2-argument arctangent function. By definition, $\mathrm{atan2}(y,x)$ is the angle measure in radians $(-\pi,\pi]$ between the positive x-axis and the ray from the origin to the point (x,y) in the Cartesian plane.

Substituting Eq. (8.35) into the Berktay's solution Eq. (8.20), the on-axis sound pressure is obtained as

$$p_a(0,0,z,t) \propto \mathcal{M}s''(\tau) + \mathcal{M}^2\left[s'^2(\tau)+\hat{s}'^2(\tau)+s(\tau)s''(\tau)+\hat{s}(\tau)\hat{s}''(\tau)\right]. \tag{8.36}$$

In Eq. (8.36), the first term $\mathcal{M}s''(\tau)$ represents the fundamental component, and the distortion is originated by the terms $\mathcal{M}^2\left[s'^2(\tau)+\hat{s}'^2(\tau)+s(\tau)s''(\tau)+\hat{s}(\tau)\hat{s}''(\tau)\right]$.

Considering the pure-tone audio signal $s(t) = \cos(\omega_a t)$, the Hilbert transform is $\hat{s}(t) = \sin(\omega_a t)$. Substituting them into Eq. (8.36), it has

$$p_a(0,0,z,t) \propto \mathcal{M}\cos(\omega_a \tau). \tag{8.37}$$

It is evident that in the SSBAM, only the fundamental component is preserved, resulting in the absence of distortions. Therefore, $\mathrm{THD}_{\mathrm{SSB}} = 0\%$. For a two-tone audio signal represented by $s(t) = \cos(\omega_a t) + \cos(\omega_a' t)$, the IMD of the demodulated audio signal can be obtained as

$$\mathrm{IMD}_{\mathrm{SSB}} = \frac{\mathcal{M}}{\sqrt{8+\mathcal{M}^2}} \times 100\%. \tag{8.38}$$

The advantages of the SSBAM include its requirement of only half the bandwidth compared to DSBAM and the absence of THD. However, Eq. (8.38) indicates that the IMD is inevitable, albeit it can be reduced by decreasing the modulation index $\mathcal{M}$.

8.3.2.2.4 Square Root (SRT) Amplitude Modulation

The *SRT amplitude modulation* was initially proposed in Kamakura et al., 1984. It can be regarded as an inverse system to the Berktay's solution. The block diagram representing this modulation scheme is depicted in Fig. 8.10. The envelope signal is [Kamakura et al., 1984, Eq. (5)]

$$E(t) = \sqrt{1 + \mathcal{M}s(t)}. \tag{8.39}$$

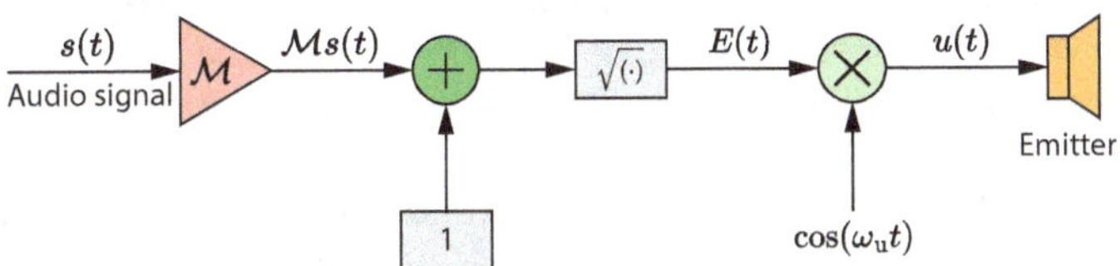

Figure 8.10 Block diagram of the SRTAM.

Substituting Eq. (8.39) into the Berktay's solution Eq. (8.20), the on-axis sound pressure is obtained as

$$p_a(0,0,z,t) \propto \mathcal{M}s''(\tau). \tag{8.40}$$

Equation (8.40) demonstrates that the SRTAM consists exclusively of the fundamental component, thereby enabling complete elimination of nonlinear distortion. Considering a pure-tone audio signal $s(t) = \cos(\omega_a t)$, the on-axis sound pressure can be obtained as

$$p_a(0,0,z,t) \propto \mathcal{M}\cos(\omega_a \tau). \tag{8.41}$$

Consequently, $\mathrm{THD_{SRT}} = 0\%$, indicating the absence of the harmonic distortion. Furthermore, when considering a two-tone audio signal represented by $s(t) = \cos(\omega_a t) + \cos(\omega_a' t)$, the IMD of the demodulated audio signal is also zero, i.e., $\mathrm{IMD_{SRT}} = 0\%$.

Comparing Eqs. (8.41) and (8.37), it is clear that the SRTAM produces an identical demodulated audio signal to that of the SSBAM with a pure-tone audio signal. However, this equivalence does not hold for wideband audio signals, as seen when comparing Eqs. (8.40) and (8.36). For instance, when the excitation is a two-tone audio signal, the SRTAM exhibits zero IMD while the SSBAM has a nonzero IMD.

As shown in Fig. 8.11, reproducing the infinite harmonics introduced by the square root operator in SRTAM requires an infinitely wide bandwidth ultrasonic transducer. However, this is not feasible in practice, as most ultrasonic transducers

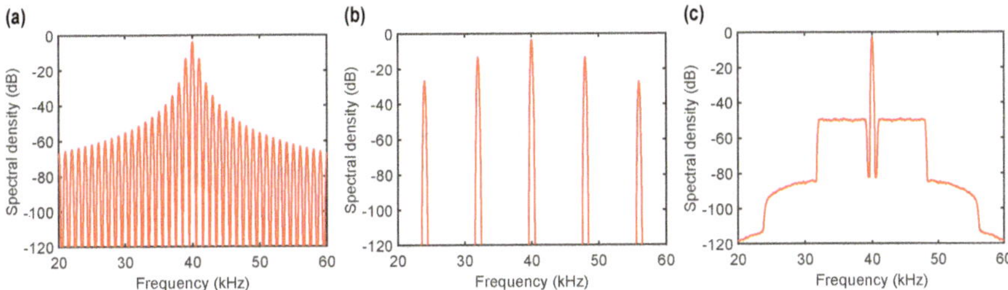

Figure 8.11 Spectral densities (dB) of different audio signals modulated by the SRTAM. The modulation index is $\mathcal{M} = 1$. The audio signal is (a) a pure-tone signal at frequency $f_a = 1\,\mathrm{kHz}$; (b) a pure-tone signal at frequency $f_a = 8\,\mathrm{kHz}$; (c) a white noise signal filtered by a bandpass filter with upper and lower cutoff frequencies of 1 kHz and 8 kHz, respectively.

have limited bandwidth. Consequently, nonlinear distortion is still present in implementations [Kite et al., 1998]. Nevertheless, the THD associated with SRTAM remains lower compared to that of DSBAM. For instance, in Pompei, 1999, an ultrasonic emitter array was used, exhibiting a frequency response deviation within ±8 dB from 40 kHz to 90 kHz. The findings indicate that SRTAM achieves a THD of less than 5% for the audio frequency ranging from 400 Hz to 4 kHz, while DSBAM generally yields THD values exceeding 20%. For a comprehensive analysis of the distortion introduced by SRTAM in a bandlimited PAL, readers are referred to Tan et al., 2010.

The truncated SRTAM introduces a modification that involves truncating the Taylor series expansion of the square root operator. To obtain the envelope of the n-th order SRTAM, a truncated Taylor series expansion is applied to Eq. (8.39), resulting in the following expression [Shi and Kajikawa, 2016a, Eq. (14)]

$$E_n(t) = \sum_{m=0}^{n} \frac{(2m)!}{(-4)^m(1-2m)(m!)^2} \mathcal{M}^m s^m(t). \tag{8.42}$$

For the second-order SRT amplitude modulation, the envelope signal is

$$E_2(t) = 1 + \frac{1}{2}\mathcal{M}s(t) - \frac{1}{8}\mathcal{M}^2 s^2(t). \tag{8.43}$$

Substituting Eq. (8.43) into the Berktay's solution Eq. (8.20), the on-axis sound pressure is obtained as [Shi and Kajikawa, 2016a, Eq. (16)]

$$p_{\mathrm{a}}(0,0,z,t) \propto \frac{\partial^2}{\partial t^2}\left[\mathcal{M}s(\tau) - \frac{1}{8}\mathcal{M}^3 s^3(\tau) + \frac{1}{64}\mathcal{M}^4 s^4(\tau)\right]. \tag{8.44}$$

Equation (8.44) implies that the SRTAM fundamentally serves as a method to shift the nonlinear distortion to higher order, specifically the third- and fourth-order terms shown in Eq. (8.44). This is achieved by introducing lower-order harmonics of the audio input, represented by the second-order term in Eq. (8.43). It is evident that as the truncated order increases, the magnitude of distortion decreases.

8.3.2.2.5 *Modified Amplitude Modulation (MAM)*

The *MAM* was initially proposed in Tan et al., 2010. The block diagram representing this modulation scheme is depicted in Fig. 8.12. To derive the MAM method,

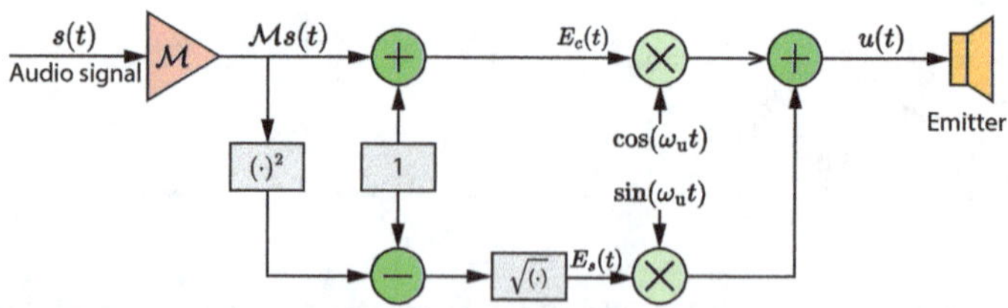

Figure 8.12 Block diagram of the MAM.

a quadrature amplitude modulation scheme is considered at first. The modulated ultrasonic signal has the form of [Tan et al., 2010, Eq. (8)]

$$\begin{aligned} u(t) &= E_c(t)\cos(\omega_{\mathrm{u}}t) + E_s(t)\sin(\omega_{\mathrm{u}}t), \\ &= \sqrt{E_c^2(t)+E_s^2(t)}\cos\left[\omega_{\mathrm{u}}t - \mathrm{atan2}(E_s(t), E_c(t))\right], \end{aligned} \tag{8.45}$$

where $E_c(t)$ and $E_s(t)$ are two components of the envelope signal. Equation (8.45) demonstrates that the ultrasonic wave undergoes a phase modification of $-\mathrm{atan2}(E_s(t), E_c(t))$, while the envelope signal is given by $E(t) = \sqrt{E_c^2(t)+E_s^2(t)}$. The advantage of employing quadrature amplitude modulation lies in its flexibility to introduce an additional orthogonal term, $E_s(t)\sin(\omega_{\mathrm{u}}t)$, for distortion reduction. For example, if the two components are chosen as

$$\begin{cases} E_c(t) = 1 + \mathcal{M}s(t), \\ E_s(t) = \sqrt{1-\mathcal{M}^2 s^2(t)}, \end{cases} \tag{8.46}$$

the envelope is then

$$E(t) = \sqrt{1+\mathcal{M}s(t)}. \tag{8.47}$$

This quadrature amplitude scheme can be understood as a DSB amplitude modulation with an orthogonal carrier modulating a pre-distortion term. The demodulated signal obtained from this quadrature amplitude modulation is equivalent to that of the SRTAM given by Eq. (8.41). Hence, this scheme is often referred to as the *root MAM* in the literature [Shi and Kajikawa, 2016a]. However, the root MAM differs from the SRTAM by a modification of the phase of the ultrasonic wave.

Since the quadrature term in the root MAM method involves the square root operation shown in Eq. (8.46), it also requires an ultrasonic emitter with an infinite bandwidth, which is not practical. Consequently, the MAMn methods have been introduced as alternatives. These methods employ the truncated Taylor series expansion of $E_s(t)$ in Eq. (8.46) as [Shi and Kajikawa, 2016a, Eq. (21)]

$$E_s^{\mathrm{MAM}n}(t) = \sum_{m=0}^{n} \frac{(2m)!}{4^m(1-2m)(m!)^2}\mathcal{M}^{2m}s^{2m}(t). \tag{8.48}$$

Here, n indicates the truncation order of the Taylor series expansion.

The quadrature terms of the first two MAMn methods are written as [Shi and Kajikawa, 2016a, Eqs. (22–23)]

$$E_s^{\mathrm{MAM1}}(t) = 1 - \frac{1}{2}\mathcal{M}^2 s^2(t), \tag{8.49}$$

$$E_s^{\mathrm{MAM2}}(t) = 1 - \frac{1}{2}\mathcal{M}^2 s^2(t) - \frac{1}{8}\mathcal{M}^4 s^4(t). \tag{8.50}$$

Applying the Berktay's far field solution yields the on-axis audio sound pressure as

$$p_{\mathrm{a}}(0,0,z,t) \propto \frac{\partial^2}{\partial t^2}\left[\mathcal{M}s(t) + \frac{1}{8}\mathcal{M}^4 s^4(t)\right], \tag{8.51}$$

$$p_{\mathrm{a}}(0,0,z,t) \propto \frac{\partial^2}{\partial t^2}\left[\mathcal{M}s(t) + \frac{1}{16}\mathcal{M}^6 s^6(t) + \frac{1}{128}\mathcal{M}^8 s^8(t)\right]. \tag{8.52}$$

The THD and the IMD for the first MAM method are given as [Wang et al., 2023b]

$$\mathrm{THD_{MAM1}} = \frac{\mathcal{M}^3}{\sqrt{8+\mathcal{M}^6}}, \tag{8.53}$$

$$\mathrm{IMD_{MAM1}} = \mathcal{M}^3\sqrt{\frac{37}{1024+37\mathcal{M}^6}}. \tag{8.54}$$

8.3.2.3 Remarks

Signal processing techniques play a crucial role in equalizing the frequency response and reducing distortion in the demodulated audio sound. While this section primarily focuses on the DSBAM, SSBAM, SRTAM, and MAM, other modulation algorithms are available in literature. For instance, a hybrid AM technique by the combination of DSBAM, SSBAM, and MAM has been proposed [Wang et al., 2023b]. The weighting coefficients of these three AMs are optimized using a genetic algorithm to minimize both total harmonic and intermodulation distortions.

In addition to modulation techniques, various preprocessing methods for the audio signal have been investigated to further reduce distortion. The Volterra filter is a widely used approach that can characterize the nonlinear behavior of PALs [Ji and Gan, 2012; Mu et al., 2014; Shi and Kajikawa, 2015b; Shi and Kajikawa, 2016b]. By designing an appropriate inverse filter, the distortion, particularly the second harmonic component, can be effectively filtered out [Shi and Kajikawa, 2015b; Shi and Kajikawa, 2016b]. Another approach involves using neural networks to model the PAL and then applying inverse control preprocessing to reduce distortion [Zhou and Chen, 2022]. This method leverages the capabilities of neural networks to learn and compensate for nonlinear distortions. These advanced preprocessing techniques contribute to the ongoing efforts in improving the quality and fidelity of the demodulated audio sound generated by PALs.

8.3.3 PHASED ARRAY PALS

The phased array technique provides enhanced flexibility in manipulating the directional audio beam generated by a PAL. The phased array PAL is commonly utilized for beam steering [Shi, 2013] and focusing [Zhong et al., 2022d] as presented in Chap. 6. In a phased array PAL, as shown in Fig. 8.13, appropriate time delays are applied to signals across different channels, which are then fed into an array of ultrasonic emitters. A simple but commonly used configuration is the uniform linear array, where ultrasonic emitters are uniformly distribution along a line. To simplify the analysis, the collimated beam is assumed in the following derivations.

8.3.3.1 Pure-Tone Audio Signal

When a ultrasonic emitter radiates two ultrasonic waves at frequencies of f_1 and f_2, the pure-tone audio wave at the frequency of $f_\mathrm{a} = f_2 - f_1$ is demodulated in air. For a specified channel in a phased array PAL, it is assumed that the time delays Δt_1 and

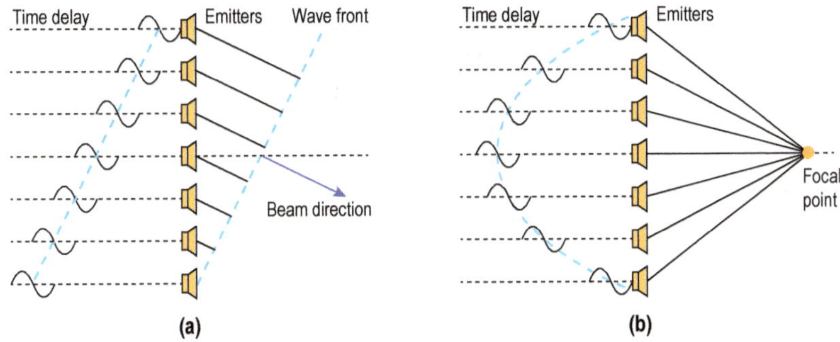

Figure 8.13 Realizing of (a) steerable and (b) focusing PALs using the phased array technique.

Δt_2 are applied to two ultrasonic waves at frequencies f_1 and f_2, respectively. The ultrasound field is then approximated by Eq. (2.57) as

$$p_{\mathrm{I}}(\mathbf{r},t) = \mathcal{A}(\mathbf{r})\left[p_{10}\mathrm{e}^{-\mathrm{i}\omega_1(t-\Delta t_1-z/c_0)-\alpha_1 z} + p_{20}\mathrm{e}^{-\mathrm{i}\omega_2(t-\Delta t_2-z/c_0)-\alpha_2 z}\right], \tag{8.55}$$

where p_{10} and p_{20} are the on-surface ultrasound pressure at frequencies of f_1 and f_2. The virtual audio source density can then be obtained by Eq. (2.58) as

$$q_{\mathrm{a}}(\mathbf{r},t) = \mathcal{A}(\mathbf{r})\frac{\beta\omega_{\mathrm{a}}}{\rho_0^2 c_0^4} p_{10}^* p_{20}\mathrm{e}^{-(\alpha_1+\alpha_2)z-\mathrm{i}[\omega_{\mathrm{a}}(t-z/c_0)-(\omega_2\Delta t_2-\omega_1\Delta t_1)]}. \tag{8.56}$$

By using the method in Sec. 2.2.4, the audio sound pressure can be obtained similar to Eq. (2.64) as

$$p_{\mathrm{a}}(\mathbf{r},t) = \frac{\mathrm{i}\beta p_{10}^* p_{20} S_0 \omega_{\mathrm{a}}^2}{4\pi\alpha_{\mathrm{t}}\rho_0 c_0^4 r}\mathcal{D}_{\mathrm{W}}(\theta)\mathrm{e}^{-\mathrm{i}[\omega_{\mathrm{a}}(t-r/c_0)-(\omega_2\Delta t_2-\omega_1\Delta t_1)]}, \tag{8.57}$$

where $\mathcal{D}_{\mathrm{W}}(\theta)$ is the Westervelt's directivity [Eq. (2.53)] as a function of the zenithal angle θ. When the time delays are the same for two ultrasonic frequencies $\Delta t_{\mathrm{a}} \equiv \Delta t_1 = \Delta t_2$, Eq. (8.57) reduces to

$$p_{\mathrm{a}}(\mathbf{r},t) = \frac{\mathrm{i}\beta p_{10}^* p_{20} S_0 \omega_{\mathrm{a}}^2}{4\pi\alpha_{\mathrm{t}}\rho_0 c_0^4 r}\mathcal{D}_{\mathrm{W}}(\theta)\mathrm{e}^{-\mathrm{i}\omega_{\mathrm{a}}(t-\Delta t_{\mathrm{a}}-r/c_0)}. \tag{8.58}$$

Equation (8.58) shows that the time delay for the demodulated audio sound is the same as for the primary ultrasonic wave.

For the beam steering presented in Fig. 8.13(a), the time delay for each channel in the phased array PAL can then be set as

$$\Delta t_{1,n} = \Delta t_{2,n} = n\frac{\Delta s}{c_0}\sin\varphi_0, \tag{8.59}$$

where $n = 1,2,...,N$ is the channel number, Δs is the pitch of the array, i.e., the separation of adjacent emitters, and φ_0 is the steering angle. It is noted that for a particular phased array PAL system, the minimum steering angle is limited by the sampling frequency. When the sampling frequency is 192 kHz, the sampling period is about 5.2 μs, and the minimum time delay is equal to the sampling period. When the emitter spacing is 5 mm, the minimum steering angle is found to be 21° based on Eq. (8.59) [Okano and Kajikawa, 2022].

For the beam focusing presented in Fig. 8.13(b), the time delay for each channel in the phased array PAL can then be set as

$$\Delta t_{1,n} = \Delta t_{2,n} = \frac{|\mathbf{r}_\mathrm{f} - \mathbf{r}_{\mathrm{s},n}|}{c_0}, \tag{8.60}$$

where $\mathbf{r}_{\mathrm{s},n}$ and $\mathbf{r}_\mathrm{f}$ are the locations of the n-th emitter and the focal point, respectively.

8.3.3.2 Wideband Audio Signal

Assume the time delays Δt_a and Δt_u are applied to the envelope and ultrasonic signals, respectively. The pressure on the source surface given by Eq. (8.13) is then modified as

$$p_\mathrm{I}(x,y,0,t) = p_0 E(t - \Delta t_\mathrm{a}) \cos[\omega_\mathrm{u}(t - \Delta t_\mathrm{u})] \mathcal{A}(\mathbf{r}). \tag{8.61}$$

The primary pressure field is then obtained by Eq. (8.14) to give

$$p_\mathrm{I}(\mathbf{r},t) = p_0 E(\tau - \Delta t_\mathrm{a}) \mathrm{e}^{-\alpha_\mathrm{u} z} \cos[\omega_\mathrm{u}(\tau - \Delta t_\mathrm{u})] \mathcal{A}(\mathbf{r}), \tag{8.62}$$

where the retarded time $\tau = t - z/c_0$ is used. The source density for the audio sound is obtained by Eq. (8.16) as

$$q_\mathrm{a}(\mathbf{r},t) = \frac{\beta p_0^2}{2\rho_0^2 c_0^4} \mathcal{A}(\mathbf{r}) \mathrm{e}^{-2\alpha_\mathrm{u} z} \frac{\partial E^2(\tau - \Delta t_\mathrm{a})}{\partial t}. \tag{8.63}$$

The on-axis demodulated audio sound can be obtained by Eq. (8.20) as

$$p_\mathrm{a}(\mathbf{r},t) = \frac{\beta p_0^2 S_0}{8\pi \rho_0 c_0^4 \alpha_\mathrm{t} z} \frac{\partial^2}{\partial t^2} [E^2(t - \Delta t_\mathrm{a} - z/c_0)]. \tag{8.64}$$

Equation (8.64) highlights a notable observation, namely that the demodulated wideband audio signal remains unaffected by the time delay applied to the ultrasonic wave, denoted as Δt_u. Instead, the demodulated wideband audio signal is only affected by the time delay, denoted as Δt_a, applied to the audio envelope signal $[E(t)]$. This conclusion has been demonstrated in Shi, 2013, Sec. 3.2. For the realization of beam steering and focusing of wideband audio signals, the necessary time delays align with Eqs. (8.59) and (8.60) correspondingly.

8.4 SPURIOUS SOUND IN MEASUREMENT

The audible sound produced by a PAL arises from the nonlinear interactions of intense ultrasound in air. However, when the intense ultrasound with two frequencies f_1

and f_2 is measured by microphones, the nonlinear intermodulation distortion of the microphone itself produces an additional unwanted signal at the difference frequency $f_a = f_2 - f_1$. This additional spurious signal does not correspond to the actual audio sound in air and cannot be perceived by the human ear. The amplitude of the spurious sound is approximately proportional to the ultrasonic sound pressure [Abuelma'atti, 2003]. Consequently, in the near field where the ultrasound field is intense, the spurious sound predominates the measured results. However, as the ultrasonic waves undergo significant attenuation over longer distances, the spurious sound becomes negligible. This observation was initially highlighted in the first experiment demonstrating the parametric effect in air [Bennett and Blackstock, 1975].

Furthermore, as the audio frequency is halved, the audio sound pressure approximately decreases by 12 dB, while the ultrasound pressure level changes a little. As a result, the presence of the spurious sound is more pronounced when measuring low-frequency audio sound fields. Hence, the primary approach to reduce the spurious sound is by minimizing the measured ultrasound levels. The presence of the spurious sound poses a major challenge in accurate measurement of the audio sound generated by a PAL. Several solutions have been proposed to mitigate the impact of the spurious sound, which are introduced in the following subsections.

8.4.1 ORIENTATION OF THE MEASUREMENT MICROPHONE

The simplest technique for mitigating the spurious sound is to leverage the sensitivity characteristics of microphones at different incident angles. For example, Bruel and Kjaer, 2017, Fig. 9 illustrates the correction curves that should be applied to the actual sound pressure measured by Brüel & Kjær condenser microphones. It is observed that when sound beams approach at larger angles, the microphone sensitivity remains relatively consistent below 10 kHz (within the typical audible frequency range), but significantly decreases in the ultrasonic range. For instance, when the wave is incident with a grazing angle (90°), where the incident beam is parallel to the microphone diaphragm, the measured ultrasound at 40 kHz is reduced by over 10 dB. This reduction is a consequence of the increased directivity of the microphone at higher frequencies, automatically suppressing the measured ultrasound pressure and resulting in a lower spurious sound. This technique was initially utilized in Ju and Kim, 2010 to measure the near field sound. For a comprehensive theoretical and experimental examination of the effects of microphone size, placement, sensitivity, distance from the sound source, and the primary ultrasound pressure on the measurement of demodulated audio sound, readers are referred to Nomura and Sato, 2022.

8.4.2 PHYSICAL ACOUSTIC FILTERS

The physical acoustic filter is a widely employed method to reduce the spurious sound. It is typically positioned in front of the measurement microphone and serves as an acoustic low-pass filter. Its purpose is to diminish the magnitude of the ultrasound pressure level while maintaining acoustic transparency for the audio sound.

In an initial experiment demonstrating the parametric effect in air, a dome-shaped acoustic filter made of a 2.5 mm thick clear plastic material was employed to enclose the microphone cartridge [Bennett and Blackstock, 1975]. This acoustic filter exhibited an insertion loss of 3.5 dB and 20 dB at audio and ultrasound frequencies, respectively. Similar acoustic filters have been utilized in various studies. For instance, a sheet of air-pad material was used providing an insertion loss over 20 dB at 40 kHz [Kamakura et al., 1984]. Four layers of felt was used providing an insertion loss over 20 dB at 30 kHz and 3–5 dB at audio frequencies [Havelock and Brammer, 2000]. Four polymer films with half-wavelength spacing provided an insertion loss of more than 30 dB between 30 kHz and 40 kHz [Toda, 2005]. An aluminum plate with a diameter of 16 mm produced an insertion loss of more than 15 dB above 30 kHz, while for audio sound below 10 kHz, the insertion loss was less than 5 dB [Ye et al., 2011].

Another effective physical acoustic filter was designed based on phononic crystals [Ji et al., 2016]. The prototype, depicted in Fig. 8.14(a), consists of an arrangement of 5×7 aluminum alloy cylinders. The simulation and experimental results, as presented in Fig. 8.14(b), demonstrate that the filter exhibits a substantial insertion loss (up to 30 dB around 40 kHz) at ultrasonic frequency while having minimal impact on audible sound (less than 3 dB below 20 kHz). Constructing a physical acoustic filter is generally straightforward, and it provides a moderate level of effectiveness in reducing the spurious sound. However, it is important to note that the demodulated audio sound may also experience some degree of attenuation when using physical acoustic filters.

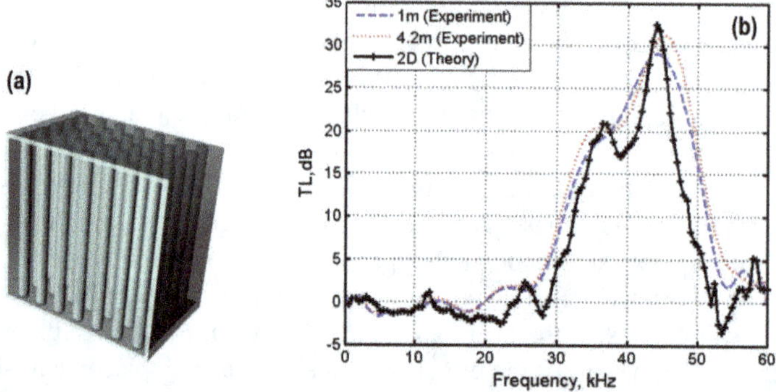

Figure 8.14 (a) Sketch of the physical acoustic filter based on phononic crystals. (b) Insertion loss of the filter on the radiator axis. Extracted from [Ji et al., 2016, Figs. 1 and 8].

8.4.3 PHASE-CANCELATION METHOD

The concept of phase-cancelation excitation for ultrasonic waves was initially demonstrated in 2008 using two strip ultrasonic emitters placed side by side [Akiyama et al., 2008] and two coaxially arranged planar radiators, one being an inner circular radiator and the other an outer ring radiator [Kamakura et al., 2008]. This technique was later employed for spurious-sound-free measurements in 2012 [Ji et al., 2012]. In the setup shown in Fig. 8.15(a), two rectangular ultrasonic emitters of identical dimensions $a \times 2b$ are placed side by side. The first emitter ($x > 0$) is driven by ultrasonic signals of $\cos(\omega_1 t) + \cos(\omega_2 t)$ while the other emits signals of $\cos(\omega_1 t + \varphi_0) + \cos(\omega_2 t + \varphi_0)$. When the phase difference $\varphi_0 = 0$, it represents the conventional rectangular PAL with a uniform radiation profile, termed in-phase excitation. In contrast, when the phase difference $\varphi_0 = \pi$, the two ultrasonic emitters are excited out-of-phase. In this case, the on-axis ultrasound pressure becomes zero due to the cancelation of the ultrasonic waves generated by the two emitters, as depicted in Fig. 8.15(b). Consequently, spurious-sound-free measurements can be conducted along the radiator axis. The phase-cancelation method effectively reduces the spurious sound without the need for installing physical acoustic filters. However, it is important to note that this method is only applicable within a specific region, which is the radiator axis in this case.

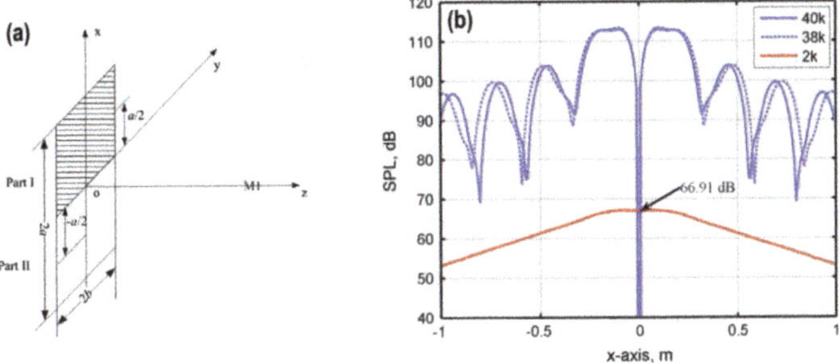

Figure 8.15 (a) Sketch of the setup of the phase-cancelation method for the spurious-sound-free measurement. (b) Sound pressure levels on the transverse direction. Extracted from [Ji et al., 2012, Figs. 1 and 2].

8.4.4 OPTICAL INTERFEROMETERY

In 2021, a spurious-sound-free measurement technique utilizing optical interferometry was proposed to measure the audio sound generated by a PAL [Ishikawa et al., 2021]. The setup, as depicted in Fig. 8.16, employs an interferometer to measure the sound pressure integrated along a laser path. Various reconstruction methods can then be employed to obtain point-wise pressure along this path [Xiao, 2022]. Optical interferometry offers a contactless and mechanical-component-free approach, ensuring that no spurious sound is introduced during the measurement process. As a result, it enables effective measurement of audio sound at any location without the presence of spurious sound. However, it is important to note that optical interferometry requires costly equipment, which can be a significant obstacle in practice. Additionally, the noise level of the measurement system based on optical interferometry is considerably higher compared to conventional microphones, posing challenges in accurately measuring low-frequency audio sound [Ishikawa et al., 2021]. Moreover, the use of Gaussian beam expansion in Ishikawa et al., 2021, is known to be inaccurate at low audio frequencies even within the paraxial region, as demonstrated in Sec. 2.4. Therefore, it is necessary to design more accurate measurement systems based on the Westervelt or Kuznetsov equations by incorporating optical interferometry.

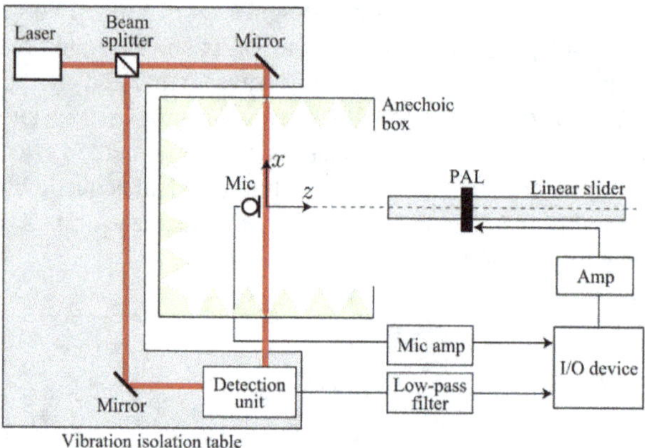

Figure 8.16 Schematic of the spurious-sound-free measurement system based on the optical interferometery. Extracted from [Ishikawa et al., 2021, Fig. 1(a)].

8.5 SAFETY ISSUES

The audio sound generated by a PAL is a result of nonlinear interactions among intense ultrasonic waves. The ultrasound pressure level produced by a PAL depends on many factors such as the transducer's dimensions, initial sound pressure, and the listener's position. Typically, in the far field, where the listener is located at a considerable distance from the PAL, the ultrasound wave attenuates rapidly due to atmospheric absorption. As a result, safety concerns are generally considered negligible. However, there is growing interest in utilizing PALs in the near field, such as focusing [Zhong et al., 2022d] and length-limited [Hedberg et al., 2010] PALs. In the near field, the ultrasound pressure level can exceed 130 dB even for a conventional PAL. Exposure to such high-intensity ultrasound may raise safety concerns for listeners.

Human body reflects approximately 99.9% of airborne ultrasonic energy due to the significant acoustic impedance mismatch between air and tissue [Ludwig, 1950]. A typical commercial PAL may produce a maximum ultrasound density of 2.9 $\mathrm{mW/cm^2}$, and consequently only around 0.0029 $\mathrm{mW/cm^2}$ at maximum actually enters the body due to the skin reflection [Mehta et al., 2015a]. The Food and Drug Administration (FDA) has established standards for internal ultrasound, which set limits of 720 $\mathrm{mW/cm^2}$ for fetal, cardiac, and vascular imaging and 50 $\mathrm{mW/cm^2}$ for ophthalmic imaging [Nelson et al., 2009]. These standards represent a safety margin of over 1/17,000 in comparison to the FDA limits for non-invasive medical applications of ultrasound. Thus, the primary risk of ultrasound exposure, in practical terms, occurs through the ear.

Exposure to loud sound can result in temporary or permanent hearing threshold shifts. Temporary threshold shift (TTS) occurs when hearing thresholds shift due to noise exposure but return to normal or baseline within 16 to 48 hours [Moore, 2013]. The extent of hearing loss caused by noise exposure is determined by the intensity and duration of the noise. Prolonged exposure to noise that induces TTS can lead to chronic or permanent threshold shifts. International guidelines have been established to ensure the safety of humans regarding ultrasonic airborne emissions. These guidelines typically provide recommendations for maximum permissible levels (MPLs) of airborne ultrasound in the frequency range of 20–100 kHz [Di Battista et al., 2022]. The current MPLs are based on the observation that no adverse effects have been observed in humans at sound pressure levels (SPLs) below 110 dB.

Conversely, SPLs above 145 dB can cause temporary threshold shifts, which, if recurrent, may lead to permanent damage to auditory function. Between these SPL limits 110 dB and 145 dB, various subjective effects have been reported, including dizziness, tinnitus, headache, nausea, stress, otalgia, vertigo, and pain, following exposure to ultrasound within the 110–145 dB range. These effects are subjective and can vary in severity among individuals, depending on the circumstances of exposure [Di Battista et al., 2022]. Currently, quantifying and analyzing these subjective effects remain challenging, and further research is needed to develop more effective methods.

International guidelines for ultrasonic emissions primarily focus on occupational health and safety, considering the well-being of industrial workers who are exposed to ultrasonic devices during activities like cleaning, drilling, and soldering. These guidelines typically assume an exposure duration of 8 hours per day. They may not be directly applicable to consumer devices like PALs, which are typically intended for casual use.

Studies investigating the auditory function and cognitive effects of short-term exposure to high-intensity airborne ultrasound have been conducted. One study examined the auditory function of individuals exposed to a 110–120 dB SPL tone at 40 kHz and found no evidence of threshold shifts or changes in behavioral or electrophysiological subclinical measures [Carcagno et al., 2019]. Another study focused on the cognitive function of participants exposed to high-intensity airborne ultrasound (40 kHz, 120 dB) [Di Battista et al., 2022]. Forty test participants were asked to perform a go/no-go task and continuous performance test under baseline conditions. The results consistently demonstrate that exposure to ultrasound at this intensity and frequency had no detectable effect on cognitive task performance. Participants' ability to select the correct response and their reaction times when responding correctly were not affected by ultrasound exposure. It should be noted that there is limited research available on the effects of ultrasound exposure above 120 dB, and further studies in this area are needed to provide a more comprehensive understanding of its potential effects.

It is important to consider that the above findings and conclusions regarding ultrasound exposure may not directly apply to the operation of PALs. The studies mentioned earlier primarily focus on the effects of exposure to pure-tone or wideband ultrasonic waves, which differ fundamentally from the narrow spectral band emitted by PALs. A relevant study investigating the safety of PALs is the one conducted with a HyperSound Audio System (HSS) under normal use conditions [Mehta et al., 2015b]. In this study, 20 adult subjects were exposed to a 113-minute movie with sound delivered through the PAL at a level of 80 dBA. Following the exposure, a post-exposure assessment was conducted. The results revealed no statistically significant changes in audiometric thresholds or pure-tone average thresholds between the pre-exposure and post-exposure measurements. Furthermore, none of the participants reported any new onset otological symptoms after the exposure.

Based on these findings and the absence of subjective or objective adverse effects, the HSS received FDA clearance in 2014 as a medical device [Mehta et al., 2015b]. This device is designed for use as a group hearing aid, catering to one or more listeners, regardless of hearing loss, and whether or not they are using hearing aids. Its purpose is to enhance clarity and comprehension of sounds originating from various sources, including a microphone, CD/DVD player, TV, stereo system, or other sound generation systems.

However, it is noted the ultrasonic output frequency from the HSS system used in the aforementioned study was above 96 kHz [Mehta et al., 2015b]. As the ultrasonic carrier frequency decreases, certain factors such as the power conversion efficiency from ultrasound to audio sound and the directivity of the demodulated audio beam can improve, as demonstrated in Chap. 3. Furthermore, it has been reported that adverse effects are more significant at lower ultrasound frequencies. For instance, ultrasound frequencies close to the range of human hearing (below 80 kHz) can generate subharmonic tones at 1/2 or 1/4 of the frequency, which can be perceived as audible noise. High levels of audible noise from these subharmonics have been associated with reports of headache and nausea [Labor, 2022]. Considering these factors, it is crucial to conduct further investigations to assess the safety of PALs with ultrasonic carrier frequencies below 96 kHz. This can provide a comprehensive understanding of the potential effects and safety considerations associated with PALs operating at lower ultrasound frequencies.

8.6 SUMMARY

The readers are introduced to the implementation of PALs in this chapter. Due to the fundamentally physical difference in comparison of conventional loudspeakers, the implementation of PALs presents unique challenges in hardware (Sec. 8.2), signal processing (Sec. 8.3), measurement (Sec. 8.4), and safety issues (Sec. 8.5). For the hardware, three important components including ultrasonic emitters, power amplifiers, and signal processors are introduced in Secs. 8.2.1, 8.2.2, and 8.2.3, respectively.

For the signal processing, the quasilinear solution of the demodulated audio sound in the time domain is presented in Sec. 8.3.1. The Berktay's far field solution with the collimated beam assumption serves as the theory basis for further signal processing algorithms. Preprocessing techniques for the input audio signal are analyzed in Sec. 8.3.2. Several commonly modulation algorithms are introduced, which include DSBAM, SSBAM, SRTAM, and MAM. The phased array technique is discussed in Sec. 8.3.3, which allows for manipulation of the directional audio beam generated by PALs. The structure and realization of a phased array using time delays are presented for both pure-tone and wideband audio signals.

The generation of audible sound by PALs involves nonlinear interactions of intense ultrasound in the air. This poses challenges in accurately measuring the demodulated audio sound compared to conventional loudspeakers. The presence of spurious sound contaminates the measured results, and different solutions to address this issue are presented in Sec. 8.4. Furthermore, the intensity of ultrasound in PALs may raise safety concerns, particularly in near-field applications. Existing studies on the exposure to intense ultrasound waves and their potential effects are given in Sec. 8.5. Overall, the implementation of PALs requires careful consideration of hardware components, signal processing techniques, measurement methodologies, and safety aspects to ensure optimal performance and user safety.

A Attenuation Coefficients of Pure-Tone Sound in Air Due to Atmospheric Absorption

A.1 ABSORPTION COEFFICIENTS

Sound waves propagating in air experience attenuation as they travel over distance due to various irreversible processes that convert acoustic energy into heat [Kinsler et al., 2000, Chap. 8]. Consequently, the pure-tone sound attenuation coefficient becomes essential in calculating the sound field generated by a PAL. Atmospheric absorption in air encompasses classical thermoviscous absorption and molecular relaxation absorption [Evans et al., 1972]. The resulting attenuation coefficient is commonly referred to as the *absorption coefficient*. Traditionally, in a classical thermoviscous fluid, the attenuation coefficient is proportional to the square of the frequency. However, in real media such as air, relaxation phenomena introduce weak dispersion, causing the attenuation coefficient to deviate from a purely quadratic frequency dependence to some extent [Hamilton, 2016]. A convenient closed-form solution that accounts for both classical thermoviscous and relaxation effects has been derived by fitting experimental data and is expressed as [ISO 9613-1, 1993, Eq. (5); Bass et al., 1990; Bass et al., 1995; Bass et al., 1996]

$$\alpha(\omega) = f^2 \left[1.84 \times 10^{-11} \frac{P_{\mathrm{ref}}}{P_{\mathrm{atm}}} \left(\frac{T}{T_{\mathrm{ref}}} \right)^{1/2} + \left(\frac{T}{T_{\mathrm{ref}}} \right)^{-5/2} \left(\frac{0.01275 \mathrm{e}^{-2239.1/T}}{f_{\mathrm{rlx,O}} + f^2/f_{\mathrm{rlx,O}}} + \frac{0.1068 \mathrm{e}^{-3352.0/T}}{f_{\mathrm{rlx,N}} + f^2/f_{\mathrm{rlx,N}}} \right) \right] \quad [\mathrm{Np/m}]. \tag{A.1}$$

It is worth noting that the unit of the absorption coefficient calculated in Eq. (A.1) is represented in nepers per meter, indicated as Np/m. The coefficient of the unit of decibels per meter, denoted as dB/m, can be obtained by

$$\alpha\,[\mathrm{dB/m}] = \frac{20}{\ln 10} \alpha\,[\mathrm{Np/m}]. \tag{A.2}$$

In Eq. (A.1), f is the frequency, $\omega = 2\pi f$ is the angular frequency, P_{atm} is the atmospheric pressure, $P_{\mathrm{ref}} = 1$ atm $= 101.325$ kPa is the reference ambient atmospheric

DOI: 10.1201/9781003354994-A

pressure, T is the air temperature in Kelvin, $T_{\text{ref}} = 293.15\,\text{K}\,(20\,°\text{C})$ is the reference air temperature, and $f_{\text{rlx,O}}$ and $f_{\text{rlx,N}}$ are the relaxation frequencies for oxygen and nitrogen, respectively, given by [ISO 9613-1, 1993, Eqs. (3) and (4)]

$$f_{\text{rlx,O}} = \frac{P_{\text{atm}}}{P_{\text{ref}}}\left(24 + 4.04\times 10^4 h \frac{0.02+h}{0.391+h}\right), \tag{A.3}$$

$$f_{\text{rlx,N}} = \frac{P_{\text{atm}}}{P_{\text{ref}}}\left(\frac{T_0}{T}\right)^{1/2}\left(9 + 280he^{-4.17\left[(T_0/T)^{1/3}-1\right]}\right). \tag{A.4}$$

The terms on the right-hand side of Eq. (A.1) account for the classical thermoviscous, the relaxation effect due to the oxygen and nitrogen, respectively.

The absolute humidity, h, is related to relative humidity, denoted as h_{rel}, through the saturation vapor pressure, denoted as P_{sat}, by [ISO 9613-1, 1993, Eq. (B.1)]

$$h = h_{\text{rel}}\frac{P_{\text{sat}}}{P_{\text{atm}}}. \tag{A.5}$$

The saturation vapor pressure can be approximated by [ISO 9613-1, 1993, Eqs. (B.2) and (B.3)]. The absorption coefficient, measured in Np/m at an atmospheric pressure of 1 atm, is depicted in Fig. A.1 as a function of frequency across different temperatures and relative humidities. It is evident that the coefficient remains

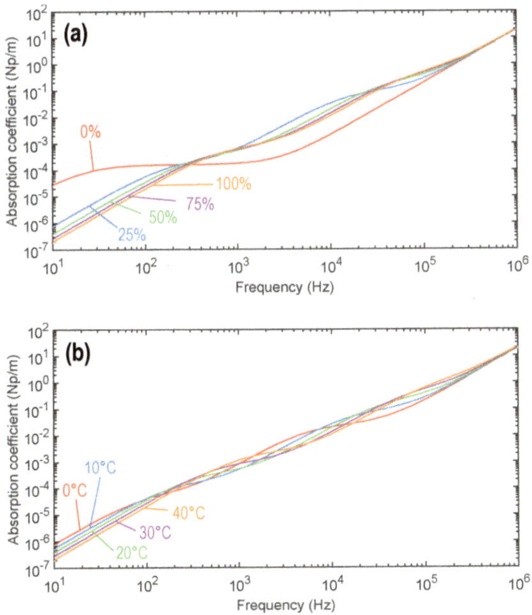

Figure A.1 Pure-tone sound atmospheric absorption coefficients, measured in Np/m at an atmospheric pressure of 1 atm, are presented as a function of frequency. (a) The temperature is 293.15 K (20°C). (b) The relative humidity is 50%.

negligible at lower audio frequencies and monotonically increases with increasing frequency. For instance, at 1 kHz, when the temperature is 20°C and the relative humidity is 50%, the absorption coefficient is only 5.37×10^{-4} Np/m. However, as the frequency increases to 40 kHz, a typical carrier frequency for PALs, it increases significantly to 1.52×10^{-1} Np/m. One of the major distinctions between PALs and conventional loudspeakers lies in the imperative need to consider the absorption coefficient.

A.2 ABSORPTION DISTANCES

The atmospheric absorption results in that the sound pressure decays exponentially with distance, expressed as $\exp(-\alpha r)$, where r is the propagating distance. It is convenient to define an *absorption distance* as the reciprocal of the absorption coefficient as [Sparrow and Kim, 2002, Eq. (15)]

$$\mathscr{D}_{\text{ab}} \equiv \frac{1}{\alpha[\text{Np/m}]}. \tag{A.6}$$

Beyond the absorption distance, the sound pressure amplitude reduces to only 36.8% (−8.7 dB). Therefore, the absorption distance can be used to characterize the absorption effect in air. The absorption distance at an atmospheric pressure of 1 atm, is depicted in Fig. A.2 as a function of frequency across different temperatures and relative humidities.

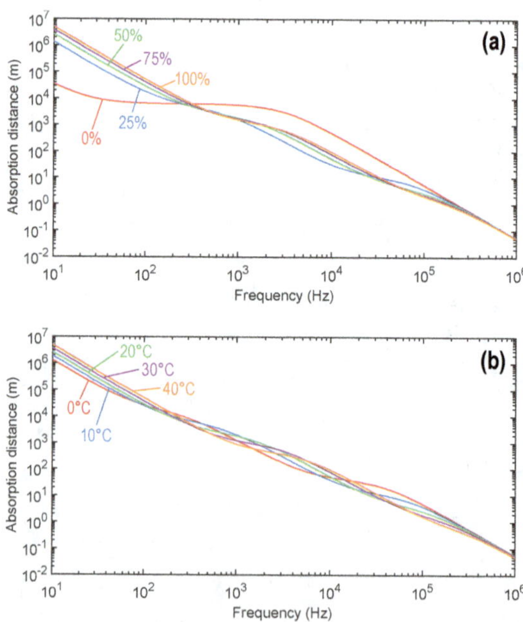

Figure A.2 Absorption distances of pure-tone acoustic waves, at an atmospheric pressure of 1 atm, are presented as a function of frequency. (a) The temperature is 293.15 K (20°C). (b) The relative humidity is 50%.

In PALs, the absorption coefficient is chosen at the ultrasound frequency f_u, so the absorption distance for PALs is defined as [Sparrow and Kim, 2002, Eq. (15)]

$$\mathscr{D}_{ab} \equiv \frac{1}{\alpha_u}. \tag{A.7}$$

When the audio sound is a pure-tone signal at frequency f_a, two harmonic ultrasound waves at frequencies $f_1 = f_u - f_a$ and $f_2 = f_u + f_a$. In this scenario, the absorption distance for PALs is sometimes defined as [Farias and Abdulla, 2015, Eq. (9)]

$$\mathscr{D}_{ab} \equiv \frac{1}{\alpha_1 + \alpha_2}, \tag{A.8}$$

where α_i is the absorption coefficient (in Np/m) of the ultrasound beam at frequency f_i, $i = 1, 2$.

B Mathematical Formulae on Special Functions

In this appendix, various mathematical formulae used in this book are summarized.

B.1 LEGENDRE POLYNOMIALS

Definitions

The *Legendre polynomial* is defined as (*Rodrigues formula*) [Zhang and Jin, 1996, Eq. (4.2.2)]

$$P_n(x) = \frac{1}{2^n n!} \frac{\mathrm{d}^n}{\mathrm{d}x^n}(x^2 - 1)^n, \quad n \in \mathbb{N}. \tag{B.1}$$

Integrals

The integral of the Legendre polynomial is

$$\int P_n(x)\mathrm{d}x = \begin{cases} x, & n = 0, \\ \dfrac{P_{n+1}(x) - P_{n-1}(x)}{2n+1}, & n > 0. \end{cases} \tag{B.2}$$

B.2 SPHERICAL HARMONICS

Definitions

The *spherical harmonics* of degree ℓ and order m are defined as [Rafaely, 2015, Eq. (1.9)]

$$Y_\ell^m(\theta, \varphi) = \sqrt{\frac{2\ell+1}{4\pi} \frac{(\ell-m)!}{(\ell+m)!}} P_\ell^m(\cos\theta) \mathrm{e}^{\mathrm{i}m\varphi}. \tag{B.3}$$

where $\ell, m \in \mathbb{Z}$, $\ell \geq |m|$, and $P_\ell^m(\cdot)$ is the *associated Legendre function*, the definition of which varies from book to book [Martin, 2006, Sec. 3.2]. The definition used in this book is

$$P_\ell^m(x) = (-1)^m (1 - x^2)^{m/2} \frac{\mathrm{d}^m}{\mathrm{d}x^m} P_\ell(x). \tag{B.4}$$

This definition differs by a factor of $(-1)^m$ (Condon-Shortley phase) from that used in some literature.

DOI: 10.1201/9781003354994-B

Properties

The complex conjugate of spherical harmonics can be represented by [Rafaely, 2015, Eq. (1.10)]

$$[Y_\ell^m(\theta,\varphi)]^* = Y_\ell^m(\theta,-\varphi) = (-1)^m Y_\ell^{-m}(\theta,\varphi). \tag{B.5}$$

Deritivatives

The derivative of spherical harmonics with respect to the zenithal angle is [Zhong et al., 2022b, Eq. (28)]

$$\frac{\partial Y_\ell^m(\theta,\varphi)}{\partial\theta} = m\cot\theta Y_\ell^m(\theta,\varphi) + \sqrt{(\ell-m)(\ell+m+1)}Y_\ell^{m+1}(\theta,\varphi)\mathrm{e}^{-\mathrm{i}\varphi}. \tag{B.6}$$

The derivative of spherical harmonics with respect to the azimuthal angle is

$$\frac{\partial Y_\ell^m(\theta,\varphi)}{\partial\varphi} = \mathrm{i}mY_\ell^m(\theta,\varphi). \tag{B.7}$$

Integrals.

Integral of spherical harmonics and complex exponential functions [Zhong et al., 2022b, Eq. (23)]:

$$\int_0^{2\pi} Y_\ell^{m,*}(\theta,\varphi)\mathrm{e}^{\mathrm{i}m'\varphi}\mathrm{d}\varphi = 2\pi Y_\ell^m(\theta,0)\delta_{mm'}, \quad m,m' \in \mathbb{Z}. \tag{B.8}$$

where $\delta_{mm'}$ is the Dirac delta function.

Integral of triple spherical harmonics [Martin, 2006, Eq. (B.2)]:

$$\begin{aligned}
I(\ell_1,\ell_2,\ell_3;m_1,m_2,m_3) &\equiv \int_0^{2\pi}\int_0^{\pi} Y_{\ell_1}^{m_1}(\theta,\varphi)Y_{\ell_2}^{m_2}(\theta,\varphi)Y_{\ell_3}^{m_3}(\theta,\varphi)\sin\theta\mathrm{d}\theta\mathrm{d}\varphi \\
&= \sqrt{\frac{(2\ell_1+1)(2\ell_2+1)(2\ell_3+1)}{4\pi}} \\
&\qquad \times \begin{pmatrix} \ell_1 & \ell_2 & \ell_3 \\ 0 & 0 & 0 \end{pmatrix}\begin{pmatrix} \ell_1 & \ell_2 & \ell_3 \\ m_1 & m_2 & m_3 \end{pmatrix} \\
&= \delta_{m_1+m_2,-m_3}(-1)^{m_1+m_2}\mathscr{G}(\ell_1,m_1;\ell_2,m_2;\ell_3).
\end{aligned} \tag{B.9}$$

Here, $\begin{pmatrix} \ell_1 & \ell_2 & \ell_3 \\ m_1 & m_2 & m_3 \end{pmatrix}$ is the Wigner $3j$ symbol as defined by Eq. (B.12). According to Eq. (B.17), it is clear that $I(\ell_1,\ell_2,\ell_3;m_1,m_2,m_3) \neq 0$ only when $\ell_1+\ell_2+\ell_3$ is even integer.

Special Values.

Special value at $m = 0$ is [Zhang and Jin, 1996, 4.4.3]

$$Y_\ell^0(\pi/2,0) = \sqrt{\frac{2\ell+1}{4\pi}} P_\ell^0(0)$$
$$= \sqrt{\frac{2\ell+1}{4\pi}}(-1)^{\frac{\ell}{2}} \begin{cases} \frac{(\ell-1)!!}{\ell!!} = \frac{1}{\sqrt{\pi}} \frac{\Gamma(\frac{\ell}{2}+\frac{1}{2})}{\Gamma(\frac{\ell}{2}+1)}, & \ell \text{ is even}, \\ 0, & \ell \text{ is odd}. \end{cases} \tag{B.10}$$

Special value at $\theta = \pi/2$ and $m \neq 0$ is [Zhang and Jin, 1996, 4.4.3]

$$Y_\ell^m(\pi/2,\varphi) = \sqrt{\frac{2\ell+1}{4\pi}\frac{(\ell-m)!}{(\ell+m)!}} P_\ell^m(0)\mathrm{e}^{\mathrm{i}m\varphi}$$
$$= \sqrt{\frac{2\ell+1}{4\pi}\frac{(\ell-m)!}{(\ell+m)!}}(-1)^{\frac{\ell+m}{2}}\mathrm{e}^{\mathrm{i}m\varphi}$$
$$\times \begin{cases} \frac{(\ell+m-1)!!}{(\ell-m)!!} = \frac{2^m}{\sqrt{\pi}} \frac{\Gamma(\frac{n+m}{2}+\frac{1}{2})}{\Gamma(\frac{n-m}{2}+1)}, & \ell - m \text{ is even}, \\ 0, & \ell - m \text{ is odd}. \end{cases} \tag{B.11}$$

B.3 WIGNER 3J SYMBOLS AND GAUNT COEFFICIENTS

Definitions

The *Wigner 3j symbol* is denoted as

$$\begin{pmatrix} \ell_1 & \ell_2 & \ell_3 \\ m_1 & m_2 & m_3 \end{pmatrix}. \tag{B.12}$$

Here, the quantities ℓ_1, ℓ_2, ℓ_3 are called *angular momenta*. Either all of them are nonnegative integers, or one is a nonnegative integer and the other two are half-odd positive integers. Then triangular inequalities hold [Olver et al., 2019, Eq. (34.2.1)]

$$|\ell_1 - \ell_2| \le \ell_3 \le \ell_1 + \ell_2. \tag{B.13}$$

The quantities m_1, m_2, m_3 are called *projective quantum number* satisfying [Olver et al., 2019, Eq. (34.2.2)]

$$m_i = \ell_i, -\ell_i, \ldots, \ell_i - 1, \ell_i, \quad i = 1,2,3. \tag{B.14}$$

The definition of the *Gaunt coefficient* as the integral of spherical harmonics is expressed as [Martin, 2006, Eq. (3.71)]

$$\mathscr{G}(n,m;\nu,\mu;q) = (-1)^{m+\mu} \iint_{4\pi} Y_n^m(\theta,\varphi) Y_\nu^\mu(\theta,\varphi) Y_q^{-m-\mu}(\theta,\varphi) \mathrm{d}\Omega, \tag{B.15}$$

where the solid angle $\mathrm{d}\Omega = \sin\theta \mathrm{d}\theta \mathrm{d}\varphi$, and the surface of the unit ball is $\Omega = \{(\theta,\varphi) : 0 \leq 0 \leq \pi, 0 \leq 0 < 2\pi\}$.

Relations

Gaunt coefficients can be represented using Wigner $3j$ symbols as [Martin, 2006, Eq. (3.73)]

$$\begin{aligned}\mathscr{G}(n,m;\nu,\mu;q) = (-1)^{m+\mu}\sqrt{\frac{(2n+1)(2\nu+1)(2q+1)}{4\pi}} \\ \times \begin{pmatrix} n & \nu & q \\ 0 & 0 & 0 \end{pmatrix}\begin{pmatrix} n & \nu & q \\ m & \mu & -m-\mu \end{pmatrix}.\end{aligned} \tag{B.16}$$

Special Values

Special value for the wigner $3j$ symbol when $m_1 = m_2 = m_3 = 0$ is [Messiah, 1962, Eqs. (C.23a) and (C.23b); Olver et al., 2019, Eq. (34.3.5)]

$$\begin{pmatrix} \ell_1 & \ell_2 & \ell_3 \\ 0 & 0 & 0 \end{pmatrix} = \begin{cases} 0, & 2\ell \text{ is odd}, \\ \dfrac{(-1)^{\ell}\ell!}{(\ell-\ell_1)!(\ell-\ell_2)!(\ell-\ell_3)!} \\ \quad \times\sqrt{\dfrac{(2\ell-2\ell_1)!(2\ell-2\ell_2)!(2\ell-2\ell_3)!}{(2\ell+1)!}}, & 2\ell \text{ is even}. \end{cases} \tag{B.17}$$

where $2\ell \equiv \ell_1 + \ell_2 + \ell_3$.

B.4 BESSEL FUNCTIONS

Definitions

The *Bessel functions* are the solutions of the Bessel's equation. The first and second Bessel functions are denoted by $J_\nu(z)$ and $Y_\nu(z)$, respectively. The first and second kind of Hankel functions are defined by [Zhang and Jin, 1996, Eqs. (5.1.5) and (5.1.6)]

$$H_\nu^{(1)}(z) = J_\nu(z) + \mathrm{i}Y_\nu(z), \tag{B.18}$$

$$H_\nu^{(2)}(z) = J_\nu(z) - \mathrm{i}Y_\nu(z). \tag{B.19}$$

In this book, only the first kind of Hankel function is used. Consequently, the symbol $H_\nu(z) = H_\nu^{(1)}(z)$ is used for simplicity.

Limiting Forms

The Hankel function at large argument $z \to \infty$ has the limiting form of [Zhang and Jin, 1996, Eq. (5.1.17)]

$$H_\nu(z) \sim \sqrt{\frac{2}{\mathrm{i}\pi z}} \frac{\mathrm{e}^{\mathrm{i}z}}{\mathrm{i}^\nu}. \tag{B.20}$$

Integral Representations

The integral representation for the Bessel function of order zero is [Olver et al., 2019, Eq. (10.9.1)]

$$J_0(z) = \frac{1}{\pi} \int_0^\pi \mathrm{e}^{\mathrm{i}z\cos\theta} \mathrm{d}\theta. \tag{B.21}$$

Derivatives

The derivative of order 0 is [Zhang and Jin, 1996, Eq. (5.1.25)]

$$\mathscr{C}_0'(z) = -\mathscr{C}_1(z). \tag{B.22}$$

Here, the symbol $\mathscr{C}$ represents Bessel ($\mathscr{C} = \mathrm{j}$) or Hankel ($\mathscr{C} = \mathrm{h}$) functions.

B.5 SPHERICAL BESSEL FUNCTIONS

Definitions

The *spherical Bessel functions* of first and second kinds of order n are denoted by $\mathrm{j}_n(z)$ and $\mathrm{y}_n(z)$, respectively. The *spherical Hankel functions* of first and second kinds of order n are defined by [Zhang and Jin, 1996, Eqs. (8.1.5–6)]

$$\mathrm{h}_n^{(1)}(z) = \mathrm{j}_n(z) + \mathrm{y}_n(z), \tag{B.23}$$

$$\mathrm{h}_n^{(2)}(z) = \mathrm{j}_n(z) - \mathrm{y}_n(z). \tag{B.24}$$

In this book, only the first kind of Hankel function is used. Consequently, the symbol $\mathrm{h}_n(z) = \mathrm{h}_n^{(1)}(z)$ is used for simplicity.

Relations

Spherical Bessel and Hankel functions can be represented using Bessel and Hankel functions as [Zhang and Jin, 1996, Eq. (8.1.3–6)]

$$\mathscr{S}_n(z) = \sqrt{\frac{\pi}{2z}} \mathscr{C}_{n+1/2}(z). \tag{B.25}$$

Here, the symbol $\mathscr{S}$ represents spherical Bessel ($\mathscr{S} = J$) or Hankel ($\mathscr{S} = H$) functions.

Limiting Forms

The spherical Hankel function at large argument $z \to \infty$ has the limiting form of [Olver et al., 2019, Eq. (10.52.4)]

$$\mathrm{h}_n(z) \sim \frac{\mathrm{e}^{\mathrm{i}z}}{\mathrm{i}^{n+1}z}. \tag{B.26}$$

Special Values

The special value of the spherical Bessel function when the argument is zero is

$$\mathrm{j}_n(0) = \delta_{n0} = \begin{cases} 1, & n = 0, \\ 0, & n = 1, 2, 3, \ldots. \end{cases} \tag{B.27}$$

Wronskians

The Wronskian between $\mathrm{j}_n(z)$ and $\mathrm{h}_n(z)$ is

$$\mathcal{W}[\mathrm{j}_n(z), \mathrm{h}_n(z)] \equiv \mathrm{j}_n(z)\mathrm{h}_n'(z) - \mathrm{h}_n(z)\mathrm{j}_n'(z) = \frac{\mathrm{i}}{z^2}. \tag{B.28}$$

B.6 ZERNIKE POLYNOMIALS

Definitions

The *Zernike polynomial* is defined as [Flusser et al., 2017, Eq. (7.126)]

$$Z_n^m(\rho, \varphi) = R_n^m(\rho)\mathrm{e}^{im\varphi}, \tag{B.29}$$

where $R_n^m(\rho)$ is called the *radial Zernike polynomial*, and $n, m \in \mathbb{Z}, n \geq |m|, 0 \leq \rho \leq 1, 0 \leq \varphi < 2\pi$.

The radial Zernike polynomial of degree n and azimuthal order m is defined as [Flusser et al., 2017, Eq. (7.127); Chong et al., 2003, Eq. (2.3)]

$$R_n^m(\rho) = \sum_{l=0}^{(n-|m|)/2} \frac{(-1)^l (n-l)!}{l!\left(\frac{n+|m|}{2} - l\right)!\left(\frac{n-|m|}{2} - l\right)!}\rho^{n-2l}, \tag{B.30}$$

where $0 \leq r \leq 1$.

Properties

Zernike polynomials satisfy the relation of orthogonality [Flusser et al., 2017, Eq. (7.128)]

$$\int_0^{2\pi}\int_0^1 Z_n^{m,*}(\rho,\varphi)Z_{n'}^{m'}(\rho,\varphi)\rho d\rho d\varphi = \frac{\pi}{n+1}\delta_{nn'}\delta_{mm'}, \tag{B.31}$$

which is resulted from the orthogonality of the radial Zernike polynomials [Flusser et al., 2017, Eq. (7.129)]

$$\int_0^1 R_n^m(\rho)R_{n'}^m(\rho)\rho d\rho = \frac{\delta_{nn'}}{2(n+1)}. \tag{B.32}$$

Zernike Expansion and Moment

Zernike polynomials form a orthogonal and complete set in over a unit circle. Consequently, a function $F(\rho,\varphi)$ can be expanded using the Zernike polynomials as

$$F(\rho,\varphi) = \sum_{n=0}^{\infty}\sum_{m=0}^{n}{}' M_n^m Z_n^m(\rho,\varphi), \tag{B.33}$$

where the prime over the summation sign indicates that the summation is performed for terms only with even $n-m$. In Eq. (B.33), M_n^m are the *Zernike moments* [Flusser et al., 2017, Eq. (2.209); Mukundan and Ramakrishnan, 1995, Eq. (8)]

$$M_n^m = \frac{n+1}{\pi}\int_0^{2\pi}\int_0^1 F(\rho,\varphi)Z_n^m(\rho,\varphi)\rho d\rho d\varphi. \tag{B.34}$$

Bibliography

Aanonsen, S. I., T. Barkve, J. N. Tjøtta, and S. Tjøtta (1984). Distortion and Harmonic Generation in the Nearfield of a Finite Amplitude Sound Beam. In: *J. Acoust. Soc. Am.* 75.3, pp. 749–768.

Aarts, R. M. and A. J. Janssen (2010). Sound Radiation from a Resilient Spherical Cap on a Rigid Sphere. In: *J. Acoust. Soc. Am.* 127.4, pp. 2262–2273.

Aarts, R. M. and A. J. E. M. Janssen (2009). On-Axis and Far-Field Sound Radiation from Resilient Flat and Dome-Shaped Radiators. In: *J. Acoust. Soc. Am.* 125.3, pp. 1444–1455.

Abbasov, I. B. and N. P. Zagrai (1995). Sphere Scattering of Nonlinearly Interacting Acoustic Waves. In: *Fluid Dyn.* 30.2, pp. 158–165.

Abuelma'atti, M. T. (2003). Improved Analysis of the Electrically Manifested Distortions of Condenser Microphones. In: *Appl. Acoust.* 64.5, pp. 471–480.

Ahn, H., K. Been, I.-D. Kim, C. H. Lee, and W. Moon (2019). A Critical Step to Using a Parametric Array Loudspeaker in Mobile Devices. In: *Sensors* 19.20, p. 4449.

Akahori, H., H. Furuhashi, and M. Shimizu (2014). "Direction Control of a Parametric Speaker". In: 2014 IEEE International Ultrasonics Symposium, pp. 2470–2473.

Akiyama, M., T. Kamakura, and C. Hedberg (2008). Parametric Sound Fields by Phase-cancellation Excitation of Primary Waves. In: *AIP Conference Proceedings* 1022.1, pp. 30–33.

Allard, J. F. and N. Atalla (2009). *Propagation of Sound in Porous Media: Modelling Sound Absorbing Materials*. 2nd ed. West Sussex, United Kingdom: John Wiley & Sons.

Allen, J. B. and D. A. Berkley (1979). Image Method for Efficiently Simulating Small-room Acoustics. In: *J. Acoust. Soc. Am.* 65.4, pp. 943–950.

Ang, L. Y. L., Y. K. Koh, and H. P. Lee (2017). The Performance of Active Noise-Canceling Headphones in Different Noise Environments. In: *Appl. Acoust.* 122, pp. 16–22.

ANSI/IEEE Std 176 (1987). *IEEE Standard on Piezoelectricity*.

Ardid, M., M. Bou-Cabo, F. Camarena, V. Espinosa, G. Larosa, J. A. Martínez-Mora, and M. Ferri (2009). Use of Parametric Acoustic Sources to Generate Neutrino-like Signals. In: *Nuclear Instruments and Methods in Physics Research Section A: Accelerators, Spectrometers, Detectors and Associated Equipment*. ARENA 2008 604 (1, Supplement), S208–S211.

Aretz, M., P. Dietrich, and M. Vorländer (2014). Application of the Mirror Source Method for Low Frequency Sound Prediction in Rectangular Rooms. In: *Acta Acust. united Acust.* 100.2, pp. 306–319.

Arnela, M., O. Guasch, P. Sanchez-Martin, J. Camps, R. Alsina-Pages, and C. Martinez-Suquia (2018). Construction of an Omnidirectional Parametric Loudspeaker Consisting in a Spherical Distribution of Ultrasound Transducers. In: *Sensors* 18.12, p. 4317.

Arnela, M., C. Martinez-Suquia, and O. Guasch (2020). "Exponential Sine Sweeps to Measure the Diretivity of an Omnidirectional Parametric Loudspeaker". In: Forum Acusticum 2020, p. 4.

Arnela, M., C. Martinez-Suquia, and O. Guasch (2021). Characterization of an Omnidirectional Parametric Loudspeaker with Exponential Sine Sweeps. In: *Appl. Acoust.* 182, p. 108268.

Arnela, M., C. Martinez-Suquia, and O. Guasch (2022). Carrier Frequency Influence on the Audible and Ultrasonic Fields Generated by an Omnidirectional Parametric Loudspeaker Excited with Exponential Sine Sweeps. In: *Appl. Acoust.* 200, p. 109073.

Assaad, J. and J. M. Rouvaen (1998). Some Considerations Concerning Replicated Spectra. In: *J. Acoust. Soc. Am.* 104.1, pp. 580–582.

Averkiou, M. A., Y.-S. Lee, and M. F. Hamilton (1993). Self-demodulation of Amplitude- and Frequency-modulated Pulses in a Thermoviscous Fluid. In: *J. Acoust. Soc. Am.* 94.5, pp. 2876–2883.

Bai, M. R., P.-J. Hsieh, and K.-N. Hur (2009). Optimal Design of Minimum Mean-Square Error Noise Reduction Algorithms Using the Simulated Annealing Technique. In: *J. Acoust. Soc. Am.* 125.2, pp. 934–943.

Bass, H. E., L. C. Sutherland, and A. J. Zuckerwar (1990). Atmospheric Absorption of Sound: Update. In: *J. Acoust. Soc. Am.* 88.4, pp. 2019–2021.

Bass, H. E., L. C. Sutherland, A. J. Zuckerwar, D. T. Blackstock, and D. M. Hester (1996). Erratum: Atmospheric Absorption of Sound: Further Developments [J. Acoust. Soc. Am. 97, 680–683 (1995)]. In: *J. Acoust. Soc. Am.* 99.2, pp. 1259–1259.

Bass, H. E., L. C. Sutherland, A. J. Zuckerwar, D. T. Blackstock, and D. M. Hester (1995). Atmospheric Absorption of Sound: Further Developments. In: *J. Acoust. Soc. Am.* 97.1, pp. 680–683.

Bates, T. P., I. C. Bacon, J. D. Blotter, and S. D. Sommerfeldt (2022). Vibration-Based Sound Power Measurements of Arbitrarily Curved Panels. In: *J. Acoust. Soc. Am.* 151.2, pp. 1171–1179.

Bellin, J. L. S. and R. T. Beyer (1962). Experimental Investigation of an End-fire Array. In: *J. Acoust. Soc. Am.* 34.8, pp. 1051–1054.

Bellin, J. L. S., P. J. Westervelt, and R. T. Beyer (1960). Experimental Investigation of a Parametric End Array. In: *J. Acoust. Soc. Am.* 32.7, pp. 935–935.

Bellin, J. L. S. and R. T. Beyer (1960). Scattering of Sound by Sound. In: *J. Acoust. Soc. Am.* 32.3, pp. 339–341.

Ben Hagai, I., M. Pollow, M. Vorländer, and B. Rafaely (2011). Acoustic Centering of Sources Measured by Surrounding Spherical Microphone Arrays. In: *J. Acoust. Soc. Am.* 130.4, pp. 2003–2015.

Bennett, M. B. and D. T. Blackstock (1975). Parametric Array in Air. In: *J. Acoust. Soc. Am.* 57.3, pp. 562–568.

Beranek, L. L. and T. J. Mellow (2019). *Acoustics: Sound Fields, Transducers and Vibration*. Second edition. London [England]: Academic Press. 881 pp.

Berktay, H. O. (1965a). Possible Exploitation of Non-Linear Acoustics in Underwater Transmitting Applications. In: *J. Sound Vib.* 2.4, pp. 435–461.

Berktay, H. O. and D. J. Leahy (1974). Farfield Performance of Parametric Transmitters. In: *J. Acoust. Soc. Am.* 55.3, pp. 539–546.

Berktay, H. O. and J. A. Shooter (1973). Nearfield Effects in End-Fire Line Arrays. In: *J. Acoust. Soc. Am.* 53.2, pp. 550–556.

Berktay, H. (1965b). Parametric Amplification by the Use of Acoustic Non-Linearities and Some Possible Applications. In: *J. Sound Vib.* 2.4, pp. 462–470.

Betlehem, T. and T. D. Abhayapala (2005). Theory and Design of Sound Field Reproduction in Reverberant Rooms. In: *J. Acoust. Soc. Am.* 117.4, pp. 2100–2111.

Betlehem, T. and M. Poletti (2012). Sound Field of a Directional Source in a Reverberant Room. In: *N. Z. Acoust.* 25.4, pp. 12–22.

Bevelacqua, P. J. and C. A. Balanis (2007). Minimum Sidelobe Levels for Linear Arrays. In: *IEEE Trans. Antennas Propagat.* 55.12, pp. 3442–3449.

Beyer, R. T. (1960). Parameter of Nonlinearity in Fluids. In: *J. Acoust. Soc. Am.* 32.6, pp. 719–721.

Bilbao, S. and J. Ahrens (2020). Modeling Continuous Source Distributions in Wave-Based Virtual Acoustics. In: *J. Acoust. Soc. Am.* 148.6, pp. 3951–3962.

Bolton, J. S., B. K. Gardner, and T. A. Beauvilain (1995). Sound Cancellation by the Use of Secondary Multipoles. In: *J. Acoust. Soc. Am.* 98.4, pp. 2343–2362.

Boodoo, S., R. Paurobally, and Y. Bissessur (2015). A Review of the Effect of Reflective Surfaces on Power Output of Sound Sources and on Actively Created Quiet Zones. In: *Acta Acust. united Acust.* 101.5, pp. 877–891.

Bowman, J. J., T. B. Senior, and P. L. Uslenghi (1970). *Electromagnetic and Acoustic Scattering by Simple Shapes*. Amsterdam: North-Holland Publishing Company.

Brooks, L. A., A. C. Zander, and C. H. Hansen (2005). "Investigation into the Feasibility of Using a Parametric Array Control Source in an Active Noise Control System". In: Proceedings of ACOUSTICS. Vol. 2005, pp. 39–45.

Browning, D. G., M. B. Moffett, and W. L. Konrad (2009). "Parametric Acoustic Array Development a the US Navy's New London, Connecticut Laboratory". In: 157th Meeting Acoustical Society of America. Portland, Oregon, pp. 045002–045002.

Bruel and Kjaer (2017). *Condenser Microphone Cartridges — Types 4133 to 4181*.

Bruno, D., A. Frezzotti, S. H. Jamali, and W. Van De Water (2023). Finding the Bulk Viscosity of Air from Rayleigh-Brillouin Light Scattering Spectra. In: *J. Chem. Phys.* 158.3, p. 031101.

Bullmore, A. J., P. A. Nelson, A. R. D. Curtis, and S. J. Elliott (1987). The Active Minimization of Harmonic Enclosed Sound Fields, Part II: A Computer Simulation. In: *J. Sound Vib.* 117.1, pp. 15–33.

Burka, A., A. Qin, and D. D. Lee (2014). "An Application of Parametric Speaker Technology to Bus-Pedestrian Collision Warning". In: *17th International IEEE Conference on Intelligent Transportation Systems (ITSC)*. 17th International IEEE Conference on Intelligent Transportation Systems (ITSC), pp. 948–953.

Cai, A., J. A. Sun, and G. Wade (1988). Adapting Parametric Acoustic Arrays for Tomographic Imaging. In: *Ultrason. Imaging* 10.3, pp. 196–203.

Cai, C. and C. M. Mak (2016). Noise Control Zone for a Periodic Ducted Helmholtz Resonator System. In: *J. Acoust. Soc. Am.* 140.6, EL471–EL477.

Campo-Valera, M., I. Rodríguez-Rodríguez, J.-V. Rodríguez, and L.-J. Herrera-Fernández (2023). Proof of Concept of the Use of the Parametric Effect in Two Media with Application to Underwater Acoustic Communications. In: *Electron.* 12.16 (16), p. 3459.

Carcagno, S., A. Di Battista, and C. J. Plack (2019). Effects of High-Intensity Airborne Ultrasound Exposure on Behavioural and Electrophysiological Measures of Auditory Function. In: *Acta Acust. united Acust.* 105.6, pp. 1183–1197.

Castagnède, B., A. Moussatov, D. Lafarge, and M. Saeid (2008). Low Frequency in Situ Metrology of Absorption and Dispersion of Sound Absorbing Porous Materials Based on High Power Ultrasonic Non-Linearly Demodulated Waves. In: *Appl. Acoust.* 69.7, pp. 634–648.

Castagnède, B., M. Saeid, A. Moussatov, V. Gusev, and V. Tournat (2006). Reflection and Transmission at Normal Incidence onto Air-Saturated Porous Materials and Direct Measurements Based on Parametric Demodulated Ultrasonic Waves. In: *Ultrasonics* 44.2, pp. 221–229.

Červenka, M. and M. Bednařík (2013). Non-Paraxial Model for a Parametric Acoustic Array. In: *J. Acoust. Soc. Am.* 134.2, pp. 933–938.

Červenka, M. and M. Bednařík (2015). On the Structure of Multi-Gaussian Beam Expansion Coefficients. In: *Acta Acust. united Acust.* 101.1, pp. 15–23.

Červenka, M. and M. Bednařík (2019). A Versatile Computational Approach for the Numerical Modelling of Parametric Acoustic Array. In: *J. Acoust. Soc. Am.* 146.4, pp. 2163–2169.

Červenka, M. and M. Bednařík (2021). Parametric Acoustic Array Lensed by a Gradient-Index Phononic Crystal. In: *J. Acoust. Soc. Am.* 149.6, pp. 4534–4542.

Červenka, M. and M. Bednařík (2022). An Algebraic Correction for the Westervelt Equation to Account for the Local Nonlinear Effects in Parametric Acoustic Array. In: *J. Acoust. Soc. Am.* 151.6, pp. 4046–4052.

Červenka, M., M. Bednařík, and P. Konicek (2009). "Numerical Simulation of Parametric Field Patterns of Ultrasonic Transducer Arrays". In: 2009 IEEE International Ultrasonics Symposium, pp. 1–4.

Chapra, S. C. and R. P. Canale (2015). *Numerical Methods for Engineers*. Seventh edition. New York, NY: McGraw-Hill Education. 970 pp.

Cheer, J. (2012). "Active Control of the Acoustic Environment in an Automobile Cabin". PhD thesis. Southampton, Hampshire, England: University of Southampton.

Chen, W., W. Rao, H. Min, and X. Qiu (2011). An Active Noise Barrier with Unidirectional Secondary Sources. In: *Appl. Acoust.* 72.12, pp. 969–974.

Cheng, J. (2019a). *Principles of Acoustics, Volume I [Chinese].* Beijing: Science Press.

Cheng, J. (2019b). *Principles of Acoustics, Volume II [Chinese].* Beijing: Science Press.

Chong, C.-W., P. Raveendran, and R. Mukundan (2003). A Comparative Analysis of Algorithms for Fast Computation of Zernike Moments. In: *Pattern Recognit.* 36.3, pp. 731–742.

Coates, R., M. Zheng, and L. Wang (1996). "BASS 300 PARACOM": A Model Underwater Parametric Communication System. In: *IEEE J. Ocean. Eng.* 21.2, pp. 225–232.

Darvennes, C. M. and M. F. Hamilton (1990). Scattering of Sound by Sound from Two Gaussian Beams. In: *J. Acoust. Soc. Am.* 87.5, pp. 1955–1964.

David, A. and S. J. Elliott (1994). Numerical Studies of Actively Generated Quiet Zones. In: *Appl. Acoust.* 41.1, pp. 63–79.

Di Battista, A., A. Price, R. Malkin, B. W. Drinkwater, P. Kuberka, and C. Jarrold (2022). The Effects of High-Intensity 40 kHz Ultrasound on Cognitive Function. In: *Appl. Acoust.* 200, p. 109051.

Dimant, Y., I. Roger-Eitan, and A. Gleizer (2021). Lens Mediated Scattering of Sound by Sound with Finite Aperture Beams. In: *J. Acoust. Soc. Am.* 149.1, pp. 207–214.

Ding, D. and J. Xu (2006). The Gaussian Beam Expansion Applied to Fresnel Field Integrals. In: *IEEE Trans. Ultrason., Ferroelect., Freq. Contr.* 53.1, pp. 246–250.

Ding, D. and Y. Zhang (2004). Notes on the Gaussian Beam Expansion. In: *J. Acoust. Soc. Am.* 116.3, pp. 1401–1405.

Ding, D., Y. Zhang, and J. Liu (2003). Some Extensions of the Gaussian Beam Expansion: Radiation Fields of the Rectangular and the Elliptical Transducer. In: *J. Acoust. Soc. Am.* 113.6, pp. 3043–3048.

Dirkse, B. (2014). "Finite Element Method Applied to the One-dimensional Westervelt Equation". Delft University of Technology.

Donley, J. (2018). "Reproduction of Personal Sound in Shared Environments". University of Wollongong.

Donley, J., C. Ritz, and W. B. Kleijn (2016). "Reproducing Personal Sound Zones Using a Hybrid Synthesis of Dynamic and Parametric Loudspeakers". In: 2016 Asia-Pacific Signal and Information Processing Association Annual Summit and Conference (APSIPA). IEEE, pp. 1–5.

Du, J. T., W. L. Li, Z. G. Liu, H. A. Xu, and Z. L. Ji (2011). Acoustic Analysis of a Rectangular Cavity with General Impedance Boundary Conditions. In: *J. Acoust. Soc. Am.* 130.2, pp. 807–817.

Duraiswami, R., D. N. Zotkin, and N. A. Gumerov (2007). Fast Evaluation of the Room Transfer Function Using Multipole Expansion. In: *IEEE Trans. Audio Speech Lang. Process.* 15.2, pp. 565–576.

Elliott, S. (2000). *Signal Processing for Active Control.* Cambridge: Academic Press.

Elliott, S. J., P. Joseph, A. J. Bullmore, and P. A. Nelson (1988). Active Cancellation at a Point in a Pure Tone Diffuse Sound Field. In: *J. Sound Vib.* 120.1, pp. 183–189.

Elliott, S. J. and J. Cheer (2015). Modeling Local Active Sound Control with Remote Sensors in Spatially Random Pressure Fields. In: *J. Acoust. Soc. Am.* 137.4, pp. 1936–1946.

Elliott, S. J., W. Jung, and J. Cheer (2018). Head Tracking Extends Local Active Control of Broadband Sound to Higher Frequencies. In: *Sci. Rep.* 8.1, pp. 1–7.

Enflo, B. O. and C. M. Hedberg (2002). *Theory of Nonlinear Acoustics in Fluids*. Dordrecht ; Boston: Kluwer Academic Publishers.

Epain, N. and E. Friot (2007). Active Control of Sound inside a Sphere via Control of the Acoustic Pressure at the Boundary Surface. In: *J. Sound Vib.* 299.3, pp. 587–604.

Escolano, J., J. J. Lopez, and B. Pueo (2007). Directive Sources in Acoustic Discrete-Time Domain Simulations Based on Directivity Diagrams. In: *J. Acoust. Soc. Am.* 121.6, EL256–EL262.

Evans, L. B., H. E. Bass, and L. C. Sutherland (1972). Atmospheric Absorption of Sound: Theoretical Predictions. In: *J. Acoust. Soc. Am.* 51 (5B), pp. 1565–1575.

Farias, F. and W. Abdulla (2015). "On Rayleigh Distance and Absorption Length of Parametric Loudspeakers". In: *2015 Asia-Pacific Signal and Information Processing Association Annual Summit and Conference (APSIPA)*. Hong Kong: IEEE, pp. 1262–1265.

Fenlon, F. (1978). "On the Maximum Achievable Conversion Efficiency on a Parametric Acoustic Array". In: *ICASSP '78. IEEE International Conference on Acoustics, Speech, and Signal Processing*. IEEE International Conference on Acoustics, Speech, and Signal Processing. Vol. 3. Tulsa, Oklahoma: Institute of Electrical and Electronics Engineers, pp. 127–129.

Filippi, P., D. Habault, J.-P. Lefebvre, and A. Bergassoli (1999). *Acoustics: Basic Physics, Theory and Methods*. Academic Press.

Flammer, C. (2005). *Spheroidal Wave Functions*. New York: Dover.

Flusser, J., T. Suk, and B. Zitova (2017). *2D and 3D Image Analysis by Moments*. Chichester, West Sussex, United Kingdom: Wiley.

Foote, K. G. (2014). Discriminating between the Nearfield and the Farfield of Acoustic Transducers. In: *J. Acoust. Soc. Am.* 136.4, pp. 1511–1517.

Foote, K. G. (2021). Reflections on "Parametric Acoustic Array," Source of Virtual-Array Sonars. In: *J. Acoust. Soc. Am.* 150.1, R1–R2.

Gallego-Juarez, J. A., G. Rodriguez-Corral, and L. Gaete-Garreton (1978). An Ultrasonic Transducer for High Power Applications in Gases. In: *Ultrasonics* 16.6, pp. 267–271.

Gan, W.-S., E.-L. Tan, and S. M. Kuo (2011). Audio Projection: Directional Sound and Its Applications in Immersive Communication. In: *IEEE Signal Process. Mag.* 28.1, pp. 43–57.

Gan, W.-S., J. Yang, and T. Kamakura (2012). A Review of Parametric Acoustic Array in Air. In: *Appl. Acoust.* 73.12, pp. 1211–1219.

Gan, W.-S., J. Yang, K.-S. Tan, and M.-H. Er (2006). A Digital Beamsteerer for Difference Frequency in a Parametric Array. In: *IEEE/ACM Trans. Audio, Speech, Lang. Process.* 14.3, pp. 1018–1025.

Garai, M. and F. Pompoli (2005). A Simple Empirical Model of Polyester Fibre Materials for Acoustical Applications. In: *Appl. Acoust.* 66.12, pp. 1383–1398.

Garner, G. (2011). "Design of Optimal Directional Parametric Acoustic Arrays in Air". Raleigh, North Carolina: North Carolina State University.

Garner, G. and M. B. Steer (2012a). Third-Order Parametric Array Generated by Distantly Spaced Primary Ultrasonic Tones. In: *IEEE Trans. Ultrason., Ferroelect., Freq. Contr.* 59.4, pp. 776–784.

Garner, G. and M. B. Steer (2012b). A Cascaded Second-Order Approach to Computing Third-Order Scattering of Noncollinear Acoustic Beams. In: *Appl. Acoust.* 73.12, pp. 1220–1230.

Garrett, G. S., J. N. Tjøtta, R. L. Rolleigh, and S. Tjøtta (1984). Reflection of Parametric Radiation from a Finite Planar Target. In: *J. Acoust. Soc. Am.* 75.5, pp. 1462–1472.

Gaunaurd, G. C., H. Huang, and H. C. Strifors (1995). Acoustic Scattering by a Pair of Spheres. In: *J. Acoust. Soc. Am.* 98.1, pp. 495–507.

Geddes, E. R. and L. Lee (2001). *Audio Transducers*. Novi, MI: GedLee Publishing.

Geddes, E. R. (2010). Comments on "Estimating the Velocity Profile and Acoustical Quantities of a Harmonically Vibrating Loudspeaker Membrane from On-Axis Pressure Data". In: *J. Audio Eng. Soc.*

Gerges, S. N. Y. (2012). Earmuff Comfort. In: *Appl. Acoust.* 73.10, pp. 1003–1012.

Goss, S., L. Frizzell, J. Kouzmanoff, J. Barich, and J. Yang (1996). Sparse Random Ultrasound Phased Array for Focal Surgery. In: *IEEE Trans. Ultrason., Ferroelect., Freq. Contr.* 43.6, pp. 1111–1121.

Gradshteyn, I. S. and I. M. Ryzhik (2015). *Table of Integrals, Series, and Products*. Eighth edition. Amsterdam; Boston: Elsevier, Academic Press. 1133 pp.

Greenspan, M. (1979). Piston Radiator: Some Extensions of the Theory. In: *J. Acoust. Soc. Am.* 65.3, pp. 608–621.

Gu, J. and Y. Jing (2015). Modeling of Wave Propagation for Medical Ultrasound: A Review. In: *IEEE Trans. Ultrason., Ferroelect., Freq. Contr.* 62.11, pp. 1979–1992.

Guasch, O. and P. Sánchez-Martín (2018). Far-Field Directivity of Parametric Loudspeaker Arrays Set on Curved Surfaces. In: *Appl. Math. Model.* 60, pp. 721–738.

Guigné, J. Y., V. H. Chin, and S. M. Solomon (1989). Acoustic Attenuation Measurements Using Parametric Arrays. In: *Ultrasonics* 27.5, pp. 297–301.

Guigné, J. Y., N. Rukavina, P. H. Hunt, and J. S. Ford (1991). An Acoustic Parametric Array for Measuring the Thickness and Stratigraphy of Contaminated Sediments. In: *J. Great Lakes Res.* 17.1, pp. 120–131.

Guo, J. and J. Pan (1998). Effects of Reflective Ground on the Actively Created Quiet Zones. In: *J. Acoust. Soc. Am.* 103.2, pp. 944–952.

Guo, J., J. Pan, and C. Bao (1997). Actively Created Quiet Zones by Multiple Control Sources in Free Space. In: *J. Acoust. Soc. Am.* 101.3, pp. 1492–1501.

Hahn, N., J. Ahrens, and C. Andersson (2021). "Parametric Array Using Amplitude Modulated Pulse Trains: Experimental Evaluation of Beamforming and Single Sideband Modulation". In: *Audio Engineering Society Convention 151*. Audio Engineering Society, pp. 1–7.

Hald, J. (2005). "Estimation of Partial Area Sound Power Data with Beamforming". In: INTER-NOISE and NOISE-CON Congress and Conference Proceedings. Rio de Janeiro, Brazil, pp. 2424–2433.

Hamilton, B. and S. Bilbao (2021). Time-Domain Modeling of Wave-Based Room Acoustics Including Viscothermal and Relaxation Effects in Air. In: *JASA Express Letters* 1.9, p. 092401.

Hamilton, M. F. (2016). Effective Gol'dberg Number for Diverging Waves. In: *J. Acoust. Soc. Am.* 140.6, pp. 4419–4427.

Hamilton, M. F. and D. T. Blackstock (1990). On the Linearity of the Momentum Equation for Progressive Plane Waves of Finite Amplitude. In: *J. Acoust. Soc. Am.* 88.4, pp. 2025–2026.

Hamilton, M. F. and D. T. Blackstock (2008). *Nonlinear Acoustics*. New York: Acoustical Society of America.

Hamilton, M. F., V. A. Khokhlova, and O. V. Rudenko (1997). Analytical Method for Describing the Paraxial Region of Finite Amplitude Sound Beams. In: *J. Acoust. Soc. Am.* 101.3, pp. 1298–1308.

Hansen, C., S. Snyder, X. Qiu, L. Brooks, and D. Moreau (2012). *Active Control of Noise and Vibration*. CRC press.

Hassani, S. (2013). *Mathematical Physics: A Modern Introduction to Its Foundations*. 2nd ed. 2013. New York: Springer.

Havelock, D. I. and A. J. Brammer (2000). Directional Loudspeakers Using Sound Beams. In: *J. Audio Eng. Soc.* 48.10, pp. 908–916.

Hedberg, C. M., K. C. E. Haller, and T. Kamakura (2010). A Self-Silenced Sound Beam. In: *Acoust. Phys.* 56.5, pp. 637–639.

Hines, P. C. (1999). Quantifying Seabed Properties in Shelf Waters Using a Parametric Sonar. In: *Meas. Sci. Technol.* 10.12, p. 1127.

Hirayama, R., D. Martinez Plasencia, N. Masuda, and S. Subramanian (2019). A Volumetric Display for Visual, Tactile and Audio Presentation Using Acoustic Trapping. In: *Nature* 575.7782, pp. 320–323.

Hogan, C. M. and G. L. Latshaw (1973). "The Relationship between Highway Planning and Urban Noise". In: *Proceedings of the ASCE Urban Transportation Division Environment Impact Specialty Conference, May 21-23, 1973, Chicago, Illinois*.

Holosonics (2019). *Audio Spotlight 24i*. URL: https://www.holosonics.com/audio-spotlight-24i (visited on 06/11/2019).

Hoshi, T., M. Takahashi, T. Iwamoto, and H. Shinoda (2010). Noncontact Tactile Display Based on Radiation Pressure of Airborne Ultrasound. In: *IEEE Trans. Haptics* 3.3, pp. 155–165.

Hu, Q. and S. K. Tang (2019). Active Cancellation of Sound Generated by Finite Length Coherent Line Sources Using Piston-like Secondary Source Arrays. In: *J. Acoust. Soc. Am.* 145.6, pp. 3647–3655.

Huang, D. and M. A. Breazeale (1999). A Gaussian Finite-Element Method for Description of Sound Diffraction. In: *J. Acoust. Soc. Am.* 106.4, pp. 1771–1781.

Humphrey, V. F. (1985). The Measurement of Acoustic Properties of Limited Size Panels by Use of a Parametric Source. In: *J. Sound Vib.* 98.1, pp. 67–81.

Humphrey, V. F., S. P. Robinson, J. D. Smith, M. J. Martin, G. A. Beamiss, G. Hayman, and N. L. Carroll (2008). Acoustic Characterization of Panel Materials under Simulated Ocean Conditions Using a Parametric Array Source. In: *J. Acoust. Soc. Am.* 124.2, pp. 803–814.

IEC 60268-5 (2007). *Sound System Equipment — Part 5: Loudspeakers*. Geneva, Switzerland.

Iijima, R., S. Minami, Y. Zhou, T. Takehisa, T. Takahashi, Y. Oikawa, and T. Mori (2019). Audio Hotspot Attack: An Attack on Voice Assistance Systems Using Directional Sound Beams and Its Feasibility. In: *IEEE Trans. Emerg.* 9.4, pp. 2004–2018.

Ingard, U. and D. C. Pridmore-Brown (1956). Scattering of Sound by Sound. In: *J. Acoust. Soc. Am.* 28.3, pp. 367–369.

Ishikawa, K., Y. Shiraki, and T. Moriya (2021). Spurious-Sound-Free Measurement of Parametric Acoustic Array Using Optical Interferometry. In: *JASA Express Letters* 1.11, p. 112801.

ISO 10534-2 (2001). *Acoustics — Determination of Sound Absorption Coefficient and Impedance in Impendance Tubes — Part 2: Transfer-function Method*. Geneva, Switzerland.

ISO 10847 (1997). *Acoustics — In-situ Determination of Insertion Loss of Outdoor Noise Barriers of All Types*. Geneva, Switzerland.

ISO 11546-1 (2009). *Acoustics — Determination of Sound Insulation Performances of Enclosures — Part 1: Measurements under Laboratory Conditions (for Declaration Purposes)*. Geneva, Switzerland.

ISO 15667 (2000). *Acoustics — Guidelines for Noise Control by Enclosures and Cabins*. Geneva, Switzerland.

ISO 3741 (2010). *Acoustics — Determination of Sound Power Levels and Sound Energy Levels of Noise Sources Using Sound Pressure — Precision Methods for Reverberation Test Rooms*. Geneva, Switzerland.

ISO 3745 (2012). *Acoustics — Determination of Sound Power Levels and Sound Energy Levels of Noise Sources Using Sound Pressure*. Geneva, Switzerland.

ISO 9613-1 (1993). *Acoustics — Attenuation of Sound during Propagation Outdoors — Part 1: Calculation of the Absorption of Sound by the Atmosphere*. Geneva, Switzerland.

ISO 9614-1 (1993). *Acoustics — Determination of Sound Power Levels of Noise Sources Using Sound Intensity — Part 1: Measurement at Discrete Points*. Geneva, Switzerland.

ISO 9614-2 (1996). *Acoustics — Determination of Sound Power Levels of Noise Sources Using Sound Intensity — Part 2: Measurement by Scanning*. Geneva, Switzerland.

ISO/TS 7849-1 (2009). *Acoustics — Determination of Airborne Sound Power Levels Emitted by Machinery Using Vibration Measurement — Part 1: Survey Method Using a Fixed Radiation Factor*. Geneva, Switzerland.

ISO/TS 7849-2 (2009). *Acoustics — Determination of Airborne Sound Power Levels Emitted by Machinery Using Vibration Measurement — Part 2: Engineering Method Including Determination of the Adequate Radiation Factor*. Geneva, Switzerland.

Jager, A. (2019). "Airborne Ultrasound Phased Arrays". Darmstadt, Germany: Technical University of Darmstadt.

Je, Y., H. Lee, K. Been, and W. Moon (2015). A Micromachined Efficient Parametric Array Loudspeaker with a Wide Radiation Frequency Band. In: *J. Acoust. Soc. Am.* 137.4, pp. 1732–1743.

Je, Y., H. Lee, and W. Moon (2013). The Impact of Micromachined Ultrasonic Radiators on the Efficiency of Transducers in Air. In: *Ultrasonics* 53, pp. 1124–1134.

Ji, P., W. Hu, and J. Yang (2016). Development of an Acoustic Filter for Parametric Loudspeaker Using Phononic Crystals. In: *Ultrasonics* 67, pp. 160–167.

Ji, P., W. Liu, S. Wu, J. Yang, and W.-S. Gan (2012). An Alternative Method to Measure the On-Axis Difference-Frequency Sound in a Parametric Loudspeaker without Using an Acoustic Filter. In: *Appl. Acoust.* 73.12, pp. 1244–1250.

Ji, P., E.-L. Tan, W.-S. Gan, and J. Yang (2011). A Comparative Analysis of Preprocessing Methods for the Parametric Loudspeaker Based on the Khokhlov–Zabolotskaya–Kuznetsov Equation for Speech Reproduction. In: *IEEE/ACM Trans. Audio, Speech, Lang. Process.* 19.4, pp. 937–946.

Ji, P. and J. Yang (2019). An Experimental Investigation about Parameters' Effects on Spurious Sound in Parametric Loudspeaker. In: *Appl. Acoust.* 148, pp. 67–74.

Ji, P., J. Yang, and W.-S. Gan (2009). The Investigation of Localized Sound Generation Using Two Ultrasound Beams. In: *IEEE Trans. Ultrason., Ferroelect., Freq. Contr.* 56.6, pp. 1282–1287.

Ji, W. and W.-S. Gan (2012). Identification of a Parametric Loudspeaker System Using an Adaptive Volterra Filter. In: *Appl. Acoust.* Parametric Acoustic Array: Theory, Advancement and Applications 73.12, pp. 1251–1262.

Jimenez, N., J. Ealo, R. D. Muelas-Hurtado, A. Duclos, and V. Romero-Garcia (2021). Subwavelength Acoustic Vortex Beams Using Self-Demodulation. In: *Phys. Rev. Appl.* 15.5, p. 054027.

Jing, Y., D. Shen, and G. T. Clement (2011). Verification of the Westervelt Equation for Focused Transducers. In: *IEEE Trans. Ultrason., Ferroelectr. Freq. Control* 58.5, pp. 1097–1101.

Johnson, S. J. and M. B. Steer (2014). An Efficient Approach to Computing Third-Order Scattering of Sound by Sound with Application to Parametric Arrays. In: *IEEE Trans. Ultrason. Ferroelectr. Freq. Control* 61.10, pp. 1729–1741.

Johnson, S. J. (2016). "Modulation of Radio Frequency Signals by Nonlinearly Generated Acoustic Fields". Raleigh, North Carolina: North Carolina State University.

Johnson-Ulrich, L., V. Demartsev, L. Johnson, E. Brown, A. Strandburg-Peshkin, and M. B. Manser (2023). Directional Speakers as a Tool for Animal Vocal Communication Studies. In: *R. Soc. Open Sci.* 10.5, p. 230489.

Joseph, P., S. J. Elliott, and P. A. Nelson (1994). Near Field Zones of Quiet. In: *J. Sound Vib.* 172.5, pp. 605–627.

Ju, H. S. and Y.-H. Kim (2010). Near-Field Characteristics of the Parametric Loudspeaker Using Ultrasonic Transducers. In: *Appl. Acoust.* 71.9, pp. 793–800.

Jung, W. (2018). "Mid-Frequency Local Active Control of Road Noise". Southampton, Hampshire, England: University of Southampton.

Jung, W., S. J. Elliott, and J. Cheer (2018). Estimation of the Pressure at a Listener's Ears in an Active Headrest System Using the Remote Microphone Technique. In: *J. Acoust. Soc. Am.* 143.5, pp. 2858–2869.

Kaduchak, G., D. N. Sinha, D. C. Lizon, and M. J. Kelecher (2000). A Non-Contact Technique for Evaluation of Elastic Structures at Large Stand-off Distances: Applications to Classification of Fluids in Steel Vessels. In: *Ultrasonics* 37.8, pp. 531–536.

Kagawa, Y., T. Tsuchiya, T. Yamabuchi, H. Kawabe, and T. Fujii (1992). Finite Element Simulation of Non-Linear Sound Wave Propagation. In: *J. Sound Vib.* 154.1, pp. 125–145.

Kamakura, T., T. Ishiwata, and K. Matsuda (2000). Model Equation for Strongly Focused Finite-Amplitude Sound Beams. In: *J. Acoust. Soc. Am.* 107.6, pp. 3035–3046.

Kamakura, T., S. Sakai, H. Nomura, and M. Akiyama (2008). Parametric Audible Sounds by Phase-cancellation Excitation of Primary Waves. In: *J. Acoust. Soc. Am.* 123 (5˙Supplement), p. 3694.

Kamakura, T., Y. Tasahide, and K. Ikeyaya (1985). A Study for the Realization of a Parametric Loudspeaker. In: *J. Acoust. Soc. Japan* 6, pp. 1–18.

Kamakura, T., M. Yoneyama, and K. Ikegaya (1984). "Developments of Parametric Loudspeaker for Practical Use". In: Proc. 10th International Symposium on Nonlinear Acoustics, pp. 147–150.

Karnapi, F. A., W. S. Gan, and M.-H. Er (2002). "Method to Enhance Low Frequency Perception from a Parametric Array Loudspeaker". In: *Audio Engineering Society Convention 112*. Audio Engineering Society, pp. 1–5.

Karnapi, F. A., W.-S. Gan, and Y. K. Chong (2004). FPGA Implementation of Parametric Loudspeaker System. In: *Microprocess Microsyst.* 28.5-6, pp. 261–272.

Kashiwase, S. and K. Kondo (2014). "Towards a Parametric Speaker System with Human Head Tracking Beam Control". In: *2014 IEEE 3rd Global Conference on Consumer Electronics (GCCE)*. 2014 IEEE 3rd Global Conference on Consumer Electronics (GCCE), pp. 22–23.

Katz, B. F. G. (2000). Acoustic Absorption Measurement of Human Hair and Skin within the Audible Frequency Range. In: *J. Acoust. Soc. Am.* 108.5, pp. 2238–2242.

Khodier, M. and C. Christodoulou (2005). Linear Array Geometry Synthesis with Minimum Sidelobe Level and Null Control Using Particle Swarm Optimization. In: *IEEE Trans. Antennas Propagat.* 53.8, pp. 2674–2679.

Kidner, M. R. F., C. Petersen, A. C. Zander, and C. H. Hansen (2006). "Feasibility Study of Localised Active Noise Control Using an Audio Spotlight and Virtual Sensors". In: Proceedings of ACOUSTICS, pp. 55–61.

Kim, C., Y. Hwang, K. Shin, and W. Moon (2022a). A Parametric Array Loudspeaker Featuring a Langevin Transducer, a Circular Aluminum Plate, and Composite Polymer Steps. In: *Sens. Actuator A Phys.* 338, p. 113466.

Kim, D. H., Y. S. Kwon, D.-H. Kang, S. Shim, J. H. Park, and B. C. Lee (2022b). "Difference-Frequency-Based Ultrasonic Contrast Imaging of Material Elasticities". In: *2022 IEEE International Ultrasonics Symposium (IUS)*. 2022 IEEE International Ultrasonics Symposium (IUS), pp. 1–4.

Kim, H.-J., L. W. Schmerr Jr, and A. Sedov (2006). Generation of the Basis Sets for Multi-Gaussian Ultrasonic Beam Models—An Overview. In: *J. Acoust. Soc. Am.* 119.4, pp. 1971–1978.

Kinsler, L. E., A. R. Frey, A. B. Coppens, and J. V. Sanders (2000). *Fundamentals of Acoustics*. 4th ed. New York: John Wiley & Sons, Inc.

Kirkeby, O., P. A. Nelson, F. Orduna-Bustamante, and H. Hamada (1996). Local Sound Field Reproduction Using Digital Signal Processing. In: *J. Acoust. Soc. Am.* 100.3, pp. 1584–1593.

Kirkpatrick, S., C. D. Gelatt, and M. P. Vecchi (1983). Optimization by Simulated Annealing. In: *Science* 220.4598, pp. 671–680.

Kite, T. D., J. T. Post, and M. F. Hamilton (1998). Parametric Array in Air: Distortion Reduction by Preprocessing. In: *J. Acoust. Soc. Am.* 103.5, pp. 2871–2871.

Kleiner, M. and J. Tichy (2014). *Acoustics of Small Rooms*. CRC Press.

Klippel, W. (2006). Tutorial: Loudspeaker Nonlinearities—Causes, Parameters, Symptoms. In: *J. Audio Eng. Soc.* 54.10, pp. 907–939.

Klippel, W. and C. Bellmann (2016). "Holographic Nearfield Measurement of Loudspeaker Directivity". In: Audio Engineering Society Convention 141. Audio Engineering Society.

Kopp, L., D. Cano, E. Dubois, L. Wang, B. Smith, and R. Coates (2000). Potential Performance of Parametric Communications. In: *IEEE J. Oceanic Eng.* 25.3, pp. 282–295.

Kreyszig, E., H. Kreyszig, and E. J. Norminton (2011). *Advanced Engineering Mathematics*. 10th ed. Hoboken, NJ: John Wiley.

Kritz, J. (1977). Parametric Array Doppler Sonar. In: *IEEE J. Ocean. Eng.* 2.2, pp. 190–200.

Kurup, D., M. Himdi, and A. Rydberg (2003). Synthesis of Uniform Amplitude Unequally Spaced Antenna Arrays Using the Differential Evolution Algorithm. In: *IEEE Trans. Antennas Propagat.* 51.9, pp. 2210–2217.

Kurze, U. J. (1974). Noise Reduction by Barriers. In: *J. Acoust. Soc. Am.* 55.3, pp. 504–518.

Kuttruff, H. (2017). *Room Acoustics*. 6th ed. Boca Raton: CRC Press.

Labor, U. D. of (2022). *Occupational Safety and Health Administration. OSHA Technical Manual, Section III: Chapter 5 - Noise Measurement. Appendix C.* The Journal of the Acoustical Society of America. URL: https://www.osha.gov/otm/section-3-health-hazards/chapter-5#appendixc (visited on 07/03/2023).

Lam, B. (2019). "Active Control of Noise through Open Windows". Singapore: Nanyang Technological University.

Lam, B., W.-S. Gan, and C. Shi (2014). "Feasibility of a Length-Limited Parametric Source for Active Noise Control Applications". In: Int. Congress on Sound and Vibration.

Lam, B., W.-S. Gan, D. Shi, M. Nishimura, and S. Elliott (2021). Ten Questions Concerning Active Noise Control in the Built Environment. In: *Build Environ.* 200, p. 107928.

Lam, B., D. Shi, W.-S. Gan, S. J. Elliott, and M. Nishimura (2020). Active Control of Broadband Sound through the Open Aperture of a Full-Sized Domestic Window. In: *Sci. Rep.* 10.1, p. 10021.

Lee, H., D. Kang, and W. Moon (2006). "Microelectromechanical-Systems-Based Parametric Transmitting Array in Air - Application to the Ultrasonic Ranging Transducer with High Directionality". In: *2006 SICE-ICASE International Joint Conference*. 2006 SICE-ICASE International Joint Conference, pp. 1081–1084.

Lee, H., D. Kang, and W. Moon (2007). "Design and Fabrication of the High Directional Ultrasonic Ranging Sensor to Enhance the Spatial Resolution". In: *TRANSDUCERS 2007 - 2007 International Solid-State Sensors, Actuators and Microsystems Conference*. TRANSDUCERS 2007 - 2007 International Solid-State Sensors, Actuators and Microsystems Conference, pp. 1303–1306.

Lee, H., D. Kang, and W. Moon (2009). A Micro-Machined Source Transducer for a Parametric Array in Air. In: *J. Acoust. Soc. Am.* 125.4, pp. 1879–1893.

Lee, S. and Y. Park (2023). Circular Active Noise Barrier Using Theoretical Control Filter Considering Interaction between Speaker and Barrier. In: *Sci. Rep.* 13.1 (1), p. 2649.

Lehmann, E. A. and A. M. Johansson (2008). Prediction of Energy Decay in Room Impulse Responses Simulated with an Image-Source Model. In: *J. Acoust. Soc. Am.* 124.1, pp. 269–277.

Li, D. and M. Hodgson (2005). Optimal Active Noise Control in Large Rooms Using a "Locally Global" Control Strategy. In: *J. Acoust. Soc. Am.* 118.6, pp. 3653–3661.

Li, M., J. Zhong, Y. Jing, and J. Lu (2023). A Note on the Audio Sound Power Generated by a Parametric Array Loudspeaker. In: *J. Acoust. Soc. Am.* 154.6, pp. 3899–3905.

Li, Y. Y. and L. Cheng (2004). Modifications of Acoustic Modes and Coupling Due to a Leaning Wall in a Rectangular Cavity. In: *J. Acoust. Soc. Am.* 116.6, pp. 3312–3318.

Li, Y., D. Polyak, E. Johnson, D. Yecies, S. Shevidi, A. de la Zerda, M. H. Gephart, and S. Chu (2020). Difference-Frequency Ultrasound Imaging With Non-Linear Contrast. In: *IEEE Trans. Med. Imaging* 39.5, pp. 1759–1766.

Liang, J., J. Gou, R. Bai, and C. Shi (2020). "Design of a Constant Beamwidth Beamformer for the Length-Limited Parametric Array Loudspeaker". In: *INTER-NOISE and NOISE-CON Congress and Conference Proceedings*. Vol. 261, pp. 3658–3664.

Lighthill, J. (1978). *Waves in Fluids*. Cambridge University Press.

Lin, Z., J. Lu, C. Shen, X. Qiu, and B. Xu (2004). Active Control of Radiation from a Piston Set in a Rigid Sphere. In: *J. Acoust. Soc. Am.* 115.6, pp. 2954–2963.

Listen Inc. (2020). *SoundCheck User Manual Version 18*.

Liu, W., P. Ji, and J. Yang (2008). Development of a Simple and Accurate Approximation Method for the Gaussian Beam Expansion Technique. In: *J. Acoust. Soc. Am.* 123.5, p. 3516.

Liu, W. and J. Yang (2010). A Simple and Accurate Method for Calculating the Gaussian Beam Expansion Coefficients. In: *Chinese Phys. Lett.* 27.12, p. 124301.

Liu, X., J. Li, X. Gong, Z. Zhu, and D. Zhang (2007). Theoretical and Experimental Study of the Third-Order Nonlinearity Parameter C/A for Biological Media. In: *Phys. D: Nonlinear Phenom.* 228.2, pp. 172–178.

Liu, Z. and C. Maury (2017). Numerical and Experimental Study of Near-Field Acoustic Radiation Modes of a Baffled Spherical Cap. In: *Appl. Acoust.* 115, pp. 23–31.

Loubet, G., F. Vial, A. Essebbar, L. Kopp, and D. Cano (1996). "Parametric Transmission of Wide-Band Signals". In: *OCEANS 96 MTS/IEEE Conference Proceedings. The Coastal Ocean - Prospects for the 21st Century*. OCEANS 96 MTS/IEEE Conference Proceedings. The Coastal Ocean — Prospects for the 21st Century. Vol. 2, 839–844 vol.2.

Lucas, B. G. and T. G. Muir (1982). The Field of a Focusing Source. In: *J. Acoust. Soc. Am.* 72.4, pp. 1289–1296.

Lucas, B. G., J. N. Tjøtta, and T. G. Muir (1983). Field of a Parametric Focusing Source. In: *J. Acoust. Soc. Am.* 73.6, pp. 1966–1971.

Ludwig, G. D. (1950). The Velocity of Sound through Tissues and the Acoustic Impedance of Tissues. In: *J. Acoust. Soc. Am.* 22.6, pp. 862–866.

Ma, Q., X. Gong, and D. Zhang (2006). Third Order Harmonic Imaging for Biological Tissues Using Three Phase-Coded Pulses. In: *Ultrasonics* 44, e61–e65.

Makov, Y. N., V. J. Sánchez-Morcillo, F. Camarena, and V. Espinosa (2008). Nonlinear Change of On-Axis Pressure and Intensity Maxima Positions and Its Relation with the Linear Focal Shift Effect. In: *Ultrasonics* 48.8, pp. 678–686.

Mangiante, G. A. (1977). Active Sound Absorption. In: *J. Acoust. Soc. Am.* 61.6, pp. 1516–1523.

Marchal, J. and P. Cervenka (2004). Modeling of the Parametric Transmission with the Spatial Fourier Formalism. Optimization of a Parametric Antenna. In: *Acta Acust. united Acust.* 90.1, pp. 49–61.

Marchal, J. and P. Cervenka (2012). Underwater Parametric Transmission with a Linear Array. In: *Appl. Acoust.* 73.12, pp. 1239–1243.

Martin, P. A. (2006). *Multiple Scattering: Interaction of Time-harmonic Waves with N Obstacles*. Encyclopedia of Mathematics and Its Applications 107. Cambridge: Cambridge University Press.

Marzo, A., A. Barnes, and B. W. Drinkwater (2017). TinyLev: A Multi-Emitter Single-Axis Acoustic Levitator. In: *Rev. Sci. Instrum.* 88.8, p. 085105.

Marzo, A., T. Corkett, and B. W. Drinkwater (2018). Ultraino: An Open Phased-Array System for Narrowband Airborne Ultrasound Transmission. In: *IEEE Trans. Ultrason. Ferroelectr. Freq. Control* 65.1, pp. 102–111.

Mater, O., J. Remenieras, C. Bruneel, A. Roncin, and F. Patat (1998). Noncontact Measurement of Vibration Using Airborne Ultrasound. In: *IEEE Trans. Ultrason. Ferroelectr. Freq. Control* 45.3, pp. 626–633.

Matsui, T., D. Ikefuji, M. Nakayama, and T. Nishiura (2013). "A Design of Audio Spot Based on Separating Emission of the Carrier and Sideband Waves". In: Proceedings of Meetings on Acoustics ICA2013. Vol. 19. Acoustical Society of America, p. 055049.

McGovern, S. G. (2009). Fast Image Method for Impulse Response Calculations of Box-Shaped Rooms. In: *Appl. Acoust.* 70.1, pp. 182–189.

Mehta, R. P., S. L. Mattson, B. A. Kappus, and R. L. Seitzman (2015a). *Speech Recognition in the Sound Field: Directed Audio vs. Conventional Speakers*. AudiologyOnline. URL: https://www.audiologyonline.com/articles/speech-recognition-in-sound-field-14901 (visited on 07/03/2023).

Mehta, R. P., S. L. Mattson, B. A. Kappus, and R. L. Seitzman (2015b). Safety of the HyperSound® Audio System in Subjects with Normal Hearing. In: *Audiol. Res.* 5.2, p. 136.

Mei, S., H. Xu, Y. Hu, M. Alkahtani, and Y. Wang (2023). "The Parametric Array Speaker: A Review". In: *Conference Proceedings of 2022 2nd International Joint Conference on Energy, Electrical and Power Engineering*. Ed. by C. Hu and W. Cao. Lecture Notes in Electrical Engineering. Singapore: Springer Nature, pp. 1254–1271.

Meissner, M. (2013). Acoustic Behaviour of Lightly Damped Rooms. In: *Acta Acust. united Acust.* 99.

Mellen, R. H. (1976). Nearfield Beam Patterns of Exponentially Shaded End-fire Line Arrays. In: *J. Acoust. Soc. Am.* 60.2, pp. 505–506.

Mellen, R. H. (1977). Nearfield Axial Levels of Exponentially Shaded End-fire Arrays. In: *J. Acoust. Soc. Am.* 61.2, pp. 599–601.

Mellen, R. H. and M. B. Moffett (1978). A Numerical Method for Calculating the Nearfield of a Parametric Acoustic Source. In: *J. Acoust. Soc. Am.* 63.5, pp. 1622–1624.

Mellow, T. and L. Karkkainen (2011). On the Sound Fields of Infinitely Long Strips. In: *J. Acoust. Soc. Am.* 130.1, pp. 153–167.

Merklinger, H. M. (1973). Fundamental-frequency Component of a Finite-amplitude Plane Wave. In: *J. Acoust. Soc. Am.* 54.6, pp. 1760–1761.

Merklinger, H. M. (1975). Improved Efficiency in the Parametric Transmitting Array. In: *J. Acoust. Soc. Am.* 58.4, pp. 784–787.

Messiah, A. (1962). *Quantum Mechanics: Volume II.* North-Holland Publishing Company Amsterdam.

Mitri, F. G. (2010). Nonlinear Acoustics in Higher-Order Approximation: Comment. In: *IEEE Trans. Ultrason. Ferroelectr. Freq. Control* 57.8, pp. 1715–1716.

Moffett, M. B. and R. H. Mellen (1977). Model for Parametric Acoustic Sources. In: *J. Acoust. Soc. Am.* 61.2, pp. 325–337.

Moffett, M. B. and R. H. Mellen (1981). Nearfield Characteristics of Parametric Acoustic Sources. In: *J. Acoust. Soc. Am.* 69.2, pp. 404–409.

Moffett, M. B., P. J. Westervelt, and R. T. Beyer (1970). Large-Amplitude Pulse Propagation—A Transient Effect. In: *J. Acoust. Soc. Am.* 47 (5B), pp. 1473–1474.

Moffett, M. B., P. J. Westervelt, and R. T. Beyer (1971). Large-Amplitude Pulse Propagation—A Transient Effect II. In: *J. Acoust. Soc. Am.* 49 (1B), pp. 339–343.

Moore, B. C. (2013). *An Introduction to the Psychology of Hearing*. Ed. by 6. Leiden, The Netherlands: Brill NV.

Morse, P. M. and R. H. Bolt (1944). Sound Waves in Rooms. In: *Rev. Mod. Phys.* 16.2, pp. 69–150.

Mu, Y., P. Ji, W. Ji, M. Wu, and J. Yang (2014). Modeling and Compensation for the Distortion of Parametric Loudspeakers Using a One-Dimension Volterra Filter. In: *IEEE/ACM Trans. Audio, Speech, Lang. Process.* 22.12, pp. 2169–2181.

Muir, T. G. and J. E. Blue (1969). Experiments on the Acoustic Modulation of Large-Amplitude Waves. In: *J. Acoust. Soc. Am.* 46 (1B), pp. 227–232.

Muir, T. G., L. L. Mellenbruch, and J. C. Lockwood) (1977). Reflection of Finite-amplitude Waves in a Parametric Array. In: *J. Acoust. Soc. Am.* 62.2, pp. 271–276.

Mukundan, R. and K. R. Ramakrishnan (1995). Fast Computation of Legendre and Zernike Moments. In: *Pattern Recognit.* 28.9, pp. 1433–1442.

Münzel, T., T. Gori, W. Babisch, and M. Basner (2014). Cardiovascular Effects of Environmental Noise Exposure. In: *Eur. Heart J.* 35.13, pp. 829–836.

Murata Manufacturing (2017). *MA40S4R/MA40S4S Ultrasonic Sensor Data Sheet.*

Murino, V., A. Trucco, and C. Regazzoni (1996). Synthesis of Unequally Spaced Arrays by Simulated Annealing. In: *IEEE Trans. Signal Process.* 44.1, pp. 119–122.

Murphy, D. T., A. Southern, and L. Savioja (2014). Source Excitation Strategies for Obtaining Impulse Responses in Finite Difference Time Domain Room Acoustics Simulation. In: *Appl. Acoust.* 82, pp. 6–14.

Naka, Y., A. A. Oberai, and B. G. Shinn-Cunningham (2005). Acoustic Eigenvalues of Rectangular Rooms with Arbitrary Wall Impedances Using the Interval New-tongeneralized Bisection Method. In: *J. Acoust. Soc. Am.* 118.6, pp. 3662–3671.

Nakagawa, K., C. Shi, and Y. Kajikawa (2019). "Beam Steering of Portable Parametric Array Loudspeaker". In: *2019 Asia-Pacific Signal and Information Processing Association Annual Summit and Conference (APSIPA ASC)*. 2019 Asia-Pacific Signal and Information Processing Association Annual Summit and Conference (APSIPA ASC), pp. 1824–1827.

Nakagawa, Y., W. Hou, A. Cai, N. Arnold, and G. Wade (1986). "Nonlinear Parameter Imaging with Finite-Amplitude Sound Waves". In: *IEEE 1986 Ultrasonics Symposium*. IEEE 1986 Ultrasonics Symposium, pp. 901–904.

Nakashima, Y., T. Yoshimura, N. Naka, and T. Ohya (2007). Prototype of Mobile Super Directional Loudspeaker. In: *NTT DoCoMo Technical Journal* 8.1.

Nakashima, Y., T. Ohya, and T. Yoshimura (2005). "Prototype of Parametric Array Loudspeaker on Mobile Phone and Its Acoustical Characteristics". In: Audio Engineering Society Convention 118. Barcelona, Spain: Audio Engineering Society, pp. 1–6.

Nelson, P. A., A. R. D. Curtis, S. J. Elliott, and A. J. Bullmore (1987). The Minimum Power Output of Free Field Point Sources and the Active Control of Sound. In: *J. Sound Vib.* 116.3, pp. 397–414.

Nelson, P. A. and S. J. Elliott (1992). *Active Control of Sound*. San Diego, CA: Academic Press Inc.

Nelson, T. R., J. B. Fowlkes, J. S. Abramowicz, and C. C. Church (2009). Ultrasound Biosafety Considerations for the Practicing Sonographer and Sonologist. In: *J. Ultrasound Med.* 28, pp. 139–150.

Nichols Jr., R. H. (1971). Finite Amplitude Effects Used to Improve Echo-Sounder. In: *J. Acoust. Soc. Am.* 50 (4A), pp. 1086–1087.

Nishiura, Y., M. Hoshiyama, and Y. Konagaya (2018). Use of Parametric Speaker for Older People with Dementia in a Residential Care Setting: A Preliminary Study of Two Cases. In: *Hong Kong J. Occup. Ther.* 31.1, pp. 30–35.

Nolan, M. and J. L. Davy (2019). Two Definitions of the Inner Product of Modes and Their Use in Calculating Non-Diffuse Reverberant Sound Fields. In: *J. Acoust. Soc. Am.* 145.6, pp. 3330–3340.

Nomura, H., H. Adachi, and T. Kamakura (2018). Feasibility of Low-Frequency Ultrasound Imaging Using Pulse Compressed Parametric Ultrasound. In: *Ultrasonics* 89, pp. 64–73.

Nomura, H., C. M. Hedberg, and T. Kamakura (2012). Numerical Simulation of Parametric Sound Generation and Its Application to Length-Limited Sound Beam. In: *Appl. Acoust.* 73.12, pp. 1231–1238.

Nomura, H. and H. Sato (2022). Influence of Microphone Characteristics on Demodulated Sound Measurement in near Field of Parametric Loudspeaker. In: *Jpn. J. Appl. Phys.* 61 (SG), SG1008.

Novikov, B. K., O. V. Rudenko, and V. I. Timoshenko (1987). *Nonlinear Underwater Acoustics*. New York: AIP.

Nutter, D. B., T. W. Leishman, S. D. Sommerfeldt, and J. D. Blotter (2007). Measurement of Sound Power and Absorption in Reverberation Chambers Using Energy Density. In: *J. Acoust. Soc. Am.* 121.5, pp. 2700–2710.

Ochiai, Y. and T. Hoshi (2018). "System and Method for Generating Spatial Sound Using Ultrasound". U.S. pat. 9936324B2. Pixie Dust Technologies Inc.

Ochiai, Y., T. Hoshi, and I. Suzuki (2017). "Holographic Whisper: Rendering Audible Sound Spots in Three-Dimensional Space by Focusing Ultrasonic Waves". In: *Proceedings of the 2017 CHI Conference on Human Factors in Computing Systems*. Association for Computing Machinery, pp. 4314–4325.

Ogami, Y., M. Nakayama, and T. Nishiura (2019). Virtual Sound Source Construction Based on Radiation Direction Control Using Multiple Parametric Array Loudspeakers. In: *J. Acoust. Soc. Am.* 146.2, pp. 1314–1325.

Okano, A. and Y. Kajikawa (2022). "Phase Control of Parametric Array Loudspeaker by Optimizing Sideband Weights". In: *ICASSP 2022 - 2022 IEEE International Conference on Acoustics, Speech and Signal Processing (ICASSP)*. ICASSP 2022 - 2022 IEEE International Conference on Acoustics, Speech and Signal Processing (ICASSP), pp. 31–35.

Olszewski, D. (2009). Targeted Audio. In: *Computers in the Human Interaction Loop*, pp. 133–141.

Olszewski, D. and K. Linhard (2006). "Highly Directional Multi-Beam Audio Loudspeaker". In: *Ninth International Conference on Spoken Language Processing*.

Olszewski, D., F. Prasetyo, and K. Linhard (2005). "Steerable Highly Directional Audio Beam Loudspeaker". In: *Ninth European Conference on Speech Communication and Technology*.

Olver, F. W. J., A. B. O. Daalhuis, D. W. Lozier, B. I. Schneider, R. F. Boisvert, C. W. Clark, B. R. Miller, B. V. Saunders, H. S. Cohl, and M. A. McClain (2019). *NIST Digital Library of Mathematical Functions Release 1.0.25 of 2019-12-15*. URL: http://dlmf.nist.gov/.

Organization, W. H. (2011). *Burden of Disease from Environmental Noise: Quantification of Healthy Life Years Lost in Europe*. Regional Office for Europe.

Ortega, D., J. Carbajo, P. Poveda, and J. Ramis (2021). "On the Use of Audible Sound from Modulated Ultrasound in Indoor Spaces". In: EuroNoise 2021. Madeira, Portugal.

Ostrovsky, L. A. (2009). "Research on Parametric Arrays in Russia: Historical Perspective". In: 157th Meeting Acoustical Society of America. Portland, Oregon, pp. 045004–045004.

Padhye, R. and R. Nayak (2016). *Acoustic Textiles*. Singapore: Springer.

Pan, J., S. D. Snyder, C. H. Hansen, and C. R. Fuller (1992). Active Control of Farfield Sound Radiated by a Rectangular Panel—A General Analysis. In: *J. Acoust. Soc. Am.* 91.4, pp. 2056–2066.

Pellicier, A. and N. Trompette (2007). A Review of Analytical Methods, Based on the Wave Approach, to Compute Partitions Transmission Loss. In: *Appl. Acoust.* 68.10, pp. 1192–1212.

Pierce, A. D. (2019). *Acoustics: An Introduction to Its Physical Principles and Applications*. 3rd. Cham, Switzerland: Springer Nature.

Poblet-Puig, J. and A. Rodríguez-Ferran (2013). Modal-Based Prediction of Sound Transmission through Slits and Openings between Rooms. In: *J. Sound Vib.* 332.5, pp. 1265–1287.

Poletti, M. A. (2005). Three-Dimensional Surround Sound Systems Based on Spherical Harmonics. In: *J. Audio Eng. Soc.* 53.11, pp. 1004–1025.

Poletti, M. A. (2018). Spherical Expansions of Sound Radiation from Resilient and Rigid Disks with Reduced Error. In: *J. Acoust. Soc. Am.* 144.3, pp. 1180–1189.

Poletti, M. A. (2019). Cylindrical Expansions of Sound Radiation from Resilient and Rigid Infinite Strips with Reduced Error. In: *J. Acoust. Soc. Am.* 145.5, pp. 3104–3115.

Pompei, F. J. (1999). The Use of Airborne Ultrasonics for Generating Audible Sound Beams. In: *J. Audio Eng. Soc.* 47.9, pp. 726–731.

Pompei, F. J. (2002). "Sound from Ultrasound: The Parametric Array as an Audible Sound Source". Cambridge, MA, USA: Massachusetts Institute of Technology.

Qian, Z.-w. (1995). Nonlinear Acoustics in Higher-Order Approximation. In: *Acta Physica Sinica (Overseas Edition)* 4.9, pp. 670–675.

Qiu, X. and C. Hansen (2000). Secondary Acoustic Source Types for Active Noise Control in Free Field: Monopoles or Multipoles? In: *J. Sound Vib.* 232.5, pp. 1005–1009.

Qiu, X. (2019). *An Introduction to Virtual Sound Barriers*. New York: CRC Press.

Qiu, X., B. Masiero, and M. Vorländer (2009). Channel Separation of Crosstalk Cancellation Systems with Mismatched and Misaligned Sound Sources. In: *J. Acoust. Soc. Am.* 126.4, p. 1796.

Qiu, Y., J. V. Gigliotti, M. Wallace, F. Griggio, C. E. M. Demore, S. Cochran, and S. Trolier-McKinstry (2015). Piezoelectric Micromachined Ultrasound Transducer (PMUT) Arrays for Integrated Sensing, Actuation and Imaging. In: *Sensors (Basel)* 15.4, pp. 8020–8041. pmid: 25855038.

Qu, K., B. Zou, J. Chen, Y. Guo, and R. Wang (2018). Experimental Study of a Broadband Parametric Acoustic Array for Sub-Bottom Profiling in Shallow Water. In: *Shock Vib.* 2018, pp. 1–8.

Rafaely, B. (2015). *Fundamentals of Spherical Array Processing*. Vol. 8. Springer Topics in Signal Processing. Berlin, Heidelberg: Springer Berlin Heidelberg.

Reddy, J. N. (2019). *Introduction To Finite Element Method*. 4th ed.

Reis, J. (2016). "Short Overview in Parametric Loudspeakers Array Technology and Its Implications in Spatialization in Electronic". In: *Proceedings of the International Computer Music Conference*.

Robertson, J. L. B., B. T. Cox, J. Jaros, and B. E. Treeby (2017). Accurate Simulation of Transcranial Ultrasound Propagation for Ultrasonic Neuromodulation and Stimulation. In: *J. Acoust. Soc. Am.* 141.3, pp. 1726–1738.

Roddy, P. J. and J. D. McEwen (2021). Sifting Convolution on the Sphere. In: *IEEE Signal Process. Lett.* 28, pp. 304–308.

Romanova, A., K. V. Horoshenkov, and A. Hurrell (2019). An Application of a Parametric Transducer to Measure Acoustic Absorption of a Living Green Wall. In: *Appl. Acoust.* 145, pp. 89–97.

Rosenheinrich, W. (2017). *Tables of Some Indefinite Integral of Bessel Functions of Integer Order*.

Rudnick, I. (1947). The Propagation of an Acoustic Wave along a Boundary. In: *J. Acoust. Soc. Am.* 19.2, pp. 348–356.

Rupitsch, S. J. (2019). *Piezoelectric Sensors and Actuators: Fundamentals and Applications*. New York, NY: Springer Berlin Heidelberg.

Saito, S. and T. G. Muir (1986). Simple Analytical Model for Parametric Focusing Sources. In: *J. Acoust. Soc. Jpn. (E)* 7.2, pp. 113–120.

Saito, S. and T. Tsubono (1998). Experiment on Parametric Focusing Source. In: *J. Acoust. Soc. Jpn. (E)* 19.6, pp. 417–419.

Samarasinghe, P. N., T. D. Abhayapala, Y. Lu, H. Chen, and G. Dickins (2018). Spherical Harmonics Based Generalized Image Source Method for Simulating Room Acoustics. In: *J. Acoust. Soc. Am.* 144.3, pp. 1381–1391.

Sauer, T. (2012). *Numerical Analysis*. 2nd ed. New York: Pearson Education Inc.

Savioja, L. and U. P. Svensson (2015). Overview of Geometrical Room Acoustic Modeling Techniques. In: *J. Acoust. Soc. Am.* 138.2, pp. 708–730.

Sayin, U., P. Artís, and O. Guasch (2013). Realization of an Omnidirectional Source of Sound Using Parametric Loudspeakers. In: *J. Acoust. Soc. Am.* 134.3, pp. 1899–1907.

Sayin, U. and O. Guasch (2013). Directivity Control and Efficiency of Parametric Loudspeakers with Horns. In: *J. Acoust. Soc. Am.* 134.2, EL153–EL157.

Schmerr Jr, L. W. (2014). *Fundamentals of Ultrasonic Phased Arrays*. New York: Springer.

Sha, K., J. Yang, and W.-S. Gan (2003). A Complex Virtual Source Approach for Calculating the Diffraction Beam Field Generated by a Rectangular Planar Source. In: *IEEE Trans. Ultrason. Ferroelectr. Freq. Control* 50.7, pp. 890–897.

Shang, J., T. Wu, H. Wang, C. Yang, C. Ye, R. Hu, J. Tao, and X. He (2019). Measurement of Temperature-Dependent Bulk Viscosities of Nitrogen, Oxygen and Air From Spontaneous Rayleigh-Brillouin Scattering. In: *IEEE Access* 7, pp. 136439–136451.

Shi, C. (2013). "Investigation of the Steerable Parametric Loudspeaker Based on Phased Array Techniques". Singapore: Nanyang Technological University.

Shi, C. and W.-S. Gan (2010). Development of Parametric Loudspeaker. In: *IEEE Potentials* 29.6, pp. 20–24.

Shi, C. and W.-S. Gan (2011). Grating Lobe Elimination in Steerable Parametric Loudspeaker. In: *IEEE Trans. Ultrason., Ferroelect., Freq. Contr.* 58.2, pp. 437–450.

Shi, C. and W.-S. Gan (2012). Product Directivity Models for Parametric Loudspeakers. In: *J. Acoust. Soc. Am.* 131.3, pp. 1938–1945.

Shi, C. and W.-S. Gan (2013). "Using Length-Limited Parametric Source in Active Noise Control Applications". In: 20th International Congress on Sound and Vibration (ICSV). Bangkok, Thailand.

Shi, C. and Y. Kajikawa (2015a). A Convolution Model for Computing the Far-Field Directivity of a Parametric Loudspeaker Array. In: *J. Acoust. Soc. Am.* 137.2, pp. 777–784.

Shi, C. and Y. Kajikawa (2015b). "Identification of the Parametric Array Loudspeaker with a Volterra Filter Using the Sparse NLMS Algorithm". In: 2015 IEEE International Conference on Acoustics, Speech and Signal Processing (ICASSP). IEEE, pp. 3372–3376.

Shi, C. and Y. Kajikawa (2016a). Effect of the Ultrasonic Emitter on the Distortion Performance of the Parametric Array Loudspeaker. In: *Appl. Acoust.* 112, pp. 108–115.

Shi, C. and Y. Kajikawa (2016b). Volterra Model of the Parametric Array Loudspeaker Operating at Ultrasonic Frequencies. In: *J. Acoust. Soc. Am.* 140.5, pp. 3643–3650.

Shi, C., Y. Kajikawa, and W.-S. Gan (2014). An Overview of Directivity Control Methods of the Parametric Array Loudspeaker. In: *APSIPA Trans. Signal Inf. Process.* 3, pp. 1–12.

Shi, C., Y. Kajikawa, and W.-S. Gan (2015). Generating Dual Beams from a Single Steerable Parametric Loudspeaker. In: *J. Acoust. Soc. Am.* 99, pp. 43–50.

Shi, C., Y. Wang, H. Xiao, and H. Li (2022). Extended Convolution Model for Computing the Far-Field Directivity of an Amplitude-Modulated Parametric Loudspeaker. In: *J. Phys. D. Appl. Phys.* 55.24, p. 244002.

Shi, D., B. Lam, S. Wen, and W.-S. Gan (2020). "Multichannel Active Noise Control with Spatial Derivative Constraints to Enlarge the Quiet Zone". In: ICASSP 2020 - 2020 IEEE International Conference on Acoustics, Speech and Signal Processing (ICASSP).

Shi, X., J. Tao, and X. Qiu (2008). Sound Insulation of an Infinite Partition Subject to a Point Source Incidence [in Chinese]. In: *Acta Acustica* 33.3, pp. 268–274.

Shin, H.-C. (2019). Time-Harmonic Convention of Exp (–Iωt) Can Be Preferable to Exp (+ Iωt)(L). In: *J. Acoust. Soc. Am.* 146.3, pp. 1851–1854.

Shmilovitz, D. (2005). On the Definition of Total Harmonic Distortion and Its Effect on Measurement Interpretation. In: *IEEE Trans. Power Deliv.* 20.1, pp. 526–528.

Silva, G. T. and A. Bandeira (2013). Difference-Frequency Generation in Nonlinear Scattering of Acoustic Waves by a Rigid Sphere. In: *Ultrasonics* 53.2, pp. 470–478.

Skinner, E., M. Groves, and M. K. Hinders (2019). Demonstration of a Length Limited Parametric Array. In: *Appl. Acoust.* 148, pp. 423–433.

Skudrzyk, E. (1971). *The Foundations of Acoustics: Basic Mathematics and Basic Acoustics*. Vienna: Springer Vienna.

Sloane, N. J. A., R. H. Hardin, and W. D. Smith (2021). *Spherical Codes: Nice Arrangements of Points on a Sphere in Various Dimensions*. URL: http://neilsloane.com/packings/dim3/ (visited on 08/22/2021).

Snyder, S. D. and N. Tanaka (1995). Calculating Total Acoustic Power Output Using Modal Radiation Efficiencies. In: *J. Acoust. Soc. Am.* 97.3, pp. 1702–1709.

Sparrow, V. W. and W. Kim (2002). "Audio Application of the Parametric Array - Demonstration through a Numerical Model". In: Audio Engineering Society Convention 113. Audio Engineering Society.

Stark, S. (2007). "Direct Digital Pulse Width Modulation for Class D Amplifiers". Sweden: Linköping University.

Sugahara, A., H. Lee, S. Sakamoto, and S. Takeoka (2017). "A Study on the Measurements of the Absorption Coefficient by Using a Parametric Loudspeaker". In: INTER-NOISE and NOISE-CON Congress and Conference Proceedings. Vol. 255. Hongkong, China: Institute of Noise Control Engineering, pp. 3743–3751.

Sugahara, A., H. Lee, S. Sakamoto, and S. Takeoka (2019). Measurements of Acoustic Impedance of Porous Materials Using a Parametric Loudspeaker with Phononic Crystals and Phase-Cancellation Method. In: *Appl. Acoust.* 152, pp. 54–62.

Sugibayashi, Y., S. Kurimoto, D. Ikefuji, M. Morise, and T. Nishiura (2012). Three-Dimensional Acoustic Sound Field Reproduction Based on Hybrid Combination of Multiple Parametric Loudspeakers and Electrodynamic Subwoofer. In: *Appl. Acoust.* 73.12, pp. 1282–1288.

Suomi, V., J. Jaros, B. Treeby, and R. O. Cleveland (2018). Full Modeling of High-Intensity Focused Ultrasound and Thermal Heating in the Kidney Using Realistic Patient Models. In: *IEEE. Trans. Biomed. Eng.* 65.5, pp. 969–979.

Tan, K., W. Gan, J. Yang, and M. Er (2004). "An Efficient Digital Beamsteering System for Difference Frequency in Parametric Array". In: *2004 IEEE International Conference on Acoustics, Speech, and Signal Processing*. 2004 IEEE International Conference on Acoustics, Speech, and Signal Processing. Vol. 2, pp. ii–193.

Tan, E.-L., P. Ji, and W.-S. Gan (2010). On Preprocessing Techniques for Bandlimited Parametric Loudspeakers. In: *Appl. Acoust.* 71.5, pp. 486–492.

Tanaka, K., C. Shi, and Y. Kajikawa (2014). "Multi-Channel Active Noise Control Using Parametric Array Loudspeakers". In: *Signal and Information Processing Association Annual Summit and Conference (APSIPA), 2014 Asia-Pacific*. 2014 Asia-Pacific Signal and Information Processing Association Annual Summit and Conference (APSIPA). Chiang Mai, Thailand: IEEE, pp. 1–6.

Tanaka, K., C. Shi, and Y. Kajikawa (2017). Binaural Active Noise Control Using Parametric Array Loudspeakers. In: *Appl. Acoust.* 116, pp. 170–176.

Tanaka, N. and M. Tanaka (2010). Active Noise Control Using a Steerable Parametric Array Loudspeaker. In: *J. Acoust. Soc. Am.* 127.6, pp. 3526–3537.

Tanaka, N. and M. Tanaka (2011). Mathematically Trivial Control of Sound Using a Parametric Beam Focusing Source. In: *J. Acoust. Soc. Am.* 129.1, pp. 165–172.

Tang, K., Y. Wang, S. Wang, D. Gao, H. Li, X. Liang, P. Sebbah, Y. Li, J. Zhang, and J. Shi (2023). *Hyperuniform Disordered Parametric Loudspeaker Array*. arXiv: 2301.00833 [physics]. URL: http://arxiv.org/abs/2301.00833 (visited on 07/08/2023). preprint.

Tao, S., J. Ren, and C. Shi (2021). "Virtual Bass Preprocessing and Carrier Frequency Optimization for the Parametric Array Loudspeaker". In: *INTER-NOISE and NOISE-CON Congress and Conference Proceedings*. Vol. 263. 5. Institute of Noise Control Engineering, pp. 1676–1682.

Tjøtta, J. N. and S. Tjøtta (1987a). Interaction of Sound Waves. Part I: Basic Equations and Plane Waves. In: *J. Acoust. Soc. Am.* 82.4, pp. 1425–1428.

Tjøtta, J. N. and S. Tjøtta (1987b). Interaction of Sound Waves. Part II: Plane Wave and Real Beam. In: *J. Acoust. Soc. Am.* 82.4, pp. 1429–1435.

Tjøtta, J. N. and S. Tjøtta (1988). Interaction of Sound Waves. Part III: Two Real Beams. In: *J. Acoust. Soc. Am.* 83.2, pp. 487–495.

Toda, M. (2005). New Type of Acoustic Filter Using Periodic Polymer Layers for Measuring Audio Signal Components Excited by Amplitude-Modulated High-Intensity Ultrasonic Waves. In: *J. Acoust. Soc. Am.* 53.10, pp. 930–941.

Tong, Y. G., S. K. Tang, J. Kang, A. Fung, and M. K. L. Yeung (2015). Full Scale Field Study of Sound Transmission across Plenum Windows. In: *Appl. Acoust.* 89, pp. 244–253.

Treeby, B. E. and B. T. Cox (2010). K-Wave: MATLAB Toolbox for the Simulation and Reconstruction of Photoacoustic Wave Fields. In: *J. Biomed. Opt.* 15.2, p. 021314.

Treeby, B. E., J. Jaros, A. P. Rendell, and B. T. Cox (2012). Modeling Nonlinear Ultrasound Propagation in Heterogeneous Media with Power Law Absorption Using a K-Space Pseudospectral Method. In: *J. Acoust. Soc. Am.* 131.6, pp. 4324–4336.

Treeby, B. E., E. Z. Zhang, and B. T. Cox (2010). Photoacoustic Tomography in Absorbing Acoustic Media Using Time Reversal. In: *Inverse Probl.* 26.11, p. 115003.

Trolier-McKinstry, S. and P. Muralt (2004). Thin Film Piezoelectrics for MEMS. In: *J. Electroceramics* 12.1, pp. 7–17.

Trucco, A. and A. Pescetto (2000). Acoustic Detection of Objects Buried in the Seafloor. In: *Electronics Letters* 36.18, pp. 1595–1596.

Tseng, V. F.-G., S. S. Bedair, and N. Lazarus (2018). Phased Array Focusing for Acoustic Wireless Power Transfer. In: *IEEE Trans. Ultrason., Ferroelect., Freq. Contr.* 65.1, pp. 39–49.

Tu, Z., J. Lu, and X. Qiu (2016). Robustness of a Compact Endfire Personal Audio System against Scattering Effects (L). In: *J. Acoust. Soc. Am.* 140.4, pp. 2720–2724.

Van Buren, A. L. (2017). "Accurate Calculation of Oblate Spheroidal Wave Functions". arXiv: 1708.07929.

Vos, H. J., D. E. Goertz, A. F. van der Steen, and N. de Jong (2011). Parametric Array Technique for Microbubble Excitation. In: *IEEE Trans. Ultrason., Ferroelect., Freq. Contr.* 58.5, pp. 924–934.

Wang, L. S., B. V. Smith, and R. Coates (1999). The Secondary Field of a Parametric Source Following Interaction with Sea Surface. In: *J. Acoust. Soc. Am.* 105.6, pp. 3108–3114.

Wang, S., J. Tao, and X. Qiu (2015). Performance of a Planar Virtual Sound Barrier at the Baffled Opening of a Rectangular Cavity. In: *J. Acoust. Soc. Am.* 138.5, pp. 2836–2847.

Wang, S., Z. Yang, J. Tao, X. Qiu, and I. S. Burnett (2023a). Transmission Loss and Directivity of Sound Transmitted through a Slit on Ground. In: *J. Acoust. Soc. Am.* 153.1, pp. 224–235.

Wang, X., Q. Xia, and J. Shi (2023b). "An Optimization Design of Parametric Array Signal Modulation Algorithm Based on NSGA-II and Entropy Weight TOPSIS". In: *2023 4th Information Communication Technologies Conference (ICTC)*. 2023 4th Information Communication Technologies Conference (ICTC), pp. 223–227.

Wang, Y., X. Li, L. Xu, and L. Xu (2009). "SSB Modulation of the Ultrasonic Carrier for a Parametric Loudspeaker". In: 2009 International Conference on Electronic Computer Technology. IEEE, pp. 669–673.

Wang, Y., R. Li, C. Shi, and Y. Lv (2019). "Formation of Local Quiet Zones Using the Length-Limited Parametric Array Loudspeaker". In: 23rd International Congress on Acoustics. Aachen, Germany.

Wen, J. J. and M. A. Breazeale (1988). A Diffraction Beam Field Expressed as the Superposition of Gaussian Beams. In: *J. Acoust. Soc. Am.* 83.5, pp. 1752–1756.

Westervelt, P. J. (1957a). Scattering of Sound by Sound. In: *J. Acoust. Soc. Am.* 29.2, pp. 199–203.

Westervelt, P. J. (1957b). Scattering of Sound by Sound. In: *J Acoust Soc Am* 29.8, pp. 934–935.

Westervelt, P. J. (1963). Parametric Acoustic Array. In: *J. Acoust. Soc. Am.* 35.4, pp. 535–537.

Westervelt, P. J. (1960). Parametric End-Fire Array. In: *J. Acoust. Soc. Am.* 32.7, pp. 934–935.

Wiedmann, K., T. Buch, and T. Weber (2012). "Parametric Underwater Communications". In: Proceedings of Meetings on Acoustics ECUA2012. Vol. 17. 1. Acoustical Society of America, p. 070021.

Williams, E. G. (1999). *Fourier Acoustics: Sound Radiation and Nearfield Acoustical Holography*. San Diego, CA: Academic.

Wu, S., M. Wu, C. Huang, and J. Yang (2012). FPGA-based Implementation of Steerable Parametric Loudspeaker Using Fractional Delay Filter. In: *Appl. Acoust.* 73.12, pp. 1271–1281.

Wygant, I. O., M. Kupnik, J. C. Windsor, W. M. Wright, M. S. Wochner, G. G. Yaralioglu, M. F. Hamilton, and B. T. Khuri-Yakub (2009). 50 kHz Capacitive Micromachined Ultrasonic Transducers for Generation of Highly Directional Sound with Parametric Arrays. In: *IEEE transactions on ultrasonics, ferroelectrics, and frequency control* 56.1, pp. 193–203.

Xiao, T. (2022). "Laser Doppler Vibrometry Based Remote Sensing for Active Noise Control". Sydney, Australia: University of Technology Sydney.

Xu, B. and S. D. Sommerfeldt (2010). A Hybrid Modal Analysis for Enclosed Sound Fields. In: *J. Acoust. Soc. Am.* 128.5, pp. 2857–2867.

Xue, H., X. Zhang, X. Guo, J. Tu, and D. Zhang (2022). Optimization of a Random Linear Ultrasonic Therapeutic Array Based on a Genetic Algorithm. In: *Ultrasonics* 124, p. 106751.

Yang, J., K. Sha, W.-S. Gan, and J. Tian (2005). Modeling of Finite-Amplitude Sound Beams: Second Order Fields Generated by a Parametric Loudspeaker. In: *IEEE Trans. Ultrason., Ferroelect., Freq. Contr.* 52.4, pp. 610–618.

Ye, C., Z. Kuang, P. Ji, and J. Yang (2008). Generation of Audible Sound from Two Ultrasonic Beams with Dummy Head. In: *Jpn. J. Appl. Phys.* 47.5, pp. 4333–4335.

Ye, C., Z. Kuang, M. Wu, and J. Yang (2011). Development of an Acoustic Filter for Parametric Loudspeaker in Air. In: *Jpn. J. Appl. Phys.* 50 (7S), 07HE18.

Ye, C., M. Wu, S. Wu, C. Huang, and J. Yang (2010). Modeling of Parametric Loudspeakers by Gaussian-Beam Expansion Technique. In: *Jpn. J. Appl. Phys.* 49 (7S), 07HE18.

Ye, C., M. Wu, and J. Yang (2012). "Actively Created Quiet Zones by Parametric Loudspeaker as Control Sources in the Sound Field". In: NONLINEAR ACOUSTICS STATE-OF-THE-ART AND PERSPECTIVES: 19th International Symposium on Nonlinear Acoustics. Tokyo, Japan, pp. 367–374.

Yoneyama, M., J.-i. Fujimoto, Y. Kawamo, and S. Sasabe (1983). The Audio Spotlight: An Application of Nonlinear Interaction of Sound Waves to a New Type of Loudspeaker Design. In: *J. Acoust. Soc. Am.* 73.5, pp. 1532–1536.

Yu, S., B. Liu, K. Yu, Z. Yang, G. Kan, and L. Zong (2023). Application of a Parametric Array over a Mid-Frequency Band (4–10 kHz) –Measurements of Bottom Backscattering Strength. In: *Ocean Engineering* 280, p. 114914.

Yuldashev, P. V. and V. A. Khokhlova (2011). Simulation of Three-Dimensional Nonlinear Fields of Ultrasound Therapeutic Arrays. In: *Acoust. Phys.* 57.3, pp. 334–343.

Zhang, D., X. F. Gong, and B. Zhang (2002). Second Harmonic Sound Field after Insertion of a Biological Tissue Sample. In: *J. Acoust. Soc. Am.* 111 (1 Pt 1), pp. 45–48. PMID: 11831820.

Zhang, D., X. Chen, and X.-f. Gong (2001). Acoustic Nonlinearity Parameter Tomography for Biological Tissues via Parametric Array from a Circular Piston Source—Theoretical Analysis and Computer Simulations. In: *J. Acoust. Soc. Am.* 109.3, pp. 1219–1225.

Zhang, J., T. D. Abhayapala, W. Zhang, and P. N. Samarasinghe (2019). Active Noise Control over Space: A Subspace Method for Performance Analysis. In: *Appl. Sci.* 9.6, p. 1250.

Zhang, S. and J. Jin (1996). *Computation of Special Functions*. New York: John Wiley & Sons.

Zhang, X., H. Zhang, and D. Li (2021). Design of a Hexagonal Air-Coupled Capacitive Micromachined Ultrasonic Transducer for Air Parametric Array. In: *Nanotechnology and Precision Engineering (NPE)* 4.1, p. 013004.

Zhong, J., B. Chen, J. Tao, and X. Qiu (2020a). The Performance of Active Noise Control Systems on Ground with Two Parallel Reflecting Surfaces. In: *J. Acoust. Soc. Am.* 147.5, pp. 3397–3407.

Zhong, J., C. Hu, K. Wang, J. Ji, T. Zhuang, H. Zou, J. Lu, H. Heo, B. Liang, Y. Jing, and J.-C. Cheng (2023a). Local-Nonlinearity-Enabled Deep Subdiffraction Control of Acoustic Waves. In: *Phys. Rev. Lett.* 131.23, p. 234001.

Zhong, J., R. Kirby, M. Karimi, X. Qiu, and J. Lu (–Oct. 28, 2022a). "Audio Sound Field Generated by a Parametric Array Loudspeaker in a Rectangular Room with Lightly Damped Walls". In: Proceedings of the 24th International Congress on Acoustics. Gyeongju, Korea.

Zhong, J., R. Kirby, M. Karimi, and H. Zou (2021a). A Cylindrical Expansion of the Audio Sound for a Steerable Parametric Array Loudspeaker. In: *J. Acoust. Soc. Am.* 150.5, pp. 3797–3806.

Zhong, J., R. Kirby, M. Karimi, and H. Zou (2022b). A Spherical Wave Expansion for a Steerable Parametric Array Loudspeaker Using Zernike Polynomials. In: *J. Acoust. Soc. Am.* 152.4, pp. 2296–2308.

Zhong, J., R. Kirby, M. Karimi, H. Zou, and X. Qiu (2022c). Scattering by a Rigid Sphere of Audio Sound Generated by a Parametric Array Loudspeaker. In: *J. Acoust. Soc. Am.* 151.3, pp. 1615–1626.

Zhong, J., R. Kirby, and X. Qiu (2020b). A Non-Paraxial Model for the Audio Sound behind a Non-Baffled Parametric Array Loudspeaker (L). In: *J. Acoust. Soc. Am.* 147.3, pp. 1577–1580.

Zhong, J., R. Kirby, and X. Qiu (2020c). A Spherical Expansion for Audio Sounds Generated by a Circular Parametric Array Loudspeaker. In: *J. Acoust. Soc. Am.* 147.5, pp. 3502–3510.

Zhong, J., R. Kirby, and X. Qiu (2021b). The near Field, Westervelt Far Field, and Inverse-Law Far Field of the Audio Sound Generated by Parametric Array Loudspeakers. In: *J. Acoust. Soc. Am.* 149.3, pp. 1524–1535.

Zhong, J. and X. Qiu (2020). On the Spherical Expansion for Calculating the Sound Radiated by a Baffled Circular Piston. In: *J. Theor. Comp. Acoust.* 28, p. 2050026.

Zhong, J., J. Tao, F. Niu, and X. Qiu (2018). Effects of a Finite Size Reflecting Disk in Sound Power Measurements. In: *Appl. Acoust.* 140.1, pp. 24–29.

Zhong, J., J. Tao, and X. Qiu (2019a). Increasing the Performance of Active Noise Control Systems on Ground with a Finite Size Vertical Reflecting Surface. In: *Appl. Acoust.* 154, pp. 193–200.

Zhong, J., J. Tao, and X. Qiu (2019b). Increasing the Performance of Active Noise Control Systems on Ground with Two Vertical Reflecting Surfaces with an Included Angle. In: *J. Acoust. Soc. Am.* 146.6, pp. 4075–4085.

Zhong, J., S. Wang, R. Kirby, and X. Qiu (2020d). Insertion Loss of a Thin Partition for Audio Sounds Generated by a Parametric Array Loudspeaker. In: *J. Acoust. Soc. Am.* 148.1, pp. 226–235.

Zhong, J., S. Wang, R. Kirby, and X. Qiu (2020e). Reflection of Audio Sounds Generated by a Parametric Array Loudspeaker. In: *J. Acoust. Soc. Am.* 148.4, pp. 2327–2336.

Zhong, J., T. Xiao, B. Halkon, R. Kirby, and X. Qiu (2020f). "An Experimental Study on the Active Noise Control Using a Parametric Array Loudspeaker". In: InterNoise 2020. Seoul, Korea.

Zhong, J., T. Zhuang, R. Kirby, M. Karimi, X. Qiu, H. Zou, and J. Lu (2022d). Low Frequency Audio Sound Field Generated by a Focusing Parametric Array Loudspeaker. In: *IEEE/ACM Trans. Audio Speech Lang. Process.* 30, pp. 3098–3109.

Zhong, J., T. Zhuang, R. Kirby, M. Karimi, H. Zou, and X. Qiu (2022e). Quiet Zone Generation in an Acoustic Free Field Using Multiple Parametric Array Loudspeakers. In: *J. Acoust. Soc. Am.* 151.2, pp. 1235–1245.

Zhong, J., T. Zhuang, M. Li, R. Kirby, M. Karimi, J. Lu, and D. Zhang (2023b). Sidelobe Suppression for a Steerable Parametric Source Using the Sparse Random Array Technique. In: *IEEE/ACM Trans. Audio Speech Lang. Process.* 31, pp. 3152–3161.

Zhong, J., H. Zou, and J. Lu (2023c). A Modal Expansion Method for Simulating Reverberant Sound Fields Generated by a Directional Source in a Rectangular Enclosure. In: *J. Acoust. Soc. Am.* 154.1, pp. 203–216.

Zhong, J., H. Zou, J. Lu, and D. Zhang (2023d). A Modified Convolution Model for Calculating the Far Field Directivity of a Parametric Array Loudspeaker. In: *J. Acoust. Soc. Am.* 153.3, pp. 1439–1451.

Zhou, H., S. H. Huang, and W. Li (2020). Parametric Acoustic Array and Its Application in Underwater Acoustic Engineering. In: *Sensors* 20.7, p. 2148.

Zhou, Z. and S. Chen (2022). Design of Parametric Acoustic Array Based on Inverse Control Processing. In: *Appl. Acoust.* 195, p. 108816.

Zhu, Y., W. Ma, Z. Kuang, M. Wu, and J. Yang (2023). Optimal Audio Beam Pattern Synthesis for an Enhanced Parametric Array Loudspeaker. In: *J. Acoust. Soc. Am.* 154.5, pp. 3210–3222.

Zhuang, T., J. Zhong, F. Niu, M. Karimi, R. Kirby, and J. Lu (2023). A Steerable Non-Paraxial Gaussian Beam Expansion for a Steerable Parametric Array Loudspeaker. In: *J. Acoust. Soc. Am.* 153.1, pp. 124–136.

Zou, H. and X. Qiu (2008). Performance Analysis of the Virtual Sound Barrier System with a Diffracting Sphere. In: *Appl. Acoust.* 69.10, pp. 875–883.

Zou, H., X. Qiu, J. Lu, and F. Niu (2007). A Preliminary Experimental Study on Virtual Sound Barrier System. In: *J. Sound Vib.* 307.1-2, pp. 379–385.

Alphabetical Index

www.ingramcontent.com/pod-product-compliance
Lightning Source LLC
LaVergne TN
LVHW020605110826
845149LV00002B/374

* 9 7 8 1 0 3 2 4 0 8 5 3 8 *